Death Rode the Rails

Death

American Railroad Accidents and Safety
1828–1965

Rode the Rails

Mark Aldrich

THE JOHNS HOPKINS UNIVERSITY PRESS · BALTIMORE

9 8 7 6 5 4 3 2 1

The Johns Hopkins University Press
2715 North Charles Street
Baltimore, Maryland 21218-4363
www.press.jhu.edu

Library of Congress Cataloging-in-Publication Data

Aldrich, Mark.
    Death rode the rails : American railroad accidents and safety, 1828–1965 /
Mark Aldrich.
        p.   cm.
    Includes bibliographical references and index.
    ISBN 0-8018-8236-2 (hc. : alk. paper)
    1. Railroads—United States—Accidents—History.   2. Railroads—United
States—Safety measures—History.   I. Title. TF23.A474 2006
    363.12′2′0973—dc22   2005013355

A catalog record for this book is available from the British Library.

To Michele,
for everything

Contents

Photo gallery appears following page 236

Figures

Tables

Preface

My interest in railroads and their safety began when I was about five years old, in 1946. My mother would take my sister and me down to the Amherst, Massachusetts, Central Vermont Railroad station to meet the train on Saturday mornings. We got to know the engineman, Jim Thurston—Engineer Jim to us—and he would let us get into the cab as he took on water in those last days of steam. I was hooked and have been a railroad buff ever since. Several years later my father took me to the Boston & Maine switch yard in East Deerfield; we got to ride in the cab of one of the new diesel switchers and the engineman even let me "run" the train. As the locomotive started to rumble and move out, I became scared: "How do you stop this thing?" I wailed, to the immense enjoyment of both men. I like to think that my interest in railroad safety dates from this encounter.

But being a railroad enthusiast by no means ensures that one will be interested in the safety of railroads, except perhaps in the avoidance of train wrecks. However, I am also an economic historian—a profession that has always been concerned with the evolution of well-being, of which personal safety is an important part. I have written elsewhere on work safety among railroaders. In doing that research I discovered that railroad safety extended beyond the obvious areas of work and travel and affected the lives of virtually all Americans as they crossed or walked the tracks. The rise and fall of railroad safety as a public concern is intriguing and important whether one is a railroad buff or not. So I hope that this story will instruct and entertain not only railroad enthusiasts but also those who are not especially railroad lovers—who wouldn't know a fishplate from a railroad frog and who think a highball is just a drink and the Johnson bar is where you get one.

The evolution of railroad safety is very much a matter of economics—of labor markets, resource allocation, information flows, innovations, trade-offs, and costs and benefits. So this book contains economic themes and economic analysis. Indeed, one of my central arguments is that early American railroad safety followed a separate path because product and labor market forces on this continent were fundamentally different from those in Europe. But if this is a work of business and economic history, I have also tried to write in English, not social science blah. The book does contain some simple statistical analysis, but that is part of the supporting cast, not the star of the show, and it has been corralled in the notes and appendixes. *Political economy* is a better term to describe my approach than *economics*, for it suggests the more interdisciplinary perspective that I have chosen. The history of railroad safety is inevitably intertwined with the history of technology and I have made a serious effort to understand how technologies actually worked. For without the nuts and bolts one cannot grasp dangers or safety or problems of diffusion. The evolution of railroad safety also reflects information problems, organizational change, and political pressures—subjects that economic historians have only recently begun to study—as well as ideology and changing attitudes toward risk and public/private boundaries, topics often reserved to cultural or intellectual historians.

All studies contain gaps and biases and this one is no exception. I have spent little time on the struggle to require air brakes or automatic couplers, for those chestnuts have been roasted and served many times, including once by me, but I have taken a fresh look at their consequences. I have done no statistical analysis of the political coalitions for and against various safety laws. There is little here on full crew bills and much other legislation. Because of a lack of data, there are no quantitative cost-benefit analyses, but I have not shrunk from qualitative assessments where the evidence seems compelling. No doubt critics will find it easy to add to this list of omissions.

Railroad safety has never been simply the stuff of vast impersonal forces, for real people chose bridge designs, and invented train rules, and switched cars, and stole rides, and real men and women died in accidents by the thousands. They too are part of the story. Inevitably, however, the available sources ensure that we see only glimpses of dangers and safety as they looked to everyday people. Although I have consulted newspapers, labor journals, archives, and reminiscences, most extant materials see events largely through the railroads' eyes—or occasionally those of their critics. Still, I have tried to understand on their own terms why workers and others behaved as they did. But I have also eschewed the reformers' easy rhetoric that described managers who failed to employ some safety device simply as heartless, murdering capitalists.

The reader should note several terminological and methodological issues. Safety and danger have a number of indicators; I focus largely on the behavior of fatalities and fatality rates, for they are reported with the least error, while nonfatal injuries were so poorly and inconsistently recorded as to be of little use. Until very recently railroading was an overwhelmingly masculine activity and I have therefore referred to railroaders in those terms but tried to employ sex-neutral terminology where that is appropriate. I have reserved the term *engineer* to describe those technically trained in civil and mechanical engineering and similar activities. I use *engineman* or *engine driver* to refer to those in charge of the locomotive.

Occasionally I employ modern terminology unhistorically; thus for stylistic reasons I have individuals describing metal fatigue before that term was widely used. I have, however, followed contemporary rather than modern practice in referring to the stress calculations for bridge members as *strain sheets*.

Throughout the text there are often places where it is important to grasp the relative magnitude of sums of money—for example, when looking at accident liability awards. Economists sometimes attempt to do this by employing a price index to convert past sums into "present" dollars. Although this is appropriate for short periods, over many decades it is problematic. For while a price index converts $450 in 1886 into $9,136 in 2005, it makes no sense to say these have equivalent purchasing power, for individuals in 1850 could buy things we cannot buy now and vice versa. My approach is to express dollar magnitudes relative to average earnings at the time. Thus to learn that a liability award of $450 in 1886 is roughly equal to a nonfarm employee's yearly earnings gives a sense of its relative magnitude.

A large number of people and organizations have improved this book, contributing time, inspiration, criticism, and money. Smith College routinely supported me with small grants, for which I am deeply grateful. Librarians and archivists at the following institutions were unfailingly helpful: the Bancroft Library, the Dodd Library at the University of Connecticut, the Colorado Historical Society, the New York State Archives, Smithsonian Museum of American History Archives, Stanford Library Special Collections, Syracuse Library Special Collections, and the Union Pacific Archives. Kurt Bell at the Railroad Museum of Pennsylvania went out of his way to make my stay there profitable. At the Library of Congress, Constance Carter found materials when others could not. At Smith, the reference and interlibrary loan librarians, especially Sika Berger, Chris Ryan, Bruce Sajdak, Pam Skinner, and Naomi Sturtevant, cheerfully answered what must have seemed like endless foolish questions and retrieved obscure works. I am deeply grateful for their assistance. At the Hagley Museum, Chris Baer and Marjorie McNinch were simply wonderful. I also spent endless hours in Cornell's Library Annex, which has a superb collection of old engineering periodicals, and Cammie Hoffmier and her staff, LuAnne Beebe, John Howard, and Michele Payne, tirelessly tracked them down for me. I am at a loss for words to express how much I appreciate their help.

A number of scholars have influenced my thinking on railroads, business history, technology, and safety. No one writing in this area can escape the influence of Alfred Chandler's works, but I have also gained insight from a reading of his critics such as Naomi Lamoreaux, Daniel Raff, Peter Temin, and Richard Langlois. Nathan Rosenberg's many writings on technology have been a rich source of ideas. Price Fishback's work provides a model of how an economist should integrate theory and history. Steven

Usselman's scholarship on railroad technology superbly contextualizes the subject. Alan Leviton of the California Academy of Sciences helped me master digital photography while Jon Harkness guided me through the medical history literature and Jack Larner found valuable materials relating to the regulation of explosives. Editors Walter Friedman at *Business History Review,* Roger Grant at *Railroad History,* Peter Lyth of the *Journal of Transport History,* Robert Post and John Staudenmaier at *Technology and Culture,* and Gert Breiger at the *Bulletin of the History of Medicine* improved earlier versions of these arguments at many points. At the Johns Hopkins University Press, my editor Robert J. Brugger provided both support and guidance. I especially want to thank William Withuhn for reading the entire manuscript with great care and for his many penetrating comments, which saved me from all sorts of errors. My wife, Michele Aldrich, read the entire manuscript and immensely improved it, which is surely devotion above and beyond the marriage contract. By long tradition the remaining errors and blunders are on my head.

Death Rode the Rails

Introduction

For nearly a century—from roughly 1830 to 1920—railroads dominated American transportation. They transformed life as perhaps only the automobile and the computer have done since, and they captured the imagination. For if Americans were often critical of the railroads, they loved their trains too. Railroad stories were a staple of popular fiction. Americans sang "The Wabash Cannonball" and "The Big Rock Candy Mountain." They named trains—the *Twentieth Century Limited,* the *Super Chief,* the *Overland Express,* and the *Zephyr*—while the Lackawanna's Phoebe Snow, in her gown of white, urged all to ride the road of anthracite. Americans wrote about their trains; for Hawthorne the Celestial Railroad was a vehicle to satirize transcendentalism, while Emily Dickinson "liked to see it lap the miles—and lick the valleys up," and thousands of lesser authors penned endless lines of doggerel.

Up until about 1950 for most Americans railroads were as visible and vital a part of life as the thumb on your hand. They brought the mail and delivered Grandma for a visit. Directly or indirectly they gave work and pride to millions of men and women. The tracks provided a shortcut to town while the road signs made wonderful targets (and still do). Railroads delivered the shotgun, or the doll, or the magic electric belt that you ordered from the Sears catalogue. They took newlyweds on honeymoons and young men off to war and brought some of them home again. Draped with bunting, from Lincoln to Eisenhower they delivered presidents to a final resting place. On Saturdays children might go to the depot to watch the train come in and perhaps meet the engineman and even get into the cab. For about a century Americans listened to the mournful whistle of the train and thought of faraway places.[1]

Yet that mournful whistle also warned: Look out for the train! For as railroad writers, musicians, and poets noted, danger also rode the rails. Railroads were a transformative technology; they were a dangerous one as well, and as nearly everyone understood, the two were in proportion. For every song like the "Orange Blossom Special" that celebrated speed there was another like "The Wreck of the Old '97." Poets wrote of "an awful human carnage—a dreadful wreck of cars."[2]

The carnage began early; 1853 is still remembered as the year the surge in train accidents began. Eleven major collisions and derailments yielded 121 fatalities. Such disasters made headlines; the far more numerous little accidents did not. Although no major disasters marred the record of the Erie that year, the company killed fifteen individuals in the course of business. On October 4, Edward P. Triffin, a passenger on the Erie's night express, fell while trying to board the moving train. He was run over by the cars and killed. The previous day had seen the death of Eli Dana, a gravel train foreman, who was run over by a derailed train. A week later, an engine struck and killed an unknown man, neither passenger nor employee. Actually, 1853 represented something of a lull for the Erie, for the next year the death list came to thirty-six. The events of 1853 were in some ways atypical; the spectacular explosion in train accidents was without parallel either then or later. But for other accidents it was just another year. The railroads, hardly a quarter century old, were taking a fearful toll, and American lines were far more dangerous than those in Europe.

The evolution of American railroad safety from the early days of many spectacular wrecks and scores of little accidents is an important subject partly because, as the truism has it, how we die reveals much about how we live. The Erie's dreary figures, like those of every other line, reflected the reality of railroading on every nineteenth-century carrier and therefore the lives of nineteenth-century Americans. The railroad journey, as one writer has noted, was how the middle classes experienced industrialism with its novel dangers. Death by railroad represented the first large-scale public experience with the dangers of new technology. Early western steamboats were also risky, but fewer people experienced them, and by the 1880s they were in eclipse, while the railroads were expanding and killing more with every passing year.[3]

By the twentieth century railroads were the largest single American industry, employing over a million men and women in 1900 and dwarfing other dangerous trades. For their blue-collar employees, work accidents were the single most important cause of death. The Brotherhood of Locomotive Firemen and Enginemen's insurance fund kept records of members' mortality. From 1882 to 1912, 48 percent of all deaths were from railroad accidents, in comparison to 7 percent from typhoid fever, the next largest cause. Outside of wartime never before had such large groups of individuals been subject to such fearsome risks from manmade causes. Railroads also killed passengers by the hundreds each year. In 1907—a peak year—fatality rates were about 110 times those of modern airlines. The reader should imagine how she or he might feel about flying today under similar conditions. American carriers were uniquely dangerous to individuals who crossed or walked the tracks or stole rides and who were killed by the thousands each year. Modern technology it seemed then, and it has seemed to many historians since, was taking a dreadful toll.[4]

The total death list, by 1907, had come to nearly 12,000 a year and no doubt the number of serious injuries was many times larger. In that year the Census reported that for registration states the death rate from railroad accidents was a little over eighteen per hundred thousand. The railroad was the largest single cause of violent death—a bit less important than diphtheria, but far more important than measles, diabetes, venereal diseases, or dysentery. To individuals such figures are most meaningfully expressed as risks—employee fatalities per thousand workers, or per capita fatalities from trespassing. An alternative approach is to see them as part of the social costs of railroading and to express them per unit of railroad output. But either way the conclusion was the same: death on the rails was a common occurrence.

While such figures underlay the widespread popular concern with railroad dangers in the decades before World War I, train accidents involving passengers dominated public concern. Modern writers on risks have stressed that public perceptions reflect more than simple accident probabilities. People especially fear risks that are unknown, involuntary, fatal, hard to control, catastrophic, and new. Nineteenth-century train accidents shared many of these characteristics, and their dangers were amplified by the popular press. A few writers downplayed the dangers. In the 1870s Charles Francis Adams of the Massachusetts Board of Railroad Commissioners pointed out that a passenger on that state's railroads would have to travel 324 million miles to get killed. Mark Twain mocked public worries, claiming that travel was less dangerous than sleeping—for many more people died in bed. Yet they were in the minority.[5]

Early European travelers, including such distinguished authors as Charles Dickens, told terrifying tales of the dangers lurking on American railroads. A standard newspaper headline of the nineteenth century was "Another Railroad Horror," and cartoonists drew "Death Rides the Rails." Such tales amplified and stigmatized rail travel, raising public fears even as safety was improving. When congressmen and journalists and regulators talked about "the railroad problem," as they did for about a half century after 1870, they were referring primarily to rates, finances, and service. But public safety was also part of the indictment.[6]

These worries, like those of epidemic diseases, have largely receded from public consciousness as both sources of mortality have declined precipitously. The results constitute a virtual risk transition that has accompanied the emergence of the modern world. Life expectancy for Americans about 1907 was less than fifty years and it had probably been rising since the Civil War. Now, about a century later, newborns can expect to live to be nearly eighty. The reasons are complex. Rising incomes have underwritten a broad range of improvements in nutrition and advances in public health and medical care and delivery. These have resulted in the conquest of epidemic diseases, at least for the time being.[7]

In advanced countries the industrial revolution has also been tamed. Environmental and workplace health risks are surely less than they were a century ago, when lead and asbestos were common, everyone burned coal, and farmers fought pests with arsenic. Accidents in the aggregate are also a far less important source of mortality, despite the contribution of the automobile. Railroad accidents, of course, have all but evaporated. This disappearance partly reflects the decline of the railroad as a force in modern life, but railroad safety largely disappeared as a public issue by 1940, when the carriers were still the dominant mode of public transportation. By that date passenger risks had declined about 90 percent from their levels in the 1890s while worker risks fell 80 percent over the same period. More broadly, the social costs of rail transportation have steadily declined for well over a century.

Railroad safety has probably been improving since the 1840s. It reflected the development of a public-private technological network that by World War I had become deeply embedded in American railroading. Other industries have developed similar systems that have helped improve both work and product safety, but as with so much else, they came first on the railroads. While historians have long emphasized that economic regulation began on the railroads, this was also true of modern social regulation. Public policies that made accidents expensive, and threatened to make them more so, have powerfully motivated carriers' safety work. Yet in the twentieth century economic regulation also helped undermine safety work.

Some of these events remain part of public memory. We all remember the story of Casey Jones and nearly everyone has heard "The Wreck of the Old '97." Train accidents such as these have usually dominated official concern, and for a long time their reduction was almost the sole focus of government regulation. There is, of course, an enormous historical literature on railroads. Insofar as this writing deals with safety, the focus has usually been on train wrecks, but none of it tries to embed them in the broader context of railroad political economy. Train accidents were actually only a small part of the railroad safety problem. Most casualties resulted from "little accidents," a term I have borrowed from Ralph Richards, dean of railroad safety workers. Like sniping, they picked off victims one at a time—workers in shops and yards, passengers at stations, and the public at large as individuals crossed the tracks or used the iron highway as a public road. Elsewhere I have written on the safety of railroad workers, and several writers have recounted the role of air brakes and automatic couplers in reducing little accidents to employees, but there is no overall history of railroad safety.[8]

Framework and Themes

The evolution of railroad safety involved the interplay of market forces, science and technology, and legal and public pressures. Economists have looked at injuries (and their converse, safety) in several ways. One approach sees injury prevention as costly and demanding resources that could be employed to produce output. Given resources and the state of technology, firms face a trade-off between safety and output.[9]

In this view, the actual amount of safety will depend on this trade-off and on the price of output and payoff to safety (the saving in accident costs). An alternative view stresses that firms could choose to produce a given level of output with varying combinations of capital, labor, resources, and accidents, with the choice depending upon their relative costs and the state of technology. These are useful conceptualizations, for railroads routinely did choose differing combinations of accidents and other productive factors, and they also traded off safety against output to improve profits. But as economists have increasingly realized, these approaches require several modifications. First, managers were not perfectly efficient, profit-maximizing automatons. They were constrained by their own history and their rationality was "bounded." Hence firms differed in organizational competence and each faced unique choices and trade-offs. Second, decision makers often lacked information on choices and its acquisition was usually costly. In addition, managers almost never acknowledged a trade-off; instead they saw accidents as evidence of organizational incompetence and spent much effort trying to increase both safety and output. Finally, over time improvements in technology shifted the safety/output relationship, potentially allowing an increase in both, and therefore yielding, as Joel Mokyr points out, "free lunches [or] . . . (more frequently) very cheap lunches."[10]

This last point deserves some explanation. Most railroad technological changes improved both safety and productivity. For example, better signals were often seen as primarily a safety improvement, but they also allowed heavier traffic, and so some of their gains might be taken as more or better output rather than as safety. Thus, the actual consequences of a "safety" innovation are socially determined and depend on, among other things, the relative prices of accidents and output. Similarly, improved rails were usually justified by their effect on productivity but they enhanced safety as well. Some productivity improvements that reduced labor requirements, such as the diesel locomotive, improved safety by shifting labor out of comparatively dangerous jobs. Better safety, therefore, was usually part of a package that also increased productivity. Both forces reduced railroads' social costs; declines in accidents and accident rates did so directly while rising labor productivity reduced worker exposure to dangerous jobs. Safety, in short, cannot be separated from the other forces that improved railroad productivity.[11]

Both the ability and desire of companies to deliver safety and productivity reflected an evolving political economy. The expansion of railroading led to organizational and institutional innovations in the form of an American railroad technological network, which was a subset of a larger technological community that was broader than railroading and international in scope. It developed in response to market forces and public policies. Its major actors have been a changing mix of railroads, individual inventors and entrepreneurs, supplier firms, technical societies, and the trade press.

By both training and institutional position the members of this community constituted a powerful force for technological improvement.

They developed and shared a fund of industry-specific and sometimes firm-specific expertise in making the technology work—and work better. Initially most were "mechanics." While few had much formal schooling and most of their early knowledge was tacit, nearly all could read and write and therefore communicate. As the century progressed an increasing number of railroaders commanded formal training in science and engineering. Technical and safety knowledge became less tacit and more formal. Engineers are trained to be technology critics—it is part of their professional ethos. Thus both mechanics and engineers possessed a fund of experience and education, and most had institutional positions that encouraged them to improve technology. In economists' terminology the accumulation of specific forms of human capital by those interested in railroads not only raised their productivity, it encouraged incremental technological progress as well.[12]

Risks are culturally constructed, and the courts, the public, and the popular press have also shaped the focus of railroad safety through legal and regulatory methods. These groups encouraged the demand for safety while the network created the supply. From the beginning, the fact that railroad accidents represented a human technological failure prevented any fatalistic view that "accidents will happen." Thomas Haskell has proposed that humanitarianism derives in complex ways from the spread of market habits and institutions. In a variant of his position, I argue that the carriers' development of the recipes for accident reduction—along with their demonstration of generalized technological competence—helped make better safety not only less costly, but also a moral imperative. Better safety, by this dynamic, induced a demand for better safety. Technological sophistication also encouraged the evolution of ideas about accidents away from an emphasis on individual to organizational or social responsibility, and this in turn has reshaped the public/private boundaries of safety. In the phrasing of economists, safety had always been a public good; increasingly it came to be seen as a public responsibility.[13]

Public control over rail safety responded to the menu of available technological choices, and sometimes shaped the evolution of that menu as well. Three forms of safety regulation evolved together: "hard" controls involving specific laws; a "softer" approach that I have elsewhere termed *voluntarism,* that relied on hectoring and publicity, always with the threat of hard controls in the background; and accident liability rules imposed by courts. In the early days court-imposed liability rules dominated. Beginning in the 1870s states evolved a mix of hard and soft regulations but there were no federal controls until the 1890s. While the Progressives passed a number of federal safety laws, by World War I voluntarism along with economic incentives became the dominant approach at the federal level until the 1960s.[14]

Safety has also been shaped by economic regulation. During the nineteenth century both rates and safety were largely unregulated. Hence, when carriers introduced technologies to improve safety and/or productivity they expected the package to return at least the cost of capital. There was, therefore, a potential conflict between safety and economic regulation. By the middle of the twentieth century stricter rate regulation helped reduce railroad returns below the cost of capital, reducing the payoff to safety-productivity packages and making their financing more difficult.

As this analysis suggests, improving safety was very much an outcome of induced technological change, which led not only to better technology but the evolution of safety organizations and institutions as well. While the inducement mechanism was sometimes the perceived rise in the cost of accidents, political and regulatory costs also powerfully encouraged a concern with safety. As noted, sometimes the rising cost of another input, such as labor, induced either technological changes or more capital-intensive methods of railroading that could reduce accidents, although that result was often contingent upon company institutions and motivation.[15]

A key feature in improved safety has always been better, cheaper information. JoAnne Yates has written of the increasing ability of managers to control large firms through communication. While her focus was on such matters as office work and accounting, her ideas have broader applicability. A central preoccupation of the railroad technological network has always been the need to monitor and control both the natural environment and workers. The study of accidents and their statistics facilitates both monitoring and control, as Henry Petroski and Theodore Porter have noted. Such information can be a focusing device to pinpoint and control physical problems. In the eyes of Adams and other regulators who urged voluntarism, the searchlight of investigation would also compel companies to follow the path of reform. In more subtle fashion, the gathering of statistics and their form of presentation can create and shape social issues. Materials testing by buyers also yields information that encourages suppliers to improve production monitoring and quality control. New technologies such as nondestructive testing reduce the costs of acquiring such information. As will be seen, automatic block signals allow managers not only to control workers but also to cope with such natural hazards as fogs, landslides, and washouts.[16]

Train wrecks, among the first large-scale technological accidents, confronted managers with unprecedented problems of information and control. In one writer's vision they led to a crisis of control on the carriers. Train accidents were early examples of what Charles Perrow and others have called "system" or "normal" accidents. They were system failures because they often involved both complex causation and tight coupling—that is, fast-acting, more or less inevitable chains of events. Because each railroad is itself an interacting technological system, and because different lines interacted with each other, problems of coordination abounded. For example, when one carrier increased freight car size the result might affect wheels and brakes and what was later termed train/track dynamics in surprising ways not only on that line but, because of freight car interchange, on other roads as well. Yet Perrow's concept of normal accidents is ahistorical, for what is normal reflects the state of technology and managerial organization, and in response, the carriers innovated industry-wide institutions to coordinate technological change. In addition, because accidents were comparatively rare events, especially on small railroads, they were difficult to learn about. Information sharing through journals and technological societies reduced its costs and was therefore crucial to progress.[17]

Employee monitoring and control are inherently difficult and costly on railroads and so "agency problems" (i.e., ensuring that workers behave in the best interests of the company) abound. Here again, the problem is one of information. Nineteenth-century American labor markets exacerbated these difficulties and worsened safety because they encouraged turnover and magnified problems of discipline and control. Agency problems might also interact with technological choices to yield either danger or safety. For example, the prolonged use of the time rather than the space method of separating trains on American railroads required brakemen to protect their trains with a flag signal, which they regularly failed to do with disastrous results. Here again, however, the railroad technological network responded with a host of organizational and physical innovations ranging from the telegraph to the standard code of rules and signals, surprise checking, employee pension plans, and organized safety work. These reduced costs of information and control and helped contain accidents due to human lapses.

As part of a broader technological network, the railroads were able to draw on the growing stock of scientific and technical knowledge to improve safety. In the nineteenth century they benefited from research on steam boiler safety and the discovery of electromagnetism, which led to development of both the telegraph and the track circuit. The twentieth century saw the introduction of such new materials and methods as alloy steels, reinforced concrete, train radio, and nondestructive testing, all of which have made important contributions to improved railroad safety. Yet the intersectoral flows of technology have gone in both directions. Carrier and supplier research generated benefits for other sectors of the economy as well. The early development of steel production in the United States was motivated in large part by the carriers' desperate need for rails that could withstand heavier trainloads. Understanding the metallurgy of iron and steel and the discovery of metal fatigue resulted from British and German investigations into puzzling failures of bridges and axles.

Railroad safety has been powerfully shaped by forces external to the industry as well, and two in particular stand out. In the twentieth century, the spread of automobiles and trucks both complicated and worsened the problem of grade crossing safety. On the other hand, automobiles changed passenger traffic in ways that reduced dangers, while the long-term rise in incomes has raised public demands for safety at grade crossings and probably other forms of safety as well. In addition, rising incomes and the spread of automobiles both contributed to the eclipse of trespassing as a serious problem.

Today railroads have nearly disappeared from our consciousness, and that is in part because the social costs in terms of accidents and casualties that they produce while generating transportation are far smaller than in the past. Their evolution from a novel, dangerous technology to near invisibility powerfully shaped the lives of generations of Americans. In 1850, at the dawn of the railway age, *Scientific American* published rules for travelers that reflected the dangers of the day: "choose a center car; travel by day, and avoid foggy weather." A century and a half later such warnings seem as old-fashioned as buttoned shoes. Today trains no longer slaughter working people forced to use the tracks as a highway. Nor do very many heavily traveled unmarked railroad crossings lie in wait for the unsuspecting motorist. Over the decades improving work safety has literally meant the gift of life and limb for tens of thousands of employees.[18]

This is a story that seems well worth understanding. Why were early American carriers so dangerous? How did the developing technological network shape safety and how did technology and economics interact? Who were the men who made railroads safer? How did labor markets affect safety and what were the ways the carriers coped with agency problems? How and why did public regulation evolve as it did, and why did the long-term improvement in safety pause in the 1950s? These are the subjects of the following chapters.

In the Beginning

American Railroad Dangers and Safety, 1828–1873

The [railway] system which has grown up under . . . [American conditions] is well adapted to the wants of the country.

—Captain Douglas Galton, Royal Engineers, 1857

When the matter of safety comes to be considered . . . America is lamentably below all other nations.

—Alfred Bunn, British tourist, 1853

JULY 4, 1828, MARKED THE BEGINNINGS of American railroading. On that date old Charles Carroll of Carrollton, last surviving signer of the Declaration of Independence, turned over the first shovel of earth to begin construction of the Baltimore & Ohio (B&O)—the first commercial railroad on the continent. Forty-five years later in 1873, American carriers operated the largest railroad network in the world, with over 70,000 miles of road. The development of American railroads during this first half century owed much to the common pool of technological knowledge shared by Europeans and Americans. Because all aspects of railroading from technology to management were new and untried, accident and casualty rates in all countries were high by modern standards. But as both contemporary and modern observers have noted, American carriers also evolved their own technological style that shaped all aspects of railroad transportation, including its safety. These choices, along with American labor conditions and the American preference for as little government regulation as possible, led to the development of a uniquely American system of railroading that was also uniquely dangerous. But safety, like every other aspect of railroading, was in constant evolution, and the very dangers of American railroads encouraged the development of a complex network of individuals and institutions devoted to improving their safety.[1]

The American System of Railroading

During the antebellum years a number of European scientists and engineers commented extensively on American railroads. These writers made it clear that a distinctly American system of railroading emerged over these years, a fact they attributed to differing European and American economic circumstances. These differences resulted from the nature of the market for railroad transportation in the United States and from the comparative expense of capital and labor in relation to natural resources such as land and wood. European travelers also commented on the distinctive characteristics of American railroading, and most were shocked by its dangers.

European Engineers View American Railroads

Among the most perceptive of the European engineers who studied American railroads was Franz von Gerstner, a German engineer who visited the United States during the late 1830s. Von Gerstner first articulated some of the fundamental differences between European and American methods. He calculated that New York railroads had far steeper grades and sharper curves than did European carriers, but had cost only $17,000 a mile, far less than most European lines. Flatter grades were avoided, he explained, "to avoid the higher construction costs," while sharp curves were employed "as a means of avoiding heavy expense." For similar reasons, in 1850 the United States with over 26,000 miles of road contained only 11 miles of tunnel, while Britain with about 8,700 miles of road had 70.[2]

Writing in 1850, the British scientist and inventor Dionysius Lardner glimpsed the same forces at work. He noted that England's railways had cost about £40,000 a mile and returned only modest profits. How, he asked rhetorically, could "this stupendous system of American railways, with a traffic comparatively so insignificant among a people where profits on capital are high . . . be made to answer?" As Lardner proceeded to explain, the American response to thin traffic and scarce capital was to construct railroad permanent way far more cheaply than was common in Europe. In the terminology of modern economics, factor proportions along with the nature of demand led American companies to choose and develop different technologies than were employed in Europe. Although American practice raised operating costs, with sparse traffic such costs were more than offset by lower interest payments. As traffic increased, improvements in permanent way became more economical, however, and in this respect, as the nineteenth century wore on, European and American railroads grew more alike (chapter 2).[3]

While most British lines were double-tracked, only a few of the early American carriers such as the Long Island and the Boston & Providence were graded for double-track, and these were sharply criticized by engineers Charles Ellett and John Trautwine for being *too* well built. "Our engineers should construct their roads with a view to *paying* well instead of *looking* well," Trautwine asserted, and he claimed that efforts to imitate English practice "must necessarily bring ruin." In 1832 the engineer Horatio Allen explained why the South Carolina railroad built part of the line on piles. To do otherwise would have doubled the cost and required use of skilled labor that was simply not available. Lardner also explained that "in laying out these lines the engineers did not, as in England, impose upon themselves the difficult and expensive condition of excluding all curves . . . nor are gradients restricted to the same low limits as with us."[4]

In addition, buildings were "cheaply constructed of timber," which was in "bounteous supply," Lardner noted. Bridges too were nearly all made of wood. Although these were short-lived, the cost of regular renewal was less than the interest on the first cost of a more substantial structure. Some early lines learned this lesson the hard way. The B&O initially followed European precedent and constructed masonry bridges until the practice nearly bankrupted the company. Wooden trestles, which avoided the need for costly filling, also characterized American railroading well into the twentieth century. Efforts to economize on first cost also ensured that ballasting was usually thin or nonexistent, while crossings were at grade and fencing was uncommon. Similarly, in America, thin traffic reduced the payoff to fixed signals. In 1853, the *American Railroad Journal* complained of a "want of signals and signal operators," while *Scientific American* concluded that the signals on all railroads are "contemptible rags." British roads, by contrast, responded to rising traffic density in the 1840s with a system of fixed signals at crossings, stations, and other danger spots. These employed boards, balls, lights, and semaphores to indicate caution as well as stop or clear. Also in the 1840s, a few British lines began rudimentary use of the telegraph and block system to control trains.[5]

However, Lardner placed the "chief source of economy in . . . the structure of the road surface" that the small volume of traffic allowed. Many of the early carriers employed stone sills and granite or wooden rails with thin strap iron nailed to them—a technique that substituted cheap wood or stone for expensive iron. Of course, American adaptations could be overdone. As early as 1834, engineers of the Columbia Railroad concluded that half-inch strap iron used by some lines "ought never to be put on a road intended for locomotive engines." They recommended five-eighths- or, better, three-quarter-inch iron instead. Even thicker strap iron could not survive locomotives, however, while stone sills and rails proved more expensive than wood and hard to keep in line and gauge, especially after winter frosts. By the 1850s such experiments were on the way out, replaced by the T rail. But unlike British lines, American carriers spiked rails directly to the ties, thereby saving the cost of iron chairs. And, as Lardner noted, "the strength necessary for the road is obtained by reducing the distance between the sleepers . . . [rather than] giving greater weight to the rails."[6]

American rolling stock also followed a different evolutionary track from that in Europe. While early American locomotives looked much like those in Britain, builders gradually added the pony truck and equalizing suspension to help them hold the track on rough American roadbeds. Cowcatchers also reflected American realities, as did the early addition of enclosed cabs. American builders mostly abandoned inside-connected driving wheels, as the crankshafts were expensive to machine and maintenance was costly where skilled labor was scarce. Lardner found American locomotives "destitute of much of that elegance . . . and luxurious beauty of workmanship" that characterized British locomotives. They also burned wood because the coke used in British engines was too expensive. Smoke was less objectionable in thinly populated rural areas than in the more urbanized European setting, but the sparks were a safety menace where bridges were made of wood, and companies endlessly experimented with spark catchers.[7]

Early American passenger cars, like those in Britain, resembled stagecoaches on rails, and might hold fifteen or twenty people. English first-class coaches retained this form well into the Victorian era, and by 1850, British and American passenger equipment differed sharply. British passengers rode in compartments that they entered from the side of the car while in America riders entered from the end of the coach and all sat together. American cars were much larger, holding fifty to sixty people, which made them cheaper to operate because they had a higher ratio of payload to dead weight. American cars were also less ornate. First-class cars made little sense where traffic was thin, and this also made them less expensive to build. As Lardner concluded, "the form and structure of the carriage is a source of considerable economy."

In all countries early freight cars were tiny, holding at most 3 or 4 tons but, again, American equipment steadily increased in size. American builders initially followed European precedent and employed two axles and wooden wheels with an iron rim on both freight and passenger equipment. In the 1850s European freight cars still ran on two axles and four wheels, but most American rolling stock employed two four-wheel swivel trucks, allowing navigation of the sharp curves on American lines. While Europeans had shifted to iron wheels with steel rims, Americans made solid wheels of cast iron with a chilled rim, which was cheaper.[8]

Captain Douglas Galton of the Royal Engineers briefly visited the United States during the fall of 1856, and in a report to the English Privy Council he provided a detailed picture of American railroads that affirmed and expanded on the findings of earlier observers. Galton's purpose was to discover the ways in which American lines might prove to be a model for railroads in other new countries. "It is to be borne in mind," he instructed the Lords, "that the American railway system must not be judged by the standard that we have established in England," and he went on to explain that "in a rapidly developing new country capital is dear. Hence a rough and ready cheap railway although it entails increased cost for maintenance is preferable to a more finished and expensive line."

Galton buttressed this conclusion with numerous details. American lines ran fewer trains than did English railroads, but they were better filled—probably because longer American journeys made reducing costs per passenger mile especially important. Light traffic and high wages ensured that an American grade crossing rarely had gates or gatekeepers. Because of this and other ways that American lines economized on labor, employment per train mile was lower than in Britain. For example, in 1856, Massachusetts carriers employed about 1.05 workers per thousand train miles compared to 1.31 for British lines. Considerations of traffic and labor costs also led American lines to place a responsible person on the train (the conductor) and subordinates at local stations while in Britain the reverse was true. Galton perceptively observed that the American system placed too great a responsibility on the conductor. American traffic patterns differed in other ways as well. By the 1850s economic development in the context of sparse population ensured that American roads were becoming dominated by freight traffic, while in Europe freight and passenger traffic remained more balanced. Although Galton expressed surprise at how little the telegraph had been employed in railway operations, he admiringly described its use on the Erie. He concluded that the American system was "well adapted to the wants of the country."[9]

European Travelers View American Railroads

European tourists were usually considerably less enthusiastic about the way Americans were adapting their railroads to local circumstances. The Frenchman Charles Oliffe who traveled in America in the 1850s found it "astonishing that an American train does not . . . leave a trail of unmitigated destruction behind it, considering the few precautions that are taken to prevent accidents." About the same time the English traveler Alfred Bunn concluded, "When the matter of safety is considered . . . America is lamentably below all other nations." Bunn indicted the flimsy permanent way, "the occurrence of accidents along shaky sleepers . . . crossings over deep waters by virtue of a few planks, misnamed bridges . . . [and] misplaced signals." All of this he found "alarming in the highest degree." Bunn lamented "the total absence of even ordinary precaution and want of the protective hand of government," and he concluded that "the whole system is wrong from first to last."

Many other travelers agreed. Rachel Felix, a Frenchwoman, also found it "quite inconceivable how few precautions are taken to avert accidents." She cited the "folly . . . [of] only one track," and on one road she claimed "the grading is not solid, the bridges . . . thrown together in haste [and] built of bad timbers." Charles Dickens also dryly remarked that American railroad bridges were "most agreeable when passed." Russian visitor Aleksandr Lakier raised his eyebrows over the "one set of tracks, bridges not finished off with guard rails, [and] road beds not raised." In 1854, the English traveler Charles MacKay suffered through four derailments on a single journey. The great distances in America led travelers to prefer speed to safety; when Charles Weld, another Englishman, complained that reckless running had caused the train he was on to derail, he found that most passengers applauded the crew's effort to make up time. "Accidents on railways are thought so little of in America it is useless to remonstrate," he concluded.[10]

British and American Railroad Safety, 1847–1900

In all countries, early railroading was highly dangerous because nearly everything was new and failure-prone. Accordingly, histories of British railroads describe many hair-raising disasters. Yet the observations of European engineers and tourists suggest that American carriers were comparatively more dangerous because they sometimes directly traded capital or labor for accidents (e.g., by eschewing gates and gatekeepers). They also substituted cheap natural resources (flimsy wooden trestles) for expensive capital and labor wherever possible and that too worsened risks. Thus American railroads terrified European tourists for precisely the same reasons that led European engineers to praise them as "well adapted."

A diagram of the process might look as follows. Suppose that capital and labor are used together and that they can be substituted for both natural resources and accidents. The curve labeled *I-I* in Figure 1.1 represents the various combinations of capital and labor and natural resources that can produce a unit of transportation output, while *AA* represents the associated accidents. In Britain, with natural resources expensive relative to capital and labor, carriers develop technologies such as point *a,* while in America it is more economical to develop *b*-type methods that use less capital and labor per unit of output but employ more natural resources and yield more accidents.[11]

Figure 1.1. **Technology and Accidents, Great Britain and America**

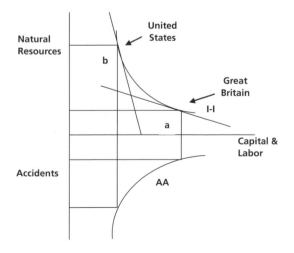

Labor scarcity also interacted with other aspects of railroading and American society to create a work culture that shaped American railroad safety in complex ways. Railroad work appealed to Americans' ideals of manliness, independence, and self-respect. It was "man's work"—hard and dangerous, and demanding strength, skill, self-reliance, and judgment. Many employees labored with little supervision. As one observer wrote in 1854, a railroad was distinct from a factory in that "a vast number of men must be employed who have to be left to themselves . . . [while] a vast consequence often hangs on the act of the humblest person employed on the road." Not only trainmen, but also station agents, track walkers, signal maintainers, and others were for a long period their own bosses. Because early track and equipment were so unreliable, trainmen were expected to be able to fix equipment in a road breakdown, or rerail a train if need be. They might also have to help dig it out of a snow bank, or help draw water from a brook, or in general do whatever was necessary to get the train in on time.[12]

This combination of freedom, responsibility, and risk contributed to a distinctive work culture. Railroad men were clannish: they drank together and looked out for each other's safety and welfare, passing on to new men the tacit knowledge needed to survive. They routinely assumed vast responsibilities and relied on their own judgment. One result was the stuff of legend: there really were brave enginemen who died at the throttle trying to save their train. But these same traits also led to chronic agency problems (i.e., problems of ensuring that workers, or agents, operate in the best interest of the company), for many men also chose which rules to obey and when to obey them. Labor scarcity protected such behavior because it reduced the carriers' disciplinary threats.

Trends in British and American Railroad Safety

The first systematic American evidence on the risk of death per worker or per passenger mile on early American railroads comes from Massachusetts. Carriers were required to file returns with the state and while these are incomplete, from 1846 on they can be used to determine trends. In New York, data on passenger risks begin about 1850. After the Civil War an increasing number of states reported information that can be used to construct casualty rates. Using Massachusetts data I also computed fatalities per million train miles to "others"—trespassers and individuals killed at grade crossings. National estimates of railroad casualties begin with Interstate Commerce Commission (ICC) data in 1889. From 1889 to 1900 they overlap data on casualties from train accidents generated by the *Railroad Gazette,* which began collecting such figures in 1873. Using the relationship between these data during the overlap period, I generated national measures of worker and passenger fatalities back to 1882, along with estimates of both passenger miles and railroad employment. British data derive from the British Board of Trade, which reported fatalities and injuries from the 1840s. Using these data and estimates of employment and passenger miles, I constructed fatality rates for 1847 on, as well as estimates of other fatalities per million train miles.

All of these data, contained in Appendix 1, require a considerable dose of salt. The ICC estimates from 1889 on are the most accurate, but even they have difficulties. The American state data are not a random sample and where they overlap ICC data they underestimate the national figures, suggesting that they probably also underestimate the risks of American railroading for the earlier period. British data also reveal considerable underreporting. With these caveats, the data are depicted in Figures 1.2–1.4 and Table 1.1. As can be seen, Alfred Bunn's worries appear to have been well founded. Throughout the nineteenth century American railroads

Figure 1.2. **Passenger Fatality Rates, America and Great Britain, 1846–1900**

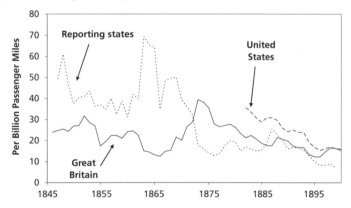

Source: Appendix 1. Data are three-year moving averages.

Figure 1.3. **Worker Fatality Rates, America and Great Britain, 1845–1900**

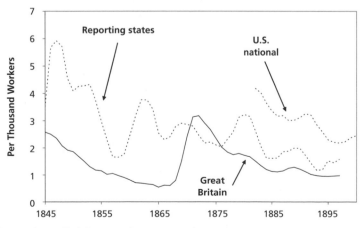

Source: Appendix 1. Data are three-year moving averages.

Figure 1.4. **Other Fatalities, America and Great Britain, 1852–1900**

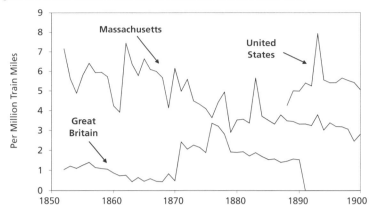

Source: Author's calculations.

were consistently more dangerous than British carriers to passengers and to workers and "others" as well. Another way of construing these figures is to think of railroad casualties as part of the social costs of the new technology. Data for Massachusetts in the 1850s suggest that American railroads killed about five times as many people per unit of transportation as did those in Britain.[13]

The Anatomy of Railroad Risks

Many features of American lines made them collision-prone, as Table 1.1 reveals. Alfred Bunn correctly observed that most were comparatively poorly signaled. A disastrous crossing collision between the Michigan Central and Michigan Southern near Chicago on April 25, 1853, that killed twenty-one people might have been prevented by proper signals. Similarly, a month later that year, forty-six passengers were drowned when a train went into an open draw at Norwalk, Connecticut. The signal was defective in two ways: it was normally clear, when normal danger would have been safer, and compounding the problem, it could not be seen in time for the train to stop.[14]

Collisions also resulted because most American lines—unlike their British counterparts—were single-track. The *Springfield Republican* reported a "Dreadful Collision" on the Western (Massachusetts) Railroad on October 9, 1841. The trains were supposed to meet and pass at a siding, but when one was late the other conductor simply went on, hoping to make the next siding. American carriers also faced increasing difficulties of coordinating passenger trains with freights, which ran much more slowly. Less than two months later on the Boston & Worcester, the 4:00 p.m. passenger train collided with a freight that had left ahead of schedule. The freight engineman said he had read in the paper that the 4:00 o'clock train had been discontinued. Clearly neither of these disasters could have occurred

Table 1.1. **Causes of Death on British and American Railroads, 1850–1852 (rates per hundred million train miles)**

Cause	New York,* 1850–1852				Great Britain, 1852			
	Employee	Passenger	Other**	Total	Employee	Passenger	Other**	Total
Collision	23	26	0	49	12	0	0	12
Derailment	47	23	0	70	13	10	0	23
Run over**	53	0	406	459	91	0	86	177
Jump off/on	35	73	6	114	20	16	0	36
Fall from train	96	35	0	131	51	8	0	59
Other	67	20	3	90	28	0	0	28
Total	321	177	415	913	215	34	86	335

Sources: Author's calculations for New York data derived from "Report of the Committee Appointed to examine and Report the Causes of Railroad Accidents, the Means of Preventing their Recurrence, &c," New York State Senate Document 13, 1853, Tables 7. For British figures, calculations are based on data from Francis Neison, "Analytical View of Railway Accidents," *Journal of the Royal Statistical Society* 16 (Dec. 1853): 289–337.

* Train miles in New York are calculated as three years at 1854 levels, as earlier figures are unavailable.

** Includes fatalities at crossings and while walking track.

on double-tracked lines—and by 1851 *Scientific American* was urging that no more single-track lines be chartered. Yet equally clearly these accidents reflected failures to follow company rules.[15]

Many accidents also resulted from passenger—not worker—foolhardiness, and jumping off or on cars and falling off the train typically took more lives than collisions and derailments combined (Table 1.1). Note also that such activities were *apparently* far more common in America than Britain. In fact, there is evidence of considerable underreporting of such accidents in Britain during these years (Appendix 1). In both countries authorities ascribed them to a "want of caution," implying that individuals were to blame. While there is obviously a good measure of truth to this, the large numbers of such fatalities in the United States also reflected engineering and managerial choices made by American carriers.

American cars had doors and platforms at each end, and passengers routinely fell off them. Most American stations were unfenced, and since one could buy a ticket from the conductor there was nothing to stop a late passenger from running after a departing train. As Table 1.1 reveals, many who sprinted did not make it. Similarly, many passengers were killed in the manner of Catherine Delaney, who jumped from a moving car on the Erie in 1855 and rolled under the wheels. In Britain stations were fenced. Also, jumping off and on cars was more difficult because most were constructed with side doors and no platform, which made it impossible to go between cars. British passengers were at risk of falling between the station platform and the car, however. Falling from trains was also a major risk for American trainmen due to the need to ride the top of freights to work brakes. In Britain such risks were reduced because trains were braked from a single car, a brake van (chapter 4).[16]

Even leaning out the window was dangerous for, as the *American*

Railroad Journal grumbled, American lines provided very little clearance. Early reports yield a sprinkling of such accidents each year. The Erie reported that one Mrs. Walters had made the mistake of leaning out a window as the train was departing to see if her baggage had been loaded. She was instantly killed when her head hit a station roof support.[17]

The prevalence of derailments on American lines reflected the flimsiness of road and equipment, lack of fencing, and harshness of the climate, as well as difficulties of enforcing speed controls on lines where fast running was unsafe. In fact, derailments typically took the lives of more passengers and employees than did collisions. Sometimes roadbed subsidence disturbed the track, causing even a carefully run train to go off the rails. Axle failures also seem to have been particularly prevalent in the United States. The heavy tonnage Philadelphia & Reading reported an average of 231 broken axles a year from 1848 to 1853. While most derailments did little damage, under the right circumstances they could be lethal. When a brake beam broke on the Boston & Worcester on November 6, 1847, it derailed a car that then struck a bridge abutment, killing six passengers. Similarly, it was a defective axle that caused a derailment on a bridge near Angola, New York, on the Lake Shore on December 18, 1867. The resulting wreck and fire killed forty-two individuals.[18]

Rails and wheels also broke because they were too skimpy for the increasingly heavy rolling stock of the 1840s and 1850s. When a broken rail threw two Erie railroad cars down an embankment near Shinn Hollow, New York, on July 15, 1858, killing six and injuring thirty, the *American Railroad Journal* concluded that the cause was a "want of correspondence between the superstructure and the locomotive." Unsupported rail joints led to low spots in track that pounded rolling stock, often shattering cast iron wheels. Axle, wheel, and rail failures also reflected the American combination of poorly ballasted roadbed and hard winters that froze it solid, for such failures were always greatest in cold weather. In the last three weeks of February 1856, the Albany & Utica broke six 6-inch locomotive driving axles. One New York railroader also concluded, "cast iron wheels do not stand the winter of this high northern latitude."[19]

Aleksandr Lakier was apparently surprised that American railroads were "unprotected against . . . animals." But, he observed, "the inevitable 'cowcatcher' substituted for the live guards and workers, of whom thousands . . . would be needed . . . to guard the thirty thousand miles of track." Cattle were always getting on the track. Sometimes they fell into bridges; sometimes they slept on the track at night; sometimes they simply walked the tracks. Engineman Henry Dawson reported one wreck when a train hit a horse a farmer was grazing on the railroad right-of-way. The *American Railroad Journal* even claimed that farmers occasionally salted the track with broken-down animals, hoping the courts might cooperate in shaking down the railroad. "Such *baseness* can hardly be credited in a Christian community," the editor fumed. Moreover, the cowcatcher did not always work, as Lakier discovered when a train he was on hit a bull and nearly derailed. Another English traveler, Charles Casey, was also on a train that hit a cow. Usually in such contests the animal lost. But not always: in December 1846, when a train on the Troy & Saratoga blew the whistle on a bull walking ahead, the animal turned and charged, derailing the train and wrecking two cars but leaving the bull with no more than a nasty headache. Sometimes the animal lost but the train still derailed, as occurred on the Rochester & Syracuse on July 5, 1852. The engine tossed a cow up an embankment but she rolled back under the train, derailing three cars, injuring several passengers, and killing a brakeman.[20]

European travelers also marveled at unfenced lines and unguarded crossings. "*They* are not required to 'look out,' but *you* are," Bunn exclaimed. He was used to crossing gates and guards, and strictly enforced rules against trespassing. Table 1.1 reveals the predictable results of American practice. The English traveler Thomas Grattan experienced them firsthand when his train hit a horse and buggy at a crossing, smashing it to pieces and killing a woman.[21]

The absence of guards at crossings was but one of many ways in which a tight labor market caused the carriers safety problems. A New York commission in 1853 blamed many accidents on the rapidly increasing demand for labor that required "the employment of incompetent or improper agents and workmen." The tight labor market also reinforced railroads' agency problems. Enforcing rules with penalties for those who caused accidents was difficult, for the rapid expansion of the rail network ensured a growing demand for experienced workers. Thus, when the B&O dismissed an agent responsible for a derailment, he was "immediately employed in the vicinity of the railroad, and at equal wages, and in fact declared that his discharge put him to no inconvenience." Similarly, attempts to enforce rules on the Erie in the 1850s led to a brief strike of enginemen.[22]

With most railroads neither guarded nor fenced, track walking also took an immense toll. "Of the hundreds who are annually maimed or slaughtered on our railroads more have certainly fallen victim from careless walking or lying on the track than from any other half-dozen causes put together," wrote the distinguished railroad journalist Henry Varnum Poor in 1855. Track walking was far more dangerous than it appeared and risks must have risen with train frequency and speed. By the early 1850s some passenger trains ran at 30 miles per hour, or 44 feet per second. At night, on a sharp curve, walking into the wind, one might never hear a train approaching from behind. Of course, older or hearing-impaired individuals were at even greater risk and state reports contain depressingly large numbers of men and women described as deaf, insane, or aged, who were killed on the track. Rolland Stebbins was one such unfortunate. He was walking the track near Deerfield, Massachusetts, on August 7, 1848, when the Connecticut River Railroad ran him over. He was deaf and had not heard the bell or whistle. Railroads also killed children by the hundreds who were playing on the track or gathering coal. *Harper's Weekly* reported two, who were run over near Newark, New Jersey, in June 1860. In vain it urged use of fences, gates, and watchmen. Finally, accident reports contain a seemingly endless list of individuals killed while under the influence. Thomas Easton was described as "intoxicated" on April 10, 1848, when he fell under a Providence & Worcester freight on which he was trying to steal a ride. James Gregg was probably drunk when he decided to take a nap on the track of the Boston & Maine on the night of April 28, 1848, and was killed.[23]

Part of the American system of railroading also included what Europeans termed a *want* of government regulation. In Britain, after 1842 the Board of Trade investigated accidents and routinely badgered the recalcitrant to install safety improvements. The board also inspected new lines, which could not open without its approval. In America, by contrast, until New York required inspection of new lines in 1855, anyone could open a railroad, and in other states this remained the case for many years. Moreover, while British and American mileage grew at about the same rate through the 1850s, thereafter growth slowed down in Britain, but not the United States. Thus, in the years after 1860 a far larger proportion of American than British mileage was of recent vintage. As a New York investigating committee commented, "many serious accidents arise from the practice . . . of opening roads . . . before the work upon them is completed."[24]

American carriers were evolving in other ways that increased dangers as well. The share of freight traffic increased rapidly during the antebellum years—far more than it did in Britain—and freight was more dangerous to workers than were passenger trains. As freight car interchange increased in the years after the Civil War, employees increasingly worked with foreign and therefore unknown equipment, which also worsened their risks. Finally, running slow freight and fast passenger trains on a single track also vastly complicated train operation and may well have raised passenger risks as well. But if some sorts of dangers increased in the antebellum decades, statistical evidence confirms that American railroading was becoming safer for both workers and passengers, because there were countervailing forces as well, and improving safety was a concern from the beginning.

Improving Safety: Management, Markets, and Technology

Just as accidents began with the first railroads, so did innovations to control them. Failure can be instructive, and in fact accidents were the main way early railroaders learned about risks. Railroad managers' interest in discovering the causes of accidents and ways to prevent them reflected humanitarian concerns, powerfully reinforced and shaped by self-interest, for accidents produced both regulations and costs. The result was a constant search for improvements and a market for those innovations that promised cost-effective gains in safety or operating practices. If the railroads were the ultimate buyers in this market, many other groups participated as a railroad technological network evolved. In the antebellum decades railroad managers and technical people, independent engineers, private inventors, and supplier industries were the major sources of safety innovations. In later years, they were increasingly joined by government agencies and by technical societies and journals, which set standards and developed, evaluated, and diffused new knowledge.[25]

Institutions and Organizations Shaping Railroad Safety

Although English travelers such as Alfred Bunn were correct that American carriers were less constrained by "the protective hand of government" than were European railroads, safety was not wholly unregulated. While nineteenth-century railroads have often been taken as a symbol of unfettered free enterprise, they were in fact creatures of the state. Early carriers were often subsidized and granted the power of eminent domain. States therefore expressed a public interest in their operation from the beginning. Moreover, the very nature of railroads encouraged the belief that "accidents" could be controlled. Writing in 1852, the *American Railway Times* argued that the railroad was "the safest mode of traveling yet discovered," because "it depends more completely . . . upon fixed principles that can be understood and . . . controlled by the intellect of man." Writing of a boiler explosion, the *New Bedford Standard* simply noted "it cannot be called an accident for the boiler was known to be defective."[26]

In response to a crossing accident in 1834, Massachusetts inaugurated the first safety legislation: in 1835 it required the carriers to mark all road crossings and the engineman to sound his bell as a warning. In 1846 it required construction and maintenance of fences and the reporting of all fatal accidents. A number of wrecks on the Western Railroad in the 1840s also led the Massachusetts legislature to inquire into the company's operating rules. This was the sort of veiled threat to improve practices that I have termed *voluntarism*, and it worked. Other states followed Massachusetts with other regulations. In 1844, New Hampshire provided for a board of railroad commissioners. One of their duties was to inspect the lines' permanent way, although they do not seem to have done so. Connecticut, in response to the Norwalk bridge disaster of 1853, required all trains to stop before a drawbridge and also set up a board of railroad commissioners. A New York statute of 1850 required fencing. New York also created a commission to investigate accidents in 1853, and in 1855 it instituted a board of railroad commissioners. Modeled after the British Board of Trade, the board included the state engineer; it inspected new lines and investigated accidents.[27]

These early regulatory efforts were of uneven value. No doubt requiring that companies guard crossings and warn of trains was desirable, but at least in New York, fencing laws seem to have had little impact. Companies routinely contracted out the requirement to landowners, who just as regularly pocketed the money and did nothing. Such behavior might prevent the farmer from recovering for a dead cow in court, but it did little for safety. The reporting of accident statistics began what would prove to have far-reaching consequences. Early state statistics, supplemented by private and, in the 1890s, federal accident information, helped establish patterns of causation that underlay most attempts to improve safety. In Massachusetts investigation of the Western Railroad probably helped spur that company's rule changes. However, no state routinely investigated accidents until Massachusetts began the practice in the 1870s. By default, accident investigation and analysis by technically trained, disinterested individuals fell to the technical journals and societies whose work helped identify error and diffuse improved practices.[28]

In both Connecticut and New York inspections and reports seem to have been modestly beneficial. For example, in 1854 the Connecticut commissioners claimed that they had recommended several improvements in roadbed, strengthening of a culvert, and stationing a flagman at a crossing, all of which the company undertook. In 1852 New York's lines had such an evil reputation that the *Ogdensburg Republican* claimed "a proposal has been made to our State Legislature to abolish hanging and substitute for punishment a ride upon some of our railroads." In 1855 that state's commissioners refused to allow a portion of the Black River & Utica to open until the company made improvements. It also requested copies of company operating rules, suggested appropriate modifications, and circulated them to other carriers.[29]

Of far greater importance than these early "hard" regulations was the system of accident liability that grew up. The common law of railroad accidents was an extension of earlier liability rules for common carriers such as stagecoaches and of the laws governing contracts between master and servant. By the 1850s this law was simultaneously simple and complex. It was simple because a few general principles governed the outcome of most cases; it was complex because cases were forever arising that were exceptions to, or modifications of, these rules. The general principles essentially all boiled down to two: railroads were responsible for injuries arising from negligence, and they were not responsible where there was contributory negligence on the part of the injured party, or where workers had freely agreed to assume the risk. As interpreted by the courts these rules made carriers strictly liable for nearly all passenger casualties that arose out of train accidents and for little else. The machinery of the law, as opposed to its content, might also protect employers. Protracted trials gave the advantage to defendants with deep pockets. In one case the New York Central ran over a ten-year-old girl in 1870, rendering her a double amputee. The courts finally awarded her $7,500 (roughly sixteen years' earnings), but the process took four trials, seven appeals, and seventeen years. Companies sued in state courts might also claim diversity jurisdiction and remove a trial to a faraway federal bench, thereby placing additional burden on a plaintiff.[30]

The courts held railroads to a high standard of care in their carriage of passengers. While they were not responsible for such acts of God as a lightning strike, in general, injury to a passenger in a train accident was regarded as prima facie evidence of negligence—in effect, a standard of strict liability. Thus, if an accident that injured passengers resulted from a mistake by railroad personnel, a broken axle, a washed-out bridge, or a cow on the track, the company would be liable. Moreover, in several cases courts ruled that railroads "are bound to avail themselves of all *new inventions* [emphasis added] and improvements known to them which will contribute materially to the safety of their passengers whenever the utility of such improvements has been thoroughly tested and demonstrated." By contrast, passengers who fell from trains, or jumped off early, or ran to board a moving train were guilty of contributory negligence and companies were not usually held liable for their injuries.[31]

These rules made most passenger train accidents expensive, and major disasters often cost several hundred thousand dollars. Compensation for the forty-six victims of the Norwalk bridge disaster of 1853 cost the New York & New Haven Railroad $252,000—about $5,478 per person at a time when common laborers might earn $324 and white-collar workers $585 a year—requiring the company to skip a dividend. Even minor train accidents could prove surprisingly expensive. In what appears to be an early example of a bogus injury (chapter 6) a jury awarded $9,045 (about twenty-nine years' earnings for a common laborer and nineteen for a white-collar worker) to Ezra Corning for injuries to his hip and spine suffered in a collision on the Connecticut River Railroad on February 11, 1848. Although he "did not appear to be seriously injured at the time," several weeks later, "his symptoms grew more and more alarming." When the company appealed, another jury tacked another $1,000 to the award. "In these class of cases," the *American Railway Times* concluded, "the railroad company always comes off second best."[32]

Such rules yielded very different results when applied to the carriers' liability for injuries to employees or others. Few employee accidents were compensated because—courts held—men had assumed the risk and had most likely been negligent themselves, or perhaps victims of negligence by fellow servants. Thus, brakemen who rode freights and were killed by hitting a bridge or falling from the car were due nothing. Nor would an engineman injured in a derailment be likely to receive compensation from the courts. Similarly, a pedestrian hit at a crossing had—almost by definition—been negligent, while someone walking the tracks was trespassing and in neither case would the company be liable. Indeed, the *American Railroad Journal* worried that automatically exonerating enginemen in crossing accidents would discourage them from taking precautions. When one court denied damages to a family whose small boy had been run over, reasoning that children "are an expense as a rule and not a pecuniary benefit," an outraged critic wondered if the carrier might then sue the parents for services rendered. Data on the compensation costs of such injuries are sparse. In Ohio, the average cost of a reported accident (injury or fatality) to a worker from 1873 to 1888 was $134 (about four months' earnings). For other persons, excluding passengers, over the same period average accident costs were $92.[33]

Such in broad strokes was the common law, but its application was tempered in a number of ways. Violation of statute law might make a company liable. For example, if the engineman failed to sound a warning before a crossing where the state required it, the company might be liable if a pedestrian were hit. Companies were also sometimes held to a higher standard of care for children and women than for adult males. As Peter Karsten has shown, the law had a heart as well as a head—especially in state courts—and juries sometimes found for the injured despite the law. For this reason, and because it was both humane and yielded good publicity, carriers always tried to settle out of court. But the broad rules remained; they influenced the level of out-of-court settlements, and they shaped railroad safety concerns in predictable ways.[34]

Concerns with passenger safety contributed to steady if unspectacular improvements in track, equipment, and operating practices that benefited workers as well as passengers. Inventors responded to the perceived railroad interest in passenger safety with a host of patented innovations. Beginning in the 1830s governments, individuals, and scientific organizations in Britain, Germany, and the United States chipped in with metallurgical investigations and studies of boiler accidents. In the 1850s the railroads and their suppliers began to employ standardized tests to evaluate quality in the purchase of axles, wheels, and rails. Railroads also experimented to improve operating procedures.

Early technical societies such as the Franklin Institute and American Society of Civil Engineers devoted much time to railroad problems, and the 1860s saw the beginnings of specialized societies devoted to railroad improvements. The Master Car Builders dated from 1864 and the Master Mechanics from 1868. Finally, during the antebellum decades a number of technical publications arose, including the *American Railroad Journal, American Railway Times, Journal of the Franklin Institute, Railway World, Railroad Record, Scientific American,* and the *Proceedings* of the British Institution of Civil Engineers. They began to publicize and analyze accidents, and assess and sometimes advertise safety devices and procedures. More broadly, these groups provided not only the information that *allowed* better safety; as courts sometimes held and as Thomas Haskell has argued, that very fact created a moral imperative to act.[35]

Of these publications, the most important during antebellum times was the *American Railroad Journal.* Founded in 1832, the *Journal* was the major source for diffusing the emerging techniques of railroad engineering. It published important company reports, reprinted articles from other technical publications both at home and abroad, and published letters on all aspects of railroad engineering and management. The *Journal* also advertised many safety devices, while the editors, especially Henry Varnum Poor, who took over in 1849 and remained until 1862, routinely commented on accidents they thought reflected poor practice and urged adoption of better equipment and methods. By such activities the *Journal* and other technical publications helped weed out bad practice and encouraged the diffusion of safer methods. Although nearly every aspect of railroading attracted the attention of inventors, six areas were of particular importance in the years before 1870: couplings, brakes, track and its appurtenances, boilers, wheels and axles, and management.[36]

Innovations in Equipment and Track

While early trains were simply chained together, the link and pin coupler was in wide use by the 1830s on both freight and passenger equipment. It worked the way it sounds. In coupling an engineman backed the train toward a waiting car while a brakeman walked between the cars holding a large horizontal chain link attached to one car that he would insert into a slot on the second car and then hold with a pin. The link and pin represented an innovation that almost certainly worsened worker safety. Its use was not only dangerous to trainmen (chapter 4); companies found it expensive as well, for the pins disappeared by the thousands. The slack that resulted from coupling with the link and pin also resulted in jerky starting and stopping that might toss passengers around the car. The link and pin freights broke in two so frequently that Edwin Price, engineman on the Nashville & Chattanooga, developed the habit of following prior trains closely up hills, so that when the inevitable break occurred the wild cars would have little momentum when they hit him. Finally, cars so coupled might telescope during a collision or derailment, with the rear cars crashing through the frame of those preceding and doing much damage.

Of the many early efforts to improve couplings, the most important was Ezra Miller's combination of a coupling and buffered platform intended for passenger service that he first patented in 1863. The device included a coupler that looked like a crochet hook. It coupled automatically, was held together by the spring-loaded buffer, and could be uncoupled from outside the car. Of equal importance was the trussed platform that was intended to prevent telescoping. The platform also reduced the likelihood that a passenger might fall between cars.[37]

In late 1863 the *American Railroad Journal* described Miller's device and urged its adoption. In 1865 the *Journal* returned to the subject, with a similar article and a notice that the Chicago & North Western had begun experimenting with Miller's device in 1864. In 1866 the Erie adopted Miller's platform, and in 1867 New York passed a vague law requiring companies to have better platforms that probably boosted Miller's device. So did some well-publicized wrecks. A train on the Allegheny Valley equipped with Miller's device slammed into a mud slide at 40 miles an hour on May 25, 1870, but it did not telescope and there were no casualties. In August of that year the Michigan Central also reported a collision with no casualties, thanks to the use of Miller's platform. Thereafter, use spread rapidly and by 1874 Miller claimed that 587 roads, or about 84 percent of all carriers in the United States and Canada, were using his device. Since these tended to be larger carriers, probably more than 90 percent of all passenger miles ran on Miller-equipped cars—a spectacular rate of diffusion for a new technology.[38]

Yet Miller's device was both less successful and more complex than early writers let on. It reduced but did not prevent telescoping, for if a rear car rose above the frame of a preceding one, Miller's platform became a battering ram rather than a safety device. The platform is also a good example of a device that, by making trains safer, changed carrier behavior. As the *Journal* enthusiastically noted, "under the old system 20 miles per hour is the maximum of *safe speed!* Under the new, 60 and more miles per hour may be run with greater safety."[39]

Thus, Miller's platform, like better brakes, steel rails, and many other innovations, did not simply improve safety. Rather, all of them expanded managers' menu of choices, allowing some combination of both heavier or faster trains *and* greater safety. This is depicted in the two sets of production possibilities in Figure 1.5. Depending on the comparative payoff to safety and service (the slope of lines 1 and 2 in Figure 1.5), carriers might, as the *Journal* suggested, choose to increase train speed so that safety is only marginally improved. Such a choice is depicted in Figure 1.5 in the movement from *A* to *B* rather than to *C*.

Brakes were perhaps the weakest link in early railroading. Initially trains simply didn't have any. By the late 1830s most American cars had hand brakes that worked on a single truck per car; the brakeman turned a wheel that tightened a brake pad, usually made of wood with a leather face, against the wheel. Because on passenger equipment the wheel was usually on the platform between cars, a brakeman who remained at his post during a collision was likely to be crushed. Such a system was not only dangerous to the brakeman; it simply didn't work very well. Moreover, since one brakeman could not usually brake more than three cars, longer trains would be more poorly controlled unless they carried more brakemen. In a

Figure 1.5. **The Effect of Managerial Choices on the Payoff to Safety Equipment**

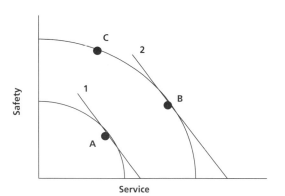

land of high wages this was an unattractive alternative, limiting its use to passenger trains.

In the 1850s a number of devices were invented and widely adopted that allowed one hand brake to work brakes on both car trucks, essentially doubling braking power. There were also endless attempts to develop some form of continuous brake. Self-acting brakes were one solution, because it seemed plausible to use the energy of the train to stop it. As early as the 1830s, the Reading used a self-acting brake on its coal cars. Another version, described in the *American Railroad Journal* in 1836, worked by allowing the car to slide forward on a frame, thereby activating the brake, but it seems to have found no favor. In the 1840s, George Griggs, superintendent of machinery on the Boston & Providence, also invented a band and pulley brake that relied on the momentum of the car for its power but could be activated by the conductor or engineman. In the 1850s William Creamer developed a spring-loaded brake intended for use only in emergencies that found some favor. When the engineman pulled a cord it released the spring and set the brakes. Patent office records reveal that in the four decades after 1830, hundreds upon hundreds of other inventors tried their hand at improving train brakes. Experimentation with a compressed air brake began in England in the 1840s.[40]

In the late 1860s Westinghouse developed the brake that would dominate railroading. The initial version, called the straight air brake, contained a steam-powered air compressor with a reservoir on the engine, a controller worked by the engineman, a train pipe with flexible connections, and a cylinder with pistons and mechanical leverage that worked the brakes. When the engineman activated the controller it emitted air into the train pipe, which pressurized the cylinders and applied the brakes.

Although superior to its competitors, the straight air brake had two important drawbacks: it was too slow for increasingly fast trains and it did not fail safely, for if a train should break, draining the air from the pipe, the last half lost its brakes. These problems led Westinghouse in 1872 to develop an automatic air brake that contained key modifications. Now each car had its own air reservoir that applied the brakes when the engineman *reduced* pressure in the train pipe, which made it failsafe should the train break apart, and in addition speeded up brake application.

The new brake was difficult to maintain and required considerable skill to operate (chapter 3). In simplified form, it worked as follows. An ingenious triple valve connected train pipe, car reservoir, and brake cylinder. With the brakes off, train pipe pressure (typically 70 psi at the time) kept the valve linking the car reservoir and cylinder closed and reservoir pressure also at 70 psi. When the engineman wished to apply the brakes, he shifted the valve handle to let (say) 10 psi from the train pipe. The lower pipe pressure opened the triple valve between the reservoir and cylinder and sent the 10 psi of air to the cylinder, which applied the brakes. In an emergency stop the engineman turned the valve handle to dump air from the pipe. But there was a catch—if the train had made several recent service applications there might be no more air in the reservoirs. To release, the engineman shifted the brake lever to "release and charge"; train pipe pressure would rise and when it exceeded reservoir pressure the triple valve shifted to connect the train pipe to the reservoir and also vented the cylinder to the atmosphere. But here again there was a catch—*there was (and is) no graduated release.* Thus he might face the choice of stopping too soon if he applied too much air, or releasing the brakes and causing a runaway. To avoid such problems the brakes came with a retaining valve that could hold pressure when the engineman shifted the lever to the "release and recharge" position, but retaining valves had to be manually set on each car, usually at the crest of a grade.[41]

Westinghouse faced competition from a host of rivals, of which the most important was the vacuum brake. The air brake triumphed in part due to its inventor's marketing skills and connections and in part because it was decisively superior in two key areas: speed and power of application. Because the vacuum had to be created to stop the train whereas the air was already under pressure with the Westinghouse brake, the latter could stop a train more quickly. While the air brake might yield pressures of 70 psi, the vacuum brake typically yielded only 8 to 10. As one speaker told the Master Mechanics in 1874, "[which brake is the safest] is a matter of much more importance than merely a few dollars in the first cost of the brake or in the cost of keeping it up . . . The brake that is the most effective is the cheapest."[42]

By 1876, carriers with about 75 percent of all passenger equipment employed some form of Westinghouse brake, and since these were on the high-density lines, like Miller's platform it probably protected 85 to 90 percent of all passenger miles.[43]

The Westinghouse brake and Miller's platform were complementary safety innovations: the platform would have little effect in collisions at high speeds, while even if a train were braked to very low speeds a wreck might take many lives if cars telescoped. An example of their value as a package comes from a collision on the Boston & Albany, near Brimfield, Massachusetts, on October 3, 1873. The eastern express was traveling about 40 miles per hour when the engineman saw a switch set for the siding some 400 feet ahead and applied the brake, slowing the train so that while it ran into some freight cars, Miller's platform prevented telescoping and no one was badly hurt. Of course, as the *American Railroad Journal* noted with respect to Miller's coupler, if the train had not been equipped with such safety devices it would not have been going 40 miles per hour. Indeed, some years later, after brakes had been applied to freight trains, one railroad manager estimated that they allowed him to double his average speed. As with Miller's platform, managers took some of the benefits of the new brake in greater speed rather than improved safety.[44]

Improvements in track and rail were less dramatic but equally important. By the 1840s most companies laid rail directly on cross ties, having scrapped the use of stone blocks or longitudinal stringers. The T rail was invented by Alexander Stevens and first appeared on the Camden &

Amboy in 1831. As trains grew steadily heavier, companies gradually substituted it for strap iron rails and they increased its size. Early versions
were pear-shaped, but as understanding of the rail as a girder grew, later
shapes were deeper with a sharper angle between the head and web to hold
a fishplate (the splice bar that joined ends of rails). But by the 1850s and
1860s wrought iron rails had reached the limit of their ability to sustain
weight increases. On the Reading the annual number of rails replaced
relative to tonnage rose steadily from about one in four hundred in 1848 to
one in 150 in the early 1860s. In 1865 the Pennsylvania Railroad explained to
the stockholders that its shift to steel had been motivated by "the rapid destruction [of iron rails] from ordinary wear . . . [and] the risk to the trains
from their frequent breakage." Similarly, on other major lines, rail breakage
and deterioration motivated experimentation with the new material in the
1860s, but in the minds of most railroad men steel remained experimental
until about 1870 (chapter 2).[45]

The weak link in rail remained the joint, however. As engineers almost immediately grasped, two unconnected rail ends, sometimes in a
chair, but often simply spiked to the tie, lacked any strength as a girder,
thus putting the full weight of the train on the tie, which soon settled,
causing problems noted above. In the 1840s the *American Railroad Journal* advertised a number of compound rails that essentially did away with
joints, and while these were tried on such lines as the New York Central
and Utica & Schenectady, they proved a maintenance nightmare and were
abandoned. Instead, companies turned to fishplates, which were also first
employed on the Camden & Amboy in the 1830s to join the rails. These
were an advance, and they were, in turn, greatly improved.[46]

If joints were danger spots, so were switches because the widely employed stub switch broke the main track; if not perfectly lined up it might
cause a car to climb the rail, leading to a wreck. "The switch, in this railway
day, is far more a terror to manhood," punned the *American Railway Times,*
than its old-fashioned ancestor had been, even in the hands of a "semi-sav-
age schoolmaster." In the late 1840s Tyler's Safety Switch appeared. The engineering editor of the *American Railroad Journal* described it as "the best
switch we have seen." Although the Tyler switch also broke the main track,
cars coming toward it from the wrong track would not derail, because the
wheel flanges would ride on a casting forming an inclined plane and land
on the right track. Carriers in New England adopted it widely and it was
used on the Lake Shore and some other lines. The Wharton Safety Switch,
introduced about 1865, worked on the same principle but did not break the
main track. By the 1870s it was widely used on such major carriers as the
Pennsylvania, the Chesapeake & Ohio, and Lehigh Valley, as well as many
smaller roads.[47]

Railroad technology was also the beneficiary of new knowledge stimulated by dangers elsewhere. In 1830, when the steam boat *Helen McGrego*r
blew up at the dock in New Orleans, killing over forty people, it moved the
Franklin Institute to investigate boiler explosions, a project that the U.S.

Treasury soon subsidized. The study refuted such popular theories as the claim that water under pressure decomposed into hydrogen and oxygen. It showed that the fundamental cause was simply the development of too much pressure for the boiler and it also demonstrated that overheating iron would weaken it, thus revealing the dangers of boiler encrustation or low water. These findings did not immediately eradicate all error: in 1849 one writer was still claiming that at high temperatures steam would turn into "stame" with disastrous results, while a decade later another claimed that his "Electrical Insulated Rod" would prevent explosions. Nor did they end the problem of boiler explosions on either steamboats or railroads. But they did focus the carriers' efforts on the need to wash boilers and maintain water levels. They also established the institute's expertise, leading a number of railroads to call on it to investigate later boiler explosions, and lighting a trail for other researchers as well.[48]

Railroads also contributed to the common pool of technological knowledge. In Britain construction of the tubular Britannia bridge about 1850 led to much experimentation with the use of thin-walled, riveted, wrought iron tubular structures. The new knowledge was of immense value in both building and ship construction. At about the same time, concerns with the safety of iron railroad bridges led to the formation of the English Iron Commission, which made a major contribution to metallurgy. Research by William Fairbairn and others amassed evidence on the breaking strength of both wrought and cast iron, as well as the concept of its elastic limit—the point beyond which, if strained, material would not return to its original shape. They also discovered that iron repeatedly strained beyond this point would break. In Germany August Wöhler's investigation into axle fractures discovered that iron repeatedly stressed always broke at sharp corners; he found a limiting stress below which it would not break. These results were widely reported in the United States and they encouraged not only more scientific design but metallurgical testing as well. In the 1850s various tests of the strength of iron were performed and publicized by the U.S. Navy. In 1861 one engineer also reported that he had the Watertown Arsenal test the iron for a bridge being built on the Boston & Providence for both elongation and tensile strength. Such tests were not limited to bridges, but spread to other structural members as well, including axles and wheels.[49]

The large number of failures of axles and wheels on early lines reflected a combination of poor road and track along with deficient understanding of metallurgy. Early axles, wheels, and rails were often proportioned largely by rules of thumb, while companies knew nothing about iron quality and purchased on trust alone. In the early years improvements were the products of both research and experience. In 1853 managers of the Syracuse & Utica revealed that they were aware of the then-accepted explanation of metal fatigue. "There is a theory," they reported, "that iron changes its fiber by long use . . . and it may break suddenly . . . when, originally, it was of best quality." The company also noted that since it increased the size of its axle journals to 3 inches "there are much [*sic*] fewer breaks of axles than before."[50]

In 1840, the dangers of broken axles also resulted in Kite's Patent Safety Beam. As Wöhler was discovering in the laboratory, metal fatigue led axles to break at the junction with the wheel. Joseph Kite invented a double beam running the length of the car from which two collars surrounded the axle near the wheel. Should the axle break it would drop into the collar without derailing the car. It was employed on a number of New York roads and on the Philadelphia, Wilmington & Baltimore and prevented many accidents. Changes in operating procedures could also offset dangers from wheels and axles. By the 1850s, some New York lines ran them only one year under passenger equipment and then transferred them to freight cars. By this time many carriers required a trainman to sound the wheels on passenger trains with a hammer at each stop—an early example of nondestructive testing. The Pennsylvania, New York Central, New York & New Haven, Michigan Southern, and probably others followed this practice while the Boston & Lowell removed wheels to check axles during the winter.[51]

In the 1850s some British railroads began to subject axles to drop tests, and by 1869, the Pennsylvania Railroad performed a drop test on every axle. These discoveries and tests were widely reported in early technical publications such as the *Journal of the Franklin Institute,* the *Proceedings* of the British Institution of Civil Engineers, and the *American Railroad Journal.* Such procedures no doubt encouraged manufacturers to improve quality. Some began to stamp the manufacturing date on wheels and axles, thereby simplifying railroads' efforts to achieve quality control. They also experimented with rolling axles instead of forging to achieve a more homogeneous product. By 1860 the Pennsylvania and some other railroads were using rolled axles under heavy cars. The impact of these changes can be traced in the reports of the Philadelphia & Reading Railroad. That company's axle failures fell from about 230 per year in the 1850s to about thirty in the 1870s, even though it owned many more cars during the latter period.[52]

Intrafirm Institutional Innovation

Within companies innovations in rules and norms were more subtle although no less important in improving safety. Their thrust, invariably, was to improve communication and enhance managers' ability to monitor and control operations. Even the tiny early railroads were quite large compared to most businesses of their day. The Utica & Schenectady, for example, was by no means New York's largest carrier in 1840, but it was about 78 miles long and yielded annual revenues of $360,000, making it far larger than most textile mills of the day. The Western Railroad was even larger when completed between Boston and Albany in 1841, and size complicated its operation because the longer road resulted in more train meetings. By 1850, the Erie operated nearly 500 miles of main line. Railroad operations involved more people, doing more complex, interrelated operations over greater distances than virtually any other human activity. Size worsened the agency problems common to all carriers because it made communication, monitoring, and control more difficult. As a number of writers have pointed out, the need to manage such far-flung enterprises with safety and accountability led the carriers to develop military-style organizational structures with detailed operating rules and to strive for rigid obedience.[53]

From the beginning companies realized that control of their operations required timely communication of vital information. As early as 1832 Alexander Black informed the directors of the South Carolina Railroad of the success of his "system of supervision, which was undertaken to protect, maintain, and keep the road in order." Black required monthly reports of maintenance work that he hoped would discover the weak spots of the road. Functional specialization came more slowly because it was unnecessary on small early carriers. In 1839 a superintendent was in charge of everything on the Tonawanda, while the Auburn & Syracuse employed as managers only a superintendent and an engineer. About the same time, the Boston & Providence employed a master of transportation, a master mechanic, and baggage masters.[54]

By the late 1840s, the evolution of larger carriers such as the Pennsylvania and Erie increased the need for information and specialization. They responded, developing complex bureaucracies with detailed rules. One of the earliest American statements of the requisites for good management came from the Erie in 1850. Superintendent Daniel McCallum emphasized that the key to good management was "*personal accountability.*" To achieve it he stated four main principles: (1) proper division of responsibility; (2) adequate power for responsible individuals; (3) a reporting system to discover whether responsibilities were executed; and (4) prompt reporting. The Erie innovated a line and staff organization that reflected these goals. Divisional and branch superintendents, a master of engines, a wood agent, and other officers each had clearly defined duties and reporting requirements. Such a system, McCallum thought, would yield the information that would allow the general superintendent to "detect errors immediately . . . [and] point out the delinquent." After the 1850s many large carriers developed and constantly tinkered with divisional and departmental structures to achieve these goals. These systems provided companies an institutional memory that kept track of such matters as axle quality, allowing them to learn from their own experience and that of others when such information was shared through technical publications. As a result they could direct investment and innovation to enhance both efficiency and safety.[55]

McCallum's principles were reflected—often imperfectly—in many of the early rules governing the movement of trains. All the early lines employed a timetable and set of rules and signals that governed its use and other aspects of train control. Initially rules were few enough to fit on a single sheet of paper; by the 1870s they would number in the hundreds and take up entire books. There were also whistle, flag, and lantern signals to communicate a host of road and train conditions such as a crossing approach, or that a train should be backed up, or that there was a second section to the train following, and many other matters as well. Some of the earliest surviving operating rules are from the Boston & Providence and were in effect by 1839. In the railroads' military-style organization the conductor was like a U.S. Navy captain: the Boston & Providence's rules emphasized that he was in sole charge of the train and that he should wear a badge. Since train meets were governed by time, he was required to compare his watch with the clock in the depot. Since the unreliability of early equipment routinely resulted in late trains, a set of rules governed such situations. Rule 15 stated that when a train waited on a turnout for another train that was delayed more than 30 minutes, the conductor should send out a brakeman. If he failed to return in 30 minutes, the train might proceed "carefully and cautiously," the definition of which was left up to the conductor. Similarly, Rule 16 required the late train to proceed with "the greatest of care." These two rules reflected the realities of early railroading: trains broke down regularly, and no one knew where they were, but both rules were vague and therefore dangerous.

About the same time, the Baltimore & Susquehanna had similar regulations and, it warned, "no person will be retained . . . who in any instance disregards them." Passenger trains ran under the direction of conductors; the enginemen governed freights, but the rules were silent on who ran mixed trains. Passenger trains would only wait 30 minutes at turnouts and again in vague language the conductor was told to "proceed slowly." By 1852, the Ogdensburg Railroad had rules requiring the engineman to carry a copy of the timetable and forbidding anyone from talking to him while the train was in motion. The latter rule was also in force on the Western & Atlantic. Its wisdom had been revealed by its absence two years earlier on the Camden & Amboy. An engineman in the midst of conversation with a companion let the water in his boiler run low, resulting in an explosion that killed both men. The Western & Atlantic, Georgia Central, and Camden & Amboy all had a truly amazing system of train rights: when trains met, the one closest to a turnout was to back up to it. Of course, this encouraged train crews to speed, both to go beyond a turnout and, if that failed, to back into it. The result, on the Camden & Amboy on August 29, 1855, was a spectacular disaster that killed twenty-three people when a train backing up too fast hit a horse and buggy and derailed.[56]

Such experiences led companies to modify their procedures and the trend was toward what Frederick Gamst has called a "web of rules" that became ever more complex. Between 1840 and 1842, the Western Railroad experienced a crossing accident, a head-on collision, and two derailments that together killed nine people. In response the company developed procedures to compensate victims and it established a committee to investigate operating practices. The committee concluded that the system of train control was far too lax and the company responded with organizational changes to improve procedures and fix responsibility. The road was divided into divisions and roadmasters were to inspect it daily; trains were to carry at least three brakemen (one per passenger car) and were under the sole control of the conductor. The rules also specified that conductors should not start trains "under any circumstances" before the time specified, and required conductors east of Springfield to set their watches by the clock in the upper depot in Worcester. They also specified where trains would meet along the line. Brakemen were to inspect journal bearings at each stop, and enginemen were to sound the whistle and bell on every curve. By the 1850s the Boston & Providence and some other carriers were also furnishing high-quality watches to trainmen and having them inspected, practices that gradually spread.[57]

Use of the telegraph to control train traffic began in Britain in the 1840s, due to problems resulting from rising traffic density. Its first use in the United States was in 1851, on the Erie. Prior to the telegraph, once a train left a station it was like a ship at sea—entirely under the command of the conductor since no one save the crew knew its whereabouts. Use of the telegraph greatly reduced the discretion of the train crew—in the words of one old-time engine driver, "the fingers of the old brass-pounders [telegraph operators] were stretching out after him, trying to harness him to the telegraph keys."[58]

The new technology also led to organizational innovation, creating the new position of train dispatcher. Each station with a telegraph would inform him when a train passed. Thus, in the event of a breakdown, the dispatcher could issue a train order, informing the waiting train to proceed to the next station and holding the late train should it arrive. In later years to improve safety companies also developed complex rules for the delivery of train orders. On most lines agents delivered two copies on special forms—one to the conductor and one to the engineman—who might be required to read them back to the agent.

By reducing train delays this system increased productivity, and because the train order system could prevent rear as well as head-on collisions, McCallum claimed that a single-track line with the telegraph was safer than a double-track without it. Indeed, when the *American Railroad Journal* reported the disastrous head-on collision at Camp Hill on the North Pennsylvania Railroad that took sixty-six lives in 1856, it excoriated the company for not employing the telegraph, claiming that because of it, the Erie had experienced a "singular exemption from accidents." By the end of the decade, not only the Erie, but also the Pennsylvania, B&O, New York Central, Michigan Central, and Michigan Southern were employing the telegraph for train control, and its use soon became universal.[59]

But no matter how well-designed the rules, even with use of the telegraph, agency problems remained because most employees worked unsupervised and even many supervisors had little incentive to follow company rules. As one observer concluded, "all experience shows that from incapacity, unfaithfulness, misconception of orders, inattention or mistake, conductors or engineers are never implicitly to be relied on." When one engineman on the Nashville & Chattanooga who routinely trespassed on the time of other trains finally caused a wreck, the company discovered that he had no watch. On September 5, 1856, a misplaced switch at a station led a New York Central train to run onto a siding, hitting a cattle train and killing six people. While the switch had been incorrectly set, the more basic problem was the failure of the engineman to control his train. Companies experimented with a variety of contractual incentives to penalize such behavior and reward those who obeyed rules. In the 1840s the B&O developed a labor contract that not only fired workers who caused accidents but withheld back wages as well. The Erie also attempted to fix responsibility and fire those such as the New York Central engineman who had accidents. In the 1850s the Illinois Central would "dismiss from the service any men . . . who disregard or neglect the rules or regulations." Some companies employed positive incentives as well, and in the 1850s both the Little Miami and the Columbus & Xenia began to reward (with a silver pitcher) those enginemen who had the fewest accidents per train mile, while the Western & Atlantic paid a wage premium to employees who didn't drink alcohol.[60]

Safety problems also resulted from the immense labor turnover that characterized nineteenth-century railroads. Much knowledge of safety and other aspects of technology was tacit—unwritten and often involving subtle tricks of the trade such as the best way to light a switchman's lamp in a high wind. In addition, knowledge was often unique to the firm. On railroads safety reflected a host of conditions, including rules and signals, the position of close clearances, the location of yard hazards, the eccentricities of locomotives, and dozens of other details. When veteran engineman Edwin Price joined the Western & Atlantic in the 1850s he spent two weeks learning the road and memorizing the location of every curve before he took out a train. A collision on the Chicago & Alton resulted when a new worker who was unaware that it used a type of switch with separate levers for the switch points (the rails) and sign threw only the first lever.[61]

As Oliver Williamson and others have stressed, where workers have firm-specific skills their loss is expensive to both workers and employers, and so both have incentives to develop long-term relationships. And so both agency problems and turnover led companies to develop internal labor markets that promised promotion to faithful, long-term employees, thereby raising the costs of both dismissal and voluntary separation. In the 1850s the Illinois Central fired those who disobeyed rules, but it ensured that the policy was applied fairly and it filled vacancies from the lower ranks. By the late 1860s promotion from within was widespread among the carriers. The object was to develop a corps of workers who were "convinced that their own welfare will best be promoted by an implicit obedience of orders." In later years companies developed medical, benefit, and disciplinary plans that were in part intended to reduce turnover (chapter 6).[62]

These incentive schemes all suffered from two flaws. Middle managers sometimes rewarded trainmen not only for following the rules, but for breaking them as well if it meant bringing a train in on time. The inability to supervise trainmen ensured that there was seldom any sanction for dangerous behavior as long as it didn't cause an accident. Thus enginemen routinely ran too fast around sharp curves, trusting to luck they would not get caught. But sometimes luck ran out. The Providence & Worcester rules enjoined enginemen to "proceed with caution" if the train was late. But on August 12, 1853, a passenger train that had been late for a meeting steamed around a curve at 40 miles per hour near Valley Falls, Rhode Island, colliding with another train and killing thirteen people. Not until the twentieth century were these problems adequately addressed.[63]

The Origins of Disaster

The Valley Falls collision was but one of a number of disasters that suddenly erupted during the 1850s. The 1830s and 1840s yielded many casualties and small-scale accidents, but few major train wrecks. Paradoxically this picture changed sharply in 1853 even as overall railroad safety was improving. As noted, on May 6 of that year an engineman on the New York & New Haven ran his train into an open drawbridge at Norwalk, Connecticut, and drowned forty-six people. There were ten other major accidents in 1853, of which the worst were a crossing collision on April 25 near Chicago involving an immigrant train that killed twenty-one, and the head-on crash on the Providence & Worcester described above. The catastrophic butting collision at Camp Hill, Pennsylvania, that killed sixty-six came only three years later in 1856.

This eruption of accidents excited much comment in newspapers and the trade press at the time and its causes have puzzled historians ever since. According to the *American Railway Times,* "some epidemic madness seems to have seized [railway] . . . management." The *Albany Evening Journal* blamed excessive train speed, a position the *American Railroad Journal* described as "superficial." The *Journal*'s editor attributed the wrecks to weak rails, rotten bridges, and generally poor maintenance. The historian Seymour Dunbar agreed with both. The early absence of catastrophe he attributed to slow speeds and absence of night travel, while the outbreak of disasters reflected fast running with flimsy cars over poor roadbeds. A modern writer, Robert Reed, similarly blames rising speed.[64]

The difficulty with these views is that they suggest that the increase in disasters reflected worsening safety but, as Figures 1.2 and 1.3 reveal, worker and passenger safety was probably improving. It is certainly true that increasing speed can turn an otherwise minor accident into a disaster, but if that were the problem passenger fatality rates should have risen. Instead, this upsurge in spectacular train wrecks almost certainly resulted simply from an increase in exposure as large trains became more common and passenger train miles increased. To explain this argument we need evidence on both the frequency and severity of train wrecks.

The first reasonably complete nationwide reporting of train accidents began in 1873, when the *Railroad Gazette* first published its monthly compilation, which it continued to issue in increasingly elaborate detail down to 1901. As its editors repeatedly warned, the *Gazette*'s data were a very deficient measure of total railroad casualties, for they ignored all individuals who were hurt in ways not involving a train accident. In addition, they were a poor measure of all train accidents, for they failed to count many minor accidents in which no one was hurt. But, the editors insisted, the *Gazette*'s data were a reasonably complete compilation of those train accidents that involved injuries, and subsequent comparisons with ICC data proved them right. Since there is no generally accepted definition of the term *disaster*, I will define it as any train wreck that kills at least six people. By this definition, the *Gazette*'s data yield thirteen disasters from 1877 to 1880, a period marked by the Tariffville bridge collapse that killed thirteen and several other major wrecks.[65]

The *Gazette* data can be employed to estimate just how common disasters were during the 1870s. From 1877 to 1880, there were 3,619 reported train accidents and I have calculated that there were about 610 million passenger train miles (Appendix 1). Thus disasters were extremely rare events, comprising only 0.36 percent of all accidents serious enough to be reported, and occurring about $13/610 = 0.021$ times per million passenger train miles. If these relationships held for earlier decades, they help explain the absence of early disasters. I estimate that American railroads traveled about 48 million passenger train miles from 1840 to 1849 (see Appendix 1). If they were as risky as railroads of the 1870s we would expect there to have been about one disaster ($0.021 \times 48 = 1.02$). In fact, there were two. On the Boston & Worcester on November 6, 1847, a derailment killed six passengers while on the Eastern a year later on November 3, 1848, a collision also took six lives. The point remains: the early freedom of American railroads from disaster was probably a matter of luck. But the upsurge of the 1850s cannot be similarly explained. There were about 190 million passenger train miles during this decade, leading us to expect about four disasters if the risks of 1877–80 had obtained. In fact, there were sixteen. The likelihood that this was a chance occurrence is miniscule.[66]

Thus the disasters of the 1850s stood out not only compared to earlier periods, but to *subsequent years as well*. Although the late nineteenth century was littered with major train accidents, no similar period reveals as many disasters relative to passenger train miles. Even the 1886–88 period, which was marked by a spate of memorable wrecks, yields an incidence of disasters of only 0.03 per million passenger train miles.[67]

One possible explanation for the eruption and then relative decline of disasters reflects both train and car size. As noted, early cars held no more than fifteen or twenty passengers, but by the 1850s they might contain sixty people. Since wrecks often severely damaged only one car, this increase in size raised the potential for disaster. Train size also increased. New York data suggest that the average passenger train carried about fifty-three people in 1839; in Massachusetts in 1846–47 the average of all carriers was fifty-five. By the mid-1850s average train size in New York had risen to sixty-four. Although the lack of data on individual carriers prevents computation of the variance of train size, it is in the nature of statistical distributions that a small increase in the average implies a disproportionate increase in the upper tail. Hence there must have been a disproportionate increase in the number of large trains. In fact, by the 1850s both immigrant and excursion trains sometimes carried five hundred to six hundred people. Excursion trains were likely to be dangerous for other reasons as well. They were extras and hence not on the regular timetable; they might also be run in sections. The first train would carry a flag warning that it was the first section but if a waiting train missed the flag, disaster could ensue. Both the Valley Falls and Camp Hill disasters involved large excursion trains, as did many later disasters.[68]

The growth in average train size ended in the 1860s, as the carriers began to run many more locals. New York data show that the average number of passengers per train hit a peak of ninety-two in 1866 and then declined more or less steadily. Similarly, national data reveal declining train size from 1882 on. Thus the number of very large trains probably grew more slowly than train or passenger miles and so therefore did disasters. While declining average train size reduced the potential for disaster, the Miller platform and the air brake also reduced the severity of accidents. The combined effect was to reduce the average number of casualties per accident in the late nineteenth century (chapter 2). As noted, this would lead to a disproportionate reduction in disasters.

In 1828 railroads were everywhere a new technology and almost everything had to be invented or at least greatly modified from past practice. One result of such growing pains was that carriers in all countries were remarkably dangerous by modern standards. In America, market forces ensured that by the 1850s railroads evolved in ways that sharply distinguished them from their European counterparts. Their various adaptations to cope with sparse traffic, high wages, cheap natural resources, and expensive capital yielded a distinctly American variation of railroading that was also especially dangerous. In the United States, railroads also remained "new" far longer than they did in Europe, and the immense additions to capacity that characterized the decades after 1850 meant that even as older carriers and regions were becoming safer, new lines constantly buoyed up the accident rate.

Economic expansion also shaped the labor market in ways that adversely affected safety. Finding skilled labor was difficult and discipline for infractions of rules held little penalty. Many men job-hopped and with much experience company-specific this too reduced safety. On all the carriers, growth meant increasing traffic density with all the dangers this implied for collisions and crossing accidents. At the same time, American lines were evolving into freight carriers, an evolution that worsened worker risks. As passenger train speed rose, sharing a single track with slow freights may have worsened passenger risks as well. Finally, the tiny cars and locomotives of the antebellum years slowly evolved into the monsters of early-twentieth-century railroading. The result was unprecedented problems for brakes, wheels, and much else.

Yet economic incentives also encouraged railroad managers and others to improve productivity and passenger safety as well, and the result was both institutional innovation within carriers and the emergence of a complex technical network. As the technical ability to improve safety increased, its very existence placed moral pressures on men to act. The statistical evidence suggests that both worker and passenger risks modestly declined from the 1840s on, although the dangers to others probably increased. Still the social costs of railroading probably fell throughout the nineteenth century, for nationwide figures in the 1880s (Appendix 2) are lower than those for Massachusetts in the 1850s. Despite these apparent safety gains, this same period also saw the beginnings of major disasters in which dozens of people were killed or injured. While disasters declined relative to train miles in the years following the Civil War, they increased in absolute number, making railroad safety newsworthy and a potential subject for public regulation. Such concerns, along with liability rules, focused carriers' concerns on passenger safety. Their efforts to reduce train accidents are the subjects of the following two chapters.

Off the Tracks

The Changing Pattern of Derailments, 1873–1900

When track is once brought up to the east of Chicago, British and German standard, derailments of serious moment tend to become evanescent.

—*Engineering News*, 1892

On many of our railways the track construction is far below the standard required for the safe and economical operation of the traffic which it has to carry.

—E. E. Russell Tratman, 1896

At 1:00 p.m. on January 10, 1888, a Boston & Maine (B&M) fast express pulling seven cars left Boston with about 150 passengers onboard. At 2:00 p.m., it rounded a curve on a downhill grade near Bradford station running about 30 miles per hour. The front truck of a smoking car derailed at a facing point switch, as did three coaches, one of which rammed into the supports of a water tower, which fell on it, killing fifteen and seriously injuring another forty-five people.

Investigation by the Massachusetts Board of Railroad Commissioners revealed the complexity of causes that had led to the tragedy. The smoker had broken a wheel that had damaged the switch, causing the subsequent derailments. The switch faced traffic, although good practice on double-track lines employed only trailing switches. It was also an old-fashioned stubb switch with a Tyler attachment, and so it broke the main track. The wheel was chilled cast iron and most railroad managers thought steel wheels were safer. In Britain, where steel rims were fixed to wheels with a retaining ring, such accidents were virtually unheard of. Finally, although the B&M employed car inspectors to check wheels, their performance was, as one writer put it "lax, if not criminally negligent [and] akin to . . . sitting over a powder mine." Yet none of this would have mattered had coach one veered 10 feet to either the left or right.[1]

The Bradford wreck captures themes common to many train accidents. Derailments were by far the most likely form of wreck throughout the nineteenth century, and they were far more common on American than British railroads. Like most derailments the one in Bradford seemed to result from a failure of technology. It was what Charles Perrow calls a system accident resulting from the interaction of track, wheel, and environment in the form of a water tower. But such apparent failures of technology were also matters of economics, as when steel wheels, heavier rail, or better ballasting that could reduce derailments would not repay the cost of capital. Derailments sometimes resulted from bad management or employee malfeasance as well—the B&M did not schedule enough time at stops for proper wheel inspection. Just as engineering problems often had economic roots, so managerial errors might also reflect economics, which encouraged companies to employ too few inspectors. Human error could also have a technological fix, perhaps in the form of better wheels or a more modern switch.

Yet Bradford was in many other ways highly atypical. By the 1880s few carriers used cast iron wheels on passenger equipment, although they remained ubiquitous on freight cars. Derailments were also becoming less common, although they would again burst into the news in the twentieth century. Most involved freights, but even passenger train derailments were rarely tragedies, and in most cases no one was even hurt. The decline in derailments had little to do with public policy; rather, it reflected market forces that encouraged the carriers to upgrade rail and track in proportion to traffic density along with increasing scientific engineering analysis that developed and diffused safer railroad technology. Yet potentially safer technology was sometimes employed to improve service instead, while new equipment sometimes worsened risks when introduced piecemeal into a complex technological system.

The Evolving Pattern of Derailments

The evolution of derailments from 1873 to 1900 mirrored the evolution of railroading. As in the antebellum decades derailments continued to reflect the problems of lightly built, new lines; the construction of flimsy railroads in the South and West worsened the problem. The drive for efficiency also led to ever-increasing train weight that caused a host of roadbed and equipment-related derailments. Yet there were countervailing forces as well. On eastern and Midwestern carriers heavier traffic encouraged investments in roadbed and more careful inspection and maintenance that helped offset the dangers of derailment from rising train weight. So did more careful purchasing, while the increasing sophistication of the railroad technological network improved and helped diffuse new techniques to counter the growing dangers. The net result was the introduction of new and more capital-intensive methods that raised railroad output relative to all forms of accident and injury, thereby diminishing the social costs of transportation.[2]

Accident Trends

Beginning in 1873 and continuing through 1901, the *Railroad Gazette* published monthly statistics of derailments and other train accidents. As noted earlier, the *Gazette*'s compilation excluded all injuries and fatalities that did not arise from train accidents, as well as most train accidents in which no one was hurt; it did include most accidents that resulted in fatalities. From 1889 on, the ICC also collected injury and fatality statistics. Comparison of the two data sources where they overlap suggests that the *Gazette* data did indeed include most serious train accidents. Together these sources provide the statistical underpinnings of this chapter.[3]

Figure 2.1 presents the distribution of accidents by severity for two periods in the late nineteenth century, derived from data in the *Railroad Gazette*. Most train accidents were minor affairs. In over half of all reported wrecks, no one was even injured, while disasters that killed at least six individuals were, as noted in chapter 1, extremely rare events occurring even in the worst of times at rate of no more than 0.03 per million passenger train miles. In 1880, the typical passenger train in a reported accident experienced 1.5 injuries and 0.25 fatalities. Since the average train carried about forty-six passengers, a passenger in a train wreck stood about a 3 percent chance [(1.5/46) × 100] of being injured and a 0.5 percent chance of being killed. One reason for such good odds was that early cars usually acted like shock absorbers, dissipating the energy of a collision or derailment. Thus in a head-on collision most casualties would be in the first car, while the last car was similarly dangerous in a rear collision. Many accidents occurred in which passengers in other cars were unaware that they had been in a wreck.[4]

In the late 1870s derailments took about as many lives as did collisions, but their comparative importance for both passengers and work-

Figure 2.1. **Percentage of Accidents by Number of Casualties, 1877–1880 and 1898–1900**

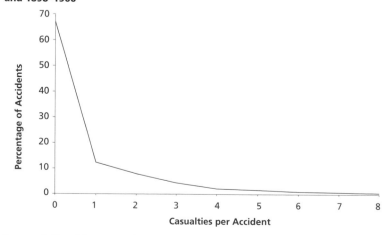

Sources: Railroad Gazette; author's calculations.

ers had faded by the century's end (Table 2.1; Figure 2.2), and this relative decline continued into the twentieth century. Passenger trains that derailed typically had fewer fatalities than trains involved in collision, and since most collisions involved two trains rather than one (typical of most derailments), a collision involved far more loss of life than a derailment. Thus while most train accidents were derailments, most disasters were collisions.

Table 2.1. **Passengers, Workers, and Train Accidents, 1878–1880 and 1898–1899 (annual averages)**

		Passengers					Workers		
	Accidents (number)	Number Killed	Rate (1)*	Passenger Trains in Accidents	Rate (2)**	Number Killed	Rate (1)*	Freight Trains in Accidents	Rate (2)**
1878–1880 Period									
Collisions	322	37	6.44	153	0.24	49	0.12	490	0.10
Derailments	545	31	4.94	191	0.18	70	0.17	347	0.20
1898–1899 Period									
Collisions	1,077	52	3.56	359	0.14	196	0.21	1,557	0.13
Derailments	1,185	27	3.20	233	0.11	147	0.16	916	0.16

Source: Railroad Gazette, various years; calculations by the author. Fatalities for 1898–99 are from the *Gazette* and differ somewhat from ICC figures. Accidents do not equal the number of trains because collisions can include more than two or fewer than one train (in a break-in-two) while derailments can include more than one train.

* Rate (1) is per billion passenger miles or per thousand employees in 1880 and 1899; passenger miles from U.S. Census and ICC; employees from U.S. Census.

** Rate (2) is fatalities per passenger or freight train involved in an accident.

Figure 2.2. **The Evolution of Train Accidents, 1873-1900**

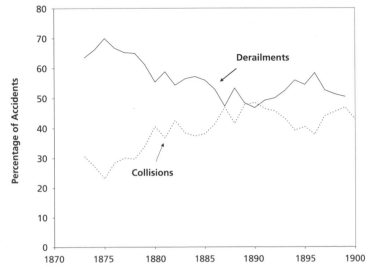

Source: Appendix 1.

Figure 2.3. **The Changing Causes of Derailments, 1873-1900**

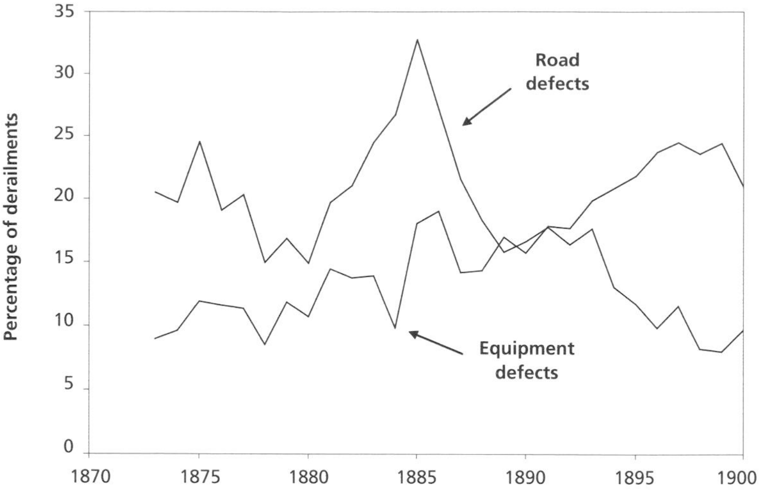

Source: Appendix 1.

If passenger train derailments were usually less lethal than collisions, fatalities per accident declined for both types of wreck. Table 2.1 indicates that fatalities per derailed passenger train declined from 0.18 in 1878–80 to 0.11 at the end of the century; fatalities per passenger train in collision fell from 0.24 to 0.11 over the same period. While some of the decline was simply a result of the reduction in average train size (chapter 1), these trends almost surely reflected improvements in car construction as well. Not only did cars become heavier, but as the *Railroad Gazette* noted in 1896, "trains now fully equipped with vestibules are quite common." These were even more effective than Miller's platform, and the journal noted that there had been a number of collisions and derailments in which the firm bracing of the cars mitigated the destructiveness of the shock.[5]

Derailments also increasingly came to involve freight trains, as Table 2.1 reveals, and after 1900 this trend continued. The evolution of derailments into a problem that largely afflicted freights resulted in part from a second trend. The causes of derailments evolved from predominantly roadbed problems and track obstructions in the early decades to equipment-related wrecks as the century ended (Figure 2.3). Collectively these patterns reflected both company investments intended to raise safety and productivity and the successes and failures of the evolving railroad technological community.

The Railroad Technological Network

Beginning in the 1870s the scientific and institutional basis of the railroad technological network widened and deepened. While its members' human capital in the form of education and experience contributed importantly to better safety, the network itself was a kind of social capital. It developed technological standards, reduced information costs, and provided peer monitoring, all of which yielded industry-wide returns. As one journal said of the Master Mechanics Association upon its founding, "the result of such meetings must be beneficial to the entire railway interest. There will be free exchange of the opinions and experience of the different men and of practice in the different shops . . . every member will be willing to add his mite of experience to the general fund of intelligence and a good deal of dust will be rubbed out of the eyes of some."[6]

Basic developments in metallurgy as well as a host of more minor and more applied advances underlay improved railroad safety and reliability. While many engineers remained empirical and self-taught, the younger men increasingly sported college degrees along with theoretical and mathematical skills. Gavin Wright has stressed the national character of technological networks during these years, and while the characterization is broadly true of railroads, nationality should not be confused with insularity. American railroad technologists read and digested European developments, making their own contributions to the common pool and borrowing selectively.[7]

As the size of the railroad technological community grew, so did the number and specialization of technical organizations and journals. Increasingly specialized technical societies overshadowed the Franklin Institute and other general scientific groups as repositories of railroad technical information. The American Society of Civil Engineers, the American Institute of Mining Engineers, the American Society of Mechanical Engineers, and various regional engineering societies were not railroad organizations, but they devoted large amounts of space to railroad scientific and engineering problems until the end of the nineteenth century.

In addition, there arose an enormous number of societies devoted entirely to railroad management and engineering. These included the Air Brake Association, the American Railway Association (ARA), the American Association of Railroad Superintendents, the American Railway Bridge and Building Association, the American Railway Engineering Association, the Master Car Builders (MCB), the Master Mechanics, the Railway Signal Club, the Roadmasters, and the Traveling Engineers, to name only a few. Beginning in the 1880s, regional railroad "clubs" that were actually engineering societies sprang up that included both railroad men and representatives of equipment suppliers and that held monthly meetings devoted to technical matters. These included, among others, the New England, the New York, the Central, the Western, and the Southern and Southwestern railroad clubs. Their growth resulted from the nationalization of railroad engineering and the difficulties that long-distance travelers still faced.

Some of these groups, such as the ARA and the MCB, developed industry-wide technological standards that would reduce transaction costs associated with interchange. Along with the large number of individual master mechanics and engineers who moved from road to road, the technical associations were also vehicles for diffusing new knowledge, and a few even sponsored research. The *Gazette* termed the *Proceedings* of the Western Railway Club "among the most useful publications sent out by scientific societies of this country." The Western also encouraged other railroad clubs to share proceedings. The clubs sometimes coordinated the review of topics serving as "focal points for thought and discussion," in the words of one writer. Such specialization was unknown in Britain, the London *Engineer* lamented, and "much good must result" from it.[8]

The 1870s saw the rise of research organizations that worked on railroad problems. The Pennsylvania Railroad opened the first corporate research laboratory in America in 1874 at Altoona, Pennsylvania, and the next year it employed Yale Ph.D. chemist Charles B. Dudley. Dudley pioneered in developing standards that the company employed in its purchasing, did path-breaking research in the chemistry of steel rails, and contributed to the development of standards for shipping hazardous materials (chapter 8). Similar organizations devoted largely to materials testing were soon founded on the Burlington (1876), Santa Fe (1883), Erie (1883), B&O (1884) Milwaukee (1886), North Western (1886), and other large carriers. Their reports and papers at technical societies ensured that the improvements spilled over to other firms as well. This routine sharing of research reflected good economic logic. Much railroad technical progress was incremental and cumulative and many innovations were therefore complementary, leading to what Robert Allen has termed *collective invention*. Moreover, the interchange of freight equipment often ensured that large carriers benefited when their findings improved track or rolling stock on other lines.[9]

Much of what was done involved equipment and materials testing; this was hardly research, but it motivated research into quality control and product improvement by suppliers of lubricants, rails, signals, wheels, and other equipment. The carriers also benefited from metallurgical investigations of rails, couplers, and other equipment at the Watertown Arsenal from 1876 on, while engineering schools and faculty also supplied contract research. Such distinguished professors as George Swain, Gaetano Lanza, and George Vose of MIT as well as Mansfield Merriman of Lehigh, W.F.M. Goss of Purdue, and others provided technical expertise to both railroads and regulators, as did respected consulting engineers such as Octave Chanute and Theodore Cooper. Although railroad engineers debated whether the Master Mechanics or MCB should found its own research facility, neither did. However, in the 1890s the MCB and other societies as well as some railroad equipment suppliers formed long-term arrangements to conduct research with the engineering schools at Purdue and the University of Illinois.[10]

Railroad technical journals helped knit together this technical community. The dominant publications date from the 1870s or later, and included the *Railroad Gazette, Railway Age, Railway Review, Locomotive Engineering, Railway Mechanical Engineer,* and, for a time, *Engineering News.*

All were edited at various times by distinguished civil and mechanical engineers, including W. M. Camp, Matthias Forney, George Fowler, David M. Stauffer, Angus Sinclair, E. E. Russell Tratman, and Arthur Mellen Wellington. The British transportation economist William Acworth noted enviously that in Britain "not one of . . . [the railroad journals] is a real railway technical paper like the . . . *Railroad Gazette.*"[11]

The journals performed two functions. First, they dramatically lowered the costs of acquiring information by publishing much of the engineering literature from both American and foreign associations, as well as government research and reams of other technical information on accidents, economics, and regulatory matters. Some developed special technical sections on air brakes or "shop kinks" that tried to formalize and communicate tacit knowledge. Second, the editors provided forceful, expert critique of both theory and practice. Matthias Forney as editor of the *Railroad Gazette* from 1870 to 1883 was at the center of these events. Although largely self-educated (he had worked as an apprentice to locomotive builder Ross Winans and had been a draftsman at both the B&O and Illinois Central), he was a tireless advocate for more scientific engineering. He was, therefore, a transitional figure in the evolution of engineering. Forney virtually invented technical journalism, gleaning and disseminating the relevant information from the meetings of the important technical societies. Forney also was very nearly the safety conscience of the industry. He not only inaugurated the *Gazette*'s monthly survey of train accidents, he analyzed their causes, blasted dangerous practices, and praised what he thought were sensible state regulations. As both an editor and a member of technical societies he pressed for safer freight cars and locomotives (chapter 4).

In 1884 Forney was replaced at the *Gazette* by the equally distinguished Arthur Mellen Wellington. Like Forney he was a transitional figure, for he was both self-taught and a powerful advocate of more scientific engineering at both the *Gazette* and the *Engineering News,* which he joined in 1887. Steven Usselman has suggested that nineteenth-century railroads represented a triumph of engineering over economics, yet Wellington saw no such conflict, for he stressed that good engineering *required* economics. He is perhaps best known for the observation that "engineering . . . is the art of doing that well with one dollar, which any bungler can do with two after a fashion." Wellington's most important contributions to safety were his campaigns for better track and bridge construction (chapter 5). Other less well-known engineer-editors such as Russell Tratman and W. M. Camp performed similar functions.[12]

Roadbed and Rails

In 1879 Matthias Forney surveyed the pattern of train accidents for the previous year. The *Gazette*'s editor was optimistic. "There has been for years a quite noticeable progress in the condition of tracks and road-beds," Forney noted. He emphasized that this was "not only in the substitution of steel for iron alone, but in the care of track and road-bed, and everything belonging to them." As a result, the number of derailments had fallen in half since 1873, while derailments from broken rails had declined from 111 to seventeen over the same period. Forney acknowledged that recent mild winters had reduced the number of accidents, as had the depression-induced decline in traffic, but he thought that some of the improvement was permanent, for winter breaks typically resulted from "inflexible, uneven roadbed."

Five years later, in 1883, Forney returned to this theme. A decade earlier, in 1873, he noted, "a very large proportion of the roads were new and badly constructed." Possibly he was thinking of a rather minor wreck on the New Haven on March 8, 1872. The southbound train broke the inside rail on a curve, near Springfield, Massachusetts, rolling several cars into a creek but resulting in only minor injuries. Investigation revealed that the outside rail had been overly elevated and the rail itself was too light, weighing only 56 pounds per yard. The ties were 3 feet on center, which was too far apart, and the roadbed only 10 feet wide, which was too narrow, and this is the reason why the train had rolled into the water. Forney believed such conditions were disappearing. "Since 1878 about 40 percent has been added to the mileage of the country, which in construction has been much superior," he concluded.[13]

Forney thought that rail failures stemmed mostly from poor roadbed that would break even good rail—whether of iron or steel. The Massachusetts railroad commissioners concurred. "Upon roads which were well ditched, drained and ballasted," the commissioners observed in 1874, "very few breakages of any kind occurred, and upon roads or sections of roads where this matter . . . of ballasting . . . had not received proper . . . care, the list of breakages was fearful." In fact, as will be seen, improvements in both rail and track contributed to fewer derailments. Better track and appurtenances also reduced derailments that the *Gazette* listed under other causes such as misplaced switches, washouts, and cattle and objects on the track.[14]

Upgrading Roadbed

As the carriers strove to reduce costs of transportation and maintenance, they began large-scale upgrading of roadbed, especially on high-density lines. Proper ballast performed a number of functions: it ensured good drainage, diffused the weight of the train, acted like a shock absorber, and held the track in place. The best material was crushed stone, followed by furnace slag or gravel. Sand, the American engineer Alexander Holley had written in 1860, "hardly deserves the name of ballast," although it was widely used, and dirt was worse. What was termed *mud ballast* caused no end of difficulties. The engineer William S. Huntington described one newly opened road in 1871: "it was in wretched condition . . . The ties were thrown down in the mud, and portions of the iron were out of sight, cov-

Table 2.2. **Accidents from Broken Rails, Winter and Summer, 1873–1880**

	1873	1874	1875	1876	1877	1878	1879	1880	Total
January–March	65	20	96	26	26	7	34	7	275
July–December	5	5	3	5	7	2	5	7	39

Source: "Train Accidents in 1880," *Railroad Gazette* 13 (Feb. 11, 1881): 85.

ered with mud, the rails warped and bent." With such ballast the roadbed could not be kept level at any time of year, thus worsening the danger
of derailments, and in the winter it froze solid. While cold weather made
high-phosphorous rails more brittle, the major problem came from frozen
ballast. In winter, with an inflexible roadbed, a blow from a flat car wheel
or a poorly counterbalanced locomotive or simply at a low spot was far
more likely to snap the rail than during any other season, as Table 2.2 reveals. Of course, not all such accidents were serious, but a broken rail in the
wrong spot could lead to disaster, as occurred on the C&O on December
28, 1889, when one caused a derailment on a high embankment near Sulfur
Springs, West Virginia, resulting in ten deaths.[15]

In 1872, the Pennsylvania began to implement a standard track that
substantially upgraded all aspects of its roadbed. Other large carriers pursued similar policies. In 1874 the Grand Trunk reported a sharp drop in
rail failures, a result of the shift to steel, the increase in rail weight, and improved support that resulted from putting ties 2 instead of 3 feet on center.
In 1878, the Pennsylvania reported only 51 out of 331,000 steel rails had broken, compared to 261 three years earlier due to "the improvement . . . in the
condition of the road-bed." By 1885 the *Railroad Gazette* reported a general
upgrading of ballast. New England roads mostly employed gravel while
the Mid-Atlantic states, especially Pennsylvania and Maryland, had about
a quarter of their lines ballasted in stone and the rest in gravel. However,
while some southern roads ballasted with stone or gravel, many used sand
or dirt. The Santa Fe still ballasted much of its road with dirt while a "large
proportion" of the newly constructed lines north and west of Chicago also
employed dirt, which probably accounts for the boom in roadbed-related
derailments during the 1880s, as depicted in Figure 2.3. A decade later a
survey of fifty-three large carriers, including many western lines, reported
that most had 6 to 12 inches of ballast under the ties, and most used stone
or gravel. In the twentieth century advances in rock-crushing machinery
reduced costs, helping spread the use of crushed stone.[16]

As chapter 1 described, early railroads often simply spiked track on
alternate ties that they placed on leveled ground, with no provision for embankments, ditching, or ballasting. Without plates, rails cut into the ties.
This was dangerous and expensive, because lateral wheel pressure might
widen gauge or tip the rail over. Ties of unequal size had unequal bearing
surface, resulting in low spots in track. Poorly spiked rail was also likely

to lose gauge after a few years when the ties had begun to rot—especially on the outside of curves where wheel pressures were greatest—and the increasing weight of locomotives worsened the problem. Hence the large number of derailments from "spreading rails," a phenomenon, the *Gazette* claimed, "really ought not to be classed as an accident. It is almost a crime." Similarly, poorly laid and spiked rails might kink on a hot day, thereby losing gauge. On one Kansas railroad, when heat bowed the rails a foot and a half, passengers had to help the crew haul the track into gauge with a chain before they could continue. Sometimes, of course, bowed rails led to derailment, as occurred to the Southbridge train on the Boston, Hartford & Erie one hot day in early April 1871. No doubt to the great surprise of passengers, "the whole track, ties and all, was suddenly lifted up at least a foot," rolling two cars down a bank into a pond and killing one woman.[17]

Tie plates dated from at least the 1870s, but early ones were expensive and too light to work well. With light trains and cheap ties there was not much motive to use them. However, as train weight increased and ties rose in price, problems of tie destruction worsened. In the 1880s heavier, better designed tie plates came into use, and by 1900 large carriers were beginning to employ them on sharp curves and on soft-wood ties. In the 1880s the cost of timber in the West made the Santa Fe the first carrier to begin chemical tie preservation but only with the twentieth-century rise in wood prices did the practice become common. Tie plates or preservatives were safety measures as well as cost-reducing investments, for both diminished risks of spread rail and the dangers resulting from maintenance work.[18]

Companies also invested in improved track appurtenances. In 1885 the Pittsburgh, Cincinnati & St. Louis found 27 percent of all fishplates broken in one 6-mile section, but by the 1890s, use of longer, heavier fishplates contained such problems. Guard rails on curves and switches to reduce derailments also came into use, and most carriers finally discarded the old stubb switch that broke the main track when set to the siding. One survey in 1896 found only split switches employed in the main track of major carriers. These could be controlled by a spring so that a train approaching from the wrong direction would push the points apart rather than derail. As a result, misplaced switches, which had accounted for about 10 percent of all derailments in the 1870s, became a much less common cause of accidents.[19]

Fencing and cattle guards also contributed to better safety. Derailments from cattle on the track had been routine events in antebellum days, and headlines such as "Bull Attacks Train" remained common well into the twentieth century, while state and company reports suggest that tens of thousands of cattle must have been killed each year. In 1876 the Missouri, Kansas & Texas reported that it had killed 1,948 animals in three western states. Still, cattle caused fewer wrecks as the nineteenth century wore on, despite the enormous expansion of the rail network into grazing lands. The growth in train size helped, as bulls found themselves increasingly overmatched by the locomotives. A few companies also gave incentives to enginemen for avoiding cattle. In 1875, the Southern Minnesota awarded a prize of $50 to the engineman who killed the fewest cattle per mile run.

Of greater importance, the use of fencing also spread. In 1870 Connecticut's railroad commissioners had been "struck with astonishment" by the miles of unfenced road; a generation later many eastern and Midwestern carriers fenced their lines. The invention and rapidly dropping price of barbed wire played a role, for unlike wooden fences it required little maintenance and was more likely to dissuade an angry or amorous bull. So did the combination of state laws and liability rules. Where laws required companies to fence they were automatically liable for any animals killed, at prices set by farmer-dominated juries. In the late 1870s the Texas and Pacific reported that the annual cost of killing livestock amounted to nearly $66 a mile. A few states encouraged fencing by placing the burden on landowners who might receive no compensation for cattle that were run over. Claim agents noted that farmers always requested payment irrespective of the circumstances of the accident, however, and that even the scrawniest animal immediately went up in value when killed by a railroad.[20]

One farmer even pressed his claim with poetry:

> My razorback strolled down your track
>   A week ago today;
> Your 29 came down the line
>   And snuffed his life away.
> You can't blame me—the hog, you see,
>   Slipped through a cattle gate,
> So kindly pen a check for ten
>   The debt to liquidate.

Instead of a check for ten, he got back a rhyme of eight:

> When farmer's swine get on the line
>   Where trains have right of way,
> And when a stake he tries to make
>   By clamoring for pay.
> He wastes his time in penning rhyme;
>   The claim's not worth a fig.
> It seems to me—and you'll agree—
>   He should have penned the pig.[21]

Cattle guards were also improved. Such guards were placed in tracks to prevent cattle from using them to get into a neighboring pasture. Early guards were simply pits under the track, but these were a menace. They would derail a train if a truck was off the rails, and they often trapped cattle, which also caused derailments. Newer guards would not trap the animals, and by the 1890s the *Engineering News* announced hopefully, if a bit prematurely, that the old pit guard was finally going.[22]

Rail Technology

Some of the decline in derailments also resulted from the substitution of steel for iron. Chapter 1 noted the Pennsylvania's first experiment with steel rails in 1864 in heavy traffic areas. Steel was twice as expensive as iron, but on high-density lines it might last seven or more years whereas iron needed to be replaced every four to six months. Safety as well as economics pushed the carriers toward steel, for the Pennsylvania also emphasized the dangers from frequent breakage of iron rails, as did the Grand Trunk when it too began to introduce steel in 1864.[23]

Early steel rails were of uneven quality; mysteriously, some batches wore well and yielded few breaks, while others performed poorly. In 1869 the *American Railway Times* termed rail purchase "a good deal like buying a lottery ticket," and a year later the *Railroad Gazette* claimed "there is still considerable distrust of the Bessemer process in scientific circles." The combination of high cost and uncertain properties led to small-scale experiments with steel: the Pennsylvania's first batch was 150 tons of cast steel rails. In an effort to discover cheaper substitutes, that company also experimented with "steeled heads" on iron rail, as did the Reading and other carriers, but these proved a failure. A few companies even tried cast iron, with equally poor results. Despite glowing public testimonials to the value of steel rails, most companies felt the need to try their own pilot studies. Most placed steel rails side by side with iron in main track to test their comparative value. The Philadelphia, Wilmington & Baltimore introduced steel rails in 1864; in 1869 it reported that one had already outworn sixteen iron rails, and soon began to lay them exclusively.[24]

By 1867 the Pennsylvania's engineers termed use of steel rails "wholly a commercial question," and it noted that "where the business of the line is small, it will still be economy to use iron rails." In 1870, in one of the first attempts to formulate the economic choice of iron versus steel in mathematical terms, Ashbel Welch concluded that steel rails were the low-cost choice and he then proceeded in a later article to investigate their optimal durability. Many companies concurred, and in 1873 nearly all would have agreed when the Chicago & Alton announced that steel was "no longer an experiment." By 1880, steel rails comprised about 35 percent of all track.[25]

The proportion of steel rail in track understates its significance in improving safety and productivity, for just as occurred with the air brake and Miller platform, economic considerations led companies to introduce it first in high-traffic areas. The following calculation illustrates the importance of this allocation. A random sample of fifty carriers from the 1880 Census Region II (the Mid-Atlantic states) had 55 percent of all track in steel. Using each carrier's proportion of steel track to weight its passenger miles implies that 66 percent of passenger travel was on steel track. By this calculation, steel rails supported about 20 percent $[(66/55) - 1]$ more traffic than they comprised of track. Of course this procedure assumes that a company with 50 percent of its track in steel had 50 percent of its traffic on such track. But as noted, companies put steel into their main road first—indeed, some lines replaced the steel rails in main lines several times before they substituted steel for iron for branch traffic. In 1880, carriers in the Mid-Atlantic states owned main track that amounted to 68 percent of their total. If steel rails and passenger miles had both been entirely on main track, then 76 percent $[(66/(55 \times 0.68)) - 1]$ of all passenger miles in this region would have been on steel rails in 1880. While these figures are hardly precise, what they suggest is surely true: in the 1870s, the impact of steel rail on safety and productivity was substantially greater than is indicated by its share of track.[26]

As the carriers shifted from iron to steel they also learned about the new metal, from their own experiences and through the technological network. The *Railroad Gazette* and other journals routinely disseminated European and American discoveries. By 1869 engineers had discovered that punching rather than drilling holes for fishplate bolts yielded stress concentrations, weakening the rail. In the 1870s and 1880s American engineers discovered that steel that was too high in carbon or phosphorus might be brittle and that slag could also result in weak metal. Improper cooling could yield an ingot with a cavity ("pipe") near its top that when rolled might yield a piped top rail. Cooling could also yield an ingot in which the carbon and phosphorous were segregated near the top and which might yield a defective top rail. Problems of both wear and breakage also led to a variety of efforts to improve quality. Engineers developed a standard taxonomy of rail failures. The most common breaks were split or crushed heads. There were also vertical breaks, cracked flanges, and split webs. As the *Gazette* stressed, these failures resulted from a complex of causes, for the rail was part of a technological system that included roadbed and rolling stock.[27]

Improving rail quality was not simply a matter of improving technology, for it involved problems of "moral hazard" as well. That is, buyers' difficulty in obtaining accurate information on quality encouraged opportunistic sellers to supply poor rails. The carriers evolved several interrelated approaches to discover rail quality. One method was to develop standards that embodied the new chemical and metallurgical knowledge and then to inspect or test the final product. Inspection of manufacturing *processes* by buyers was not common until the twentieth century, however. Alternatively or additionally, carriers might keep records of rail failure and modify purchases according to experience.

Large carriers such as the Pennsylvania developed purchasing specifications, as did some mills, and many small roads adopted these as their own. By the 1870s specifications included such important details as rail section (shape), weight, and chemistry, as well as tests and inspections to be employed. Some railroads also contracted with engineering firms such as Frank Ward & Brothers and Robert W. Hunt & Co. for their rail specifications and inspection.[28]

Discoveries in the 1890s revealed that steel's molecular structure might be as important as its chemistry and depended on both the temperature and duration of its working. Research by Albert Sauveur of Illinois Steel, who went on to become one of the leading metallurgists of his day, confirmed that working rails at too high a temperature resulted in large grains that made the metal fracture-prone. Rail engineers began to complain that newer, faster mills worked the rail too fast. Yet metal worked at too low a temperature might be "blue short," and by the 1890s specifications commonly forbade working at less than red heat. Enforcing proper working proved difficult, however.[29]

Physical testing remained haphazard and rare. In 1869 the *American Railway Times* reported that most companies that tested employed one ton dropped 18 feet, but the Philadelphia, Wilmington & Baltimore Railroad used a one-ton weight dropped only 15 feet while Pennsylvania Steel dropped 1,640 pounds 20 feet. In 1876 the *American Railroad Journal* urged the carriers to undertake far more testing, not only of rails but of wheels and axles as well, claiming the result would improve materials, design, and construction as well as diminish accidents. But as late as 1888 Frank Ward noted "of all the railroads for whom I have so far inspected but one road has required physical tests to be made of its rail steel," and he thought chemical tests inadequate.[30]

Testing always involves an analogy between the test circumstances and actual product use, and the *Gazette* pointed out that many companies tested rails under compression and tension the same way they did bridge members, which made little sense, for what was needed was evidence of impact resistance. As the editors later concluded, "no one has yet been able to frame specifications on which rails may be made and bought with confidence." In part this was a matter of monitoring, for few companies checked to see if tests were performed as specified, and in part it reflected the use of small samples, which ensured that even appropriate, properly conducted tests were unreliable. Still, testing may have spurred manufacturers to improve quality anyway. In 1885, the *Gazette*'s Arthur Wellington tartly informed the Michigan Central (a road that did not test): "roads which make . . . tests . . . [get] very good rails"; those that don't "complain loudly."[31]

An alternative to specifications and tests was simply to trust a manufacturer's reputation for quality. By the 1870s, the Philadelphia, Wilmington & Baltimore, the Pennsylvania, and some other lines kept records revealing the maker, date of purchase, position in ingot, and location in track for each rail, all of which helped pinpoint problems stemming from manufacture or use. Even the *Gazette*'s editors couldn't decide whether testing was useful, concluding at one point that keeping careful records and then buying according to company reputation was the best approach. This was especially attractive when sellers offered a guarantee, but not all makers would guarantee their product, for here another moral hazard problem arose. As one rail maker complained, "the manufacturer has no control of the condition of the tracks and they may put his rails into some miserable mud bed . . . or the business of the road may be increased . . . or the . . . weight of locomotives and rolling stock [may increase]."[32]

A survey in 1899 revealed that rail testing had spread, with a number of carriers adopting the Pennsylvania's procedures or those of Robert W. Hunt. But many, including such large carriers as the B&O, the Lehigh Valley, and the Great Northern followed manufacturers' specifications, which did not include physical tests. Neither procedure yielded entirely satisfactory results. In 1891 a Pennsylvania inspector wrote the Pennsylvania Steel Company complaining of the "great number" of broken rails. Two years later the Pennsylvania's chief engineer W. H. Brown warned the Maryland Steel Company: "[if you] cannot make any better rail than this we will have to get it from some other mill. We cannot afford to run such risks as this." At about the same time bad rails from Lackawanna Steel plagued the Great Northern, leading James J. Hill to threaten to cancel his orders. The problems of rail contracting and testing remained unsolved, contributing to an upsurge of rail failures in the early years of the twentieth century (chapter 7).[33]

While the carriers struggled with ways to ensure manufacturing quality, they found other methods to improve rails. One obvious solution was simply to increase their size. In 1872, writing in the *Railroad Gazette* Matthias Forney observed that locomotive weight had risen from 3 to 5 or 6 tons per driving wheel. "One prolific cause of accidents is the fact that the weight of rails is insufficient for that of the rolling stock with which roads are now equipped," he concluded. On the eve of the Civil War, rails weighed 56 pounds to the yard or less and many of them were still in use in 1872, as in the above-described wreck on the New Haven. But by 1885, a survey found a shift toward heavier rail, although southern and southwestern lines still laid 56-pound rail. Thereafter the carriers steadily increased rail weights, and by the end of the century eastern lines such as the B&M were generally laying 85- to 95-pound rail in main track, while southern carriers such as the Louisville & Nashville employed 70- to 80-pound rail.[34]

Engineers also endlessly experimented with the rail section. As noted above, the conception of rails as girders focused engineers on their depth, and in 1865 the Pennsylvania claimed that shifting to a 4.5-inch rail would increase its strength by 30 percent. The leading exponents of "strength through depth" were the respected engineers Arthur Mellen Wellington and Plimmon H. Dudley.

In 1879 Dudley invented a dynagraph to study track deflection on the New York Central that recorded a detailed graphical picture of the track contour. By the mid-1880s it was also used on the Eastern Railroad, Maine Central, and Lake Shore. Later Dudley also invented a stremmatograph to obtain the first actual measurements of rail stress in track. On Dudley's advice, the Central led the procession toward heavier, deeper sections with a 5-inch, 80-pound rail in 1884 and a 6-inch, 100-pound rail in 1892. By the 1890s the number of shapes had become so bewildering (and expensive to manufacture) that the American Society of Civil Engineers (ASCE) developed a standard section that was widely employed well into the twentieth century.[35]

Wellington emphasized the economic value of heavier rails. When

Figure 2.4. **The Trade-off Between Rail Weight and Cross Ties**

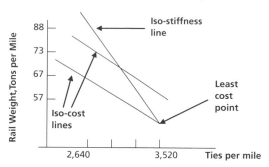

Sources: "Cross Ties," *Railroad Gazette* 17 (Apr. 10, 1885): 230–31; author's calculations.

he joined the *Gazette* in 1884, Wellington had acquired broad experience in railway construction and had published his opus *Economic Theory of Location of Railways*. In it he stressed that good engineering was also good economics. Engineers had analyzed the economics of rail composition and shape since the 1870s when Ashbel Welch had assessed the payoff to steel and tried to compute the optimal level of durability. Wellington, however, was interested in broader issues. "In buying rails . . . we do not care to buy [steel]," he instructed. Instead, strength, durability, and stiffness were the goals. Wellington pointed out that rail weight varied with its breadth × height, while stiffness varied with breadth × height *cubed*. Hence the average cost of stiffness declined with heavier rail and light rail was false economy.[36]

An even better way to increase rail stiffness, Wellington argued, was to employ ties of equal bearing surface and increase their number per mile. In 1885, he explained the engineering-economics, and his analysis is reproduced in Figure 2.4. Wellington claimed that the combination of 2,640 ties to the mile and 50 pound-to-the-yard rail (88 tons per mile) would have the same stiffness as 3,520 ties to the mile and 32-pound rail (57 tons per mile). Thus the issue became whether the additional 880 cross ties cost more or less than the saving of 31 tons of steel. With steel at $30 per ton, substitution of ties was economic even if they cost $0.50 each. "The *first* essential is to use ties as freely as possible," Wellington emphasized. While carriers commonly employed around 2,200 ties per mile in the 1880s, by the end of the century many lines employed between 2,800 and 3,000 ties to the mile.[37]

Inspection and Maintenance

Routine inspection and maintenance of rail in track could reduce the potential for derailment by improving the line of the road, and by discovering obstructions, washouts, misplaced switches, loose spikes, bad ties, and broken rails and joints before they caused a wreck. Daily track inspections had been instituted on the Western Railroad in the 1840s, but this practice was exceptional; many carriers did not inspect tracks more than once a week in the antebellum decades. Experience, in the form of many wrecks, soon suggested the need for more careful monitoring and other lines began daily inspections. By the 1870s the Pennsylvania required each section to be inspected daily by the foreman. On the B&O the inspector was the foreman or another "reliable and experienced" man. His job was to "do anything and everything to protect the railroad from accident."

In track work as elsewhere companies faced moral hazard or agency problems because monitoring track workers was difficult. In response, in 1873 the Pennsylvania pioneered an annual inspection that awarded substantial prizes for the best-maintained section. This system slowly spread to other lines; in the 1880s, the Savannah, Florida & Western, the Charleston & Savannah, the Boston & Albany, the Erie, Louisville & Nashville, and other companies began to award premiums for the best-maintained track. After the New York Central rather belatedly began to award premiums in 1899 it reported a sharp increase in the quality of maintenance.[38]

Yet, maintenance itself was risky. A passenger train derailed on the Old Colony on August 19, 1890, because a maintenance crew had failed to send a flagman far enough out to warn the oncoming train, and because the crew could not remove a track jack in time. Like Bradford and most other serious wrecks, a whole host of other circumstances conspired to raise the death toll. The train had bad brakes; the jack happened to be on the inside of the rail; the one car with the most passengers slammed into the locomotive; a steam pipe broke, scalding many to death. Such well-publicized accidents were instructive and they revealed the value of the technical press: learning from the Old Colony's experience, B&O and other companies soon came to require that jacks be used only on the outside of rails.[39]

While well-motivated, sharp-eyed inspectors might catch potential sources of derailment, what the *Gazette* termed "malicious obstructions" were more difficult to detect. Some of these came from small boys who—then as now—delighted in placing obstacles on the track to see if the train would crush them. More dangerous were obstructions placed on tracks by individuals with a grievance against the railroad, for they not only knew how to wreck a train; they sometimes knew how to avoid an inspector. Companies were likely to find track obstructions if they enforced antitrespassing rules, and there were outbreaks of deliberate derailments during the tumultuous years of labor unrest during the depressions of the 1870s and 1890s.

Robbery was a regular although minor source of derailments. The *Gazette* claimed there were sixty-one wrecking attempts in the first six months of 1893, twenty-one of which involved robbery. There were three ways to hold up a train: thieves could board it at the station, they could stop it by piling an obstruction on the track, or they could simply derail it. The carriers countered the first two approaches in several innovative ways. They armed trainmen and sometimes sent along a posse in an armored

car for good measure. Thieves responded, sensibly enough, by simply removing a rail or otherwise wrecking the train, which usually put both the
trainmen and posse out of action. Such activities continued well into the
twentieth century. The Southern Pacific reported a robber had removed a
rail near Bon, Arizona, on September 8, 1924, causing a wreck that killed
a fireman. The Union Pacific reported two similar wrecks in November
1929. In yet another example of the use of economic incentives to improve
safety, company detectives paid $5,000 for the tip that resulted in capture
of the bandit, who spent the remainder of his days in jail.[40]

The Economics of Improvement

Nearly all improvements, from better rails to better inspection, were, as
the Pennsylvania put it, "commercial matters." That is, they were expected
to earn at least the cost of capital by cheapening maintenance or transportation, or by increasing capacity. Thus they rarely measured up to the
standards of high-density European lines. E. E. Russell Tratman surveyed
European and American practice in 1896. Tratman was an expert; he had
emigrated from England in 1884 and was an editor of *Engineering News*.
He found that while the best American lines matched European standards
for rail weight and ballast depth, the Europeans ran much lighter rolling
stock. "On the whole, they have a much better relation between track and
traffic," he concluded. Nearly a decade later another writer agreed, claiming that "the track of many of our main lines still compares unfavorably
with that of leading European railways." Economic considerations ensured
that upgrading, like steel rail and other investments, was concentrated on
high-density lines where most workers labored and passengers traveled.
Thus, if market forces sometimes encouraged companies to take the gains
from safety investments in better service, the market also allocated these
investments in ways likely to maximize safety as well as economic gains.

In 1892, Wellington, now at the *Engineering News,* concluded: "when
track is once brought up to the east of Chicago, British, and German standard, derailments of serious moment tend to become evanescent." As Table
2.3 suggests, he was essentially correct. Although large eastern carriers still
had some bad track on main lines (in 1895 the Lake Erie & Western had 163
of its 423 miles of road ballasted in dirt), most of it was concentrated in
the South and West. As late as 1903 one authority noted that "a very large
mileage of track in this country is dirt ballasted."[41]

What travel was like where "commercial matters" made upgrading
and maintenance uneconomic is suggested by the 1900 report of Missouri's
railroad commissioners. The Kansas City, Excelsior Springs & Northern
had about 15 percent of its ties rotten and the bridges were "quite dangerous." On the Omaha & St. Louis, 50 percent of the ties were rotten.
The Southern Missouri & Arkansas yielded trestles with bunched ties and
rotting pilings, while one section of road had 75 percent of its ties rotten.
With a few exceptions, "no ballasting has been done." Eastern branch lines

Table 2.3. **Passenger and Worker Fatality Rates from Derailments, by Region of the Country, 1890**

	North	South	West
Passenger fatality rate	0.45	1.79	1.48
Worker fatality rate	0.12	0.44	0.28

Source: ICC, *Third Annual Report on Statistics of Railways, 1890* (Washington, 1891).

Notes: Rates are per billion passenger miles and per thousand employees. The North is ICC regions 1–3. These include New England, New York Pennsylvania, Maryland, Delaware, Ohio, Indiana, and Michigan. The South is ICC regions 4–5. These include Virginia, Kentucky, Tennessee, the Carolinas, Georgia, Florida, Alabama, and Mississippi. The West is ICC regions 6–10 and includes all other states.

also left much to be desired. The roadbed of the Brattleboro & White-hall (Vermont) was so narrow that if the engine lurched it would bump into projecting rocks. In 1903 a local newspaper called it "the flimsiest, the most dangerous piece of public conveyance imaginable." Thus, while eastern carriers had indeed engineered a great improvement in mainline track conditions by the 1890s, rickety roadbed continued to characterize a significant fraction of the country's railroads. In less than a decade all the carriers would be struck by another bout of rail failures.[42]

Equipment Failures and Derailments

As Figure 2.3 reveals, equipment failures became an increasing source of derailments as the nineteenth century wore on. These were of compara-tively little moment to passengers because most involved freight trains. Thus in the 1888–92 period passenger fatalities from equipment failures amounted to about 37 percent of those from road defects, while for work-ers the proportion was 87 percent. Equipment failures rose in part simply because the number of freight cars grew from roughly 300,000 in 1870 to 1.4 million at the turn of the century. But of course track also increased while roadbed-related accidents declined. Nearly every part of a freight car from drawbars to trucks might cause a derailment, but the following analysis focuses on efforts to reduce wheel and axle failure and on two problems that interfered with these efforts. These were the complexity of the freight car as a technological system and the economics of freight car interchange.

The Problem of Freight Car Size

The rising number of equipment failures after 1870 resulted from a head-long increase in equipment size without adequate attention to the propor-tioning and strength of running gear. It provides an example of a techno-logical improvement that worsened at least some forms of safety. While railroad engineers understood that cars and their relationship with track were complex systems, predicting the impact of any given change was initially beyond their capabilities. At best individual parts might be tested, but for the whole system, experience was the only teacher. As the carriers increased the size of their equipment they learned many painful lessons.

The freight car of 1870 had not been designed so much as it had evolved by dint of cut-and-try methods. Everything that could be was made of wood. Box cars weighed about 10 tons, and they carried about 10 tons, although they were often overloaded. But as builders were discovering, freight cars could transport goods more cheaply if they were bigger. Thus, by the 1880s, cars were typically designed to carry 20 tons and a committee of the Master Car Builders was pondering whether their size could safely be increased. Some of the advantages the committee stressed are depicted in Table 2.4.[43]

As can be seen, the economics of larger cars were compelling. Not only was the capital cost of cars less, but fewer sidings would also be needed, and there would be fewer parts to inspect and to break. Having set out the advantages of larger cars, the committee then did a curious thing. Claiming the need to obtain information "from practical men," it polled members on how such cars should be designed.

Like most other technical organizations of the time, the MCB was evolving from a craft to an engineering organization; it had begun in 1864, and grew out of the problems associated with the increasing interchange of freight cars. One motive was the need to standardize axles, wheels, and other parts used in repair. Initially many members of the MCB were craftsmen with little formal technical training. Hence much of their knowledge was tacit; surveys and discussions decided engineering matters, and scientific analysis played a minimal role. In the case of freight cars, most of those polled thought that 30-ton cars would require both a larger wheel and axle although a significant minority saw no need for any increase. The committee concluded with a recommendation that builders of 30-ton cars should increase wheel weight from 550 to 575 pounds.[44]

Table 2.4. **Characteristics of Freight Cars Required to Carry One Thousand Tons, by Car Size**

Characteristics	10-Ton Car	20-Ton Car	30-Ton Car
Number of cars	100	50	34
Weight of cars	1,000 tons	550 tons	412 tons
Weight of trucks	450 tons	250 tons	175 tons
Train length	3,100 feet	1,550 feet	1,440 feet
Cost of cars	$57,000	$30,000	$21,000
Number of wheels	800	400	272
Number of axles	400	200	136
Number of drawbars	200	100	68

Source: Master Car Builders *Proceedings* 16 (1882): 49.

This, in fact, was how all early technical societies did most of their business. In 1880, when the Master Mechanics wanted to know the best means to prevent smoke, it sent out a questionnaire. The original MCB standard axle for 10-ton cars, adopted in 1873, had been the outcome of immense discussion. A committee had proposed 3.5 inches diameter at the journal—the crucial part on which car weight rested. This was countered with a minority recommendation of an increase to 4 inches, and the result was a compromise on 3.75 inches. As a later critic observed, in the whole discussion there had been no mention of fiber stresses or elastic limit; it had been purely empirical. The later decision to recommend an increase in journal size to 4.25 inches for 30-ton cars was similarly based.

Engineers with technical training found this personal experience-based approach maddening. "Average practice is not a safe guide for proportioning axles any more than it would be for proportioning an iron bridge," Matthias Forney complained in 1879. That same year the *Railroad Gazette* reported thirteen different axle sizes used on seventeen roads—many of them smaller than the MCB standard. Few lines tested or bought according to specifications and many car builders thought that a journal worn down to 2 inches was perfectly safe. Some years later the engineer David Barnes surveyed the whole problem of car axles. Barnes had studied at MIT; he was one of a new generation of technically trained engineers and had little patience for rules of thumb. He estimated that by the time axles wore to the MCB limit of 3.75 inches, on many cars they sustained fiber stresses of 16,000 psi. "In no other country are such loads carried," Barnes claimed.[45]

Wheels too had evolved largely though trial and error and knowledge of their strength was almost wholly empirical, Forney noted in 1881. Safety concerns that were heightened by wrecks such as that at Bradford led most companies to shift to steel or steel-tired wheels for passenger equipment. In Britain both freight and passenger cars rode on steel-tired wheels, and where retaining rims held on tires, failure would not cause a wreck. Hence Britain reported *no* fatalities from wheel failures in the decade after 1878. By contrast, one-piece chilled cast iron wheels might simply shatter, although such breaks were less likely to cause derailment under American trucks than on four-wheeled English cars. Thanks in part to a campaign by the *Gazette,* some carriers equipped trucks with safety chains to keep them straight if they did derail. Still, wheel failures could cause disaster, as Bradford and other wrecks revealed. Freight cars still rode on chilled cast iron wheels and American cars carried far greater loads relative to the size of their wheels and axles than did British rolling stock. The first cost of cast iron was far less than steel, although steel wheels lasted longer. While partisans argued that the longer life of steel more than offset the extra cost, the evidence was sufficiently unclear that few companies chose steel.[46]

Practical men ran the early wheel foundries, largely without benefit of chemistry. They chose iron ores based on experience, added scrap, and hoped for the best. The carriers, who knew even less about wheels, bought on price or by maker's reputation. By the 1870s this process had yielded a 550-pound wheel that performed well on 10-ton cars where each carried a total load of perhaps 5,000 pounds. Thereafter, however, car weight sky-rocketed to 30, 40, and, by the end of the century, 50 tons, with still larger cars in the wings. Cars were also routinely overloaded, sometimes by large amounts. In 1887 the editor of *Railway Master Mechanic* told of a 30-ton car that had its capacity increased by the simple expedient of stenciling it "40 tons." He wondered if the stenciling also improved the wheel and axle strength. In response, companies slowly increased wheel weight; by 1900, wheel loads exceeded 20,000 pounds, and wheel failures were of epidemic proportions. One company reported that failures on 50-ton cars were three times as great as on 30-ton cars.[47]

Increasing weight raised wheel stresses in many ways. There were greater shocks when wheels struck frozen roadbed. The need to brake heavier equipment caused many problems. Well into the twentieth century, some carriers controlled trains in part with hand brakes on a few cars, which generated enormous amounts of heat on their wheels, leading to flange and tread cracks. Partially air-braked trains caused similar problems, and air brake pressures, which were usually set at 70 percent of the weight of the empty car, rose much faster than wheel weight, also generating increased heat.

The newer cars were also simply too heavy for their wooden support members. A report to the MCB in 1893 estimated that there wasn't a single safe truck bolster (the beam on which the car sat). The result was that on heavily loaded cars the bolster sagged and the car sat not only on the center plate of the truck but on the side bearings as well. This increased friction and made the truck corner more poorly, thereby rubbing the wheel flange against the rail. Of course this raised the cost of haulage; it also broke some flanges and caused others to square off, so that their greater friction made wheels more likely to climb the rail. Some companies placed side bearings too close to the center. When rounding a corner, especially on track with the outer rail elevated for fast passenger traffic, a slow freight might place enormous pressure on the bearing, preventing the truck from turning and leading to flange wear and derailment.[48]

Information and Contracting Problems

Information problems and the economics of freight car interchange compounded the technical difficulties associated with improving the quality of wheels and axles. Critics complained that many carriers bought the cheapest wheels on the market, irrespective of quality, and so, as the Pennsylvania's superintendent of motive power Theodore Ely put it, "the market was flooded with wheels that were not safe." As was true with rails, this resulted in part from asymmetric information between buyers and sellers. Writing in the *Gazette* in 1883, Forney complained that there was no sure way to test wheels; hence buyers could not easily ascertain quality, while price proved no guarantee. This explains why competition failed to weed out bad makers. Probably contracting problems were what led the Pennsylvania, the Canadian Pacific, and the Milwaukee to integrate backward into wheel manufacture. However, the failure of other companies to follow suggests both that information problems were inherent in wheel manufacture and that contracting was improving (see below). For carriers that bought wheels, the only sure approach was to keep careful records of failure and use them to modify specifications and the choice of suppliers, but few carriers followed this approach. Perhaps the reason was that even if buyers could discover quality, freight car interchange reduced their incentives to pay for it.[49]

Car interchange generated a host of perverse incentives that were never adequately solved. In 1872 the MCB adopted rules governing all sorts of repair to foreign cars. When the wheel on such a car broke due to "fair use," the company that owned the car could be billed at an MCB-set price. Part of the problem was that such prices were often too low, especially for western carriers that had higher labor costs. Later, in the 1890s, low prices discouraged cleaning air brakes, which led to sticking brakes and overheated wheels. But the more basic difficulty was that the repairing company had a strong incentive to use the cheapest parts on the market, for it saw none of the "costs" of poor quality.

These third-party costs bedeviled repair of all sorts. Sometimes companies got back their own cars with wheels of different diameter on the same axle, or improperly gauged wheels that might cause a wreck, or out-of-square trucks, which might wear flanges, and there were endless complaints of shoddy work. A later study revealed that a company's car repair costs rose with the proportion of its cars on foreign lines. Similarly, any company building a new car that would soon enter the interchange system would receive per mile (or later per diem) payments that were independent of quality, while its car might well return with cheap replacement parts. Such a system undermined incentives for quality improvement. As one editor described the problem: "No amount of care and no price will keep the best wheels under the cars . . . Under the interchange system they come back shorn . . . And the tendency is to cease, at least, the effort to put good wheels under other people's cars while getting poor ones in return."[50]

Not surprisingly, bad wheels and bad makers abounded. In 1884, the Erie reported that around one percent of the wheels of its regular makers failed annually, compared to 25 percent of the wheels of some makers it had employed in an emergency. Another company also reported that about 4 percent of all wheels from its six best makers failed or wore out compared to 20 percent for the twelve worst makers. A year later the *Gazette* estimated that the Erie's wheel failure rate was 50 percent greater than those on British roads. As with rails that broke in track, inspection could catch some bad wheels before they caused problems. But as the Bradford wreck demonstrated, inspection might fail, for defects were hard to spot and the

Railway Review claimed that inspections were usually slap-dash. Similarly, inspectors of freight cars received in interchange were overworked and hard to monitor. Companies such as the Pennsylvania invariably found a disproportionate number of wheel failures occurred under foreign cars, but it couldn't catch them all. In 1887 that company experienced a serious derailment that resulted from a poor-quality broken wheel on a foreign car. Between 1892 and 1897 it replaced 6,446 bad wheels; of these 87 percent were on foreign cars, although such cars accounted for only 55 percent of all wheels.[51]

On the Track of Better Wheels and Axles

Although the *Railroad Gazette* complained about wheel and axle failures from time to time during the 1880s, reforms were slow in coming. As it often did, the Pennsylvania led the way in the increasing application of scientific and engineering analysis to axle design. The likely reason was that with research a fixed cost, as the largest carrier the Pennsylvania experienced the lowest per unit costs. In addition, while freight car interchange reduced the benefits of investing in car quality, it encouraged sharing of research, for the researching company might capture some of the external benefits when connecting lines implemented its findings.

The Pennsylvania had experimented with steel axles as early as 1861, and it subjected both wheels and axles to physical and chemical tests. The move from iron to steel axles caused many problems. While iron axles had been built up, steel axles were hammered down from billets, and early efforts employed hammers that were too light, leading to unhomogeneous steel that would fracture; this problem caused most carriers to shy away from steel until the end of the century. The mechanical engineer Angus Sinclair was a British immigrant and editor of *Locomotive Engineering*. He thought "the antipathy which railroad men in America have displayed toward the use of steel one of the most extraordinary things known in the history of engineering." Sinclair attributed it to loose record keeping and lack of tests and inspection. Use of heavier hammers and annealing to relieve stresses improved steel quality, but in the 1890s the Pennsylvania was still plagued by axle failures resulting from detail fractures at the journal. It began an investigation that led to a redesign: the new axle was heavier, employed steel with much higher tensile strength, and "put it where it belonged." At the same time, the company began fatigue tests.[52]

In 1896, when the MCB next took up the matter of axle specifications, the chair of its committee—E. D. Nelson—came from the Pennsylvania, thus ensuring a national payoff to that company's research. Other members included distinguished mechanical engineers William Forsyth of the Burlington and George Gibbs of the Milwaukee. Both carriers had test departments and both men had strong technical training. Gibbs was a graduate of Stevens Institute, had been an assistant to Edison, and was a distinguished consulting engineer. Forsyth graduated from Pennsylvania Polytechnic College and would become a professor at Purdue and mechanical editor of *Railway Age*.[53]

Their methods could not have been more different from the earlier "practical" approach. The authors acknowledged the need to consider both safety and cost. The report began with a theoretical discussion of stresses. It went on to consider steel chemistry, fiber stresses, the role of the elastic limit of the metal, and a number of other technical issues, drawing on the expertise of A. J. DuBois of the Yale Sheffield School and citing such experts as Wöhler and others. Based on this analysis the MCB adopted new, larger axles for 50-ton cars—one of the first instances of a proactive approach to equipment design. Although the *Railroad Gazette* complained that companies often ignored the standard, better design and the eventual shift to steel curtailed axle failures in the early years of the twentieth century.[54]

Nelson's report provides a good example of Robert Allen's "collective invention" that characterized much of railroad technological progress. It was also symptomatic of broad changes that were sweeping through railroad engineering. As more carriers supported research laboratories, and as better-trained engineers came to the fore, professional norms and reciprocity led to a pooling of knowledge, and the MCB adopted an increasing scientific approach toward equipment standards. Initially the association relied upon the laboratories of such member lines as the Pennsylvania and the Burlington for its analysis, but in 1900 the MCB made its first arrangements with Purdue University to perform drop tests on couplers. In 1902 Professor W.F.M. Goss, dean of engineering at Purdue and an MIT graduate, became an associate member of the MCB and was soon chairing its committee on brake shoe tests. Other engineering faculty from Purdue and the University of Illinois also joined. By 1910, the association had purchased a machine for testing air brake triple valves that it located at Purdue. It also tested brake shoes and hoses, couplers, and drawbars. This arrangement proved nearly cost-free to the carriers. The university supplied the laboratory and paid the researchers while suppliers furnished most of the materials.[55]

The path to better wheels also led through the laboratory, but it was a complex journey. The wreck on the Pennsylvania in 1887 that resulted from a bad foreign wheel led the company to petition the MCB. "I do not know of any one thing that needs closer attention than this matter of car wheels," the company's Theodore Ely wrote, and he enclosed a copy of the Pennsylvania's specifications. The result was a committee that met with wheel makers and then developed specifications including a standard guarantee and drop tests that reflected practice on the Pennsylvania and the Milwaukee. The tests were not very demanding; one required a wheel to withstand five drops of a 140-pound weight from 12 feet on its center, or 1,680 foot-pounds. In contrast, some European carriers tested with a blow as great as 28,000 foot-pounds. Alternatively the buyer might choose thirty drops of 100 pounds from 7 feet on the rim. In either case, if the wheel failed, the inspector was to try another wheel, and if it failed, amazingly enough, he was allowed to try a third.[56]

Apparently, use of the specifications spread slowly. Although the tests were far from stringent, committee member J. N. Barr of the Milwaukee claimed "they would exclude nine-tenth's of the cast iron wheels made in this country," and perhaps that was the reason. In 1889 a writer to the *Gazette* claimed that "generally purchasing and using car wheels is [*sic*] conducted in such a careless manner." However, in 1887 *Railway Master Mechanic* reported that laboratory testing had led some western roads to specify the chemical composition of wheels. By 1890 both the B&O and the Milwaukee had laboratories and routinely applied physical and chemical tests to wheels and a host of other materials. On the B&O tests rejected about 15 percent of all axles and 2 percent of wheels. The Burlington, Great Northern, and some other lines also tested wheels. In 1892, a report to the MCB claimed that companies buying about half of all wheels required testing.[57]

As testing spread, it was improved and companies discovered the need to pay much more attention to such test details as the type of base and shape of the weight and anvil. Manufacturers, the *Gazette* reported, sometimes employed wood bases that cushioned the blow from a drop test, allowing poor material to survive the test. There were complaints that the drop tests on the hub encouraged manufacturers to put the metal there, resulting in wheels that would pass a test and fail in practice. Such problems led to tests that allowed buyers to choose the impact point and to the development of a thermal test, intended to mimic the effects of brake heating. Developed by P. W. Griffin of the New York Car Wheel Works, probably around 1891, the thermal test was soon adopted by the Pennsylvania and the Southern Pacific and was recommended by the MCB in 1897.[58]

Makers such as Griffin had always prided themselves on producing a quality product. Although wheel making was more art than science, that company marked each wheel, kept careful records, employed only charcoal iron, and tested regularly. Gambling that at least some carriers would pay for quality, in the early 1890s, the company experimented with new ores that led to substantially stronger cast iron. In 1895 Griffin tried to signal his improvements by publicly subjecting his wheels to European tests. They passed and were soon rolling under cars belonging to the Pennsylvania and the Carnegie Steel Company. The Burlington also reported that testing was resulting in better wheels. One maker's 585-pound wheels that had withstood only twenty-four blows in 1893 could take 165 by 1900. In the early years of the twentieth century most manufacturers performed chemical analysis of their metals whereas a decade earlier wheels had been made "without the aid of chemistry." When the Canadian Pacific introduced chemical analysis and a number of other scientific methods it cut wheel failures by 80 percent.[59]

As the nineteenth century ended a burgeoning number of derailments seems to belie any claim of progress on this front. Yet the reality was more complex, for some of the increase came about in response to the solution of earlier problems. In 1870, weak rails and lightly constructed, poorly maintained roadbeds characterized even the best of American carriers. During the next three decades the response to these conditions was a massive investment in improved rail and track where traffic was expected to justify the expenditure. Because better inspection and maintenance could also avoid derailments, employers also experimented with various incentive schemes to deal with agency problems. The result was a safer roadbed and reduced costs—and of course these set the stage for larger, heavier trains. Unsurprisingly, therefore, some of the gains of better rails and roadbed came not in fewer derailments, but in better and cheaper transportation. Still, if heavier equipment contributed to further rail and equipment-related derailments, heavier passenger cars, complete with vestibules, also made derailments less dangerous.

The search for solutions to derailments also led the railroads away from the older "cut and try" approach. Led by Forney, Wellington, and the newer college-trained generation of engineers, they began to apply economics and science-based engineering to the construction of track and the design and testing procedures for rails, axles, wheels, and other equipment. These results emerged from scientific research done by the Pennsylvania and other large carriers, and by equipment suppliers, sometimes in loose cooperation with universities. The resulting new knowledge reflected a kind of collective innovation; it was evaluated and diffused through railway clubs, the trade press, and the MCB, and it helped push the latter organization toward a more scientific, proactive stance toward equipment standards. This in turn led both the association and the wheel manufacturers to formalize long-term research ties with major engineering schools that institutionalized the process of product improvement.

Yet contracting problems continued to exacerbate the scientific uncertainties surrounding both rail and wheel quality. Moral hazard undermined the use of guarantees, and buyer ignorance allowed bad makers to flourish, while in the case of wheels freight car interchange discouraged all carriers from learning about or investing in better quality. The solutions involved a mix of vertical integration, tests, record keeping, and better specifications. In the case of wheels, the perverse incentives generated by interchange required industry-wide specifications, which emerged in the 1890s, but which proved difficult to enforce.

Collisions and the Rise of Regulation, 1873–1900

There is no crop raised in this country which can be relied upon with . . . the certainty that the next few months will be prolific in what are known to railroad men as "tail-enders."

—Arthur Wellington, 1884

The . . . railroads kill . . . [300 passengers a year, but] 987,631 corpses die naturally in their beds! You will excuse me from taking any more chances. The railroads are good enough for me.

—Mark Twain, 1871

Few years during the nineteenth century brought a more dreadful crop of train accidents than 1887. The *Railroad Gazette* recorded 207 passenger fatalities and as usual collisions took a goodly toll. The year began badly. On January 3, B&O freight train No. 26, under command of Conductor L. F. Fletcher, left Garrett, Indiana, heading east toward Republic, Ohio. The engine was not steaming properly and it finally broke down just west of the station. The road to Republic was a single track, and as Fletcher well knew, the westbound express was due at the station in 10 minutes. In fact, the express was running 3 minutes late, but unaccountably Fletcher failed to send out flag protection for his train, as company rules required, and the express plowed into the freight, killing thirteen people.

At first glance, the disaster was but one more example of employee nonfeasance: had the conductor done his job there would have been no accident. But as both the coroner's jury and the railroad press pointed out, the B&O was far from blameless. The freight engine was unfit for service and road breakdowns were invariably highly dangerous. Moreover, the B&O was one of the few companies that did not employ Westinghouse brakes on passenger equipment. It still used the Loughridge straight air brake that was slower than the Westinghouse automatic and had been obsolete since 1872. The engineman of the express had seen the freight about 1,000 feet away while he was running about 43 miles per hour; an automatic brake should have stopped the train. The accident also revealed the need for a "proper system of clearly visible fixed signals in conjunction

Figure 3.1. **The Rise of Collisions, 1873–1900**

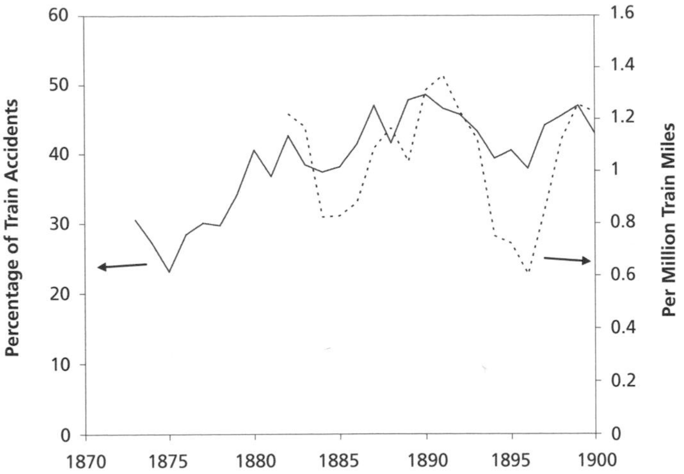

Source: Appendix 1.

with the block system," the *Railroad Gazette* noted. Finally, the B&O heated the express with old-fashioned stoves, although various safety heaters were then on the market, and most of the casualties had been caused by the fire, not the collision.[1]

Republic reveals much about the nature of collisions. They frightened the public out of all proportion to their likelihood, for to passengers they were uncontrollable, mysteriously caused, and, where fire was involved, especially dreadful. Like Bradford (chapter 2), Republic also reflected a complexity of causes. While the tragedy seemed to result from human error by the conductor, it also stemmed from managerial decisions and technological choices that reflected economic considerations. For if the maturation of American railroads after 1870 lessened derailments, it increased collisions. Rising traffic density in the context of an expanding labor market overwhelmed the old train order system. In addition, increasing passenger train speed and weight largely offset beneficial effects of better brakes, while the introduction of very large freight cars created a bulge of freight collisions from break-in-twos. Pressured by the trade press and public outcry, the carriers responded tardily with new brake and signal technology and improved company employment practices. Yet neither private nor public efforts proved adequate and by 1900 collisions again became a public issue.

Regulatory Beginnings, 1871–1880

A rear collision on the Eastern Railroad on August 24, 1871, near Revere, Massachusetts, that killed twenty-nine people was responsible for the first serious state accident investigation. In 1869, partly at the behest of Charles Francis Adams Jr., Massachusetts established a Board of Railroad Commissioners that included Adams and that he soon dominated. Modeled after the British Board of Trade, it could investigate and make recommendations, but otherwise had no power. Adams was no enthusiast of regulation. An elitist, he had little sympathy for passengers killed through "their own fault" (chapter 4), and while he favored state action to control train accidents, he believed that more might be accomplished by negotiation and public pressure than by laws and rules. This is the sort of "soft" regulation I have termed *voluntarism*.[2]

Under Adams's guidance, the board pressed for numerous reforms in response to the Revere disaster. The commission's investigation found a veritable smorgasbord of obsolete and dangerous practices. Neither train had air brakes, nor did the company use the telegraph. The board also found "defects of management," many of which were common to other roads. It discovered that only three roads used the same rules and signals, and that on many others they were cumbersome and confusing. Adams skillfully used public outrage over the collision to force Massachusetts carriers to draw up a general code of rules and signals. Problems remained, however. In 1890 the board again complained that the "lack of uniformity on different roads in rules governing the train service is a definite danger." It also noted "the alarming diversity of [signal] practice," observing that "the arrangement of lanterns which means safety on one line means danger on another."[3]

The Adams approach represented a conscious alternative to "hard" regulation, but it worked best with expert commissions and where regulation remained a credible threat—where there was a "gun behind the door," in the phrase of Thomas McCraw. Under Adams, the Massachusetts board soon gained a reputation for expertise and good judgment and its findings were widely and respectfully reported. Because Massachusetts's lines ran throughout New England, that state's policies improved safety region-wide. In the 1880s, New York also began to investigate accidents thoroughly. In 1877 Adams urged voluntarism on the national stage, authoring a bill introduced by Representative James A. Garfield and modeled on his own and British experience that would have authorized federal railroad inspection employing men from the Corps of Engineers. While it failed to pass, the Accident Reports Act of 1910 finally gave similar powers to the ICC; thereafter, although safety laws multiplied, Adams's approach largely prevailed at the national level (chapter 7).[4]

In most states, however, until 1910 the carriers investigated their own wrecks, usually hushing up the results, or coroners' juries sometimes looked into accidents, but their reports were typically superficial. With so few accidents thoroughly investigated it was difficult to learn from experience. In Britain, by contrast, the Board of Trade routinely examined a large number of accidents and their findings were broadly disseminated. The *Gazette* under both Forney and Wellington subscribed to Adams's view that publicity was the best remedy for railroad dangers, and its own accident statistics and commentary were intended to be a private substitute for this public good. From the 1880s on, all the major railroad journals urged federal investigation of train accidents and they pressed the carriers for a host of improvements, including better brakes.[5]

Collisions, Air Brakes, and Passengers

Collisions such as Revere and Republic had been a regular feature of American railroading since the 1850s. After 1873, when the *Railroad Gazette* began to keep score, they rose sharply, from around 300 a year to about 1,200 at the end of the century, surpassing derailments and far outpacing the growth of train miles (Figures 3.1 and 3.2). The surge of "crossing and miscellaneous collisions" reflected the relative rise in freight traffic, for most of them involved yard work and switching. As a result, collisions increasingly involved freight trains (recall Table 2.1). But butting and especially rear collisions rose sharply as well.

Despite the increasing prevalence of collisions, passenger fatality rates from collisions were lower in the late 1890s than in 1880, while ICC data reveal uneven decline from 1889 to 1896, followed by a sharp increase during the traffic upswing to about 1907, and then long-term decline. Nor were individual accidents themselves becoming more severe. Here the decline in passengers per train was part of the explanation, but the spread of heavier cars and vestibule-equipped trains also reduced the lethality of passenger train collisions, just as it did derailments. Fatalities per passenger train involved in a collision declined from about 0.24 in 1878–80 to 0.11 in 1898–99—and the decline continued into the twentieth century. Often the only casualties were the train men.[6]

Figure 3.2. **The Changing Nature of Collisions, 1873–1900**

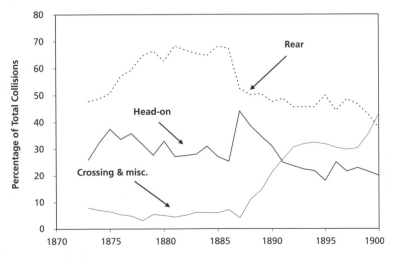

Source: Appendix 1.

Wooden coaches were battered,
    Train was torn in two
And yet no passenger injured,
    Nobody killed but the crew.
None but the men in the engine
    Hogger and fireman too;
Didn't we get off lucky?
    Nobody killed but the crew.

It is tempting to see these improvements in passenger safety as the result of better braking, yet better brakes seem to have had little impact on safety.[7]

Better Brakes, Higher Speeds, Heavier Cars

The introduction and rapid diffusion of the air brake on passenger trains in the decade after 1869 has already been described. Along with Miller's platform it contributed to the large decline in accident rates reflected in state data from 1860 to the early 1870s. As noted (chapter 1), the first straight air brake was made obsolete in 1872 by the introduction of the automatic, which was much faster and failsafe. But having invested in the former, companies were far slower to install the latter even as they increased train weight and speed—as the tragedy at Republic revealed. As late as 1881, Westinghouse estimated that about 48 percent of passenger cars equipped with his brake still used the old straight air version. Both the Boston & Providence and the Southern Pacific employed the Westinghouse straight air brake as late as 1887–88. About the same time, the Long Island was still wheezing along with vacuum brakes.[8]

A moving train contains kinetic energy measured in foot-pounds; to stop it requires an equal number of foot-pounds of resistance. Kinetic energy in a train varies with directly with its loaded weight and the square of velocity. An approximate formula is $E_k = (wV^2)/30$, where E_k is in foot-pounds, w is weight in pounds, and V is velocity in miles per hour. Thus a train weighing 200,000 pounds traveling 20 miles per hour will require 2.7 million foot-pounds to stop it, while the same train traveling 40 miles per hour requires 11 million foot-pounds. By implication, given weight and brake pressure, increasing speed disproportionately increased stopping distance, a conclusion confirmed by both experience and experiments. In 1892 the *Gazette* distilled the results of a number of tests on trains of fifty empty cars running at speeds up to 40 miles per hour as $D = 2.94S + 0.3S^2$, where S is speed and D is the distance run from the moment the brakes are applied. Thus, at 20 miles per hour the train would stop in 179 feet while at 40 the stop took 539 feet.[9]

While no national figures exist, fragmentary data reveal that passenger trains were becoming longer, heavier, and faster during these years, making them much more difficult to stop. In Ohio, for example, the number of cars per train rose from 1.97 in 1873 to 4.59 in 1888. Passenger car weight rose from 20 tons in 1870 to 30 by the 1890s, while Pullmans of that era tipped the scales at 45 to 50 tons. To haul such monsters required heavier locomotives. Moreover, some of this added weight was unbraked: Pullmans had six-wheel trucks, only four of which had brakes, and many locomotives had no driver brakes. In 1890, for example, only 28 percent of passenger locomotives in Massachusetts had engine brakes. Trains pulled by two engines might therefore be particularly poorly braked, and in addition, since the second engineman usually controlled brakes, they took longer to apply as well. The worst rear collision of the nineteenth century, on the Lehigh Valley at Mud Run, Pennsylvania, on October 10, 1888, involved a double header on which the lead locomotive did not control the brakes. It telescoped the rear two cars of an excursion train, killing sixty-six.[10]

The increasing use of heavyweight cars shaped safety in other ways as well. Most companies tried to protect passengers in the event of a collision by placing the baggage car next to the tender, and the dining and sleeping cars, which tended to be heavier, toward the end. The idea in each case was to use the heavier cars as shock absorbers. In fact, the large cars often became battering rams and as the railroads increasingly mixed heavy dining cars and sleepers in with lighter coaches, the dangers for the lighter equipment worsened. In scores of accidents the heavy cars would sustain little damage while demolishing the lighter coaches with much loss of life. The head-on collision at Silver Creek, New York, on September 14, 1886, was "decidedly mild," and neither locomotive sustained much damage. But the frame on the baggage car was higher than that on a following coach, which it telescoped. At the *Gazette,* Wellington described telescoping as a wonderful way to absorb energy: "a gentler way of stopping a train . . . could not well be [imagined]." But it killed nineteen people. Thus while contemporaries sometimes worried whether they should sit in the front or rear of a train, the solution was to sleep in a Pullman. As noted, the general use of heavier cars along with vestibules that held the train together reduced the number of casualties per collision.[11]

Train speeds also increased. As late as 1869, the Erie's express trains made 30 miles per hour, while locals ran at about 20. Later, as roadbeds improved, companies increasingly began to advertise speed. By the 1890s, some fast trains averaged 55 miles per hour, which must have meant that they traveled 60 to 65 between stations, and the express to Atlantic City *averaged* 70 miles per hour. Of course, many locals remained much slower. In a series of experiments in Britain conducted with Captain Galton of the Board of Trade, Westinghouse discovered that the coefficient of friction between brake and wheel varied inversely with speed, declining moderately as speeds rose from 20 to 40 miles per hour, and then more rapidly. Thus, given brake pressure, faster trains took a *much* longer distance to stop.[12]

Of course, such stops were under test conditions, and trains in service would be likely to take even longer to stop. In 1887, at the *Engineering News* Wellington noted that brake tests on the Burlington had revealed that "the actual condition of the brakes on the average passenger train is very far inferior to what it might be." He pointed out that sleeping and parlor cars had only eight of twelve wheels braked. In addition, sleeping car companies reduced their brake leverage to prevent wheel sliding and since they were interchanged, this encouraged the carriers to do the same with home cars, as the most heavily braked car in the train would be the first to slide wheels. Poor maintenance allowing excessive piston travel could also reduce brake power. A thirteen-car passenger train on the Southern Pacific made it over the Tehachapi range on July 1, 1888, with brakes on five cars inoperative. One writer claimed that in 1890 inspections discovered 2,316 sleeping cars that passed through Windsor, Ontario, that had "absolutely no braking power" due to excessive piston travel.[13]

Because the rate of deceleration of a train increased as velocity decreased, it was "the last 25 or 50 feet run that kills," as the *Railroad Gazette* observed, and many collisions can be traced in part to the failure of brake performance to keep up with train weight and speed. The collision at Republic is as good an example as any. Brake performance also depended on properly trained enginemen. For example, as noted earlier, if an engineman made several service stops, thereby depleting the air in reservoirs, he might have too little for an emergency. The need for training was nicely illustrated by the rear-end collision near Mays Landing on the West Jersey & Atlantic that killed twenty-three people on August 11, 1880. The second train was newly equipped with the automatic brake, but the company had not bothered to instruct the engineman in its use. As he candidly admitted, "that's the first day I ever used them."

The disaster reflected a broader problem: enginemen, as skilled workers, typically learned their trade on the job as firemen, and companies provided very little formal instruction of any kind. Thus they were unprepared to deal with air brake complexities. As late as 1882, only the Erie and New York Central had equipped brake instruction cars. Thirteen years later, in 1895, the editor of the *Railway Review* complained that he had recently traveled over several eastern roads that had no brake instruction car or plant and it was plain that the engineman did not know how to use the air brake.[14]

Of course, most of the time bad brakes simply resulted in a near miss. On April 26, 1889, the Boston & Albany 4:00 o'clock express, seven cars in all weighing 597,000 pounds with three twelve-wheel coaches and an unbraked locomotive, going 45 miles per hour, hit the brakes just outside Springfield. The track was wet and the grade descending about one percent per mile. With only 57 percent of its weight braked (good practice would have braked 77%), the express plowed into the rear of a freight at 6 miles per hour, *a little over a mile* down the track. Fortunately, no one was hurt.[15]

Improving Passenger Train Braking

By this time improvements were already in the works. The Pennsylvania had pioneered in equipping its locomotives with driver brakes in the 1870s, and by the late 1880s most carriers were ordering them on new engines.

The Old Colony wreck at Quincy, Massachusetts, in 1890 (chapter 2) would have been harmless had the train stopped only *45 feet* sooner. The investigation by George Swain of MIT for the state railroad commission discussed the importance of locomotive truck as well as driving wheel brakes, and later experiments on the Old Colony by another MIT professor, Gaetano Lanza, revealed their value, especially on short, fast trains. A four-car train traveling at 60 miles per hour could be stopped 111 feet more quickly with such a brake. The Old Colony and the New York Central promptly began to install such brakes, but their use spread slowly. About this time companies did begin to apply brakes to all wheels of twelve-wheel cars, however. Collectively, by the late 1890s, these changes probably increased passenger train braking power 20 to 40 percent, depending on the length of train and number of sleepers.[16]

Both Westinghouse and the carriers also began to improve the quality of brakes and equipment. As will be discussed in chapter 4, in 1886 and 1887 at the behest of the MCB, the Chicago, Burlington & Quincy hosted the first careful tests of brakes made in the United States. These resulted in the Westinghouse quick-acting brake. In emergency application, the new triple valve vented air from the train pipe into the brake cylinder, thereby both speeding up brake application and increasing pressure. The new brake almost certainly saved lives during a derailment on the New York Central near Garrison, New York, in 1897. Although seventeen people were killed when the engine and five cars plunged into the river, the brake stopped the remaining five cars in time.[17]

In the 1890s, the MCB continued to improve brake performance. In 1891 it reported laboratory tests undertaken at the Burlington to determine the friction coefficients of various shoe materials and it began road tests several years later. These tests continued well into the twentieth century and their findings were incorporated into association standards. The association also began to test triple valves to ensure that all makes were compatible, and in 1893 the Pennsylvania turned over to the MCB a brake-testing facility at Altoona, Pennsylvania; in 1898 the facilities were moved to Purdue University. The association also investigated brake hoses and the design of brake beams necessary for newer, heavier cars.[18]

But the carriers were increasing train speed as well, and so in 1894 Westinghouse brought out the high-speed brake. It incorporated the lessons learned at the Galton trials in 1878. Since the coefficient of brake shoe friction declined sharply at speeds of 50 to 60 miles per hour, the new brake increased train pipe pressures so that braking power equaled 125 percent of the light weight of the vehicle upon initial emergency application. As velocity declined and shoe friction rose, a reducing valve diminished pressure to prevent wheel sliding. In well-publicized trials on the Pennsylvania at Shiproad, Pennsylvania, in 1894 and then in Absecon, New Jersey, the new brake demonstrated its value, in one case stopping an engine and six cars from about 80 miles per hour in 1,000 feet less than the old, quick action brake. It was soon applied to a few fast passenger trains; after 1898 diffusion was more rapid but many high-speed suburban trains still ran without it. The *Gazette*'s editors thought the high-speed brake might have prevented the New York Central rear-end collision in the Fourth Avenue Tunnel on January 8, 1902, that killed fifteen. In 1904 the *Gazette* estimated it was on only one-third of all passenger equipment, and four years later *Locomotive Engineering* complained that some railroad companies were still operating "heavy passenger trains at the highest possible speed with the ordinary quick acting brake."[19]

The Payoff to Better Braking

Clearly one of the most important features of the Westinghouse brake was its improvability. For while the high-speed brake and later modifications in the twentieth century were lineal descendents of the old automatic brake of 1872, they were incomparably more efficient. Yet it is hard to see their imprint in the collision statistics. Rather, passenger casualty rates from collisions in 1880 seem little different from those twenty or twenty-five years later, and Table 3.1 suggests the reason. Although the high-speed brake was an enormous improvement, it merely offset the deterioration of brake power from increasing train weight. Yet speed as well as weight had increased and in the early twentieth century a representative of Westinghouse concluded that the air brake "had not kept pace with the developments of locomotion."[20]

Better brakes did not cause higher speeds, heavier cars, and denser traffic, but they *allowed* these developments to occur without increases in collision rates. That is, better brakes provided carriers with the chance to increase not only safety but also train speed and weight with more and heavier cars, which they did (see Figure 1.5), presumably because they thought the payoff to better service more than offset their detrimental effect on liability costs.

Table 3.1. **Comparative Passenger Train Brake Performance, 1875 and 1907**

					Work in Foot-Pounds Performed by Brake			
Date	Speed (mph)	Length of Stop (feet)	Time of Stop (seconds)	Weight of Train (tons)	Total	Per Second	Per Shoe per Second	Brake Power*
1875	56	1,020	22	227	23,900	1,086	14.7	88
1907	58	954	19	559	61,300	3,278	28.8	150

Source: Walter Turner and S. W. Dudley, *Development in Air Brakes for Railroads* (Westinghouse Air Brake Co., 1909), Figure 45.
* Brake power as a percentage of the unloaded weight of cars.

The Demise of the Deadly Car Stove, 1886–1892

While passengers valued the savings in time that came from faster trains they surely did not wish to be roasted in a wreck, and the Republic collision revealed a feature common to many train accidents: the ensuing fire sometimes accounted for more fatalities than the collision. Fires also made such accidents seem especially dreadful, magnifying their perceived risk. Republic also followed hard on the heels of the equally awful conflagration at Rio, Wisconsin, on October 28, 1886, in which seventeen were burned alive in one car. The *Gazette* reported the conductor's testimony: "the victims struggling in the wreck could be largely seen by the light of the flames. One woman . . . could be seen making frantic efforts to free herself from the broken timbers . . . She was unsuccessful and her struggles and awful screams were soon stopped by death."[21]

The Problem of Fire Safety

These tragedies were part of a long procession of such fires. Before Rio and Republic there had been Angola in 1867 (forty-two dead), Ashtabula, Ohio, in 1876 (eighty dead; chapter 5), and Spuyten Duyvil, New York, in 1882 (eight dead). In 1887 there were fires from a collision at Kout's Station, Indiana, that killed nine, and a bridge accident at White River Junction, Vermont, that killed thirty-two (chapter 5). The fires resulted, of course, because railroad cars were made of wood. A fire might be touched off by flames from the locomotive, by the lights, or from a stove that tipped over. In fact, while locomotives did cause some fires, even the old candles or oil lights were comparatively safe, the reason being that the shock of the accident usually extinguished them, while one journal bluntly asserted that "there is no danger from oil of the standard quality [300 degree test] used in car lamps." Later, the introduction of Pintch and other gas light systems were even safer and the dangers from lighting all but disappeared long before the advent of electricity. As one survey concluded, the main culprit was the car stove.[22]

By the mid-1880s, the stove had long been under attack, for while fires took relatively few lives they seemed especially horrifying. As Thomas Haskell's work implies, the horror must have seemed all the worse because it was so avoidable. As early as 1871 the *New York Evening Post* had urged the carriers to shift to steam, and subsequent holocausts led to both popular outcry and inventive activity. As modern writers on risks have shown, such rhetoric can amplify public worries, turning a fairly minor risk into an important public issue. In 1882, the nearby collision at Spuyten Duyvil stirred up the *New York Times,* which dramatically reported the "charred victims," while the *American Railroad Journal* termed winter "the season of railroad barbecues." In 1886 and 1887, Rio, Republic, and White River Junction led the *Times* to editorialize on "The Terrible Car Stove," terming the dead a "burnt offering to the spirit of recklessness." Inventors also responded to these events. The number of patents granted for heaters rose sharply with well-publicized fires. In the mid-1870s there were six or seven a year, but in 1877—the year after Ashtabula—there were eleven. The number then dropped until 1882, when Spuyten Duyvil yielded sixteen. Another decline ensued until 1887–89, which yielded 116.[23]

As Wellington pointed out in the *Gazette,* simply substituting boiler iron for cast iron would have been a step forward. Many innovations such as Spear's forced hot air system were primarily intended to improve heat distribution, for not only were stoves dangerous, they made cars miserably uncomfortable as well. Forney claimed that actual measurement found a 50 degree temperature difference between floor and ceiling in some cars. Other inventors placed the stove under the car, which saved space and may have been safer as well. William C. Baker's early hot water heaters, which employed a stove to heat water that ran in pipes throughout the car, were also intended to distribute heat more evenly, but in 1887 Baker invented a safety car heater enclosed in a metal safe. By this time many railroads were routinely employing a drop test to evaluate rail and wheel quality and its use on a Baker's safety heater led the company to strengthen the doors to prevent them from bursting open from the impact of a collision. The safety heater seems to have been widely used, but the real breakthrough came with the shift to steam. Experiments began as early as 1851 on the Old Colony and a bit later on the New York Central, but neither seems to have been successful. In 1878 Baker invented a system that drew steam from the locomotive and was used on the New York El, and in 1881, the Connecticut River Railroad fitted a train with the Emerson system that used a separate auxiliary boiler. Despite such small-scale experimentation, most carriers did little until the disasters of 1886–88 sparked public outcries.[24]

The reasons for such lethargy were partly economic—steam heating promised to add to costs but to do little for revenue—and partly technical, for a workable system had to solve a host of problems. An obvious difficulty was couplers; another was how to deal with condensation, and whether to draw steam from the locomotive or an independent source. If steam came from the engine, would it reduce power significantly, and what would it add to the fuel bill? Could steam heat cars well enough to stand extreme temperatures? Should steam be used with a heat exchanger, or directly, and if directly, would it be dangerous if a pipe ruptured?

The answers to these questions came once again through much experimentation, the results of which were widely diffused through the railroad clubs and journals and in state reports. Heating problems and systems were extensively discussed at the meetings of the New England, New York, and Western Railway clubs in 1887 and 1888. The trade press reported the discussions in full and also described technical details of the various heating systems. They also featured stories of the heaters in actual practice during blizzards on the Milwaukee, the Maine Central, and the Canadian Intercolonial.

State legislatures and railroad commissions also joined the fray. As early as 1882, Massachusetts required various "safeguards" on stoves, and in 1887 it banned the common stove and required all methods of heating to be approved by the Board of Railroad Commissioners. The board that year recommended that the carriers adopt steam, and in May 1888 it informed the legislature that it intended to require steam heat beginning in the fall of 1889. It hoped this warning would give the carriers time to adjust. It did: they petitioned the legislature to rein in the board, and in June, that body obediently instructed the commissioners to study the matter further. Finally, in 1891, the legislature banned all stoves, effective November 1, 1892. In 1887 Connecticut also threatened to require steam heat, but caved in under pressure from the carriers. Michigan and New York banned individual car heaters in 1888, over much protest from the carriers. In 1890, Maine's railroad commissioners reported that in response to a law of 1889 its carriers had discarded common stoves, and the other New England states also banned stoves and required that the railroad commission approve other forms of heating. Maryland also banned stoves in 1891.[25]

In 1889 the Massachusetts board published a report on steam heating by Gaetano Lanza of MIT that the *Engineering News* publicized and termed "of great value in furthering the reform." Lanza surveyed the existing systems, and performed important experiments, demonstrating that steam heating added about 6 percent to steam usage, and that even at 80 psi, a burst line in a car carried too little steam to do much damage. The *Gazette* promptly fired Lanza's findings at Michigan's roads, which it suspected were inventing technical problems when their real objection to steam heat was economic. Similarly, in 1891, when New York Central President Chauncy DePew told a reporter how much he loved the warmth of car stoves, cartoonists had a field day. The same year after a wreck and fire on the New Haven killed two people, President Clark explained that his company had ignored New York's steam heat law due to the dangers of steam escaping in the car. Wellington sarcastically shot back that unfortunately Clark failed to describe the New Haven's experiments that allowed him to speak with such confidence.[26]

Despite such foot-dragging, by 1891 about 29 percent of all cars nationwide were equipped with steam—a remarkable rate of progress considering that probably no more than a handful had steam heat as late as 1885. About 27 percent had water heaters, leaving about 44 percent equipped with stoves. As usual, most of the new equipment went on the high-density, high-risk lines, and so most of the stoves were consigned to branch lines with light traffic or to the South, where they were rarely used. Thus the stove warmed comparatively few passenger miles and the conflagrations of the 1880s faded from public memory.

This story reveals some themes that led to such other innovations as block signals (see below) and automatic couplers (chapter 4). Economists have usually emphasized the role of prices in inducing technological change, but steam heat, as well as signals and couplers, were partly acci-dent-induced. The railroad technical community played a key role in de-veloping and diffusing information on both accidents and the safer steam heating technology that made humanitarian reform seem necessary to a range of groups. It is incorrect to dismiss public policy in this trans-formation, as one writer has, or to term the behavior of the New Eng-land commissions and legislature "discreditable," as has another, for in reality, steam was a politically induced innovation. In fact, as chapter 4 notes, the episode reflected a new willingness of some states in the 1880s to press the carriers to improve a range of safety equipment, includ-ing brakes, interlockings, couplers, and bridges. The *Railway Review* noted that while nationwide only about 27 percent of all cars were steam heated, in New York and New England the proportions were 80 percent and 47 percent, respectively, a result it attributed to "the effect of state laws and also of active railroad commissioners."[27]

Train Brakes and Train Breaks, 1890–1900

Although public concerns focused on disasters involving passenger trains, the explosion of collisions depicted in Figure 3.1 increasingly involved freights. While many of these were yard collisions, toward the end of the century an epidemic of "tail enders" arose due to freight train break-in-twos. These could be highly dangerous, for the rear section of the train (or sections, for sometimes the breaks were multiple) typically had no air brakes. Accordingly, it might collide with the front, causing a general smashup, or a section might be left for a succeeding train to hit. Even more possibilities existed. On the Reading on September 19, 1890, a 150-car coal train parted. While the engineman was trying to reconnect his train, a fol-lowing freight collided with the second section, knocking some cars onto another track in front of an express, coming at 50 miles per hour. Twenty-one people perished in the ensuing wreck.[28]

No national figures exist on the number of trains that parted, but sta-tistics from individual companies suggest the problem was widespread by the 1890s. Thus, in 1896 the Rock Island reported twenty-two train break-in-twos a month, or about 3 percent of all trains. These events alarmed the MCB, which formed a committee on break-in-twos in 1896 that promptly surveyed the problem. Its findings suggested that in 1897 there were per-haps 270 break-in-twos a day nationwide, or nearly 99,000 a year. The number of collisions due to broken trains recorded by the *Railroad Gazette* rose after 1885 to about 13 percent of all collisions in 1899.[29]

The Causes of Train Breaks

This upsurge of broken trains resulted from the interaction of three simul-taneous and interrelated changes in railroad technology: the introduction of automatic couplers, the gradual shift to air brakes, and the introduction of 50-ton cars. The ability of anyone to get into the coupler market combined with the absence of any MCB tests and company incentives to pass off inferior technology through freight car exchange (see chapter 4) resulted in a plague of poor couplers. In addition, because trainmen no longer needed to go between cars to make couplings, enginemen took less care in switching, which damaged couplers and led to break-in-twos. As the *Gazette* observed, "the automatic coupler has done its work all too well . . . The use of the humane and economical automatic coupler has developed a new source of loss and danger." The new couplers were also attached to old, spring-loaded drawbars that had insufficient strength and buffing power for the newer, heavier cars.[30]

In addition, the introduction of air brakes on freights generated a blizzard of unanticipated problems involving brake rigging and maintenance and employee training. The large number of stories about wrecks when air brakes failed were almost always traceable to improper operation or inadequate maintenance. In one example, a break-in-two that led to a collision with a second train was caused by a defective brake that always went into emergency application.

Control of long freights using the air brake required a deft touch. Too rapid an application of air might buckle a train while too rapid acceleration could pull it apart. An engineman who depleted his air in service applications might cause a runaway. Once he had chosen a service application of, say, 10 psi, he could increase it if necessary or he could release the brakes, but as noted, there was no graduated release. Thus, releasing the brakes might lead to a runaway unless the retaining valves had been set to hold the air (chapter 1).[31]

Surprisingly, given their long experience with air brakes on passenger trains, few carriers were equipped to provide adequate instruction to trainmen. The Santa Fe was an early user of air brakes, and in 1884 it employed Benjamin Johnson as a traveling instructor. Johnson, who had been trained at Westinghouse, traveled about lecturing on the air brake, and sometimes riding with engine drivers, which was all the training the men got. Air brakes made obsolete the older lighter rigging, which had so much play it destroyed their efficiency. Still, such rigging remained in wide use. Finally, air brakes required a much higher standard of maintenance than carriers were accustomed to providing for freight cars, while car interchange reduced incentives to maintain brakes because prices were too low. Not surprisingly, ICC inspectors invariably found the foreign cars to be in the worst shape.[32]

If triple valves were not cleaned, or piston travel not adjusted, or pipes leaked or were plugged, or the rigging was bent, or hoses were worn, the brakes would not work properly. They might not come on, or fail to release, or apply unevenly, any of which might cause a wreck. G. W. Rhodes noted that the Burlington even discovered mud wasps had taken up residence in the retaining valves of its brakes, which prevented release. Yet companies seem to have been as slow to establish yard testing facilities as they were to train workers. In 1896 Angus Sinclair in *Locomotive Engineering* described the "miserable condition" of air brakes, and three years later one writer claimed that 20 to 25 percent of freight cars in interchange had inoperative air brakes.[33]

These difficulties with brakes and couplers interacted with train makeup in complex ways. Consider a hypothetical but not atypical fifty-car freight of 1898 making a run in hilly territory. Sometimes companies put air-braked cars in the rear, where they could not be used. Typically, however, trains were made up with loaded air-braked cars first; loaded 50-ton cars would have only about 16 percent of their total weight braked, while for 20-ton cars the figure would be nearer 25 percent. Empty air-braked cars would be next, which were braked at 70 percent of their weight. Perhaps half the cars would be braked but many of these would be cut out and not working or partly working. The older rolling stock would still have link and pin couplers, while of the rest, some MCB couplers would be worn and others of poor quality and all would attach to weak drawbars. Such a freight would have an almost infinite capacity to break in two.[34]

Simply running out the slack by using steam might break the train; a light, loaded, link-and-pin car in front of a loaded 50-ton car might let go, or if a spring-loaded drawbar extended sufficiently, it might pull the pin and unlock an MCB coupler. Going over the crest of the hill might also break a coupler or drawbar. Or suppose the engineman slowed his train in a service application. The empty air-braked cars would slow much more rapidly than the loaded heavy cars, perhaps causing a weak MCB coupler to snap. Or if the engineman slowed, and then released the brake, the front of the train would release before the rear, especially if the train pipe was partly clogged. If he then accelerated too rapidly the train might break. This latter problem may have been especially prevalent, for a survey by the air brake association revealed that many breaks occurred near the last braked car.

Efforts to remedy these problems began almost immediately. In 1889, Westinghouse established an air brake instruction car that toured companies providing gratis instruction on air brake use; in a decade it could count about 127,000 graduates. In 1891 the MCB developed a code of rules for air brake use along with complex questions and answers companies might use to test employees. In the 1890s major carriers such as the Pennsylvania, Southern, Michigan Central, and the Big Four also began to equip instruction cars and so did the International Correspondence School, which sold its services to smaller carriers that could not afford a car.

Clearly, maintenance was a network problem. Led by G. W. Rhodes and A. M. Waitt, Master Car Builders of the Burlington and Lake Shore, respectively, the Western Railway Club galvanized the technical societies in a campaign to improve maintenance, while the Airbrake Association developed lists of recommended practices for starting and stopping trains. All of these efforts were widely circulated by the technical press. The journal *Locomotive Engineering* began to publish a regular section on air brakes while labor journals such as the *Locomotive Firemen's and Enginemen's Magazine* also devoted many pages to brake operation and maintenance. In 1901, the MCB raised the prices for cleaning air brakes, which made it profitable to service foreign cars and, according to one ICC inspector, led a number of companies to establish cleaning and testing facilities in their yards. Westinghouse also began to do an increasing repair business. Finally, in 1903 Congress modified the Safety Appliance Act to require that the air-braked cars be operable. The net result of these developments was that, as another inspector asserted, "a steady improvement is taking place, especially of the air brake equipment."[35]

The work of manufacturers, the MCB, the Western Railway Club, and individual carriers to improve coupler quality and maintenance will be detailed in chapter 4, and by the early twentieth century both brake and coupler problems were declining. In addition, Congress helped to re-duce break-in-twos in 1905 when it modified the Safety Appliance Act of 1893. The new law gave the ICC power to increase the use of air brakes; it promptly raised the required proportion of braked freight cars in a train to 75 percent. It also required that if freight cars had air brakes they had to be usable *and used*. By 1906 about 95 percent of all freight cars had air brakes and maintenance had sharply improved.

In the late 1890s Westinghouse also introduced a friction drawbar that was much stronger than the older equipment. The *Railway Age* termed it "one of the most important recent railroad inventions." Companies such as the Butte, Anaconda & Pacific, which hauled heavy ore cars, promptly installed it and reported marked reductions in breakage. About this time research at the Western Railway Club using a dynamometer car reported that buffing shocks even from ordinary movement were much larger than had been expected. By the end of 1902, Westinghouse had equipped 62,000 freight cars with the new equipment.[36]

The epidemic of break-in-twos peaked in 1902–3, accounting for about 18 percent of all recorded collisions, and then declined. The epi-sode reveals the difficulties of integrating new technologies into an exist-ing system and is a warning that the effectiveness of safety devices is often contingent upon behavior—all of which has a surprisingly modern sound. Moreover, safety equipment can worsen some risks, depending on the sys-tems in which it is embedded and on how it affects behavior.

Train Control and the Labor Market

Train control requires rules, communication, and signaling. The introduc-tion of the telegraphic train order system in the 1850s was described in chap-ter 1. It worked by maintaining a time interval between trains. In contrast, the block system, which originated in Britain in the 1840s, separated trains with a space interval. As noted, in addition to these broad systems of train control, there was also an increasingly complex book of rules governing train operation as well as a large number of whistle, flag, and lantern signals.[37]

Under the block system, the line was divided into sections of perhaps one or 2 miles separated by a signal, and no more than one train would be allowed in a block. In fact, most American lines operated block signals "permissively" for following trains. That is, if an engineman found a block protected by a stop signal he must stop, but after a brief wait, even if the signal failed to clear he might proceed with his train "under control."[38]

In what was called the manual block system, each block had an operator who would set the signal to stop. Before the operator of block (A) would clear the signal to allow a train to enter he would telegraph the next station at (B) to see if the previous train had cleared and, on a one-track line, to see if a train was incoming. If all was clear he signaled the (B) operator to stop entrance, let his own train in, and set his signal to stop again. The telegraph block system arrived in the United States in the 1860s, being first installed by Ashbel Welch on the Camden & Amboy in response to a rear-end collision, and a bit later by the New Jersey Railroad. The Pennsylvania took over these lines in the 1870s and by 1875 it had blocked both the New York division and the main line from Philadelphia to Pittsburgh. Welch's version of the block system was distinctive in one important aspect: it contained the first clear enunciation of the failsafe principle. As he put it: "the thing should be presumed to be wrong until the engineman has affirmative evidence that it is right—that is to say . . . safety signals should be used and never danger signals."[39]

While the telegraphic block system was far less failure-prone than controlling trains by time interval, it was not foolproof. Operators might mistakenly let two trains into a block at once. Indeed, the ICC reported thirteen collisions from this cause in the two years after July 1, 1904, leading it to conclude that "American signalmen are not so carefully selected nor so well trained as those in England." To prevent this inventors developed various "controlled manual" systems. In one version, signals were interlocked so that the operator at (A) could not clear his signal until the (B) operator electrically released it. Moreover, the (B) operator was prevented from releasing the (A) signal until the train in the block passed his tower and released the lock.[40]

The automatic version of the block system was an American invention usually credited to Thomas S. Hall, and seems to have appeared first on the Eastern Railroad, about 1871. With Hall's device, each end of the block contained a track instrument connected by a wire. When a train entered the block it tripped the instrument, thereby breaking the circuit and setting the signal to stop until the train left the block, which closed the circuit, setting the signal to clear. This system violated the failsafe principle; if a train broke in two the signal would automatically clear when the first section left the block.[41]

The form of automatic block signal that ultimately found favor relied on the failsafe track circuit invented by William Robinson in the early 1870s. In operation on a double-track line early automatic block systems based on the track circuit worked roughly as follows. The track in each block completed a circuit and power from a battery held the signals to clear. The entrance to the block was protected by a "home" signal, while lower down on the signal mast a second "distant" signal told the engineman whether the *next* block was occupied. By the 1890s these signals might be semaphores for daytime use with lights at night (usually red for danger, green for caution, and white for clear). If, on approaching a block, the engineman found the top semaphore in the vertical position it told him that the block was clear. When his train entered it would short the circuit and shift the signal to a horizontal position, which would stop following trains. It would also shift the distant signal on the previous block to "caution," giving any following train time to stop. Similarly, if the distant signal for this block showed "caution," the engineman would know that the next block was occupied and accordingly begin to slow his train preparatory to a stop; otherwise he would maintain speed. This system was failsafe, for if the battery failed a counterweight automatically shifted the signals to stop. Similarly, equipment left on the track, or a pile of rail placed by train wreckers, or a broken rail, or anything else that broke the circuit would protect the train.[42]

While in Britain use of manual block signals was all but universal by 1880, American carriers largely stuck with the train order system. Control with train orders worked best with light traffic and an experienced, well-disciplined labor force. But American carriers faced a labor market characterized by rapid growth, high turnover, and agency problems. As traffic density rose the flaws in the train order system became increasingly apparent, because it placed too much responsibility on dispatchers, telegraph operators, and especially train crews.

The Problem of Labor Control

Reviewing the *Gazette*'s record of 220 collisions in 1878, Matthias Forney commented, "What most needs doing now is the improvement of the men . . . [both] officers and employees." Only two years later, now staring at a total of 437 collisions, he noted the great increase due to "negligence and blundering," problems he blamed on the enormous increase in mileage that led to the hiring of new men with "insufficient experience and training." Forney also noted that the expansion drew experienced men from older to newer lines and that "the best of men have much to learn on new lines."[43]

As Forney observed, the use of inexperienced workers reflected in part the enormous expansion of the rail network. Desperate for labor, the carriers often placed inexperienced boys in positions of responsibility. In the 1870s Ace Corson became a substitute telegraph operator on the Kansas Pacific at the age of thirteen. Herbert Pease recalled being given the job of night telegraph operator at the Chicago & Alton's Atlanta, Illinois, depot in April 1900 when he had precisely two years' experience as a relief operator. Sometimes such inexperience had disastrous results. In one instance in

1884, a fifteen-year-old station operator with one year of experience caused a wreck on the New York Central when he failed to hold a train as ordered. The footloose qualities of trainmen were also legendary and their turnover worsened risks because, as Forney implied, much safety information was company specific. In 1854, when Charles B. George left the Boston & Maine to try his hand on the Chicago & Aurora he noted that "it would be impossible to name even a fraction of those who left . . . to take positions on lines tributary to Chicago." Between 1868 and 1901 J. Harvey Reed worked on eleven different roads. Over a similar span, Henry Clay French worked as a conductor, engineman, fireman, switchman, brakeman, yardmaster, and telegrapher, changing employers twenty-three times on railroads from Missouri to Oregon.

Such wanderlust was by no means atypical; it resulted not only from the increasing demand for labor but also from the cyclical and seasonal character of rail employment. Thus in the 1880s the Long Island, which did an immense summer business, routinely hired men in the spring and laid them off each fall. In 1889 a survey of sixty large carriers by the U.S. commissioner of labor found that the average duration of employment for engine runners was 8 months; for conductors and flagmen, 7 months; for firemen and telegraph operators, 5 months; and for brakemen, 3.8 months. Since these were averages, probably a majority of employees had even less company-specific experience.[44]

Changing jobs faced even an experienced railroader with a bewildering variety of novel train rules and signals. In addition, as companies increasingly obtained trackage rights over other lines trainmen found signals that changed meaning along the line. Nearly all companies with double track ran on the right, but not the Old Colony, which preferred running left. The train that carried President Garfield's remains to Washington traveled through four different signal systems. As one writer described the problem, "on one road a lamp moved up and down is the signal to stop, on a connecting road the same signal means go back." Similarly, "on one a single note of the whistle means go a-head, on the other go back; on one road red is the standard signal for danger, and yet, at three points on the same road, that color means that the road is clear." In 1881 the U.S. commissioner of railroads claimed that on two hundred roads only one whistle signal always had the same meaning while some had up to forty different meanings. Two years later the New York Railroad Commission surveyed the use of color signals by carriers in that state, finding "much lack of uniformity." Most companies used red for caution, but some used red and white, others green, and still others blue and white. An extra following a regular train with the same rights usually carried a red flag, except on a few carriers that used green. As Forney summarized signaling in 1882, "diversity like that which prevailed on Noah's Ark exists everywhere." Later the editor also complained of the "lack of system" and ambiguity of rules on many carriers that would bewilder a "Philadelphia lawyer."[45]

Inconsistent and faulty rules and signals contributed to a number of collisions. As a later writer described matters, "the early rules were crude and full of loop-holes through which many have leaped to their deaths." One such leap occurred on the New York, West Jersey & Buffalo. The company used red flags to signal both danger and to stop for passengers. On October 1, 1882, the engineman of Train 72 thought he had been flagged for passengers at the St. Johnsville station. He soon pulled out onto the main line and promptly collided with another train, killing two people and injuring several more. Such accidents led the carriers to develop general principles of signaling, and in 1884 the American Railway Association developed the Standard Code of train rules. Under Wellington's editorship the *Gazette* urged its adoption, terming the existing situation one of "complete confusion," and it was indeed widely, but partially adopted. By 1889, ninety carriers with about 66,000 miles of road had adopted it. Signal systems, however, remained uncodified until the twentieth century.[46]

The value of the code was sometimes revealed in the breach, as occurred on the Reading on October 24, 1892. A station agent received a train order from the superintendent to hold a freight, as a passenger train was running south on the north track. A few moments later the agent received a telegram from the yardmaster to have the freight take some empty cars up the line. He mistook this to be a train order annulling the first one; the ensuing head-on collision killed seven and injured about thirty. The Reading system was defective and differed from the Standard Code in that train orders were delivered on ordinary paper. Moreover, the code required that duplicate orders be delivered to train crews. Had this been done the freight crew would have known why they had been detained and been on their guard.[47]

Trainmen who simply violated the rules were a fruitful source of disaster, for supervision was difficult and expensive. Occasionally a traveling engineer might ride in the cab, but that simply put everyone on good behavior. Moreover, probably because early technology was so unreliable and breakdowns were so frequent, most carriers developed an ethos of "getting the train over the road." Thus, running too fast, or trying to make the next siding, or not bothering to slow for a torpedo (an explosive device put on the track to be set off by the locomotive warning of danger ahead) were all ways to get the job done, and all traded safety for production. *Railway Age* concluded that trainmen violated the rules daily in order to "make time," and in a kind of implicit contract, minor officials agreed to wink at most violations—if they didn't cause a wreck. Arthur Wellington sarcastically observed that the system of train rules required "some sort of an interlocking apparatus, so that it would be mechanically impossible for a superintendent to 'fire' the rules at his men without at the same time going personally to see whether or not he had hit the mark."[48]

Engineman Henry Clay French described several instances in which he violated the rules. In one case he took his train out ahead of another train rather than after, as his train order required. To French such actions were good railroading, and they may indeed have been, but the point was that men expected to judge for themselves which rules to obey, and as the collision record reveals, not all were good judges. A tight labor market also reduced the penalties of dismissal, making it more difficult for companies to enforce the rules. French recalled committing the worst sin of railroading: violating "Rule G," or getting drunk on the job. He was fired, but soon got another job. Discipline, Forney thought, "would be almost like a revolution." So, despite the rules, well into the twentieth century work trains were routinely run under "smoke orders"—which meant run until you see smoke and then look for a siding—and laborers were accordingly "slaughtered like sheep."[49]

Rule 99 of the Standard Code required the conductor to send a flagman to protect his train whenever it stopped on the road. Thus the safety of two trains was hostage to the fidelity and good judgment of the least experienced, worst-paid member of the crew. Yet flagmen could not be monitored, and since most of the time protecting the train was unnecessary, the consequences were predictable. Writing in the *Gazette* in 1884, Wellington summarized the results of relying on Rule 99: "Our railroads have been trying this experiment now for nearly sixty years, only to prove by each year's melancholy results that when flagging is depended upon there is no other word than FAIL."

Wellington also observed that rear collisions were more prevalent in winter than summer, because there were more reasons for trains to stop and because flagmen were less inclined to go back in bad weather, especially at night. A good example of the genre occurred at Spuyten Duyvil, New York, on January 13, 1882. At about 7:30 that evening the *Atlantic Express,* loaded with politicians from Albany, stopped on a sharp curve apparently because someone pulled the air brake cord. Alcohol may have been involved. Predictably, the flagman failed in his duty and the Tarrytown Special plowed into the rear of the *Express,* telescoping cars and starting a fire that incinerated eight passengers, including Senator Webster Wagner. Accidents that killed politicians were clearly serious; the legislature established a committee to look into such matters and the next year reorganized the railroad commission, broadening its powers.[50]

On single-track lines, disaster in the form of a head-on collision also lurked when watches were off or train orders incorrect, or forgotten, or misread. On October 20, 1893, the conductor of an eastbound excursion train on the Grand Trunk had orders to wait at the end of a double-track section near Battle Creek, Michigan, to let the westbound *Pacific Express* pass. But he failed to do so and the resulting collision and fire killed twenty-six.[51]

The carriers tried to cope with their labor problems in a variety of ways. As discussed in chapter 1, as early as the 1850s they began to develop rules governing dismissal and promotion by seniority to encourage fidelity to rules and to discourage turnover. By the 1880s the railroad unions were contributing to this process as they pressed for contracts that required trials and allowed appeals before men were dismissed. In the 1880s companies also began to develop accident and illness compensation plans (chapter 6).

The first of these appeared on the B&O in 1880, presumably making that company a more attractive place to work. The Pennsylvania began an accident compensation plan in 1886 and a company spokesman soon noted that "it has had a wonderful restraining influence on our men [reducing labor turnover]." The Reading, the Burlington, and other large carriers adopted similar plans, and by the early twentieth century fifty carriers had some form of benefit fund.[52]

Concern with the influence of labor quality on accidents also led the Illinois Central to inaugurate the first physical exams in the late 1870s. These tested only for color blindness and were confined to trainmen. The Pennsylvania followed in 1882, testing for hearing as well. The Reading, Burlington, Santa Fe, and other lines began similar examinations in the 1880s. Companies with benefit plans that needed to guard against adverse selection began pre-employment physical exams for employees beginning with the B&O in 1884. State laws requiring tests for color blindness in Connecticut (1880), Massachusetts (1882), Ohio (1885), and Alabama (1887) boosted testing. By the 1890s accident worries led companies to broaden the physicals; the North Western began testing not only for vision and hearing but also for hernias, loss of fingers, and heart and lung problems, and rejected about 13 percent of all applicants. By 1900 the American Railway Association had a standard physical examination that was widely used.[53]

In the 1890s, physical tests for employees blended with a broader program to tighten rules and revamp methods of hiring and firing. Joseph Bromley, a locomotive engineer on the Lackawanna, recalled the increasingly strict application of the Book of Rules on that line. It included in addition to physical examinations and mechanical courses for trainmen, injunctions against tooting the whistle to girlfriends. This took a lot of romance out of railroading, Bromley recalled, "but on the other hand, a railroad man had a much better chance of living to tell about his adventures."[54]

There were also new procedures for hiring and firing. Traditionally, a standard punishment for any number of infractions had been suspension or dismissal, thereby making the carriers a major cause of the high labor turnover from which they suffered. In about 1886 the Fall Brook Railroad originated what became known as the Brown system of discipline ("record discipline") that in most cases substituted demerits for suspensions. While a sufficient number of demerits might result in termination, they could also be offset by merits. By cutting both suspensions and terminations, the program reduced the carriers' need for extra men, and its use spread widely.[55]

The Crisis in the Train Order System, 1880–1900

In Britain the combination of high traffic density along with low wages for telegraphers encouraged use of the manual or controlled-manual block system. By 1873, nearly 40 percent of all mileage was blocked. As noted, by 1880, the year before Parliament made blocking mandatory, the figure had risen to 80 percent. The spread of blocking contributed to a rapid decline in passenger collision casualties, which sank to very low levels during these years—a fact that profoundly influenced the American trade press. In 1882 on the occasion of the rear-end collision on the New York Central at Spuyten Duyvil, Forney at the *Railroad Gazette* reviewed the progress of blocking in Britain and urged its extension in America. Six years later in 1888, Arthur Wellington of the *Engineering News* pointed out that only eleven passengers had been killed in all forms of British train accidents and that passenger risks had fallen by two-thirds in a dozen years. Wellington calculated that American lines were at least five times as dangerous, and he too called on carriers to begin blocking.[56]

The Slow Spread of Block Signaling

Despite such urging, American railroads mostly stuck to the old train order system. In 1892 a survey by the *Gazette* estimated that 3,000 to 6,000 miles of road, or less than 4 percent of the total, were blocked. The reasons were partly economic and partly technological. The block system reduced accidents and train delays, and it expanded track capacity more cheaply than double-tracking. But these gains were not compelling on most lines until the rise in traffic density during the 1890s. Modern writers on risk perception have noted that people tend to be overly optimistic in assessing personal risks, and companies found it hard to learn from others' experience. For while accidents were common, disasters were rare, fostering an "it can't happen here" mentality and sometimes only a homemade disaster would spur action. Moreover, manual blocking required skilled telegraphers who were more expensive in the United States than in Britain, while application of automatic blocking was hindered by problems of technology as well as cost.[57]

On single-track lines automatic blocking was technically complex because of the need to stop trains in both directions. Moreover, early signals did not work well. The difficulty of transmitting power over long distances led the early Hall automatic block systems to use banjo signals that could be run by an electromagnet that controlled display of a red or green silk cloth inside a weatherproof compartment. An alternative was the target signal, which worked like a Venetian blind on a vertical spindle. It employed an electromagnet that actuated a pendulum that powered the signal. Neither was satisfactory; both were hard to see and the pendulum mechanism had to be wound regularly. By the 1880s, battery-controlled pneumatic signals that used air to power a semaphore were available. But since they required running compressed air the length of the line they were too expensive to find wide favor. Later signals powered by electrically activated CO_2 cartridges reduced costs, but not until about 1900, with the development of small electric motors, did all-electric systems became feasible.[58]

In contrast to the carriers' tepid response to block signals, they rapidly adopted the use of switch and signal interlocking where railroads crossed each other, in good measure due to state actions. Like the block system, interlocking also had British origins. In the 1840s, British inventors developed rudimentary techniques to control signals at a distance. The idea of interlocking switches and signals also dates from the 1840s, but John Saxby developed the first successful mechanical interlock in 1858. Beginning with Michigan in 1873 states began to regulate crossings of railroads by other lines. Usually the laws required trains to come to a stop and that the lines maintain a flagman at the crossing, but such requirements might be waived if the crossing was interlocked.

Such laws in Ohio (1882), New York (1884), Massachusetts (1885), Illinois (1887), and some other states transformed interlocking from a money loser into a paying proposition at high-density crossings. One signal engineer on the North Western later estimated that it cost about $0.45 to stop a train and that a simple two-track interlocking cost about $2,600 a year to set up and maintain. Thus once crossing density rose to sixteen trains a day interlocking became cost-effective (16×365×$0.45=$2,628). Better technology also helped. Interlocking had initially been entirely mechanical but in the 1890s first pneumatic, then electro-pneumatic, and finally all-electric systems appeared, and these too cut costs because they allowed operators to control switches at much longer distances. Such economics did not always yield interlocking due to contracting difficulties, however, for where many lines crossed and benefits were unequal, agreement might be difficult and state laws failed to compel agreement. New York was one of the few states to collect statistics on interlocking and in 1902 it reported that 72 percent of all crossings were so guarded.[59]

In the western states most interlocks included derailing switches; if the engineman ran a stop signal his train would be shunted off the track before it could do any damage. One might suppose that such extreme precautions would rarely be required but such was not the case. In 1902, W. H. Elliott, signal engineer on the Milwaukee, presented a record of 141 recent derailments at interlocks on that line. The crossing of the Reading and Pennsylvania near Atlantic City was interlocked with both home and distant signals, but it had no derails as they were particularly expensive to install at that crossing. On July 30, 1896, a Pennsylvania excursion train bound for Atlantic City was stopped by the signalman and then given the signal to proceed. A Reading express approached the crossing at a high rate of speed, ran both the caution and the home signal, and collided with the excursion train, killing forty-two people, including the engine driver of the Reading.[60]

The Rise in Traffic Density

It was rising traffic density that finally tipped the balance in favor of the block system. While the rise in average traffic density was quite modest in the years following the Civil War, by the 1880s, some main lines were carrying over a hundred trains a day. Such density put enormous pressure on single-track roads. It increased the number of potential meeting points, which magnified the possibilities for human error and train collisions. The *Railway Age* illustrated part of the problem with a formula for the likelihood of collisions on a single-track line: $(N/2)^2 + N-2$, where N is the number of trains per day in both directions. Thus on line with 10 trains a day (5 in each direction) there were 25 potential meetings that could cause a butting collision and 8 possible rear collisions, or 33 potential meetings to be guarded. Double the number of trains, and there were 118 potential meetings. For example, on October 6, 1888, an eastbound freight on the B&O had orders to wait by a switch to meet two sections of the *Pittsburgh Express* and one of the *Cincinnati & St. Louis Express* to pass, but the crew fell asleep. They awoke to mistake the second train for the third, and, pulling out onto the main track, collided with the third train, killing three and injuring seven.[61]

Rising traffic density also snowballed the consequences of train delays, requiring increasingly complex rescheduling and increasing the chances for error and disaster. Thus, on the Old Colony in April 1882 a minor accident that was quickly cleared up immediately threw fifty-five trains off schedule. The most common cause of delay was equipment failures. A seemingly endless number of mechanical failures might cause a breakdown. Locomotive detentions were so common that George Fowler wrote a book on the subject and one carrier reported them under sixty-eight separate cause classifications plus "miscellaneous." In 1888 one engineman on the Southern Pacific described the problems with such equipment.

> Our passenger engine no. 1 is in very bad condition . . . The flange to center bearing on saddle has worn down so that it is broken off . . . and I am afraid she will drop down so the truck will not curve. The stack is also in very bad condition. We will have to have another air pump very soon. The pistons and valves are blowing very badly and the tank trucks are in very bad shape.

But locomotives were only part of the problem. Records of the Southern Pacific for the 1880s reveal train delays nearly every day on every division from broken air brake rods and pipes, pulled draw heads, hot boxes, cracked wheels, poor steaming, derailments, break-in-twos, overloaded trains, and dozens of other causes. Matthias Forney termed detentions "the primary cause of a large proportion of accidents" because they threw trains off schedule, and complexity was the Achilles heel of the train order system, causing it to fail when most needed.[62]

By the 1890s train density on some lines in the East and around Chicago was becoming unworkable and the carriers' critics increasingly began to call for use of the block system. A rear-end collision near Hastings, New York, on the New York Central on Christmas Eve 1891 that resulted from the usual combination of a traffic stoppage plus failure to flag precipitated a burst of criticism. The *New York Times* featured the disaster on page 1 and urged wider use of block signals. Arthur Wellington at the *Engineering News* called on management to "block signal the entire line." The Central ran 168 trains a day on its main line. This was three times the density of the little 44-mile Providence & Worcester, which had used block signals since 1881, he noted. While the P&W had been collision-free since that time, the Hudson River division of the Central had collisions "by the dozens." Angus Sinclair, the distinguished mechanical engineer and editor of *Locomotive Engineering*, also blasted the company. Sinclair observed that the Central's publicity termed it "America's Greatest Railroad," and suggested they add "—for Rear Collisions," which he claimed "are of almost daily occurrence." "Trains follow each other 'in sight,' sixty-car freights without air brakes . . . locals . . . and . . . sixty-mile-an-hour expresses," Sinclair fumed, and he wondered, "Can the absence of the only known means of protecting trains on a road like this be called stinginess, lack of foresight or carelessness? Is it not a crime?"[63]

The Central was by no means the only carrier drowning in traffic. About two years after the Hastings wreck, on January 14, 1894, when fog slowed a suburban express on the Delaware, Lackawanna & Western another suburban train rammed it, killing fifteen and injuring forty. Again the *Times* wondered why such a rich company could not afford block signals. Angus Sinclair pointed out that the signals on the division were the same as they had been in 1840. Yet fourteen trains were scheduled from Newark to New York City from 7:30 to 9:30 each morning, and the two that collided were scheduled to arrive in Hoboken only *3 minutes* apart. Under such conditions, flagging was unworkable. A flagman who left the moment a train stopped walking at 4 miles per hour could get back 1,056 feet in 3 minutes. A four-car train going 45 miles per hour took about 860 feet to stop: should the flagman hesitate one minute, disaster was inevitable. After the wreck, the company announced it would invest in block signals "right away."[64]

In response to complaints such as these from the trade and popular press that accompanied nearly every accident, as well as the carriers' need to raise train density still further, they began to introduce the block system during the 1890s. By 1901, the *Gazette* reported that 25,000 miles of road, or about 11 percent of the total, were blocked. Unfortunately for the railroads and their passengers, this was too little and too late, for the economic upswing that began in 1896 led to an enormous expansion of traffic, which ballooned all forms of accident.

Compounding the problem was the failure of blocking to provide in America the level of safety it delivered in Britain. As its use spread a small but troubling number of collisions began to occur in which enginemen simply ran signals. To their surprise both the carriers and their critics discovered that block signals suffered from some of the same labor problems that plagued the train order system. On the Fitchburg Railroad in 1893 a freight engineman running distant and home signals crashed into the rear of a passenger train. Subsequently the engine driver explained that he never paid any attention to distant signals. His fireman didn't even know where they were. Another engineman on the line testified that even if he missed a distant signal he never slowed for the home signal. The lesson, it seemed, was that even the block system could not fully remedy the carriers' agency problems. The *Engineering News* noted that there was no reason to believe that these employees were any more poorly trained or disciplined than those of other roads. Perhaps sensing what the twentieth century would bring, the editor concluded, "the wonder is that accidents are so few."[65]

In the years after 1870 the pattern of train accidents shifted away from derailments and toward collisions, while both forms of accident increasingly came to involve freights. For passengers, while overall the safety of train travel increased from the 1880s on, collision casualty rates fluctuated sharply, exhibiting no trend; after 1896 the upswing in traffic caused a sharp deterioration in safety. In Britain, by contrast, casualties from train accidents fell sharply while collisions nearly disappeared.

American carriers led the world in their adoption of air brakes on passenger trains, but the main effect of their subsequent improvement was to allow greater size and speed, not to improve safety. On freights the interaction of heavier cars, better brakes, and weak couplers, along with management failings precipitated an epidemic of break-in-twos. Heavier traffic on lines governed by train orders combined with a system of lax management and a chronically tight labor market also contributed to collision risks.

The public perceived collisions, especially if they involved fires, as especially dreadful. The popular press also amplified these risks, while the increasing ability to control them made the horror seem all the worse. In response, two forms of public policy began to shape these results. Massachusetts under Charles Francis Adams pioneered the use of publicity as a path to better safety, while several states employed regulations to ban stoves and force interlocking. In the 1890s the carriers also responded to rising dangers, introducing record discipline and a host of other new employment practices. They also extended the block system, but its impact on accident rates was obscured by rising traffic and blunted by continued agency problems. The result, as chapter 7 discusses, was an outpouring of criticism after 1900 that led to major reforms.

4 The Major Risks from Minor Accidents, 1873–1900

The whole system of American institutions is based upon the principle that . . . people can take . . . good care of themselves.

—Charles Francis Adams, 1871

It is not in the interest of the state to have . . . [these coupling injuries] continue.

—New York Board of Railroad Commissioners, 1884

The great majority of casualties on railroads were not the result of spectacular collisions or derailments. While such disasters might injure or kill dozens of people all at once, most of the carnage resulted from "minor" accidents that picked them off one or two at a time. These were passengers who fell while running for the train, or employees who were crushed while coupling cars, or working men and women who took a shortcut across the tracks or tried to beat a train to the crossing.

Throughout the nineteenth century, such accidents were usually invisible. In part this was simply because they were little: a train accident that killed a dozen people became a headline; a dozen accidents that killed twelve people became obituaries. But as modern researchers have noted, ideology can also attenuate as well as amplify risks and the ideology of individualism also colored views on such accidents. The standard memorial for a person killed in a little accident was: "it was his own fault." Since legal liability for accidents depended on fault, the carriers had less economic interest in preventing such accidents than they did train wrecks that harmed passengers. While similar conceptions also dulled public humanitarian concerns, this fault-based view gradually receded. As understanding of accident causation became more complex, some officials and reformers began to assert that accidents had social as well as individual causes, and that the community had an interest in preventing them. Changing attitudes and improved technology reduced the number of little accidents as the nineteenth century ended, but crossing and trespassing accidents remained major problems.[1]

Taming the Iron Horse

In both Britain and the United States locomotives accounted for many accidents. Boiler explosions usually killed only trainmen, typically one or two at a time. In 1887, a typical year, the *Railroad Gazette* reported fourteen explosions that killed fourteen trainmen. American locomotives were more likely to blow up than those in Britain, for American carriers routinely exceeded designed pressures, which overstressed pins, rods, and other moving parts, and in a rickety old locomotive, might bring disaster. In 1868 the *American Railway Times* reported a boiler explosion on a Grand Trunk locomotive that was twenty years old, and which had iron near the seams corroded to one-third its original thickness. In 1870 one engineman claimed his road ran such long trains that if he followed the rules for steam pressure he could not start the train. From 1873 to 1886 the *Gazette*'s data yield an average explosion rate of 0.8 per thousand locomotives, which was admittedly a considerable undercount. Locomotives were long lived; over two decades such an annual risk implied the sobering fact that nearly 2 percent of all locomotives would explode. In Britain, by contrast, much more complete reporting yielded an explosion rate of 0.46 per thousand over the same period.[2]

Locomotives had many other ways of killing or maiming trainmen. Gilbert Lathrop reported that a fellow engineman on the Rio Grande was killed when the Johnson bar (the reversing lever) failed to latch and hit him in the stomach when he opened the throttle. Main or side rods broke with such frequency that the *Railroad Gazette* classified such accidents separately. Engineman J. Harvey Reed reported that when the main connecting pin broke on a Rogers Mogul locomotive he was on, the side rod acted like a flail, tearing up the cab. In 1887 the *Gazette* reported seventeen

Figure 4.1. **Locomotive Accidents from Explosions and Broken Rods, 1873–1900**

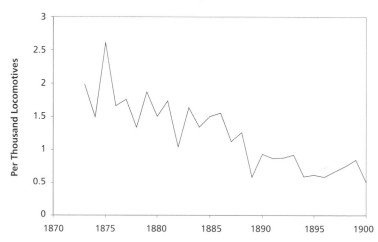

Sources: Railroad Gazette; author's calculations.

such accidents; one of them to Locomotive No. 72 of the Rome, Watertown & Ogdensburg on July 16 killed the engineman as it was drawing a special train containing President Cleveland and his wife.[3]

The Master Mechanics and the Locomotive

The men in charge of the steam locomotive were the master mechanics and their main interest was in increasing its efficiency and hauling capacity. To this end they built ever-larger machines with greater steam pressure, infinitely tinkered with their many parts, experimented with compounding, and eventually embraced the superheater. Just as the master car builders organized the MCB, so the master mechanics organized their own association—the American Railway Master Mechanics. Many men in fact joined both organizations and they eventually met together. Like the MCB, the Mechanics became increasingly scientific as the nineteenth century wore on and younger, more technically trained men took over the reins. In the early days members decided technical matters through discussion and voting; testing was all but unknown and each man appealed to personal experience as his evidence. Their results were widely reported and yielded important design improvements. Safety was an indirect concern, for a locomotive that broke down or blew up was inefficient as well as dangerous. Hence, while improving safety was not the dominant motive, most innovations in locomotive technology and operation reduced risks.

Boiler design was one topic where safety and efficiency met. In the 1870s Matthias Forney was both the editor of the *Railroad Gazette* and a member of the Mechanics. He had designed a tank engine that bore his name, and was therefore both an expert on locomotive design and thoroughly familiar with the *Gazette*'s accident data.

Forney thought many boilers poorly designed and their construction "wretchedly defective." Although explosions also resulted from low water, he opined that this cause was often a "convenient scapegoat for the sins of the boilermakers." At the Master Mechanics' 1880 meeting Forney proposed a discussion of how to secure boiler crown sheets. He explained that British builders employed staybolts attached from the crown bars to the boiler shell, concluding that "boiler construction in England is ahead of American practice," and he urged other members to study British methods. A representative of the Lake Shore agreed, saying they followed British practice, while another member claimed that it "was fast coming into use in this country." The journal *Railway Master Mechanic* also wondered "how long will it be before locomotive boilers will be properly stayed," and it complained of "long rods running from head to head with the threads . . . [not upset], just as if the strength was not measured by the weakest part." American locomotive construction was defective in other ways, Forney noted. American builders employed lap joints, and the normal expansion and contraction during heating and cooling led to metal fatigue and failure. British builders employed butt joints with plates over top and bottom that were less prone to bending and also placed less shear on the rivets. British locomotives also had fireboxes and steam domes with fewer, stronger joints, and they used heavier boiler iron than did the Americans.[4]

Incremental Improvement

Gradually, methods and materials improved. Few of these changes were either dramatic or patentable; as with much of railroad technology many reflected a kind of collective invention. As master mechanics experimented with incremental changes they shared experiences, and in response designs and practices slowly evolved. In the 1870s companies began to experiment with steel in boiler shells, fireboxes, flues, axles, and drive gear, but its use spread slowly. Early steel plate was often brittle and failed—a result of excessive phosphorous, sulfur, or carbon—while forged steel axles for locomotives had the same problems as those for cars. In 1886 one master mechanic recalled that steel had once been "a bewildering variety of substances with highly differentiated qualities." Despite difficulties, steel slowly replaced iron in axles and drive gear. In 1887 one superintendent reported that since he had begun to use steel in 1876 not a single parallel rod had broken. Even so, mechanics' distrust of the metal died hard, and as late as 1911 *Railway Age* complained that some laggards were still using wrought iron in reciprocating parts, making them both inefficiently heavy and dangerous.[5]

The use of steel castings in locomotive parts proceeded even more gradually. A report to the Master Mechanics claimed that they were little used in 1885. The reasons were partly their cost and partly quality control, for casting steel and iron required sharply different foundry practices, and early steel castings tended to have blowholes. By the mid-1890s, however, use of silicon and manganese additives and proper venting of molds improved quality. As locomotives became larger and heavier, they began to bump up against weight limits for track and bridges. As a result, in 1896, American Steel Castings produced the first cast steel locomotive frame, and steel began to be employed in wheel centers and elsewhere. Cast steel was safer than wrought iron, which might contain defective welds, and it allowed companies to reallocate weight to the boiler, thereby increasing locomotive hauling capacity.[6]

Companies such as the Pennsylvania, the Burlington, and others that developed testing laboratories and began to develop specifications and tests for rails and wheels applied the same procedures to locomotive materials. By the 1880s railroad engineers had assimilated research findings demonstrating that stress well below a metal's elastic limit might lead to failure if repeated often enough. In 1895, the B&O began to require vibration tests when it bought staybolts to guard against metal fatigue, and the Pennsylvania soon followed. In 1891 W.F.M. Goss of Purdue University developed the first locomotive-testing laboratory to optimize all aspects of locomotive design. This was followed by test facilities on the North Western in 1895, at Columbia University in 1899, on the Pennsylvania Railroad in 1904, and finally at the University of Illinois in 1914.[7]

By the 1890s the Master Mechanics belatedly formed a committee to develop specifications and tests for iron and steel. In 1893, their report, which was skeptical of the value of specifications and tests, urged expert inspection instead. The committee concluded with the observation that bread could also be made to specifications, but a culinary expert could tell far more by simply chewing it. The report provoked a sharp exchange that highlighted the contrast between the older tacit knowledge and the newer, science-based engineering. William Forsyth of the Burlington, one of the new breed of technically educated men, claimed that his company had been employing physical tests of steel for fifteen years. Samuel Vauclain of Baldwin Locomotive believed that "when you get away from the scientific inspection of boiler plate . . . you are losing ground." Buying boiler steel was not like eating bread, he thought: "you might chew boiler plate until you wore your teeth out and not know any more about it when you got through."[8]

The opponents of testing were not quite as reactionary as their critics believed. They pointed out that there were many different specifications, not all of which could be correct, and while they stressed the value of expert inspection, they also acknowledged the importance of keeping careful records. "I think if you go to the scrap pile you will tell more about steel than the laboratory," one claimed. This was a losing cause, however, and the next year the mechanics proposed specifications for boiler steel. By 1897 most major carriers and builders bought steel subject to both physical and chemical testing.[9]

A host of other incremental improvements in locomotive design, operation, and maintenance methods shaped safety. As late as the 1880s, a fireman had to walk out on the running board of a moving locomotive carrying his tallow pot to lubricate valves. By the 1890s, sight-feed lubricators operating from within the cab replaced this dangerous custom. Companies improved locomotive lubrication by shifting from animal- to petroleum-based oil, by systematic testing of lubricants, and by shifting from oil to greased bearings. Better lubrication reduced downtime and lessened the chance of dangerous failures of crank pins and other pieces of running gear. Traditionally locomotive tires were simply heated and then shrunk onto the wheel. But these might come off if the tire cracked, and in the 1890s companies began to fasten them more securely. Bad water was also a major source of boiler and flue failures. Boiler scale clogged injectors and acted like insulation. This not only wasted fuel, it caused the metal to overheat and weaken, which might burst a flue, stopping the train on the road where it was in danger of collision, or even cause a boiler explosion. Companies experimented with various additives to improve boiler water but cost-effective methods remained elusive until the twentieth century. During routine maintenance companies also washed boilers to remove scale that resulted from hard water. The Master Mechanics, following rules developed on the Pennsylvania, recommended washing every six months.[10]

Failure to fix broken staybolts was probably what caused the engine *Jupiter* to blow up on the Fall River Railroad on November 3, 1880—in fact, the company sent it out knowing that there was a leak. By the mid-1890s, a survey that covered about 12,000 locomotives found that 58 percent tested staybolts at least once a month. But testing was an imperfect art; a workman tapped the bolt with a hammer and tried to determine from the sound if it was broken. Companies began to test under pressure because they found that was more accurate, but nothing worked well and as boiler pressures increased so did expansion and contraction. The problem of broken staybolts worsened. Some companies were especially disaster-prone. The Reading had six explosions in 1892, a rate of 4.29 per thousand locomotives, and it killed twenty people, while nationwide the failure rate that year was 0.37 per thousand.[11]

Inventors responded, first developing partially drilled and then hollow staybolts, either of which would leak steam when they broke, thereby warning of danger. But bad maintenance practices could defeat the best of technologies. One carrier that had locomotives with drilled staybolts simply plugged the leaks. In an editorial in *Locomotive Engineering* Angus Sinclair termed it amazing that an engineman could be found who would take out such a locomotive. In 1878 the first flexible staybolt was invented, but it worked poorly. Only after improved versions came out about 1900 did their use become common.[12]

In a few instances builders modified designs simply to improve safety. In 1881 Matthias Forney read a paper to the Master Mechanics in which he listed a dozen accidents taken from the *Railroad Gazette* in which steam from the boiler killed or maimed trainmen or passengers. Forney pointed out that it was rare for the boiler shell to rupture; instead, the damage was done by an attachment that broke off, perhaps in an otherwise minor collision or derailment. In one wreck the engineman died when a broken gauge cock driven into his side steamed him alive. There were twenty-three openings in the boiler with attachments that might break off, Forney noted; the design resembled a porcupine, he later remarked. Nor were such accidents always the result of a collision or derailment. Henry Clay French claimed that he fired for one line that experienced two broken water glasses a week, in one case scalding the engineman so badly he had to retire. Forney's plea got results. By the 1890s the usual practice was to place all the attachments on a single stand attached to the boiler. In addition, some roads also installed check valves to the attachment inside the boiler so that an abnormal flow of steam would automatically close it. Some locomotives also began to come through with other safety attachments. In 1888 the Mechanics resolved that "all engines should have steps placed on the front end for the safety and convenience of brakemen." In 1890 the Mechanics passed a weak motion in favor of using water glasses as a precaution against boiler explosions but many locomotives lacked them until the Locomotive Inspection Act of 1910 made them mandatory.[13]

By 1900, steam locomotives were far larger and carried steam at much greater pressure than a generation earlier. They were also far better designed and built, employed much more steel of higher quality, and were subject to more careful inspection and maintenance. These incremental improvements owed nothing to public policy; their combined effect was not only a more efficient locomotive but a much safer one as well. The *Gazette* data on boiler explosions and broken running gear are surely an undercount of all such accidents, but as usual they probably caught most of those that caused casualties. From 1873 to 1900, before any state inspected locomotives, such accidents declined about 75 percent from two per thousand locomotives per year to one for every two thousand (Figure 4.1). Once again, the drive for efficiency led to major safety improvements.

The Trainman and the Freight Car

Trainmen's work has always been dangerous, and the little accidents have always done most of the killing and maiming. The earliest data from New York, Massachusetts, and Ohio from the 1850s through the 1870s reveal that collisions and derailments rarely accounted for as much as half of employee injuries or fatalities (see Table 1.1). Instead, most of the risks derived from working around cars and locomotives, from the need to ride freights to work the brakes, and from coupling cars. In 1890 the ICC published the first detailed national estimate of employee casualties by cause and occupation. There were a total of 2,451 fatalities and 22,396 injuries severe enough to be reported. Of these, about 369 fatalities and 4,750 injuries resulted from coupling cars. About 59 percent of all casualties occurred to trainmen although they comprised but one-fifth of total railroad employment. Some of the particulars of the carnage are contained in Table 4.1.

 Clearly, by modern standards all forms of railroad work were extraordinarily dangerous, while trainmen's mortality was simply staggering. The death rate from occupational hazards for members of the Brother-

Table 4.1 **Casualties to American Railroad Employees, by Occupation and Cause, 1890 (per thousand workers)**

	Coupling Cars	Riding Cars*	Train Accidents	Other	Total
All workers					
Fatality rate	0.49	0.87	0.69	1.22	3.27
Injury rate	10.48	3.61	3.45	12.36	29.90
Trainmen					
Fatality rate	1.75	3.54	2.55	1.78	9.62
Injury rate	40.10	14.18	13.27	19.34	86.89

Source: ICC, *Third Annual Report on the Statistics of Railways of the United States 1890* (Washington, 1891).
*Falling from trains and overhead obstructions.

hood of Locomotive Firemen and Enginemen in 1890 amounted to 3.38 per thousand, yet for all trainmen it was 9.62. By implication, the mortality rate of brakemen and conductors (i.e., trainmen other than firemen and enginemen) was a breathtaking 14.52 per thousand. It is no wonder that when Henry Clay French first went to work in the yards of the Hannibal & St. Joseph in 1873 his sister set aside one clean sheet in which to wrap his remains.[14]

While some of these dangers were inherent to nineteenth-century railroading, as chapter 1 explained, American railroads evolved in ways that made them more dangerous to operate than were British lines. After 1850, American carriers became far more freight-intensive than those in Britain. By 1880 freight train miles were nearly double passenger train miles, and the risks of freight work exceeded those on passenger trains by a wide margin. Some accidents were simply bizarre. On the Oregon Short Line a cow knocked the step off a caboose, and when a brakeman stepped down on it he fell under the wheels and lost an arm. But most were more predictable. The two leading dangers to trainmen, riding and coupling cars, reflected a peculiarly American evolution of the freight car.

The Evolving Dangers of Freight Work

From at least the 1840s most American cars employed link and pin couplers (chapter 1). As noted, these were a form of technical change that probably worsened worker risks. Coupling required not only that one go between moving cars, but also that the brakeman insert his hand between the couplers as they came together. In the twelve months from October 1867 through September 1868 the Erie reported twelve coupling accidents. As usual, injuries exceeded fatalities. Only one man was killed—a switchman who was crushed between cars in Jersey City, New Jersey, on November 5, 1867. Of the injured, a typical report was that of switchman George Rogers, whose thumb was taken off while coupling cars in Buffalo, New York, on November 24, 1867, or brakeman Johnson Cole, who crushed his hand near Port Jervis on January 23, 1868.[15]

> In the rain or in the sunshine
>   He must mount the speeding train
> Ride outside at post of duty,
>   Heeding not the drenching rain.
> Suddenly there comes a messenger;
>   God have mercy hear them pray,
> As they hear the fearful story—
>   Killed while coupling cars today.

So common were such accidents that in 1900 E. E. Clark of the Order of Railway Conductors reported to the Industrial Commission that "these minor injuries, such as loss of a finger, have never been considered serious by the trainmen." British freight cars, by contrast, were coupled by hook

Table 4.2. **Casualties to British Railroad Employees, by Occupation and Cause, 1890 (per thousand workers)**

	Coupling Cars	Train Accidents	Other	Total
All workers				
Fatality rate	0.05	0.03	1.36	1.44
Injury rate	0.94	0.42	7.66	9.02
Brakemen and goods guards				
Fatality rate	—	—	—	5.56
Injury rate	—	—	—	71.45
Shunters*				
Fatality rate	—	—	—	5.75
Injury rate	—	—	—	62.50

Source: Great Britain Board of Trade, *General Report Upon the Accidents that have Occurred in the United Kingdom in the Year 1890* (HMSO, 1891).
* Calculated using 1884 employment.

and chain, which could be attached from outside the cars. Accordingly, as Table 4.2 reveals, coupling cars was not a major source of risk for British trainmen.[16]

Similarly, while trainmen rode cars for many reasons, until the twentieth century riding freights to work the hand brakes was part of the job and a man might be thrown off, be knocked off, or fall when the brake wheel let go. A late-nineteenth-century freight car was about 14 feet high. A man falling from a 14-foot car going 10 miles per hour would hit the ground at about 23 miles per hour—easily enough to kill him. Death was even more likely if he rolled under the wheels as did Erie brakeman Thomas Aldridge, who fell from a freight and lost an arm and leg on January 14, 1868, and died. Well into the twentieth century when a trainman died the wreath on his casket was shaped like a broken brake wheel. By contrast in the days before power brakes, British trains employed a brake van to stop. The overall mortality of British railroad employees, while still high for those who worked around freight equipment (shunters and brakemen and goods guards), was dwarfed by the risks to American trainmen.[17]

While risks for American workers had probably been declining since before the Civil War, as Table 4.1 indicates, freight work remained breathtakingly dangerous for a very long time. It was not just the act of coupling cars or riding the tops of freights that was dangerous but, more broadly, simply being around rolling stock. Yard switchmen were forever dodging moving equipment, and to fall might mean death or dismemberment. Poling, by which a switch engine could push a car on a separate track employing a long, diagonal rod, was a highly dangerous but highly productive practice. As late as the 1890s an ARA survey revealed that except for Massachusetts and Michigan, no state required companies to block frogs (the joint where two rails cross), and many companies didn't bother. Edgar Herenden was one of the many young men who paid the price for this failure. He died at the age of twenty-three on Saturday, October 11, 1873, when he caught his foot in a frog and was run over while switching near Norwich, Connecticut, on the Norwich & Worcester.[18]

Tight clearances in yards and roads routinely killed men, one of them being Engineman O. M. Wilmot of the Boston & Maine. He mistakenly stuck his head out of his cab on January 21, 1894, just as the locomotive neared a bridge at Lyndonville, Vermont, where the clearance was less than the legal limit. Flying switches—where a brakeman dropped off a moving engine, cut out some cars, and threw a switch to direct them to another track—were as dangerous as they sound and took the lives of many men. Among them was Station Agent H. G. Grover of the Norwich & Worcester who tried to make such a switch on July 10, 1872, but he slipped "and the whole of the car passed over him."[19]

Again, some dangers may have increased. Initially carriers had built bridges 20 feet above the rail—high enough to allow a man to pass under them safely while standing on a car, but as car size increased to 14 feet and more after 1870, bridges became more dangerous, requiring the use of telltales for safety. Predictably, good and bad designs evolved. One form involved a bar across the track above the car top and on a hinge. Sometimes the wind would blow it out of the way, making it useless; sometimes ice would freeze the hinge, ensuring that it would knock a brakeman off.[20]

The risks of riding and coupling cars may also have worsened due to freight car interchange, which created strong incentives to pass along cars with faulty ladders and handholds to other lines because MCB repair rates were too low for western railroads. In 1887 when one company (probably the Boston & Albany) announced it would refuse cars in interchange that did not fasten ladders with lag screws, it found one where nails (which have a head that looks like a lag screw) had been used. They pulled out when the inspector tested them. Interchange also raised risks because it increased the heterogeneity of the freight car stock. Drawbars ranged in height from 33 to 35 inches. As car interchange spread, brakemen increasingly faced the dangerous task of trying to couple loaded 33-inch cars with empty 35-inch cars. Thus in 1888 the master car builder of the Southern Pacific complained that loaded fruit cars from the Michigan Central sat so low they could not be coupled with his own equipment.[21]

Similarly, buffer blocks, which were to protect the cars from collision during coupling, often differed from line to line. The effect of such diversity on trainmen's safety led the men to call them deadblocks or sometimes mankillers. Some lines used a single block above the coupler. When it hit a similar block it protected both the cars and the brakeman. Other carriers used double blocks on either side of the coupler. Sometimes these were so deep that reaching around them to couple was nearly impossible. In 1884 G. W. West, master mechanic of the Burlington, claimed that foreign cars with overly deep buffer blocks caused most of the coupling injuries on his line. Sometimes double blocks were too far apart to hit a single block, which would result in the trainman being crushed to death. Finally, some lines also employed blocks only 3 inches deep that would allow a man to be crushed even if the blocks did meet. When such thin blocks killed a man on the New York Central a court found them "unnecessarily dangerous" and fined that carrier $3,500 (a bit over seven years' annual earnings).

Such accidents, railroad managers invariably intoned, resulted from the men's carelessness. This construction of risk—in which those most able to improve safety blamed dangers on those least able—powerfully appealed to American ideals of individualism and self-reliance. It was also convenient at law, but it blunted interest in reform. In a survey of that state's accidents for 1883, New York's railroad commissioners also described the many coupling injuries as the result of "the carelessness of the employees themselves." Yet the commissioners concluded that because the injured often became public charges "it is not in the interest of the state to have this continue."

Moreover, everyone familiar with railroading understood that coupling accidents also reflected technology and institutions. New York's commissioners found that nearly all carriers had rules requiring men to use coupling sticks. These would guide the link and thereby protect the fingers during coupling. Yet, supervisors usually winked at the rules, for the sticks slowed a man and the carriers encouraged productivity, not safety. Hence, while a few carriers such as the Atlantic & Great Western required use of sticks, on most "the rule was very laxly enforced," and companies admitted as much. Even with a coupling stick, men still had to go between cars to pull or drop the pin, a fact that focused attention on unguarded frogs and switches that might catch a foot and on car construction. Thus the employees' careless behavior was only half of the problem, for casualties were also an outcome of management rules and technological choices that emphasized productivity over safety.[22]

The real function of the coupling stick, as employees well understood, was not to protect their hands at work but to shield employers' pocketbooks in court. Workers sometimes vented their rage at such sham safety procedures by breaking the equipment. In 1897, Joseph Cummin was a bridge superintendent on the Long Island and he marveled that when companies placed telltales above the tracks to warn of bridges, brakemen "as a rule will not allow the opportunity to go by without doing something to injure the tickler." Another superintendent was more perceptive: "these bridge warnings are not put up for the purpose of protecting brakemen alone, but to protect the railroad company as well," he explained.[23]

Building a Safer Freight Car

Matthias Forney was a member of the Master Car Builders as well as the Mechanics. Forney was well aware that coupling accidents reflected more than just carelessness and he used his position in the Car Builders and as editor of the *Gazette* to campaign against coupling injures. In the 1870s, since he thought that existing automatic couplers were "utterly useless," his approach was to improve freight car design. He especially urged companies to get rid of projecting bolt heads and standardize the position of buffers. In 1871 the MCB recommended 33 inches as the standard height of drawbars, and in 1878 the organization set up a committee on accidents to trainmen to look into car construction. The next year it reported a bewildering variety of dangers due to inconsistent heights of drawbars, dangerous dead blocks, and poorly constructed and inconsistently positioned hand brakes. Although the committee recommended abolition of double dead blocks, little was accomplished. In 1881 the MCB was still trying to standardize the position of brake shafts to the left hand corner of the car and in 1886 it was still trying to standardize dead blocks and drawbars. Two years later replies to an MCB circular asking for suggestions on ways to improve freight car safety told of a host of dangerous practices. In 1892 Wellington's *Engineering News* summarized the situation, observing that that standards for protection were "generally neglected" with the exception of the standard for dead blocks, which was "wholly neglected."[24]

Although the federal Safety Appliance Act of 1893 mandated a common height for drawbars, the efforts of Forney and the MCB to standardize freight car safety appliances largely failed until a federal law of 1910 forced it on the carriers. The episode is instructive of the difficulties in agreeing on a common technology in the absence of strong incentives to do so, and it foreshadowed the problems the carriers would face in improving coupler technology.

The link and pin coupler had one great merit—simplicity. Otherwise it was unsatisfactory as well as dangerous. Pins disappeared, which was expensive and reduced productivity, while the coupler broke regularly, leading to accidents. By the 1880s there was no shortage of alternatives; the *Gazette* estimated there were nearly a thousand patented car couplers on the market, many of which promised greater safety. These divided into three broad classes: variants on the old link and pin coupler, hooks such as that of Miller, and the newer vertical plane couplers, which worked like two hands with fingers curled. Most would not couple with each other.

Companies' interest in self-couplers reflected both humanitarian concerns and the prospect that they would increase the productivity of yard work. But their enthusiasm was tempered by the modest cost of killing and injuring brakemen. As previously discussed, in the 1870s and 1880s, the average compensation cost of a worker casualty was about $134 in Ohio. Fragmentary evidence suggests that worker risks also caused higher labor turnover and forced employers to pay a wage risk premium as well. My findings indicate that each increase in the death (injury) rate of one per thousand workers raised their annual income by 0.13 (0.05) percent. But there is no evidence that employers were *aware* that risk raised either turnover or wage payments, and therefore that such payments created safety incentives. In the *Gazette* Forney concluded, "it is economy to prevent accidents to passengers. If the much more numerous accidents to employees could be made equally costly to the companies there is good reason to believe that much more pains would be taken."[25]

The modest costs of accidents were reinforced by problems of freight car interchange. By this time most carriers hauled a significant proportion of foreign cars and saw their own rolling stock scattered in interchange. Thus in 1890 17 percent of the B&O's rolling stock was away from home; for the North Western 29 percent and the Lake Shore 45 percent of rolling stock was on foreign carriers. This yielded the same disincentive to investment in better couplers as it did to better wheels (chapter 2). Writing in 1887, Arthur Mellen Wellington thought the chief obstacle to the introduction of automatic couplers was that "owing to the continuous interchange of cars no real benefit would be derived from such a coupler until it had come into almost universal use." Wellington correctly concluded, "it may be well on toward the close of this century before automatic couplers come into use."[26]

Despite such disincentives the record reveals considerable experimentation with alternatives to the link and pin. Spooner's automatic link and pin coupler, which was used on the Eastern Railroad in the 1870s, contained a stop on the drawbar that prevented the (single) buffers from meeting when the engine backed, thereby protecting the hand of the brakeman. At about the same time, the Burlington, Maine Central, and Milwaukee used the Potter drawbar, which worked about the same way. Miller's coupler and platform was of the hook variety; it has already been discussed and was almost universal on passenger cars. But while the Southern Pacific and a few other companies employed it in freight service, few other carriers found it satisfactory for such duty. The Pennsylvania Railroad had long employed Eli Janney's vertical plane coupler on passenger equipment, and about 1880 at their urging Janney strengthened it for freight work.[27]

Regulatory Beginnings

These rather leisurely experiments were interrupted in 1882, when Connecticut became the first state to mandate use of safety couplers. The only opposition to the rule, according to the *Hartford Courant,* was from "railroads and undertakers." Massachusetts followed in 1884, Michigan and New York in 1886, and Iowa in 1890. Their actions reveal the importance of regulatory commissions that could gather information and comment on safety matters. In each of these states the railroad commission had advocated coupler safety and had also pressed for other safety rules such as switch guarding and car heating. Their actions, however, threatened the carriers' ability to make rational technical choices. Should a coupler happen to meet one or more states' requirements and be chosen by several large carriers, the need to interchange might encourage others to standardize on it, thereby locking all the roads into a poor technological choice. This was the sort of industry-wide technology problem that had led to the

MCB in the first place and the organization reacted to these threats with coupler tests that began in 1884 and ended with the Burlington air brake trials of 1886 and 1887.[28]

Bad brakes, like weak couplers, were also an efficiency problem for they constrained freight train weight and length. The Burlington, being one of the leaders in equipment testing, was a logical place for the trials. Although Westinghouse brakes predominated in passenger service, a few carriers used buffer or vacuum brakes or the Loughridge air brake. But few employed any train brake on freights. The reason again was partly freight car interchange, which reduced the payoff to brake investments. The Pennsylvania, which faced the steep grades of the Alleghenies, put air brakes on its stock cars, which weren't interchanged, while the Denver & Rio Grande, a narrow-gauge line that also faced steep grades and didn't exchange cars, used air on all its cars. Most of the other early users were also western carriers with steep grades such as the Southern and Central Pacific, the Burlington, and the Santa Fe. Other companies also felt the need for better brakes if they were to reap the gains from longer, heavier freights. Yet since freight cars were interchanged, brakes had to be standardized.[29]

The Burlington trials were intended to test the ability of various brakes to control long (fifty-car) freights, and initially none of the contenders worked very well. For example, because of the slow speed of application of the Westinghouse brake an emergency stop generated disastrous shocks in the rear of long freights. Ultimately, however, the trials solved both the coupler and the air brake problem. After the 1887 trials Westinghouse modified the automatic brake and developed the quick-acting brake. (Recall: in an emergency it vented air from the train pipe directly into the cylinder, thereby not only increasing brake pressure but also speeding application and preventing shocks.) The new brake cut the time before the brakes began to apply at the end of a fifty-car freight from 17 seconds to 6. The trials also made it clear that slack was undesirable, which automatically disqualified all the link and pin couplers. As the MCB concluded in 1887, "the Janney type of coupler . . . is the type to which the evolution of the subject has brought us," and in 1888 after holders of the Janney patents ceded some of their rights, it became the MCB standard.[30]

Thus the Burlington trials resulted in both a workable train brake and automatic coupler. Moreover, although freight car interchange dulled the incentives to install such equipment, companies that wanted the benefits of heavier trains would install air brakes and these virtually *required* automatic couplers. Despite their failings, economic incentives were therefore leading the companies to install safer appliances. By 1893, the last year in which the installations were voluntary, about one-quarter of all freight cars had some form of automatic coupler.

This process was not fast enough to suit the carriers' critics, however, and state commissioners, the ICC, and the railroad unions agitated for a federal law mandating air brakes and automatic couplers. In 1888 the ICC published its first accident data, revealing that there had been 2,000 men

killed and 20,000 injured on the railroads that year. In his first State of
the Union message President Benjamin Harrison termed such statistics a
"reproach to our civilization," and called on Congress to act. In 1889 for the
first time the commission separately reported coupling accidents, revealing 300 men had been killed and 6,757 injured coupling cars. The numbers
stunned Henry Cabot Lodge, then a congressman from Massachusetts,
and he responded with an article for the *North American Review,* "A Perilous Business and the Remedy." Like Harrison's message, it reveals the impact of such statistics in combination with known remedies on individuals' moral sentiments. Lodge quoted Robert Burns—"Man's inhumanity to
man makes countless thousands mourn." While Lodge acknowledged that
Burns probably had direct inhumanity in mind, "the failure of Congress to
act . . . is a kind of inhumanity indirect and unintentional," he claimed. He
urged that "if anything can be done, it is little short of criminal not to do
it," and he urged Congress to require automatic couplers.[31]

Unions representing the brakemen and conductors employed
Lorenzo Coffin to make their case to Congress. Coffin had been a chaplain
during the Civil War and later an Iowa railroad commissioner, and so he
brought both religious fervor and technical expertise to his job. They could
not have had a better advocate. The railroads strongly opposed legislation
that might have required specific couplers, for existing couplers were both
highly imperfect and rapidly evolving, and such a law might freeze their
technology. But there was less opposition to Coffin's bill mandating that
men need not go between cars to couple them. Such was the language of
the Safety Appliance Act, passed March 2, 1893; as initially written it required self-couplers on all equipment by January 1, 1898, and air brakes
on a sufficient number of cars to allow the engineman to control his train.
After several extensions it finally became operative in August 1900.[32]

The campaign to require automatic couplers is instructive for several
reasons. As usual, for the carriers safety was in good measure a byproduct
of the quest for productivity. As Theodore Porter has suggested, the development of statistics was a precondition for reform. Initially, information
on coupler and brake accidents was limited to the labor unions, carriers,
and state commissions, and they moved men like Matthias Forney and
Lorenzo Coffin to press for change. But the publication of national accident statistics in 1889 helped create railroad work safety as a national
problem. The episode also reflects Thomas Haskell's argument that the
emergence of humanitarian concerns required a recipe for intervention
(although Haskell's formulation ignores the work of men like Westinghouse and Janney who *invented* the recipes). By the mid-1880s the availability of workable brakes and couplers made the need for legislation seem
a moral imperative even to conservatives such as Henry Cabot Lodge.[33]

The law had a number of important consequences. There is little
evidence that it speeded up the diffusion of brakes and automatic couplers
but contemporaries certainly thought that it did. Since uniformity was a
benefit to the carriers, reformers could claim to have been better judges of

the railroads' needs than were the companies themselves. In the twentieth century this claim would be used to dismiss railroad opposition to other safety legislation (chapter 7). The law also represented the first large-scale effort by the rail brotherhoods to use federal power to improve their safety. But by focusing on specific pieces of equipment and accidents the law also switched railroad safety legislation on a track that led nowhere, for no other type of equipment yielded so many casualties and was so easily regulated. The effectiveness of the new equipment was also contingent upon its quality, on maintenance incentives, and on worker training. In short, far from being a simple technological fix, the new appliances revealed that better safety required much organizational change and learning.[34]

Learning by Using Couplers and Brakes

When the MCB adopted the vertical plane type of coupler to replace the old link and pin, they did so with some misgivings, noting that "its most serious defect is in strength," a problem they thought could be remedied. In fact, when the manufacturers of the Janney coupler ceded their patent rights, many new couplers appeared on the market, resulting in a replication of the problems that plagued car wheels, and for precisely the same reasons. The MCB had established no tests or quality standards and few companies tested. The combination of asymmetric information and the externalities associated with freight car interchange soon led to a flood of junk couplers and parts. Some roads switched low-priced for high-priced couplers on others' rolling stock, and some bought good couplers for their cars and poor ones for foreign cars and then charged for good couplers. So long as such conditions prevailed, the *Gazette* wondered how "the law of evolution can be depended upon to eventually weed out the poor couplers." Some companies even used cast iron knuckles in repair, which Wellington at the *Engineering News* termed a "moral crime."[35]

In addition, except for the contour lines (the parts of the coupler that connected with another coupler), none of the couplers had interchangeable parts, and even makers of quality couplers failed to follow contour lines. The lack of interchangeability resulted in a repairman's nightmare, while differing contour lines sometimes caused couplers to come apart, and led to uneven wear and breakage. These difficulties were compounded for a time by the need for a hole in the knuckle of a vertical plane coupler, which allowed it to couple with the old link and pin but reduced its strength. Even worse, while the MCB had thought the basic design of the coupler weak in 1887, both the design and weight remained essentially unchanged into the twentieth century—a problem to which the unchanging MCB contour lines surely contributed—even as car weight doubled.[36]

Some manufacturers strove to improve their product from the beginning. McConway & Torley, makers of the Janney coupler, guaranteed it against defects and used the returned material as a guide for product improvement. As a result the company abandoned use of malleable iron, trying cast steel, which initially proved unreliable, and then wrought iron before again returning to steel. The Burlington catalogued types of coupler failure, and in 1892 its mechanical engineer William Forsyth reported some of the first coupler tests to the Western Railway Club. These disclosed problems in design and material. By 1892 *Engineering News* was calling for standardized tests to weed out poor couplers. As with brake maintenance, the Western Railway Club took the lead in these efforts, forming a committee in 1893 chaired by G. W. Rhodes of the Burlington that sponsored tests. They found, among other things, "complete disregard" of the MCB contour lines by most manufacturers. The Western also circularized the other clubs, urging them to take up the issue. In 1893, Rhodes chaired a committee of the MCB that sponsored coupler tests at the Pennsylvania's testing plant and by the government at the Watertown Arsenal. The *Gazette* claimed that the mere prospect of testing had already led manufacturers to improve quality.[37]

In 1899 the MCB finally adopted a set of specifications and tests that it urged on members—something critics charged should have been accomplished a half decade earlier. The strategy, the Committee on Testing later explained, was to make the requirements increasingly rigid to do away with weak and poorly designed couplers. Yet the MCB had no power to compel use of its specifications and one journal complained that as long as some railroads failed to test, the rejected couplers would still find their way into circulation.

ICC freight car inspection began in 1900 and revealed an increasing problem of defective safety appliances until 1904, when inspectors reported 30 percent of all cars had defects, while coupler defects were running at a rate of 200 per thousand cars. Prodded by the ICC, in 1905 the carriers began to refuse to accept cars in interchange that had defective safety appliances. Thereafter matters improved, and by 1910 only about 5 percent of all cars had defects and coupler problems had declined to about twelve per thousand. Designs were improved, knuckles became stronger, and couplers would no longer open should the pin break. Steel casting also improved and most producers shifted to that metal. The final step—adoption of a standard coupler—proved a Herculean task, however, involving much work on the part of the MCB as well as individual carriers and manufacturers, and was not achieved until 1916.[38]

These improvements help explain not only why coupling injury and fatality rates fell as sharply as they did, but also why they continued to decline well into the twentieth century, long after use of the MCB coupler had become universal (Figure 4.2). When couplers failed to mesh, or stuck, or broke, trainmen went between the cars, putting them at risk. Thus in 1903, *after* the link and pin coupler had almost entirely departed, there were still 253 fatalities and 2,788 injuries that resulted from a host of coupler defects as well as other dangers attendant to coupling. Vermont's railroad commissioners reported the death of Maynard Ryan, brakeman on the Central Vermont, on July 12, 1903. When an automatic coupler failed to couple, Ryan stepped between the moving cars to adjust it, caught his foot in an unblocked frog, and was killed. Such casualties reflected the poor state of coupler technology combined with the absence of company safety procedures. The proper procedure when a brakeman encountered a defective coupler was to stop the train. But not until after 1910, when the Safety First movement began to interest railroad managers in worker casualties, would such procedures be common.[39]

As chapter 3 described, problems of maintenance and operation also bedeviled the early air brake (although junk air brakes never plagued the carriers, thanks to Westinghouse's monopoly) and in addition, most trains were only partially air-braked. For both reasons men still had to ride freights to stop trains well into the twentieth century. In 1907, the ICC inspected practice on a number of carriers whose lines included steep grades. Not surprisingly, brake maintenance on these lines was quite good. On the Northern Pacific it found that "hand brakes are sometimes used." On the Southern Pacific, although air brake conditions were "exceptionally good" it found that on some grades "hand brakes are used on all trains." On the Santa Fe, the general practice was to control trains going over the Tehachapi pass with hand brakes. Similarly, in the East the B&O, the Erie, the Pennsylvania, and the Reading still controlled some trains wholly or largely with hand brakes. The *Gazette* blamed this on "the present condition of the air-brake . . . the lack of thorough training and discipline . . . and the insufficiency of the forces assigned to inspection and repair." It also reflected a decision to trade safety for cost. Well into the 1920s the Pennsylvania routinely required men to help control trains on steep downhill grades with hand brakes because it was cheaper than improving mainte

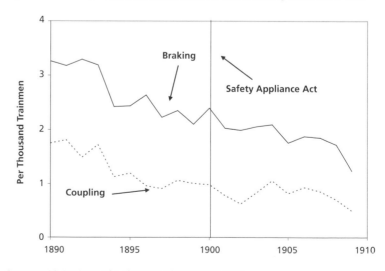

Figure 4.2. **Trainmen's Fatality Rates, Braking and Coupling Cars, 1890–1909**

Sources: ICC, Statistics of Railways; author's calculations.

nance on cars the company received in interchange. Moreover, with trains only partly equipped with air, and with some cars being well braked and others not, application of the air might generate nasty jolts that one ICC inspector thought increased the number of men falling from trains.[40]

But even where air brakes were used exclusively and worked perfectly, men sometimes had to get out on the cars. Sometimes front brakemen had to inspect the train as it went past, then hop the caboose and walk over the train to the locomotive. Local freights that picked up and dropped cars along the way also required men to ride car tops. Classification yards also needed men to ride cars. Such work was more likely to generate an injury than a fatality when a man fell from a car because it was not moving very rapidly. Charles Brown recalled being pitched off a car in the Union Pacific's Green River, Wyoming yards when a rusty brake staff broke in his hand. He was merely shaken up. Another time, when a brake ratchet wheel unwound, it tossed him next to another track, where he was knocked flat by the pilot beam of an oncoming engine. In 1903 (the first year such data are available), the ratio of injuries to fatalities from falls was 41 percent higher in yard work than over-the-road work. Moreover, yard work almost certainly increased relative to over-the-road employment as productivity gains in the form of longer, heavier trains restrained the demand for road trainmen. Thus after 1900, as companies equipped an increasing proportion of freight cars with air brakes and as better training and maintenance became widespread, the dangers from riding freights over the road diminished (Figure 4.2). But the shift of employment to yard work along with more complete reporting of injuries masked the effects on injury rates.[41]

The application of air brakes and automatic couplers remains one of the most spectacular safety investments ever made by American industry. They increased productivity, and contemporaries agreed that their costs were fully justified by the payoff in longer, heavier freights and faster yard work. But they reduced fatalities per unit of output even more than they raised productivity and so worker risks declined. Rough calculations suggest that from 1894 to 1900 they may have prevented 1,116 fatalities and 22,000 injuries from coupling and 1,329 fatalities from falls. But as noted, the Safety Appliance Act proved a poor blueprint. No other such major sources of accidents existed. Nor could the law stimulate a broader interest in accident prevention among the carriers, which remained a dead letter until the Safety First campaigns after 1910.[42]

Killing Passengers One by One

In 1885 the New York Board of Railroad Commissioners proudly reported the accident statistics for that year. "Probably for the first time since railroads have been run within the State of New York can it be said that a year has elapsed without a single passenger being killed *from causes beyond his control*," the commissioners crowed. They immediately went on to take

Table 4.3. **Casualties to Railroad Passengers in New York, by Cause, 1884–1885**

Cause	Fatalities	Injuries
Falling from trains	7	26
Getting on/off trains	15	61
Collisions	5	100
Derailments*	3	43
Other	8	56
Total	38	286

Sources: New York Board of Railroad Commissioners, *Second Annual Report, 1884* (Albany, 1885), xxvi; *Third Annual Report, 1885* (Albany, 1886), xix.

* Includes bridge accidents.

credit for this triumph, attributing it to their inspections, accident investigations, and rules, and ignoring the thirteen passengers who were killed "by their own misconduct or incaution."[43]

Table 4.3 presents the data on passenger injuries and fatalities for New York during 1884 and 1885. Collisions and derailments accounted for only a minority of the casualties. Such findings were typical of those from other states and other times; while a particularly disastrous collision might boost the toll of train accidents from time to time, in most years in most states little accidents claimed a majority of passenger casualties.

While the figures are clear that "minor" accidents routinely accounted for a majority of all passenger fatalities and injuries, their dismissal as a matter of public concern by New York's commissioners was also typical. Neither public officials nor any of the railroads' numerous critics wasted much sympathy on individuals apparently so foolish as to fall or jump off trains. Charles Francis Adams of the Massachusetts railroad commission played a central role in regulating safety (chapter 3). But Adams favored public action only for train accidents where the individual could not protect himself or herself.

In 1870 at the behest of the legislature, the Massachusetts commissioners investigated car and station safety. The report, probably written by Adams, noted that the Miller platform diminished the likelihood that one might fall off a train (chapter 1), but that it had not been widely adopted in New England. Adams pointed out that in Europe railroads behaved as if "all passengers are bent upon self destruction . . . They are fenced in until a train arrives; they are then made to pass through a particular gate to get into it, and, when in, they are locked up." Such requirements were "opposed to the habits, and, indeed, what may be called the genius of the American people," he felt. "The whole system of American institutions is based upon the principle that . . . people can take . . . good care of themselves."[44]

Adams's bracing appeal to individual freedom and responsibility ensured that nothing would be done, and it reflected a widely shared view that was enshrined in the common law of liability. Passengers killed or injured, apparently through their own carelessness, had no one to blame but themselves. In practice, of course, juries sometimes did impose liability for passengers so injured, and hence the carriers were not entirely indifferent to such accidents. Nor were they entirely blameless. Attributing the deaths of passengers who jumped on or off trains or were killed in stations to their own carelessness may have dulled humanitarian sensibilities; it also conveniently ignored train and station design. Just as American freight equipment led to a large number of worker casualties, so American station and passenger car design led to many passenger accidents. Just as active public policy required a social construction of coupling accidents, so the treatment of passenger accidents as a personal problem rather than an engineering failure must also have blunted reform.

The Passenger, the Car, and the Station

In the early days of railroad travel moving between cars required the intrepid passenger to jump across an open space between two cars lurching down a rickety track at 20 miles per hour. It was like jumping from one boat to another in a heavy sea. The introduction of Miller's platform in the 1850s and 1860s diminished these dangers by cinching the cars tightly together, but it did nothing to protect those who jumped from the train or ran to catch it. Passengers routinely jumped from trains before they were fully in the station. Sometimes the train stopped early, to wait for another to pull out, or because the engineman misjudged his air brake, and then started with a lurch, throwing off a passenger who had begun to depart. But many were killed simply because they wanted to get off early.

As traffic density rose, passengers increasingly began to get run over crossing tracks at stations. In small cities and towns most American stations were built at grade with the track entirely unfenced. Passengers and those coming to meet or see them off crossed the tracks wherever and whenever it suited them. By the 1890s the railroads running into major cities had evolved into high-density commuter lines, picking up and discharging many passengers at multiple stations just outside the city. On a double-track road passengers might have to cross a heavily traveled track to board or leave a train and most carriers ran trains through stations at high speed. After the Boston & Maine killed a passenger in a station, the *Gazette* editorialized: "It is one of the wonders of modern civilization, that so many hundreds of passenger trains daily discharge their passengers at stations where express trains pass by within four feet of unprotected steps and at speeds of from fifteen to thirty miles per hour or faster, without killing more passengers than they do."[45]

To reduce such dangers required redesign of stations and cars. In 1889, with Adams gone and two more decades of accidents to contemplate, the Massachusetts commissioners evidenced more interest in such matters, publishing reports on European station design. These revealed that in both Britain and Germany stations employed platforms, fences, bridges, and subways to prevent passenger access to the tracks. The board urged these forms of construction on American carriers. The new interest in safe design was symbolic of changing conceptions of freedom and liberty. To Adams freedom meant freedom *to* act as one wished and take the consequences. To the new board freedom meant freedom *from* risk.[46]

Companies responded, and new stations in suburbs and large cities began to be constructed with such safety features designed in. As a later survey implied, the motives were largely economic, for "the arrangements needed for convenience and speed in handling crowds also tend to safety." Separation of inbound from outbound passengers, for example, not only speeded up operations but made them safer. In 1894 the *Railway Age* reported that the Illinois Central was substituting station platforms for car steps. This not only enhanced safety and efficiency, it removed the "temptation to brakemen to exhibit unnecessary ardor in assisting attractive ladies off and on." Despite such progress, in most small towns and rural areas the station remained as dangerous as ever.[47]

Accident rates involving passengers at stations fell about 60 percent from 1892 to 1900 (Table 4.4). The decline probably reflected both the construction of safer stations in large cities and the long-term relative shift of population and traffic away from stations in rural areas and small towns. Station accident rates also declined as a result of longer journeys, which led to an increase in passenger miles relative to trips to and from stations. Throughout much of the nineteenth century the average length of a passenger journey fell due to the rise of urban and suburban traffic, but the decline reversed in the 1890s. As Table 4.4 reveals, the average journey increased about 20 percent from 1892 to 1901, and it continued to rise thereafter. The origins of this turnaround are obscure; it occurred within all ten ICC regions and so it does not simply reflect a shift of traffic from east to

Table 4.4. **Passenger Fatality Rates, by Cause, 1890–1902 (per billion passenger miles)**

Type of Accident	1890–1891	1901–1902	Change	Percent of Total Change
Train accidents	9.03	7.56	−1.47	23
All other accidents	14.41	9.37	−5.05	77
Station accidents	3.32	1.35	−1.97	30
Falling/jumping from train*	10.85	7.94	−2.91	45
Total	23.44	16.93	−6.51	100
Length of journey (miles)	24.14	29.52	+5.38	

Source: ICC, *Statistics of Railways in the United States.*
* Available from 1901 on only; figures for earlier years are a residual category and include a small number of other kinds of accidents.

west. Probably it reflected declining fares and improved quality of service as well as rising incomes and the increasing integration of the rail network. After 1900 the length of haul would continue to increase as the carriers lost their short-haul business to the electric interurbans, busses, and the automobile.

The rate of accidents from getting on and off trains also fell sharply in the 1890s (Table 4.4). Like the decline in station accidents, these resulted from growing urbanization, improvements in station design, and the increased length of journey. In addition, improvements in car design reduced accidents as a byproduct of companies' efforts to increase passenger comfort and revenue yield. As discussed in chapter 2, in the 1890s the vestibule became increasingly employed on passenger trains; it reduced casualties in train wrecks and also made falling from trains while moving between cars nearly impossible. A second improvement was the use of gates on passenger cars, pioneered by the Chicago & Alton in 1893. Like many lines, the Alton was plagued by what the *Railroad Gazette* termed "the no-fare evil." The evil actually took two forms. Sometimes conductors simply pocketed the cash paid by passengers, perhaps charging the passenger less than full fare as a bribe to keep quiet. Sometimes, however, especially on suburban lines with many stops, one could board and leave without ever surrendering a ticket. On local trains between St. Louis and Kansas City, the Alton installed the Wood Safety Gate. It also hired gatemen—who were independent of the conductor—to check for tickets prior to boarding. When a train entered the station some gates were locked and at the others passengers had to show a punched ticket in order to leave. In the first month the company paid out $300 for gatemen's wages—and sold an additional $1,200 in tickets. It later reported "a perceptible decrease in the number of damage suits on account of personal injuries." Success led the company to extend the service to Illinois as well.[48]

The Missouri railroad commission promptly sued the Alton for this infringement on passenger rights, but later backed down. The Southern Pacific, which shared the Alton's predicament, installed a similar system on its local trains in Oakland. Some Massachusetts carriers also employed gates on trains in the early 1890s, and the New Haven routinely protected passengers and revenue with both gates and vestibules. By 1898, the Burlington, the Missouri Pacific, the Missouri, Kansas & Texas, and the Chicago & West Michigan all used gates on at least some trains. George De Haven, general passenger agent of the latter company, stressed that the gates were "a decided safeguard," protecting the railroads from liability suits and passengers "against their own ignorance of the laws of gravitation."[49]

As Table 4.4 reveals, the net impact of all these developments sharply reduced passenger fatalities. From 1890–91 to 1901–2, reductions in accidents at stations, from falling or jumping from trains, and other nontrain accidents accounted for about 77 percent of the improvement in passenger safety. Because fatalities from train accidents were highly variable these percentages varied from year to year, but the general point remains: al-

Figure 4.3. **Passenger Fatality Rates, by Cause, 1890–1910**

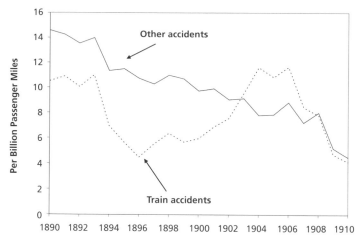

Sources: ICC, Statistics of Railways; author's calculations.

though largely ignored by the press or regulators, the decline in nontrain accidents resulted in major improvements in passenger safety, and these gains continued well into the twentieth century (Figure 4.3).

Trespassers

If many passengers were killed in little accidents, their numbers were dwarfed by the deaths of trespassers. Early state statistics divide accidents into those happening to employees, passengers, and others. Investigation of the last category invariably reveals a respectable number killed at crossings (usually nontrespassers), but far more trespassers who died walking the track or hitching a ride. Thus in New York, in 1885, 261 of the 411 individuals killed on railroads (64%) were "others." Of this group, 189, or 72 percent, had been walking the track. Most of the rest died stealing a ride, and only 30 were killed at grade crossings. National data in Table 4.5 reveal that trespassers accounted for nearly half of all fatalities in 1890, and

Table 4.5. **Fatalities to Passengers, Employees, and Others, 1890–1900**

Category	1890	Percent of Total	1900	Percent of Total
Employees	2,451	39	2,550	32
Passengers	286	5	249	3
Others	3,598	57	5,066	64
Trespassers	3,062	48	4,346	55
Nontrespassers	536	9	720	9
Total	6,335	100	7,865	100

Source: ICC, *Statistics of Railways in the United States.*

the proportion rose as both railroad work and travel became safer in the twentieth century.

The very word *trespass* is, of course, freighted with moral meaning. It is an unlawful act, a wrongful entry, a sin, an invasion of rights. Thus when a train killed or injured a trespasser who by definition had no legal right to be on the track, the company was rarely liable. Here again, there were exceptions. The *Railroad Gazette* noted that trespassing was not a capital offense, nor were the railroads appointed as executioners: the law would hold carriers liable if trainmen failed to take all reasonable precautions to avoid trespassers. Sometimes juries found for the plaintiff or his next of kin no matter what: in 1887 one Alabama court awarded damages to a trespasser who had been walking the track and was injured when a locomotive tossed a cow that landed on him. In Ohio in the 1870s and 1880s the average compensation cost to the railroads of such a casualty was about $92 (about three months' earnings in 1880). Trains also had to stop to offer aid to the injured, whom railroad doctors usually treated, at least briefly. For these, as well as humanitarian reasons, the carriers were not indifferent to the slaughter of trespassers. But neither was the problem a high priority for them, or the state commissions, or anyone else.[50]

Hopping Freights

The absence of public concern with accidents to trespassers reflected not only the ideology of individualism but the absence of any workable solution, for the problem mirrored the realities of life for working-class Americans. Many casualties involving those who stole rides on trains resulted from failed attempts to hop freights or to detrain at too high a speed. These individuals, mostly men and children, who stole rides divided into two groups. First were those who wanted a ride to the next town and could not afford a ticket. Throughout the last half of the nineteenth century fares ranged from 2 to 4 cents per mile. In Massachusetts in 1882 fares averaged 2.32 cents per mile and wages (nationwide) averaged $1.16 a day. Thus a 20-mile round trip would cost 40 percent of a day's wages. The reader who makes $200 a day should try to imagine paying $80 in order to commute 10 miles to and from work. As late as 1900, such a trip would still have taken one-quarter of a day's pay.[51]

Similar calculations contributed to the rise of a second group of travelers—the hobos. In the early twentieth century social workers began to distinguish tramps from hobos: tramps were essentially traveling bums who worked no more than necessary, while hobos were migrant workers. Migrant workers reflected the economics of certain highly seasonal industries as well as the cyclical rise and fall of employment throughout the economy. Agricultural employment followed the seasons; the demand for labor rose at plowing and harvest time, and agricultural processing was also fall work. Ice harvest was winter work as was logging, for the snow facilitated moving the logs, while construction, including railroad building and maintenance, boomed in the warm months and in the North mostly shut down in the winter.[52]

While these seasonal industries required a pool of itinerant workers who routinely traveled long distances in search of work, during a major economic contraction, workers in many other industries might hit the road looking for a job. In neither case were the men inclined to pay passenger fares. A 500-mile trip might cost nearly $12—almost two weeks' pay for some. Railroad men who traveled long distances often got preferential treatment. Both Henry Clay French and Charles Brown reported that flashing a union card got them a ride as a passenger or in a caboose, but most hobos were not so lucky. To catch a ride they sometimes had to jump a traveling freight. They rode the bumpers (dead blocks) between cars, or the truss rods or trucks, any of which was as least as perilous as the brakeman's position. Sometimes they got into cars where they might be crushed if the load shifted. Many died running for trains or jumping off, perhaps to avoid the railroad police. Jack London recorded that on one of his first trips as a hobo in the 1890s, a companion, the French Kid, lost both legs when he fell beneath a train. Of course there was the risk of derailment or collision. On October 22, 1898, one break-in-two on the Rock Island, near Fort Worth, Texas, caused a collision. No trainmen were hurt but the wreck killed eight hobos.[53]

The railroads hated the hobos, for they sometimes accidentally set fire to cars or intentionally cut air brake hoses if they wished to make an unscheduled stop. As a result, railroad police sometimes pitched them off moving trains. In 1901 the Rio Grande even fitted one locomotive with steam pipes fore and aft so that the engineman could play a stream of hot water and steam on anyone trying to steal a ride. Thugs also robbed and murdered hobos, and threw their bodies off, to be counted as railroad accidents.[54]

Just as the number of hobos reflected cyclical rhythms in the economy, so did hobo casualties. The depression of 1893–97 resulted in a sharp upturn in trespassers killed riding the rails in Massachusetts, and in 1894, the various armies of unemployed that sometimes commandeered whole trains caused a bulge in the statistics nationwide. The *Railroad Gazette* reported 104 "others" killed in train accidents that year—up from eighty-nine the year before. Logic and fragmentary evidence also suggest that there may also have been a long-term increase in hobo casualties relative to the population during the late nineteenth century. Railroad construction expanded into the South and the West while logging, agriculture, and other seasonal industries began to sell on national rather than local or regional markets and the geographic scope of labor markets also broadened. Before the Civil War state railroad commission reports contain few examples of hobos killed in train wrecks, but by 1887 their numbers had so increased that the *Railroad Gazette* began to list them separately. Its data show that fatalities to "others" rose from 13 percent of those to trainmen in 1887–89 to 17 percent in 1897–99. In 1901 the ICC reported 1,038 trespassers killed in train accidents or falling or jumping from trains.

Walking the Track

The majority of trespassers killed on the railroads were not riding trains; they were walking or crossing the tracks. Massachusetts reported that only one-fifth of trespassers killed in 1888 were stealing rides. New York's railroad commission reported in 1887 that "the most serious cause of death to others, not employees or passengers, was walking or being on the track." Obviously track walking was far more dangerous than it appeared. While many who were killed were drunk or deaf, most were simply victims of bad judgment. Double-tracks around cities, where trains were sometimes scheduled only 3 minutes apart, might be exceptionally lethal because urban noise masked the sound of the locomotive and a person could be hit by one train while avoiding another.[55]

Again, the reasons reflected the exigencies of working-class life and, as Adams might have put it, the "genius of American institutions." In Europe fences and platforms at stations prevented not only passengers, but also those coming to greet the train or merely near the station from cutting across the track. In America, by contrast, open stations took a toll on trespassers every year. In addition, British railroads had been constructed in a small country that already boasted a network of highways, while in rural and small-town America until the twentieth century the tracks were working-class highways. While some who wished to go to town might hop a freight, many walked. Since country roads were often winding and choked with dust in the summer and mud in the spring and fall, a railroad was likely to be straighter and more pleasant. Sometimes individuals walked long distances. Vermont's railroad commission reported that Barnard Cassidy had set out to walk from White River Junction to Hartland, some 8 miles, on November 2, 1904, when he was struck and killed by a train. In cities, on the other hand, neighborhoods grew up around the tracks; men and women had to cross to go to work or the market, while children walked the track to school or played on it or picked or stole coal.[56]

Trespassing, in short, although illegal, was through long custom a working-class property right. In 1890, a member of the Massachusetts commission took the 4:10 train from Beverly to Boston—a distance of perhaps 15 miles—and counted 100 trespassers on the track: 63 men, 5 women, and 32 children. "It seems not improbable that, on the eight lines of railroad running into Boston, and within a radius of twenty miles, there are on pleasant days as many as 40,000 people using the tracks as a footway," the commissioners concluded. The United States seems to have been unique in the extent of trespassing. The Massachusetts commission observed that there was one trespasser killed for every 0.21 million train miles while in Britain the number was one per 0.73 million. In Germany, such accidents were "so rare they are not separately classified." In Britain and Germany police regulations reinforced gates and fences to keep people off the tracks. A German report to the Massachusetts Board of railroad commissioners noted that all railroad employees were vested with police powers and that any form of trespassing was "severely punished." In America, by contrast, not only was trespassing rarely punished, railroads that tried to enforce the law experienced mysterious fires and found obstructions on the tracks.[57]

Grade Crossing Accidents before the Automobile

From the beginning of railroading European countries required grade separation at most rail-highway crossings and they guarded the few remaining crossings with gates and watchmen. In contrast, from the beginning American carriers and public highways routinely crossed at grade. In 1900 a team of German inspectors toured American railroads. Like European tourists a half century earlier they marveled that "a regular track guarding system is unknown in America." To guard American switches and crossings according to the German standard would require an increase in staff of about 586,000 men, they estimated. Predictably, therefore, crossing accidents were rare in Europe and common in the United States. Throughout the nineteenth century most Americans wanted railroads and did not burden them with the extra cost of grade separation. Thus while states and localities asserted a public interest in grade crossing safety, they usually required only that enginemen sound the whistle and ring the bell before a crossing rather than the more expensive gates or watchmen. In 1882 New York State boasted 6,901 crossings; only 703 had gates or flagmen, although most had warning signs. Six years later a survey found 2,229 grade crossings in Massachusetts, of which 1,374 were unprotected by gates, flags, or bells.[58]

When a locomotive struck a pedestrian or a team at a crossing, the injured party (or next of kin) was often unable to recover damages. Courts ruled that pedestrians and drivers had to exercise due care when approaching a crossing and look both ways. Those who failed to do so were guilty of contributory negligence and might not recover damages. Injured parties could usually recover little at law except where the crossing was poorly marked, or the engineman failed to sound a warning, or some other extenuating circumstances existed. In the five years between September 30, 1877, and September 30, 1882, New York railroads killed or injured 264 individuals, which generated lawsuits and other damages averaging $358 each (about one year's annual earnings in 1880). Thus, to the carriers crossing accidents were not very expensive, reducing their incentives to improve crossing safety.[59]

Abolishing Grade Crossings

As trains and traffic density increased, so did the number of accidents. Crossing accidents reflected in part people's inability to estimate risks, for trains covered distance much faster than anyone was accustomed to. Imagine a train and a horse and buggy approaching a crossing, with the train traveling at 30 miles per hour, or 44 feet per second, and the horse at 3, or

4.4 feet per second. The horse and buggy is 20 feet long and the horse's nose 20 feet from a 10-foot wide crossing. The rear of the buggy must travel 50 feet to avoid the train, and if the train is within 500 feet of the crossing, unless the horse picks up the pace, there will be a collision. Sometimes horses became ungovernable, running away at the sound of the train or bolting in front of it or occasionally standing frozen and immovable on the crossing.

The dangers of a crossing depended not only on the degree of protection that companies provided, but also on visibility, and crossings near curves in the road or railroad, or with otherwise obstructed visibility were especially treacherous. Moreover, in a disquieting development, states discovered that protected crossings usually averaged more accidents than those that were unprotected. Of course, this did not necessarily mean that protection caused accidents because protected crossings typically had heavier traffic. But individuals routinely ignored signs, whistles, watchmen, and even gates. And sometimes protection *did* contribute to the problem, for carriers might put up gates that were only operated by watchmen during the day, leaving the gate open at night to act as a trap for the unwary. Indeed, one Maine court understood that crossing accidents reflected more than simply individual carelessness. It held that unmanned gates were so misleading that they reduced the contributory negligence of the injured party who crossed. In the 1870s Massachusetts recorded about twenty-six casualties a year at crossings and by the 1880s the number increased to forty-six, nearly half of which occurred at protected crossings. Other states experienced similar increases and as a result crossing safety gradually became a public issue.[60]

The straightforward solution to crossing accidents was to get rid of grade crossings, and both the carriers and local governments had some incentive to do so. Since municipal governments could usually require that crossings be guarded or impose speed limits on trains, they could raise a carrier's payoff to grade separation just as state laws had done for interlocking (chapter 3). Municipalities, by the same token, gained both safety and fewer traffic tie-ups. One might suppose that such incentives would lead both parties to the bargaining table and result in the abolition of crossings for which these combined benefits exceeded costs. But in practice few crossings were eliminated without state encouragement, apparently because in most cases the safety and travel benefits to towns seemed inadequate or because they felt that abolition was a railroad duty.

The Fitchburg Railroad eliminated twenty-five crossings in Massachusetts between 1875 and 1890 by paying most of the costs with the towns contributing relatively little. In the late 1880s the New York Central and the city of Rochester agreed to elevate the railroad in that city with the city to bear the expenses of road grading up to the overpasses. The company spent about $1.7 million in the elevation, but lowered its operating costs $103,000 a year, which yielded a 6 percent return on investment. Similarly, in 1888, the Pennsylvania Railroad spent $1 million to eliminate crossings in Jersey City, with the municipality paying nothing. Ever the economist, Arthur Wellington at *Engineering News* estimated that the carriers' annual savings from reduced costs of crossing protection and lower legal and surgical fees amounted to about $40,000—a 4 percent return on investment. He also noted that the separation would avoid 300 injuries and fatalities a year, which if valued at $133 each would also yield 4 percent even if there were no other benefits at all.[61]

While private action eliminated a few crossings in the 1870s and 1880s, it created even more, and so states began to act to both discourage new crossings and encourage grade separation. In 1884 Connecticut passed a law allowing either municipalities or railroads to initiate crossing elimination. The railroad commission was to allocate costs with the municipalities to pay no more than half. The problem of course was that it was impossible to match each party's costs with the benefits they received from eliminating any given crossing. Suppose, for example, a yard crossing costs $100 to eliminate and generates $85 of benefits to the railroad and $10 of benefits to the town. Since total benefits are less than costs, private action will not eliminate this crossing. But if the commission allocates $80 of the cost the railroad and $20 to the town, the municipality will oppose elimination but the carrier will favor it. Such "free riding" seems to have occurred. The New Haven especially viewed the law as a windfall; by 1887 it had petitioned to eliminate seventy crossings. Although the towns were assessed only about 11 percent of costs, they nearly always objected.

A new law of 1889 apportioned all the cost to the carriers where they proposed elimination and 75 percent when the towns proposed elimination. This quieted complaints, but it also generated perverse incentives, as Table 4.6 suggests. Eliminating crossing one in Table 4.6 generates net benefits but the railroad will not propose it for it would have to bear all the costs, and neither will the municipality, for it would then incur one-quarter of the costs. Crossing two, on the other hand has net costs but because only one-quarter of the costs go to the town it may propose elimination. In any event Connecticut managed to shrink the number of crossings from 1,223 in 1884 to 997 in 1900.[62]

Massachusetts also became more interested in crossing accidents in the 1880s. In 1884 the legislature asked the railroad commissioners to report

Table 4.6. **Incentives for Crossing Elimination, Connecticut, 1889**

	Railroad	Town	Total	Eliminate?
Crossing One				No
Benefits	$80	$20	$100	
Costs	$90 or $67.50	0 or $22.5	$90	
Crossing Two				Yes (Town)
Benefits	$40	$50	$90	
Costs	$100 or $75	0 or $25	$100	

on the problem. As in Connecticut, however, initial efforts to encourage crossing reduction backfired. A law of 1885 allowed county commissioners to petition to abolish crossings, but the costs were to be apportioned by a special commission. Apparently most counties were risk-averse and so little was accomplished. The city of Springfield, however, did petition to force the Boston & Albany Railroad to provide a better station, and the hearings provide a glimpse of how serious urban crossing problems had become. The tracks crossed Main Street at grade, and one survey over a 14.5-hour period recorded that the gates were lowered 201 times to allow trains to proceed, closing the crossing approximately 20 percent of the time. Over a similar period another observer counted 2,035 teams and 8,973 individuals crossing the tracks. Because delay was expensive, such congestion explains why individuals sometimes bolted around gates and watchmen, into the path of an oncoming train.[63]

In 1888 the state appointed a board of engineers chaired by the civil engineer Augustus Locke to investigate the issue of crossing elimination. Their report, which won high praise from the engineering press, was one of the first analyses of the problem that grappled with the issue of costs and benefits. The board divided crossings into seven classes according to the cost of removal. The total, they estimated, amounted to about $42.6 million. Although they made no systematic effort to estimate benefits, the commissioners did calculate the delay at some crossings which, valued at current prices of labor, amounted to $35 a day in one case. "It must be . . . worth a large sum to these cities to be relieved of such a hindrance," the board concluded.[64]

Writing for *Engineering News,* Wellington carried the analysis another step. He noted that at 4 percent the interest on the investment amounted to about $1.68 million a year. This came to roughly $30,000 each (about sixty-four years' earnings) to avoid the average of forty-five injuries and fatalities a year at Massachusetts crossings. Wellington thought this "pretty dear . . . [and] more than the community is in the habit of paying." However, he made a rough estimate that there might be as many as 26 million teams exposed at crossings a year, and few that would not be willing to pay 5 cents to avoid the annoyance. These benefits, amounting to $1.3 million annually, in combination with savings on crossing guards and the reduction in injuries and fatalities, made the investment a bargain.[65]

Of course, the problem that both Wellington and the original report avoided was priorities: if all of the more than two thousand crossings could not be removed, which would yield the highest net benefits? The next year the state responded with a bill which, while it ignored priorities, encouraged action by requiring that localities pay only 10 percent of costs, with the carriers picking up 65 percent and the state the rest. As a result, the number of crossings, which had been growing slowly in the late 1880s, fell from 2,239 in 1888 to 2,001 in 1902.

Other states and cities also began to act. Led by its local business community, Buffalo, New York, began to press for crossing elimination in

1887 but the railroads balked and not until 1895, after much complex political maneuvering, did work begin. The carriers eliminated seventy crossings by 1905, and crossing fatalities, which had averaged about twenty-one
per year, fell to nine. In 1897 New York State passed a law similar to that
of Massachusetts but requiring the carriers to foot only half the bill, with
cities and the state splitting the remaining 50 percent. The *Railroad Gazette*
complained that the state had "no general plan" and that the choice of
crossings eliminated had more to do with politics than safety, but by 1908
220 crossings had been removed.[66]

Philadelphia began systematic elimination of crossings in 1888 and
by 1908 the Pennsylvania Railroad had eliminated all its crossings in the
city, with the carrier paying all the construction costs. In 1901 the Cleveland chamber of commerce produced a study of grade crossings, employing traffic surveys to estimate and value the loss of time from delays at several crossings. Between 1905 and 1907 the city and the railroads spent about
$2.86 million to eliminate thirteen crossings, with the carriers paying 65
percent of the total. Chicago undertook the most ambitious program. By
1908, its railroads eliminated 573 crossings with another 678 on the list.
In the process they elevated 681 miles of track and planned to eliminate
another 872 miles. The cost of $46 million down to 1908 was borne entirely
by the carriers. Between 1899 and 1908, crossing fatalities in the city fell
from 113 (6.95 per hundred thousand of population) to twenty (0.92 per
hundred thousand).[67]

Clearly these public and private works saved many lives. Although
none of the states seems to have undertaken any systematic analysis of
which crossings should be eliminated, leaving the choice up to localities
and railroads probably worked tolerably well. While loading most of the
costs on railroads may have generated perverse incentives, the focus on
crossings in cities and towns with heavy train and pedestrian traffic, combined with the penuriousness of municipalities, probably made most eliminations a paying proposition. But after 1910, the automobile transformed
the crossing problem. The many unguarded crossings that had seen little
activity in the horse and buggy age experienced a surge of traffic. Although
states and later the federal government would continue to pursue complete
elimination of crossings as the goal, the automobile brought a new emphasis on choosing priorities for elimination, on guarding, and on safety-first
exhortations.

The gradual reduction in "minor" accidents to passengers and workers
throughout the late nineteenth century made a major contribution to
improved railroad safety. Their decline was in large measure a byproduct
of the railroads' quest for better technology and more efficient operating
practices. The work of technical societies and the trade press enhanced
this process, although it was impeded by the economics of freight car interchange. With brakes and couplers and locomotives, safety resulted not

simply from the adoption of the new equipment, but also from a host of incremental improvements in technology and in maintenance and operating practices. Little accidents to passengers also declined sharply, mostly as a byproduct of other forces, although by the 1890s companies increasingly designed and operated cars and stations to make them safer and more efficient. However, better safety sometimes reflected public prodding as well. Railroad critics increasingly came to reject the idea that accidents were the inevitable result of carelessness. More complex understanding of the engineering and managerial causes of injuries helped generate a moral imperative to prevent them.

During these same years accidents to trespassers rose sharply, but they elicited little concern from either the carriers or public officials, for economic and ideological reasons and because they were so deeply embedded in working-class life. The number of crossing accidents also rose despite increasing public concerns. They reflected the American penchant for constructing railroads as cheaply as possible, and as cities grew up around the railroads, urban crossings took an increasing toll. In the 1880s state regulators, some cities, and the railroads all began to press for grade separation and by the twentieth century a few had made a significant dent in the number of crossings. The absence of third-party funding also ensured that most were high payoff crossings. In the twentieth century the advent of the automobile and the increasing role of state and federal funds required major changes in procedures, however.

Engineering Success and Disaster
Bridge Design and Failure, 1840–1900

Does anyone advocate the designing . . . of bridges to withstand the impact of a railroad train . . . ? Are such accidents . . . bridge failures or . . . failures of management?

—Theodore Cooper, Consulting Bridge Engineer, 1889

I believe . . . that the time will come when the failure of an iron bridge from an ordinary accident of train service will be regarded as discreditable to its builder and not excused as a fault of the management.

—Charles Stowell, New York State Bridge Engineer, 1889

Sometime after 8:00 p.m. on December 29, 1876, the Lake Shore & Michigan Southern *Pacific Express,* traveling west out of Erie, Pennsylvania, approached the Ashtabula, Ohio, bridge, just east of the station. The train consisted of three express and one baggage cars, three passenger coaches, and three sleepers hauled by two locomotives—the *Columbia* and the *Socrates.* The weather was foul, with snow and sleet driving off of Lake Erie, and the *Socrates*'s engineman Daniel McGuire slowed his train and cautiously approached the bridge. When he was nearly across, it gave way, dropping all of the train save his own engine about 65 feet into the chasm, where car stoves ensured that the wreckage immediately caught fire. The death toll was eighty-nine, including two officers of the company. It was the worst railroad disaster of the century.[1]

Ashtabula was what civil engineers called a "square fall"—that is, the bridge failed for lack of strength. Immediately after the event the Ohio legislature commissioned an investigation by a panel of engineers. The bridge was an iron Howe truss that Amasa Stone had designed for the Cleveland & Erie railroad after he had become its president. Stone was an old-time bridge builder who had worked with Howe and patented modifications of the original Howe truss, but who lacked scientific training. The engineers' report, which the *Engineering News* termed "full and satisfactory," concluded that Stone's bridge was in sound condition but suffered from numerous design errors. Among the many such flaws was that the top chord had a factor of safety of only 1.6 when sound engineering practice at that time employed a factor of at least 5. Moreover, the chord consisted of five unconnected beams, which implied that it had little lateral strength. Reviewing this evidence, the eminent bridge engineer Theodore Cooper concluded, "its failure has taught us nothing that we did not know before."[2]

Cooper was a member of an emerging fraternity of professional bridge engineers. Educated at Rensselaer Polytechnic Institute, Cooper worked as a railroad engineer and instructor of engineering at the U.S. Naval Academy and then for several bridge companies before launching his long distinguished career as a consulting bridge engineer. For such men, Ashtabula symbolized the problem of bridge failures as they saw it. The old rules of thumb that had dominated American practice since the beginning led to unscientific designs that resulted in square falls. Thus the central problem of bridge safety, in this view, was the need for scientifically informed bridge designers. Of course, bridges collapsed for many other reasons, but these were not engineering failures as Cooper and many other engineers saw matters. Even bridges that were knocked down by a train, for example, did not "fail."[3]

A few critics subscribed to a broader conception that saw bridge failures not just in terms of scientifically uninformed engineering but also as the result of design choices that distinguished American from British bridge-building technology and made the former inherently prone to knockdown. Yet even this was too narrow a lens with which to view the problem. Bridge disasters resulted not only from errors, or design choices, but also from a broad array of construction and management practices that characterized nineteenth-century American railroad economics, and from the same problems of monitoring, communication, and control that made all forms of derailment common.

The American System of Bridge Construction

As chapters 1 and 2 described, the high cost of capital and comparatively thin traffic induced early American railroads to choose relatively inexpensive construction methods for the permanent way, resulting in adverse consequences for safety. So it was with bridges. In the aftermath of Ashtabula the *Engineering News* observed, "there are many cases . . . where railroads have been pushed to completion with scanty means, and temporary structures have been erected to be replaced later; but . . . bad times necessitate postponement." While British railroad builders turned from stone to iron bridge construction in the 1840s, in the United States, with wood cheap, familiar, and easy to work, the early bridges were nearly all wooden trestles or Howe trusses. American carriers also routinely skimped on bridge approaches, foundations, and abutments, and instead of filling cuts, they built trestles.

"The wooden trestle is emphatically an American institution," Arthur Wellington proclaimed in *Engineering News* in 1887. In fact, he stated, "in no other part of the world have the conditions favored the use of timber so much as here," and he noted that "few of the roads . . . west of Ohio use anything but wooden trestle for their structures in first construction." Writing in 1893, the distinguished civil engineer George S. Morison explained why. "For immediate results, nothing equal to it [wood] has ever been known," Morison claimed. Although wood bridges were "very short lived," for many years, "the cost of frequent renewals of timber was less than the interest on the additional cost of iron." Hence, just as flimsy permanent way was an appropriate engineering choice for early railroads, so "it was good engineering to build wood superstructures."[4]

In the 1850s and 1860s, as train weight rose and iron prices declined, American builders also began to construct iron truss bridges. As Morison put it, in Europe bridge superstructures evolved from masonry to metal and in the United States from wood to metal. But whether they chose wood or iron, American bridge builders used less material than did their European counterparts, and they chose designs that could be factory-made and quickly assembled in order to save labor. Until the mid-1880s virtually all these choices were largely unconstrained by regulatory forces, and they made American railroad bridges far more disaster-prone than those in Britain.[5]

Rates of Failure

Assessing the prevalence of nineteenth-century bridge failures requires information on the number and type of bridges in existence. The best available evidence on that topic was collected by Theodore Cooper, and presented to the American Society of Civil Engineers in 1889. To assemble his data Cooper relied on his wide contacts in the profession, writing the chief engineers of dozens of carriers and asking for information on their bridges. By this process he was able to obtain hard information from lines with about 37 percent of all track that he then extrapolated to estimate totals. His results are presented in Table 5.1.[6]

As can be seen, the great majority of bridges were short, under 20 feet or less, and wood predominated as construction material. But for longer spans iron had become the material of choice. These patterns obtained well into the twentieth century, with the share of iron (and, after 1880, steel) gradually rising and wood still employed for large numbers of short spans wherever it was plentiful and cheap. In 1895, an engineer of the Boston & Maine explained that while his company used iron for longer spans, building short wooden bridges was still a "live business." In Massachusetts only about 2 percent of all bridge miles were iron in 1872; by the end of the century iron and steel constituted half the total.[7]

Just how many of these bridges were likely to fail during a year cannot be assessed with any precision. While a reading of the popular press reveals that bridge accidents date from the 1840s, no one seems to have collected any statistics on either their number or causes until the civil engineer Thomas Appleton began the task. Appleton derived estimates for 1873–77 from the list of train accidents published in the *Railroad Gazette* that he had checked and verified. Charles Stowell, a bridge engineer for New York State, later supplemented his work, and he too relied on the *Gazette*'s data, which he checked and supplemented from other sources. Their findings are presented in Table 5.2.[8]

These data are not entirely comparable, for while Appleton tried to include all bridge accidents, Stowell ignored trestle collapses as well as failures in culverts and cattle guards—all of which were common. Both writers also omitted certain failures such as washouts and fires where no trains were involved, as well as other failures that involved train accidents, but did not, in the authors' judgment, contribute to them. Stowell's col-

Table 5.1. **American Railroad Bridges and Trestles, 1889**

	Number of Bridges			**Miles of Bridges**		
Span (feet)	**Iron**	**Wood**	**Total**	**Iron**	**Wood**	**Total**
Under 20	5,100	722,100	727,200	17	2,407	2,424
20–50	12,900	5,250	18,150	86	35	121
50–100	4,600	4,500	9,100	66	64	130
100–150	3,900	4,100	8,000	93	97	190
150–200	2,100	1,200	3,300	69	40	109
Over 200	950	200	1,150	49	7	56
Total	29,550	737,350	766,900	380	2,650	3,030

Sources: Theodore Cooper, "American Railroad Bridges," ASCE *Transactions* 21 (1889): 1–59; "The Bridge Failures of Eleven Years," *Engineering News* 23 (Apr. 19, 1890): 373–74.
Note: Includes only wood and iron structures.

Table 5.2. **Number of Bridge Accidents by Cause, 1873–1889**

Cause of accident	**1873–1877**			**1878–1882**			**1883–1887**			**1888–1889**		
	Wood	**Iron**	**Total**	**Wood**	**Iron**	**Total**	**Wood**	**Iron**	**Total**	**Wood**	**Iron**	**Total**
Square fall	1	3	4	5	6	12	19	3	23	8	2	10
Fire	4	0	4	11	0	11	18	0	18	5	0	5
Washout	5	2	19	0	0	14	1	4	20	5	0	13
Repair	1	1	7	0	0	0	4	0	4	2	1	4
Knocked down	33	8	66	9	6	26	4	10	32	6	8	17
Unknown	2	3	26	24	2	52	5	1	43	1	0	5
Total	46	17	126	49	14	115	51	18	140	27	11	53

Sources: Thomas Appleton, "Railroad Bridge Accidents," *Engineering News* 5 (Feb. 21, 1878): 59–61; "The Bridge Failures of Ten Years," *Engineering News* 2 (Mar. 30, 1889): 288–89; "The Bridge Failures of Eleven Years," *Engineering News* 23 (Apr. 19, 1890): 373–74.
Note: The totals column includes bridges of unknown material, probably wood.

lection criteria ensure that his figures provide an undercount of *all* bridge accidents. For the class of bridges on which he focused, however, the data probably account for most accidents that resulted in casualties, for when Stowell published his data he always requested readers to supply additions and corrections. Since Stowell's criteria largely excluded accidents on short bridges, his data should be compared to Cooper's estimates of bridges over 20 feet long. Assuming that bridges of unknown construction were wood, such a calculation implies an annual failure rate in 1888–89 for iron bridges of one in 4,445, and for wooden bridges one in 726. Unfortunately, these data cannot be compared to British experience. When Cooper tried to obtain data on English railroad bridges the Board of Trade informed him that no such figures existed. However, virtually all contemporaries believed that bridge accidents were far more prevalent in the United States than in Europe.[9]

Design and Error in Bridge Failures

One obvious inference from Table 5.2 is that most bridge failures were not simply collapses. That is, square falls, where bridges went down from the weight of trains, were comparatively rare, even for wooden bridges. By implication neither unscientific engineering nor rising train weight was a major cause of accidents. Of greater importance were design choices, such as the widespread use of wood that traded safety for cost, and management procedures that made American railroad bridges disaster-prone.[10]

   Square falls arose from three quite different causes. Some, like Ashtabula, were indeed the result of uninformed engineering. Since the 1840s the railroad technological network had slowly made bridge engineering more scientific. With the publication of works by Squire Whipple and Herman Haupt, competent engineers computed strain diagrams that calculated the load on each member, whether in tension or compression, from the dead load of the bridge and some assumed live (train) load. Experiments by William Fairbairn for the English Iron Commission published in 1850 and by the German engineer August Wöhler established the ultimate strength of wrought iron at around 50,000 psi and its elastic limit at about half that level, although both might vary substantially with the quality of the iron. Similar calculations were done for cast iron and wood. Thus, by the Civil War, a skilled engineer armed with such information could proportion all bridge members to be able to withstand the assumed stress with a safety factor of 4 or 5 (relative to the breaking strength) to allow for the uncertainties in the process. As will be seen, bridge engineers constantly refined and improved these techniques throughout the nineteenth century.[11]

   But not all bridge builders pursued the path of science. Thomas Appleton told the Boston Society of Civil Engineers that "in view of the prevailing lack of system, or 'rule of thumb,'" he was surprised how few Howe trusses had collapsed. He had once computed the stress on some Howe truss bridges and found that the iron needed to be much stronger.

He showed his calculations to one of the pioneer builders, who dismissed them with the comment "we never use such heavy iron as that." Another writer told the British Institution of Civil Engineers of early American timber bridges in which the iron tension rods carried a stress of 46,000 psi when accepted procedures limited stresses to 10,000 psi. As late as 1893 the superintendent of bridges and buildings on the Big Four Railroad sent the *Railway Review* photographs of two trestles that he termed "death traps" because they lacked either lateral or diagonal bracing.[12]

The results of such practices were nicely illustrated in 1850 on the Erie when one of Rider's Patent Iron Bridges gave way, moving the railroad to abandon iron altogether for a time. Although Rider's bridge had been praised by the *American Railroad Journal* and won an award from a committee that included the distinguished engineers Horatio Allen and John B. Jervis, it was fatally flawed. Three years before the bridge fell, Squire Whipple had pointed out that Rider's design placed stresses of 26,000 psi on some of the wrought iron. John Roebling claimed that he too had denounced Rider's plan. To him it symbolized the "necessary consequence of that total want of scientific knowledge on the part of those who superintend these structures." He called upon the civil engineers of the United States "in view of their professional standing" to denounce the "wholesale veto" which the Erie had passed "*indiscriminately* upon *all* iron bridges."[13]

Square falls also resulted from overloading flimsy bridges and trestles beyond their rated capacity. Train and engine weight increased steadily throughout the nineteenth century. On the Baltimore & Ohio, for example, the early "grasshoppers" of 1835 weighed about 10.5 tons, but by the 1890s locomotives weighed 80.4 tons. Since many bridges were "cheap and nasty," as one committee of American engineers put it—which is to say, designed to carry a load only slightly greater than immediately necessary—they soon became overloaded. Nimrod Bell, conductor on many post–Civil War Southern railroads, described crossing a trestle so poorly built that the bents sank, and the water rose, putting the locomotive fire out. Another so terrified the engineer that he sent the locomotive across unmanned, and followed his train on foot.[14]

Arthur Wellington thought this the result of economic, if not engineering design errors, and in 1886 in the *Gazette,* he denounced the policy of "sailing so close to the wind," which was not only uneconomic but introduced a "constant element of danger" as well. The editor pointed out that the carrying capacity of a bridge rose more than proportionately with its weight, while costs rose less than proportionately. Under such circumstances, he argued, investing in a heavier bridge was both safer and more economic, as it postponed the costs of replacement.[15]

Several years later, Albert Robinson, bridge engineer on the Rock Island, expanded on the *Gazette*'s claim. Robinson calculated that for spans of 100 feet or less, built about 1882, the extra cost of building them to withstand 1897 train loads when compounded at 5 percent, would have been less than the cost of reinforcing the bridge in 1897 (Table 5.3). Yet given the

Table 5.3. **Extra Cost of Building 1882 Bridge to Meet 1897 Loads versus Cost of Reinforcement in 1897**

	Length of Span in Feet							
	40	**60**	**80**	**100**	**120**	**140**	**160**	**180**
Added cost for heavier load (1882)	$86	$254	$531	$714	$1,377	$1,744	$2,148	$2,601
At 5 percent for 15 years	$179	$528	$1,105	$1,485	$2,864	$2,628	$4,466	$5,410
At 8 percent for 15 years	$273	$807	$1,684	$2,265	$4,368	$5,532	$6,814	$8,251
Cost of reinforcement (1897)	$377	$713	$1,408	$1,926	$2,857	$3,574	$4,388	$5,159

Sources: Albert Robinson, "Relative cost of Heavy vs. Reinforced Bridges," *Engineering News* 30 (Sept. 21, 1893): 273–38, Table 1; calculations by the author.

apparently large number of lightly built bridges, it is hard to believe that such behavior was uneconomic. Robinson's own figures demonstrated that if he had chosen an interest rate of 8 percent, most of the heavier bridges would not have paid off. For many railroads, the prospect of earning 5 to 7 percent on bridge investments must have paled in comparison with expected returns elsewhere—even when the additional safety was factored in. In short, overloading was probably the outcome of rational engineering-economic design choices.[16]

In any event, old bridges were routinely overstressed—sometimes by large amounts. In 1885 *Engineering News* reported a bridge on the Central Railroad of New Jersey in which tension bolts with an allowable stress of 10,000 psi were subjected to loads of 22,000 psi and another bridge with portions overstressed 200 percent. In 1890 that journal described some "old Fink and Bollman bridges [on the B&O that] . . . are carrying heavier loads than they were ever designed for . . . and ought to be taken down before they fall down." Five years later a speaker to the American Society of Civil Engineers claimed that "of the bridges built during the past 15 years . . . the greater proportion are carrying loads in excess of their specification requirements." Not surprisingly, some of these overloaded bridges failed. One of many such disasters occurred on August 18, 1886, when a rickety old wood bridge on the Brattleboro & Whitehall (Vermont) collapsed, wrecking ten cars and killing two people. It had been designed for a moving load of 1,000 pounds per linear foot and was "badly overloaded."[17]

The final cause of square falls was poor maintenance, which constituted a peculiar problem for wooden bridges, and hence, a peculiar problem for American as opposed to British railroads. Here again, danger arose not from error, but from design choices that put a premium on continuous and careful inspection, for as the American engineer Zerah Colburn explained to the British Institution of Civil Engineers, "timber bridges were always rotting." Colburn claimed that when a wood bridge collapsed on the New York Central, killing nine passengers in 1858, some of the timbers were found to be so decayed that a walking stick could be pushed through them. In 1859 when a wooden bridge on the Vermont & Canada that the management had known was rotten fell in, killing thirteen, the *American*

Railway Times termed the superintendent "unfit for his position" and "either stupidly ignorant or willfully malignant." In the 1860s Nimrod Bell watched pieces of rotted wood drop off some of the trestles he passed over. But Howe trusses not only rotted, their design also used threaded iron rods as tension members and these required adjustment, for if a nut on one rod loosened, all the load would be carried on those remaining.[18]

To cope with such difficulties, some companies followed the same practices they employed to protect the permanent way. They developed elaborate monitoring procedures, requiring weekly and monthly bridge inspections by various officials. By the late 1880s, the Erie employed ten inspectors who went over bridges monthly, while roadmasters inspected bridges quarterly and the bridge engineer annually. The Buffalo, Rochester & Pittsburgh kept detailed records on the dates each bridge was inspected and its condition. For wooden bridges, inspectors were admonished to pay particular attention to "the condition of the chords . . . around the angle blocks . . . to see that . . . there are no evidences of decay in the timber."[19]

But if some companies maintained their bridges, as Table 5.2 suggests, many did not. Writing in 1891 one expert condemned the "pernicious" practice that some roads followed of letting bridges deteriorate to the point of collapse before undertaking any repair. Poor maintenance and inspection probably contributed to the collapse of the Tariffville, Connecticut, bridge, which occurred a little over a year after Ashtabula. The bridge over the Farmington River on the Connecticut Western Railroad was a double-span wooden Howe truss, about 333 feet long, with vertical iron tension rods. It had been built in 1870 and was uncovered and unpainted. An excursion train, returning from a Moody and Sankey religious meeting in Hartford, went through the bridge about 10:00 p.m. on January 15, 1878, killing thirteen people and injuring forty-six.

Mansfield Merriman, then an instructor of civil engineering at Yale and at the beginning of his long and distinguished career, investigated the disaster. He found that the upper chord of the bridge was rotten. But, like Ashtabula, the Tariffville bridge had so much else wrong with it that Merriman could not pinpoint the precise cause of failure. He calculated that the stress on the tension rods was 22,000 psi, which he thought exceeded their elastic limit. Finally, both chords were also skimpy. Merriman calculated that the upper chord needed a cross section of 396 inches but only contained 284. He concluded that the bridge "was not properly built, it was not properly kept in repair, and it was not properly inspected." At the *Railroad Gazette* Matthias Forney took the lesson of the disaster, concluding that "eternal vigilance is the price of safety."[20]

Fire—like maintenance—was also a peculiar problem for wooden bridges. Trestles and Howe trusses were always burning—the result of locomotive sparks, or lightning, or careless burning, or smoking, or arson. This too was a peculiarly American problem, for in Britain coke was the fuel of choice, which yielded few sparks. Many companies took extraordinary precautions to protect their bridges from fire, providing handy water pails (containing saltwater in the winter) and requiring constant inspection from bridge watchmen or trackwalkers, but fires continued to cause disasters at the rate of one or two a year. In 1887, a fire-weakened bridge caused the Chatsworth, Illinois, disaster, which was one of the worst in American history. About midnight, on August 10, a double-headed, sixteen-car excursion train on the Toledo, Wabash & Western bound for Niagara Falls and carrying six hundred people approached a tiny bridge over a culvert near Chatsworth. The bridge was only 15 feet long and 6 feet deep, but it was on fire. Unable to stop in time, the train broke the bridge, killing 73 and seriously injuring 374 people. Ironically, the bridge had been inspected by section men about six o'clock the previous night; they had been burning weeds and probably set the fire.

Chatsworth provides another example of a characteristic common to nearly all disasters and noted in chapter 2—the coincidence of a series of tightly coupled, independent, low-probability events. Here too, disaster reflected problems of information, communication, and control. In this case operating practices were a contributing factor because the train was a double header with the air brakes controlled in the *second* engine, which slowed their application. As in the Mud Run disaster (chapter 3), had the brakes been applied when the first engineman saw the danger, the train might have stopped in time. In addition, Wellington thought that Chatsworth revealed the dangers of running double headers. The first engine almost entirely escaped the wreck but the second engine and tender "were the true resisting force" that telescoped the following cars, causing most of the casualties, he reported.[21]

If fire did not get a bridge, a flood often (in fact, more often) did. Some washouts were virtually unavoidable, but many were not, reflecting instead companies' efforts to economize on construction costs. An early example occurred on June 27, 1859, when a culvert washed out on the Michigan Southern, causing a train of six cars to fall into a ravine and killing thirty-nine. In the 1880s Charles Folsom of the Boston Society of Civil Engineers claimed that in two years of clipping newspaper articles he had assembled a list of eighty railroad bridges and 125 culverts that washed out, killing thirty-four people. On many of his consulting jobs, Folsom found the need to "increase the number and size of spaces for waterway," and stressed that in grading, "six feet above freshet mark is a great deal better than one." Culverts were an even worse problem than bridges; they were often too narrow, causing the stream to deepen the channel and wash out the abutments.[22]

Trestles were particularly prone to washouts because their supports were vulnerable to debris during high water. A washed-out trestle bridge near Eden, Colorado, just north of Pueblo, caused the second worst train accident in American history on August 7, 1904. Denver & Rio Grande Train No. 11, the *Denver, Kansas City & St. Louis Express,* carrying 162 passengers, broke through Bridge 110-B and fell into a rain-swollen creek, drowning eighty-eight people. The bridge was a simple timber frame trestle; it was "weak and in bad condition," and had been further weakened when a county bridge upstream let go, crashing into one of the bents.[23]

Bridges that collapsed while under repair provide yet another example of how the absence of management controls contributed to bridge failures. On August 31, 1893, one of the worst of such disasters occurred near Chester, Massachusetts. At about 12:30 p.m. that day, Boston & Albany Train No. 16, the *Chicago Express*, carrying 135 passengers plunged through Willcutt's Bridge over the Westfield River, killing fourteen people and outraging Wellington at the *Engineering News*. It was "the least excusable bridge disaster of magnitude which has ever occurred," he exploded. The investigation showed that the bridge was under repair at the time by an outside contracting firm, the R. F. Hawkins Iron Works, and under the immediate supervision of a foreman. The railroad, it appears, exercised no supervision over the project. The workmen were repairing one of the truss chords, which was built up from two beams with a riveted top plate that gave it lateral stability. At about noon, they quit for lunch, leaving the top plate off, which led the chord to buckle under compression when Train 16 arrived. Such an accident, Wellington concluded, reflected "not only merely gross individual carelessness but [also] certain radical underlying errors of practice and method . . . extending through quite a chain of officials from the president down."[24]

But knockdowns were the most common single reason why American railroad bridges failed. These also resulted from design choices that diverged sharply from European practice. In 1874 the *Railroad Gazette* contained a long exchange that highlighted the distinctions between British and American bridge-building techniques. The British engineer Ewing Matheson explained that the divergence in practice resulted in part from economic circumstances. "American [iron] bridges are lighter," he noted, thereby saving in material and also transportation, both of which—given the vast American distances—were far more important in the United States than in Britain, where iron was cheaper and distances shorter.[25]

Differing contracting methods also encouraged contrasts in design. By the 1860s, American bridge companies were assuming responsibility for bridge design, thereby facilitating both standardization and mechanization. Faced with the demand for large numbers of small bridges, American producers turned to the mass production of standard designs. By contrast, in Britain even small bridges were custom-made. As Matheson put it, "the builder gets his drawings from the railway engineer, who designs every rivet and bolt . . . A.B. wishes one style, C.D. an entirely different type." A later writer suggested that the American system, which located design with the contractor, generated strong incentives for economy, while in Britain design by the user emphasized safety.

American designers also used pins or bolts to connect the main members, while British engineers made such connections with rivets. Pin connections were ideally suited for American conditions, because they allowed all the bridge members to be factory-made and rapidly field-assembled. Such procedures speeded up construction, thereby allowing the railroad to open quickly and earn revenue. One writer claimed that British methods would require ten or twelve days to erect a 160-foot riveted lattice span while Phoenix Iron Works could put up a pin-connected bridge of similar size in 8.5 hours. Factory construction was also a way of employing semiskilled labor under close supervision "where skilled labor for erection is unobtainable or where, because of the lack of supervision, there is a risk that the riveting at the site would be carelessly or badly effected."[26]

American builders also skimped on bridge floors, sometimes omitting them entirely, a practice unheard of in Britain, and they employed deeper trusses with fewer panels than did their British counterparts, which saved metal but required more lateral bracing. In the 1890s, the British journal *The Engineer* described American designs as "birdcage structures," and it pointed out that they had "all the lateral stiffness of the side of a suburban garden fence." Matheson concluded that American practice was economical but very unsafe, claiming that "Fink trusses, and others of similar character . . . would not, in the interest of public safety, be allowed by the Government Inspectors in this country." He summarized the superiority of British practice by observing that nineteen bridges had failed in the United States in 1873, while in Great Britain, "not one accident per annum happens from any failure of a railroad bridge."[27]

American practices were defended by the civil engineer Charles Bender, who rested his case on economic and technical grounds and largely ignored the matter of safety. The *Gazette*'s editor Matthias Forney summarized the debate. Forney acknowledged the need for better floors on American bridges, laconically observing that "the American engineer seems to assume that derailment will never take place on a bridge," but he challenged Matheson's claim of superior safety for British practice. Forney asserted that "there is as yet no instance on record of an iron bridge designed by any of our leading engineers or engineering firms which has ever given way." As for the nineteen bridges cited by Matheson, all were wood. Forney then reviewed failures of six iron bridges built by "reputable firms." Three had been overloaded, while the other three "did not give way under a load but . . . were knocked down by a train." Forney was a tireless critic of bad engineering and one of the most safety-conscious engineers of his day. Yet by his lights, a bridge that collapsed from a train accident did not constitute an engineering failure.[28]

In fact, the very techniques that made American iron bridges distinctive also made them peculiarly susceptible to knockdown—a point that was implicit in Forney's argument but that he did not explicitly articulate. Others did, however. In 1878 Thomas Appleton concluded: "by far the most frequent causes of accidents are derailed trains." The solutions, he thought, were simple. Echoing Forney, Appleton noted that "an all important point in any bridge, and one that has been sadly neglected, is the floor. If we can make a floor that a derailed train will not break through . . . we shall have gained a great desideratum."

The second problem was that, unlike riveted bridges, pin-connected bridges were nonredundant structures in which the integrity of each member was necessary for the integrity of the whole. Knock one single post or truss rod out and the bridge was liable to fold up like a hinge. The solution, Appleton claimed, was to "let your chords, tension members, posts, etc., all be properly proportioned . . . but *fasten them together* [with rivets]; let your structure be one integral bridge, not a conglomeration of disconnected parts."[29]

Nineteenth-century American railroad bridge failures therefore reflected not only engineering error but also the consequences of engineering-economic choices with respect to design, materials, and construction methods. Such choices put a premium on careful inspection and operation, which might offset some of the dangers resulting from such choices, but which was often lacking. Most American bridges were designed with little margin for overloading, while the choice of wood made them susceptible to fire and rot. Trestles were inherently more dangerous than embankments, grading and waterway space were often inadequate, and floors were skimpy at best. Finally, light pin-connected metal bridges were prone to knockdown from derailments at a time when other engineering-economic characteristics of equipment and roadbed made American carriers derailment-prone (chapter 2).[30]

The Campaign against Bridge Accidents

If American railroad bridges were disaster-prone, these same disasters also generated powerful impulses toward reform. Public concerns dated from the 1850s but only in the 1870s did a combination of railroads, bridge companies, the American Society of Civil Engineers, the technical press, and state regulatory commissions, all in varying degrees, begin to press for change. Their actions, along with largely independent changes in materials technology, brought about a gradual reduction in bridge failures.

Reform Beginnings

State governments and railroad commissions began to exercise some jurisdiction over railroad bridges as early as the 1850s. As chapter 1 noted, in response to the Norwalk drawbridge disaster of 1853 that killed forty-six people, Connecticut promptly passed a law requiring all trains to stop before crossing such bridges. Many state railroad commissions performed their own inspections to monitor bridge safety. Their focus was mostly maintenance rather than design; few had any special expertise for the job and most lacked powers of enforcement. In 1859, Vermont's single railroad commissioner unsuccessfully requested authority to require repair of bridges. Two years later he was complaining of bridges that lacked guards and others that were not "sufficient to afford the security, which the public have a right to require."[31]

In 1872, Ohio's railroad commission employed the civil engineer William S. Williams to inspect railroad bridges. He found many "unsound and imperfect structures." The next year Williams reported that he had inspected all the important carriers in the state and claimed that "in most cases" his suggestions were complied with. Given that Ohio then had nearly 1,700 railroad bridges and trestles, his inspections cannot have been very thorough and he seems to have found no fault with the Ashtabula bridge before it fell. Similarly, two months prior to the Tariffville disaster an engineer employed by the Connecticut Board of Railroad Commissioners had examined that bridge and pronounced it safe.

Inspection in other states was no better. In 1872, when the Massachusetts railroad commission hired civil engineer James Laurie to examine that state's bridges, he found many dangerous structures but the commission did nothing. In 1878 George Vose was professor of civil engineering at Bowdon College. He had studied at the Lawrence Scientific School, edited the *American Railway Times,* authored a *Handbook of Railroad Engineering,* and was committed to the cause of better bridges. He described one bridge that was "old, rotten, [and] worn-out," and which had "been condemned for four years," that Maine's railroad commission termed "safe for present use." Vose was thunderstruck. "How . . . this bridge . . . has suddenly become safe . . . would puzzle any one but a railroad commissioner," he snorted. Of course most states neither inspected bridges nor investigated accidents, and when the Cahaba Bridge fell in on the Louisville & Nashville, in 1896, the company conspired to cover the incident up to forestall lawsuits.[32]

With early state regulation focused on monitoring maintenance, and largely ineffective, the technical press led by *Engineering News* and the *Railroad Gazette* began a campaign to improve the safety of bridge design. They not only published the technical papers presented at engineering societies and railroad clubs, they also reported debates such as that between Matheson and Bender in which ideas about best practice were hashed out. The engineering journals also reported on and discussed the lessons to be learned from disasters, diffused the ideas of reformers, and championed specific reforms and better regulation.

Sometimes these activities yielded immediate, concrete results. In one instance, the *Gazette* published an article, complete with illustrations, describing a bridge on the Kansas City, Memphis & Birmingham Railroad. A reader noticed that some crucial members had been omitted from the trusses and promptly notified the railroad and the bridge company asking for copies of the strain sheets, which they refused to provide until he threatened to publish his conclusions. The companies then acknowledged the blunder and made the corrections.[33]

In 1873 the American Society of Civil Engineers evidenced its first concern with bridge safety. Responding to the collapse of a highway bridge in Dixon, Illinois, "and other casualties of a similar character that . . . are constantly occurring," the society appointed a seven-member committee that included some of the leading bridge engineers of the day to report "on the means of averting bridge accidents." In 1875 the committee delivered its report. In fact, it delivered four reports. The first, signed by the distinguished James B. Eads and C. Shaler Smith, developed a model bill that its authors thought could easily be embodied in a law. It included standards for loads per linear foot of track for bridges of varying length as well as maximum stresses for wood, wrought iron, and cast iron. Noting that derailed trains knocked down many bridges, it proposed standards for floor construction. The report also recommended that each state appoint a bridge expert subject to an examination by the society, and that every railroad be required to inspect its bridges once a month. Other members in separate reports objected to many of the technical details. Some also disliked the proposal that states appoint inspectors who, they thought, would be political hacks. Compromise failed and the committee was dissolved in November 1876, a little over a month before the Lake Shore's *Pacific Express* kept its appointment with the Ashtabula bridge.[34]

As noted, Ashtabula precipitated an investigation by a committee of the Ohio legislature. The committee drafted a bill to regulate bridge construction but it never passed. The ripples of Ashtabula extended beyond Ohio's borders, however. Liability suits stemming from the tragedy cost the Lake Shore over $600,000 (about seventeen years' earnings for each of the eighty-nine fatalities)—enough to get the attention of railroad managers everywhere. Both the Erie and New York Central went over all bridges as a result of the accident. "I venture to say that there is hardly a railway in the country that has not been inspected in some way as to its bridges since last December," Alfred Boller told the ASCE in February 1877. Wisconsin's railroad commissioner reported that as a result of Ashtabula, "numerous letters were received at this office inquiring as to the safety of certain bridges." In response, he circularized the carriers, urging them to employ competent engineers to inspect their bridges. "As the result of my labors . . . the railroad bridges are in better condition by far than ever before," he crowed. On the national level Representative James A. Garfield, whose district contained Ashtabula, submitted Charles Francis Adams's bill that would have required railroad inspection by army engineers (chapter 3). The Tariffville disaster of 1878 also improved bridge inspection in Connecticut, at least for a time. The commission urged carriers to compute strain sheets while the state's engineer began to inspect tension members with more care.[35]

The *Engineering News* opined that Ashtabula provided a good excuse for the ASCE to revive its committee on bridge inspection, complaining that "at the present time there is scarcely any hindrance to parties building any structure they see fit." At the April 1877 meeting Eads's report was again debated and a resolution to form a committee to draft a model law was submitted to letter ballot, but it failed. It was the last effort of the ASCE to develop a code of bridge specifications.[36]

But the ASCE and the regional engineering societies did play an important role in developing and diffusing new knowledge on proper bridge design. Before the 1870s, engineers often employed cast iron in compression. They compared the tensile stress of wrought iron to its breaking strength to estimate the safety factor, ignoring its elastic limit. Nor did designers always distinguish between live and dead loads, or how the live load was distributed. Little was known about the effect of wind on structures, or the role of impact, or the actual behavior of full-size members under load.

As in the MCB and Master Mechanics, new men with college training in mathematics and engineering increasingly dominated the ASCE and other engineering groups. The societies' papers and discussions clarified many of these technical matters. Cast iron was abandoned, and in the 1880s, the implications of Wöhler's experiments on metal fatigue were gradually digested, leading to an increased emphasis on the elastic limit of metal rather than its breaking strength. Engineers discovered that heavy loads on light bridges might lead to stress reversal, putting the same member first in tension and then compression, and they gradually discovered that impact might generate much greater stress than a static load. Various formulas were developed to incorporate these insights into designs, and by the twentieth century engineers were actually measuring the impact of live loads. Materials testing—including that of full-sized members—improved and the procedure of overloading a new bridge to see if it held up was abandoned as unsound and unsafe. These and many other improvements in technique contributed to both safer design and a better understanding of maintenance and replacement needs.[37]

Improvements in Contracting and Inspection

Changes in the methods of bridge contracting during the 1870s also contributed to improvements in design and construction. Prior to the Civil War most bridges had been constructed either by itinerant self-trained builders or by the railroads, and neither method ensured expertise in design or accuracy in calculation. In 1849 Isaac Hinckley, superintendent of the Boston & Providence, complained that the bridges built by one contractor "are in some respects inferior in dimensions of material and quality of workmanship." Some of the piles were small and poorly driven and "some of them have settled six or eight inches, thus permitting the permanent roadbed to deflect in same degree."[38]

During the 1860s as the rail network expanded, companies such as Keystone Bridge and Phoenix Iron arose that specialized in bridge construction. The carriers would contract with the bridge manufacturers, paying by the linear foot and leaving the design, construction, and factor of safety to them. Such contracts generated moral hazard problems because they provided no oversight by the railroad and no incentive save reputation for the contractor to perform honestly or competently. During the 1870s, however, the railroads began to exercise much more control over the process, hiring their own or consulting bridge engineers. In 1873, George Morison of the Erie developed the first set of detailed printed bridge specifications. The Cincinnati Southern, the Milwaukee, and other large carriers soon followed. In the 1880s prominent consulting engineers published their own specifications that were widely publicized and debated. Theodore Cooper developed a set of standard loadings for proportioning bridges that became the industry standard well into the era of diesels. Typically such specifications left design and construction details to the contractor, but specified maximum stresses for individual members, set out specific formulas to be used in computing loads and the concentration of load to be assumed, and required blueprints, strain sheets, and materials testing. By this time at least some bridge companies maintained materials testing equipment, and by the mid-1880s, Keystone Bridge and the Pennsylvania Railroad were performing compression tests on full-size members.[39]

In the absence of any group such as the MCB to develop industry-wide specifications, these efforts by companies and consulting engineers came to define best practice. All were reported from time to time in the engineering press, thus ensuring that they would be widely known. They helped prevent shoddy construction, and ensured that each bridge was independently checked at least twice. Despite such improvements, most critics continued to urge better public regulation of bridge safety. In 1880, George Vose penned two articles on bridge accidents in the *Railroad Gazette* that blamed Ashtabula, Tariffville, and similar disasters on bad design, sloppy construction, and incompetent inspection by both the railroads and state officials. Vose urged states to begin inspection by ASCE-certified engineers. At the *Gazette,* Forney went even further, advocating that states develop specifications that could become the basis for a law.[40]

In 1883, the New York State railroad commission initiated the first comprehensive program of railroad bridge inspection in the country in response to a bridge that collapsed on October 22, 1883, on the Rensselaer & Saratoga Railroad, killing three and injuring twenty-two people. It was the usual story of misfeasance and nonfeasance. The inquiry revealed that a truss rod had broken under a stress of about 25,000 psi, and, as was often the case, that no one in the company had clear responsibility for bridge inspection.[41]

In January 1884, New York's commission requested drawings and strain sheets on every railroad bridge in the state, and it hired Charles T. Stowell as bridge inspector to assess the submissions and report needed changes to the companies. Stowell symbolized the increasing role of newer scientifically trained men. He had graduated with a degree in civil engineering from Rensselaer Polytechnic Institute and been a practicing bridge engineer. This was the first time in which any public authority had required the recording of accurate strain sheets on every member of every bridge and then employed an expert to verify the calculations. It was, *Engineering News* reported, a procedure that "well deserve[s] to be widely copied."[42]

Two results followed from New York's action, the first being a widespread upgrading of the state's railroad bridges. The engineering press reported that many companies reinforced their bridges before submitting plans in order to forestall bad publicity. Still, the final report, which ran to 1,600 pages, contained an astonishing number of unsafe bridges that had been revised only in the light of Stowell's criticisms. The second result was to provide Stowell with both the information and the bully pulpit that he would use to influence bridge design.[43]

Charles Stowell, *Engineering News,* and American Bridge Design

Stowell's position made him familiar with bridge design on every railroad in New York and he soon rediscovered the argument that Appleton and Matheson had made a decade before. The pin-connected truss bridges common on most railroads were subject to knockdown from derailed locomotives or cars. To use the terminology of Charles Perrow, collapse was a "normal accident" after derailment on such bridges whereas on the riveted lattice trusses used by the New York Central it was not. To build his case, Stowell began to assemble failure statistics, ignoring trestles and simple stringer bridges since his concern was with trusses. Stowell had powerful allies in the railroad press. His statistics came mostly from the *Railroad Gazette*'s annual compilation of train accidents. Both the *Gazette* and *Engineering News* also gave Stowell's statistics and views wide publicity and at both journals Arthur Wellington chimed in with editorial support as well.[44]

Stowell's first salvo appeared in the *Railroad Gazette* in November 1885. He presented his own and Appleton's statistics on bridge accidents from 1873 through 1885, pointing out that while knockdowns were the leading cause of failure of pin-connected iron bridges, there were *no* such accidents to lattice truss bridges. Stowell reminded the reader of the earlier debate over pin versus riveted bridges and the *Gazette*'s claim that a knocked-down bridge did not really "fail." Stowell was unimpressed. That "must be small consolation either to the maimed survivors, the bereaved relatives, or the company which pays for the damage," he observed.[45]

As soon as the figures for 1886 were available Stowell struck again, this time in the pages of *Engineering News,* for by then Wellington had joined that journal. Wellington, the reader will recall, defined engineering as "the art of doing that well with one dollar which any bungler can do with two after a fashion." Bridge building, however, required a caveat— "doing well *safely.*" Of course bridge safety needed to be achieved in the least costly way, and so while Wellington also concluded that the ease with which American bridges could be knocked down revealed a design flaw, it was a different flaw than Stowell stressed. The problem, he thought, was with the floor, not the truss, perhaps because flooring was a cheaper fix.[46]

As Forney had observed a decade earlier, floors were a weak spot in many American bridges. Sometimes bridges had no floor at all, and the rails were laid on stringers that rested on the cross-braces. If ties were used they were often unattached to the stringers, did not extend the width of the bridge, and were widely spaced. In all such cases, derailment usually meant disaster. Without a floor, a derailed car might plunge into the cross-braces, carrying away the bridge, or if ties were spaced more widely than 8 inches apart, the wheel would fall between them, causing them to bunch and leading to disaster. Even if the ties were properly spaced, however, there was usually nothing to prevent a derailed car from plowing into the truss or going over the side of a deck truss bridge or trestle.

The usual results of an accident with such flooring were starkly illustrated on Saturday, February 5, 1887, on the Woodstock bridge near White River Junction, Vermont. The Central Vermont's Boston-Montreal express apparently hit a broken rail and derailed about 450 feet before the bridge, which was a deck truss. The last three cars pitched over the edge, falling about 42 feet to the ice below where one of the car stoves set them ablaze, incinerating twenty-nine victims, injuring many more, and burning the bridge down.[47]

Woodstock illustrated the ambiguities inherent in the definition of bridge failures. To most engineers, Woodstock was a derailment—which was how the *Railroad Gazette* categorized it—for the bridge didn't fail, and its subsequent burning had nothing to do with the accident. Even Charles Stowell, concerned as he was with truss failures, did not include Woodstock in his list of bridge failures. But it was meat and drink for Wellington at the *Engineering News*, who saw such disasters as the predictable outcome of derailments on bridges with unsafe floors. In all events, it had lasting consequences, contributing to the nationwide campaign to banish the "deadly car stove" that ultimately resulted in the introduction of steam-heated passenger cars (chapter 3). It also led the Vermont legislature to pass a bill authorizing the annual inspection of railroads, by experts if necessary, and it led Wellington and the *Engineering News* to launch a crusade to improve bridge floors.[48]

As soon as Vermont authorized inspections, the railroad commissioners promptly employed Hiram Hitchcock and Robert Fletcher, both of whom were civil engineers at Dartmouth, to inspect all bridges in the state. Their reports revealed the usual assortment of horrors and death traps. Hitchcock reported that many railroads were improving their bridge floors, but on others he noted "the absence of re-railing devices, improperly spaced ties, [and] inadequate guard timbers." In September 1889 he reported on the Central Vermont's Bethel Bridge, a Howe truss just then celebrating its forty-ninth birthday. The bridge had been "horsed up" (i.e., turned into a trestle), without which Hitchcock thought it would have collapsed. He found some tension rods 100 percent overstressed and some wood posts with safety factors of only 2.5.[49]

Such scrutiny got results, and several years later the commissioners reported much improvement as "the bridges of first construction on all the early built lines have almost wholly disappeared." Later they noted that the Central Vermont had adopted a standard bridge floor consisting of long ties, outside guardrails, and sometimes inside guards as well. Yet Vermont's inspection system remained inferior to that of New York. Companies were not required to file bridge plans and strain sheets with the commission, and in 1896 the commissioners complained that with few exceptions, they had no records that gave the condition, material, and safety factor of railroad bridges. When they requested such information, some companies claimed to be unable to give the safety factor on their bridges—"a confession of incompetency that merits criticism," the commissioners rather lamely concluded.[50]

At the *Engineering News* Wellington also tried to ensure that the lessons of Woodstock did not go unheeded. In February 1887, immediately after the tragedy, the journal launched a campaign for safe bridge flooring. An editorial reproduced a letter dated February 18, 1874, from the general manager of the Chicago & Michigan to Charles Latimer, inventor of a rerailing device, which described an accident it had prevented. Latimer's invention was exceedingly simple, consisting of two rails that came to a point in the center of the track and led back to a ramp up to the main rails. A derailed truck that hit the device would automatically be guided back to the ramp and up onto the rail. The editorial claimed that the device was not patented and was standard on many lines. "We could readily give a list of a dozen more similar occurrences in which trains were without doubt saved from running off bridges," it concluded.[51]

As the editorial noted, efforts to improve bridge flooring were not new. In 1881, the Massachusetts railroad commission had sent a circular to that state's carriers describing various safety systems and urging their adoption. In 1885, Ohio's commission published a letter from J. E. Childs, chief engineer on the New York, Lake Erie & Western, describing a derailment on a bridge that would have resulted in a "Second Ashtabula" but for the use of Latimer's guard. Childs also modified Latimer's device to make it part of a guard system that the *News* described in a second editorial. The system included, in addition to the rerailer, two sturdy timbers bolted or notched to the sleepers and flared at each end of the bridge that would prevent the ties from spreading and also act as outside guard rails to prevent a derailed car from hitting the truss. Finally, sturdy end-posts also protected the truss. The whole arrangement could be installed for $120 to $160 per bridge.[52]

Stowell promptly responded with a list of recent accidents, most of which involved some form of collision on the bridge that would not have been prevented by the Childs-Latimer system. He admitted it was wise to provide all bridges with guards. "But is it not also a pretty good idea," he queried the *News*, "to build your bridge [so] that if all these safe-guards fail, and the truss does happen to get struck, it will not fall down?" Stowell, who managed to retain his sense of humor while reporting these tragedies, concluded with the story of an "old bridge builder" who always claimed to tremble when riding over some bridges lest an elderly lady inadvertently stick an umbrella out the window and hit the truss.[53]

For the next three years the pages of the *News* contained a lively debate among Stowell, the editors, and various others on whether pin-connected bridges or inadequate floors were the worst danger and which of the several sorts of guard rails was the best. The *News* emphasized disasters that resulted from bad flooring. As noted, this was probably because floors were cheap and Wellington found the economics of pin-connected iron trusses too appealing to criticize. Stowell, on the other hand, bombarded the editors with examples, complete with photographs, of riveted bridges that had withstood the loss of end-posts and webbing without collapsing and of pin-connected bridges that had been knocked down under similar circumstances. Stowell breezily suggested that all such bridges carry a sign "Warning: Don't Touch the Trusses Under Penalty of a Wreck."[54]

In 1888 George H. Thomson, bridge engineer on the New York Central, joined the fray with a paper titled "American Bridge Failures: Mechanical Pathology Considered in Its Relation to Bridge Design." The Central had been constructing riveted lattice truss iron bridges since the 1860s, a type frowned on as uneconomic by the engineering establishment, and his argument constituted a self-justification as well as a critique of common practices. He delivered the paper to the British Association for the Advancement of Science and later published it in the British journal *Engineering*—both of which the American technical press considered little short of treason. Like Stowell, Thomson claimed that pin-connected spans were easily knocked down—a kind of mechanical pathology. "As long as reputable bridge practice is satisfied in making no provision for the contingencies of railroad operation—catastrophes due to broken axles etc.—just so long must we expect to hear of railroad bridge failures," he lectured.[55]

Such arguments by Stowell and Thomson amounted to a direct attack on American engineers' conception of their professional responsibility for safe bridge design. In his famous "American Railroad Bridges," presented to the ASCE in 1889, Theodore Cooper rather testily responded to such critics. "Does anyone advocate the designing and building of bridges to withstand the impact of a railroad train or the bursting effect of piling two trains on one another inside of the trusses? Are such accidents to be classed as bridge failures or as failures of management?" Cooper asked rhetorically.[56]

In the long discussion that followed Cooper's paper no one took the bait except Charles Stowell, who responded simply: "as long as trains run into bridges or cars pile up inside the trusses, a bridge to be safe must be designed and built to withstand just those things. I believe . . . that the time will come," Stowell concluded, "when the failure of an iron bridge from an ordinary accident of train service will be regarded as discreditable to its builder and not excused as a fault of the management." Another discussant, the respected J.A.L. Waddell, defended Cooper. The shift to riveted lattice truss design that Stowell advocated "would be retrograde," Waddell claimed. Like Wellington, he favored stronger floor systems instead, and concluded that "general managers and superintendents are much to blame for the improper styles of floor system used on many American lines." Thus, Cooper and Waddell (and apparently most of the others present, for no one disagreed) saw knockdowns, not as problems of engineering design but as the result of managerial choices and blunders that resulted in bad floors, collisions, and derailments. Just as attributing certain passenger accidents to personal carelessness initially blinded regulators to the need for improved train and station design, so the inability of engineers to see bridge knockdowns as engineering failures retarded reform.[57]

As noted, at the *Engineering News* Wellington was also reluctant to follow Stowell in condemning pin-connected bridges, and in response to his claims the journal usually pointed out that most bridges that failed had widely spaced ties and lacked any system of guards. Sometimes, however, the facts pushed the *News* into Stowell's camp. The journal reported a knockdown on the B&O on April 30, 1887, that resulted when the locomotive hit a cow on a bridge. This was too much for Wellington. "Whether the cow was actually . . . flung against one of the posts, or whether the cow just happened to swing her tail against one of the compression members just as it was taking strain from the locomotive we cannot say, but the internal evidence rather favors the latter theory," he thundered. "The day is near at hand," Wellington went on to pronounce, "when it will ruin a man's professional reputation to have either designed or accepted such a bridge." By 1888 the *News* admitted that "a riveted structure of the same span and strength [as a pin connected bridge] would . . . [be] much more likely to escape collapse."[58]

As this debate was proceeding, disaster gave the cause of reform another nudge. Forty days after White River Junction, on March 14, 1887, a Boston & Providence train out of Dedham, Massachusetts, with nine cars and about 275 passengers and crew plunged through the Bussey Bridge, killing twenty-three people. "The Second Ashtabula," Wellington called it, although compared to the Bussey Bridge, "the Ashtabula bridge was a masterpiece of engineering," he fumed. If anything, this was an understatement. The bridge had started life as a wooden Howe truss. It had been tinned to prevent rot, thereby winning it the title "the tin bridge," and had been rebuilt with one iron truss in 1869. A second—of different design—was added in 1876. It was not the trusses that had broken, however, but the hangers for the floor beams that held the track. The offending hangers were "of far from good iron . . . bad design . . . imperfectly welded . . . with old, deeply rusted breaks." Subsequent analysis revealed that normal train loads placed a stress on them of 48,000 psi, roughly equal to their breaking strength.

But while the hangers caused the bridge to fail, as the *News* and the official investigation pointed out, the disaster also revealed a potpourri of defective management controls and operating practices. The railroad had exercised no supervision over the contractor, who built the bridge under what may have been fraudulent conditions. It had never been tested, nor had a trained engineer ever inspected it. Once again disaster reflected the coincidence of a series of tightly coupled, independent, low-probability events. The train that broke through was equipped with old-style Westinghouse straight air brakes that lost pressure when the train parted, and had been obsolete for a decade. The train was traveling no more than 25 miles per hour when the bridge broke. Assuming standard braking efficiency, Wellington calculated that with automatic brakes the last three cars would not have gone over the edge while those that preceded would have settled with the truss rather than crashing into the abutment, thereby causing many fewer casualties.[59]

In its investigation, the Massachusetts railroad commission heard testimony from George Vose, now at MIT. In response, it recommended that the railroads be required to have all bridges inspected biennially by a competent engineer with the reports, including plans and strain sheets, going to both the railroad and the commission, which was empowered to employ its own engineer. The legislature promptly obliged and the commission employed George Swain, another MIT engineer, who was also an MIT graduate. Swain was to inspect bridges and go over the carriers' plans and strain sheets. In addition, the board sent out a circular urging the carriers to choose one of the various systems of bridge floors and guards that it described. As in New York and Vermont, the new procedures apparently generated some spectacular results, although Massachusetts, in a gesture the carriers no doubt appreciated, did not publish individual inspection reports. However, in his yearly statements to the board, Swain reported a sharp increase in the number and quality of bridge guards and a widespread upgrading in many bridges. In the early 1880s, Massachusetts carriers spent an average of about $1.1 million a year on bridge repair and renewal; in 1888, the year after the Bussey Bridge fell in, they spent $1.8 million.[60]

The Decline in Bridge Failures

These efforts by the engineers, regulators, and reformers to improve bridge safety gradually bore fruit. With the technical press and the state commissions in Massachusetts, Vermont, and Ohio strongly advocating better flooring, companies increasingly addressed the problem. Shortly after the disasters at Chatsworth, White River Junction, and Bussey, the *Railway Review* reported that "guard rails and re-railing devices are being more extensively employed." In 1893 in the pages of *Engineering News,* Wellington claimed "within the last few years there has been a decided tendency on the part of some of our larger railways to adopt solid floors, the ballast and roadways being continued on the bridge itself." By 1899 that journal concluded—with some overstatement—"it is the exception to find a steam railway bridge or trestle which is not thoroughly protected by guard rails."

In fact, reports to *Engineering News* in 1909 and to the American Railway Engineering Association from 1912 to 1914 reveal that use of guardrails, while widespread on heavily traveled lines, was by no means universal where traffic was light. The reasons, of course, were once again "commercial considerations," as C. E. Smith, bridge engineer on the Missouri Pacific, explained. That carrier had about 10,000 bridges and to guard all of them would cost $1,350,000, which at 6 percent resulted in an annual cost of $80,000, he calculated. Even if guardrails eliminated all wrecks "the sav-ing to the railway would not have equaled the increased interest charge," Smith claimed.[61]

Improvements in engineering understanding of materials and de-sign also contributed to the changes. In the 1890s, engineers abandoned light, pin-connected truss bridges for short spans in favor of more heavily built, riveted structures. In 1904 one engineer admitted that "there can now be no question that the English engineers were pretty much in the right in their old contention in favor of riveted bridges—at least for spans of less than 200 feet, which cover the bulk of ordinary railroad structures." An-other engineer writing in 1907 agreed: "the pin connected truss is ranked last [in degree of safety] . . . because of its greater flexibility . . . and the greater chance of failure through rupture of a single member."

These changing professional norms may also have reflected evolving economic circumstances, for the exit of the bridges that Stowell decried was hastened by largely independent changes in technology. The arrival of compressed air field rivet guns sharply raised labor productivity and reduced the need for skilled riveters. As iron and then steel prices fell, companies increasingly began to use steel plate girders, which had become cheaper for spans of less than 100 feet and were nearly indestructible. Rein-forced concrete made its appearance and it too yielded safety gains. Many companies also simply replaced bridges and trestles with embankments. In 1901 A. S. Markley of the Chicago & Eastern Illinois recalled that fifteen years before washouts had been common, but "we have been continuously renewing our trestles with permanent structures and . . . in the past ten years I cannot call to mind that we have had a single washout." In addition, the spread of steam heat, better brakes, and steel passenger cars also pared the casualty list from bridge disasters.[62]

Technological change gradually reduced the domain of the old wooden Howe truss. Statistics for Illinois, Ohio, and Iowa reveal that from the 1870s to 1900, the number of wooden bridges declined slightly while iron and steel structures increased sharply. The figures in Tables 5.1 and 5.2 suggest that metal bridges had about one-seventh the failure rate of wood. Thus it was not the introduction of steel, but rather the substitution of metal (and later concrete) for wood that was probably the most important improvement in bridge safety. Such a conclusion also throws the debate between Stowell and the advocates of pin-connected iron bridges in a different light. Stowell was surely right: pin-connected metal bridges were less safe than those that were riveted. But to the cost-conscious, light-traffic carriers of the 1870s, the alternative to pin-connected iron bridges was not a more expensive metal bridge: it was wood. Seen that way, pin-connected bridges improved safety because they speeded up the transition from wood to metal.

Management and operating practices also improved in response to disaster. Wooden structures became safer as companies experimented with fire protection, employing fire-retardant paint and graveled floors. As noted, contracting methods evolved that yielded better monitoring and more appropriate incentives. In the late 1880s, one journal reported that "the railways have taken hold of the matter of inspection in a manner that is not generally appreciated." By the twentieth century, most large carriers ensured that each bridge received two separate types of inspection. They required "current inspections" quarterly or monthly by foremen or section hands to look out for routine maintenance and repair. A bridge engineer or someone with similar qualifications conducted a "general inspection" of all bridges at least annually to check on maintenance and mark bridges that needed major upgrades or renewal. On the Northern Pacific these reports employed forms containing forty questions covering all aspects of the structure.[63]

Beginning in the 1890s, these efforts contained the problem of bridge failures. Many companies described instances of derailments that did not become disasters due to improvements in floors and trusses, but the only available statistical evidence derives from the *Railroad Gazette*'s compilation, for the ICC accident statistics did not allow separate tabulation of bridge failures. The *Gazette*'s data, while providing an undercount, can be used to spot trends. They reveal an annual average of twenty-four bridge accidents during the 1873–77 period, rising to thirty-eight per year in the 1888–92 period, and then declining to twenty-five from 1896 to 1900, even as the number of train miles and bridges was sharply increasing.[64]

These efforts to improve bridges reveal much about the engineering economics of nineteenth-century American railroad safety. The technological network, especially the *Engineering News* and the *Railroad Gazette*, played a central role, gathering, interpreting, and diffusing information. Once again statistics powerfully shaped reform. The rise and decline of bridge failures as a "problem" was partly an outcome of the particular form in which the *Railroad Gazette* collected accident data, for they provided Stowell and other critics with ammunition. In the twentieth century, as the gathering of accident statistics shifted to the ICC, hard evidence on bridge disasters simply disappeared. The commission did collect information on collisions, however, and as noted, these were rising sharply. As a result, both the engineering press and public outrage shifted to these more pressing problems. Concern with bridge failures thus declined more rapidly than the failures themselves.

The problem of bridge safety also reveals how accidents can induce institutional change as they lead to new rules for bridge design and inspection within companies and regulatory bodies. Bridge accidents also demonstrate the need for a historical approach to the idea of technological or normal accidents, for what was normal in 1880 reflected managerial and technological choices and as these evolved so did both the number and type of bridge failures.

The story also shows that while failure can be instructive, its teachings depend on how it is interpreted. Both professional self-interest and broader economic considerations encouraged Cooper, Waddell, and most members of the ASCE to adopt a narrow construction of bridge failures. The emphasis on engineering errors was professionally advantageous because it focused on the mistakes of outsiders rather than their own design choices that implicitly traded safety for economy—trades that engineers were loath to acknowledge. These involved wood versus metal, reinforcement versus greater initial strength, weak versus heavy floors and approaches, complete versus partial guarding, and pin versus riveted connections.[65]

These choices were not inherently bad designs. Rather, they and the debate over them reflected the changing economic forces that shaped nineteenth-century American railroad safety. The early flimsy bridges, like flimsy roadbeds were, as George Morison put it, "good engineering," because they were good economics, while their critics such as Stauffer and Wellington were the midwives of a new age struggling to be born.

6

Coping with the Casualties
Companies, Workers, and Injuries, 1850–1900

The higher the order of railway surgery, the greater the protection to the employee, the passenger, and the company.

—Motto, American Academy of Railway Surgeons

The justification for the relief department must be that it pays . . . That it does pay we know from the figures.

—Charles Elliott Perkins, President, Chicago, Burlington & Quincy Railroad

Early on the morning of May 12, 1871, a westbound Erie freight headed for Buffalo, New York, broke in two. The rear of the train, to which was attached an immigrant car, came to a stop just outside the Griswald station, near Attica. It was struck almost immediately by a following freight, demolishing the immigrant car. Six of the passengers were killed and thirty-five injured. One was badly scalded, three had fractures of the leg, and another fractures of both legs. There were also several broken arms and one fractured clavicle. Eight physicians soon appeared; the injured that could be transported safely were moved to Buffalo General Hospital. The most severely injured were removed to a recently vacated local house where one, a boy of fifteen with a compound comminuted fracture, had his leg amputated. All recovered, and the boy's father waived his right to sue for a payment of $5,000.[1]

What distinguishes this sad story from thousands of similar tragedies is that we know the medical attention that the injured received. Therein lies its interest, for in the years following the Civil War, American railroads and their unions pioneered in the development of complex medical and beneficial organizations to care for workers, passengers, and others injured by the new technology. These innovations occurred a generation before manufacturing companies showed serious interest in such matters, and involved imaginative efforts to deliver emergency and preventive medicine as well as income security to large numbers of working-class men and women. Railroad workers also responded to these dangers, primarily through the brotherhood beneficial societies that provided income security against injury or death. Collectively these suggest a significant and neglected private-sector response to the dangers of industrialization. Yet it is wrong to see these organizational innovations as entirely induced by accident costs. Companies also hoped they would increase worker loyalty and reduce turnover while both employers and unions viewed them as weapons in the struggle over organization.[2]

The Origins of Railway Medical Organizations

As previous chapters have demonstrated, railroads began generating casualties from the beginning, and they also started to devise methods to cope with the injured and dying. The details of the first arrangement between a railroad and a physician to care for those injured in accidents are obscure. The Baltimore & Ohio (B&O) may have employed a surgeon as early as 1834. In 1849, as soon as the Erie Railroad came through the Delaware Valley, there was a "railroad doctor" who cared for the inevitable casualties. In the late 1850s, the Chicago & Galena Union, the Illinois Central, the Chicago & Milwaukee, Michigan Central, Michigan Southern, and Lehigh Valley all had some form of surgical service. In 1858, President J. Edgar Thompson appointed Dr. John Lowman the first surgeon of the Pennsylvania Railroad. Perhaps the idea for a company surgeon was inspired by a naval example, for like naval vessels that carried a ship's surgeon, early railroading was dangerous, while many of these early lines went through wild and sparsely settled regions as far from doctors as ships were from seaports. But economic circumstances also suggest that some arrangement between railroads and local doctors must have been common. Physicians gained a predictable clientele while most carriers found it expeditious to take care of those injured in the course of business almost irrespective of legal liability. As noted, proper medical care was not only humanitarian, it lessened the likelihood of expensive lawsuits and ensured goodwill among both passengers and workers.[3]

Some of these early arrangements between railroads and physicians were remarkably informal. As late as the 1870s, the Philadelphia & Reading apparently had no formal contracts with physicians to care for those injured in accidents. Rather, physicians simply sent in bills for their services, which General Superintendent J. E. Wooten honored as he saw fit. Such a system presented obvious difficulties. First, as the carriers grew in size, the number of claims grew to flood proportions and it was clearly inefficient to have each one scrutinized by top management. Moreover, review by nonspecialists also tempted doctors to bilk the company; yet physicians risked having the fees for their services disallowed.

The Reading's records reveal both sorts of difficulties. On January 8, 1873, Dr. A. P. Carr presented Wooten with a bill for fourteen dressings he claimed to have made on the mashed finger of John Welch, a Reading brake-man. Wooten was suspicious and upon inquiry discovered that Carr had performed only eight dressings. Yet when Wooten disallowed a bill submit-ted by Dr. William McKenzie that he thought had been padded, he promptly got a hot note. McKenzie announced that "until quite recently I have been attending your accidents and sending no bills, but it is too much to do for nothing. I often spend two or three hours with a patient, losing my time and often getting besplattered with blood." He claimed he had been sum-moned by the employees and told Wooten to pay up or "I shall *not attend any more* of your cases and *will* also proceed to collect." Wooten called his bluff, for nearly a year later he was still trying to collect, still caring for those injured in the Reading's accidents, and still threatening to see no more such casualties.[4]

Developing Formal Medical Institutions

Most carriers found such piecemeal arrangements unsatisfactory and de-veloped more formal structures to deal with medical needs. By the 1880s, the railroads had evolved three different models of railroad medical ser-vice that reflected in part the differing economic and geographic circum-stances within which each carrier operated. Western carriers originated hospital associations: in 1867 or 1868, the Central Pacific (CP) became the first railroad to set up such an organization, which was modeled on the U.S. Marine Hospital Service. The CP and later the Southern Pacific cre-ated a formal medical service that employed a chief surgeon, and divi-sion and district surgeons (all of whom were salaried), along with local "emergency" surgeons who were compensated according to a fee schedule. In 1869 the company opened a temporary hospital in Sacramento and in 1870 it built a new structure with six wards, eight private rooms, and space for 125 patients. The service was largely funded by charging all employees $0.50 a month. All were required to participate except for Chinese who were, apparently, barred. The company contributed free transportation, as well as bookkeeping and other overhead services, and it paid for treatment of passengers and others injured on the road. Employees, but not their families, received free medical care for all injuries or illnesses provided the ailment was not due to "venereal infection . . . vicious acts . . . or previous infirmity."[5]

 The Central Pacific organization reflected the circumstances of west-ern railroad construction. With population density low the local medical market was virtually nonexistent. Thus humanitarianism combined with the need to recruit and keep a labor force prompted the carriers, as re-gional "first movers," to build hospitals and provide other medical services. For similar reasons, mines and lumber camps sometimes provided similar services. It was vertical integration to cope with what Richard Langlois has termed "dynamic transactions costs"—the problems of change when re-lated markets are undeveloped. Not surprisingly, therefore, the CP model was widely emulated by other large western carriers. The Missouri Pacific and Texas Pacific set up almost identical services in 1879. The Northern Pacific established a hospital association in 1882 and built hospitals in Brainerd, Minnesota, and Missoula, Montana. The Union Pacific system dated from 1882. The Denver & Rio Grande was next in 1883, building a hospital in Salida, Colorado. These were followed by the Santa Fe, Wabash, Milwaukee, and Great Northern, among others. A survey by the U.S. Railroad Administration revealed that one-quarter of all railroad workers were in such hospital plans by World War I (Table 6.1), and by 1930, the number was up to one-third.[6]

The Lackawanna, which both mined coal and ran a railroad, was one of the few eastern carriers to operate a hospital association, but many supported private or community hospitals. In the 1880s and 1890s, the Pennsylvania Railroad systematically contributed to the operating expenses of a number of hospitals along its line, basing its contribution on the usual number of hospital days times the hospital's estimated average daily cost per patient. The Pennsylvania also sometimes made capital contributions of various sorts, occasionally providing land or endowing beds. One advantage of such a system was that the company was able to allow injured employees and others to choose their own hospital. These procedures continued at least into the 1920s and other carriers, such as the Delaware & Hudson, had similar arrangements. On neither railroad did the company's contribution cover the cost of treating its casualties, however.[7]

The second form of medical organization—common among eastern and Midwestern carriers—involved more or less formal contractual arrangements with local physicians to care for accident victims, but did not include company-run hospitals. As noted, the Pennsylvania began a medical organization before the Civil War. By the 1890s on its lines east of Pittsburgh it employed about two hundred surgeons on contract who reported directly to the superintendent. The Chicago & North Western organized a medical service in 1865 comprised of a superintending surgeon and assistants, all of whom were salaried. No contributions were required

Table 6.1. **Railroad Medical Organizations, circa World War I**

	Number of Carriers	Total Employment	Average Employment	Physicians per Thousand Workers
No organization	8	15,988	1,999	0
Hospital	41	483,009	11,500	24.7
Other medical	108	871,871	7,999	19.7
Mutual benefit	26	577,053	22,194	12.3
Total	183	1,947,921	10,529	18.7

Source: Author's calculations from data presented in U.S. Railroad Administration, *Survey and Recommendations of the Committee on Health and Medical Relief* (Washington, 1920).

of employees and the plan provided care only to injured (not sick) employees. Most of these programs compensated local doctors on a per case basis at about half the going rate, and provided free passes.[8]

The Problem of Income Security

To understand the development of the third form of railroad medical service—the beneficial organization—requires some background, for such plans emerged in part out of the failure of other institutional arrangements to provide income security to injured workers. Compensation for injuries and fatalities could take two forms: preaccident wage premiums or postaccident insurance payments. As chapter 4 noted, while others have found higher wage premiums, my estimates suggest that each point increase of the fatality (injury) rate increased wages by 0.13 percent (0.05%). Thus, in 1903, for example, when trainmen ran death risks of eight and injury risks of 101 per thousand workers, the annual wage premium for a hypothetical brakeman earning $630 a year would have amounted to about $40.[9]

Yet this hypothetical $40 (more for higher-paid, more risky jobs) did not come with a label saying "save me in case of injury," and in fact few workers had adequate savings to fall back on. A Michigan survey of about 9,200 railroad workers in 1893 found that only 2,792 had any savings at all and these families averaged $168, or a bit over three months' average pay. A 1900 study of 222 Kansas railroad workers revealed some improvement; 74 percent had some savings, which averaged $287, or between four and five months' income. In addition, half these workers owned a house with an average value of $724.[10]

While passengers injured in an accident could usually count on a settlement from the company, legal liability rules largely precluded most workers or others from winning damage suits. Moreover, no one could sue for minor injuries because even if the case were won the payoff would not justify the legal costs. Ohio data on the amounts paid by railroads for casualties over a sixteen-year period reveal that the average compensation paid to employees reported killed or injured amounted to $132. Such figures contain an upward bias due to the potentially large number of injured who were not reported and likely received nothing.[11]

Other sources yield even more modest estimates of worker compensation. The Boston & Maine reported 149 workers injured in Massachusetts in 1890 and the first half of 1891. Of these, thirty-one received some compensation averaging $101 each and so the average for all 149 workers was $21. The New York, Providence & Boston reported fourteen casualties in 1891, two of whom were killed and none of whom received any compensation. These low averages implied that legal liability provided little security for workers. Yet the averages concealed occasional cases in which workers won large sums. Paradoxically, then, the liability system contributed to the income insecurity of both workers and their employers. As a result, both had strong incentives to contract out of the system.

With compensation unlikely and with many workers having little or no savings, a serious accident might mean catastrophe. The companies' initial response to such situations involved ad hoc payments to workers and sometimes others that they injured or killed. Again, such payments reflected humanitarian concerns, but they were also good business, for they might also reduce labor turnover and avoid large lawsuits. But their generosity was no doubt influenced by the expected value of court awards. In the twentieth century, after workers' compensation required payment for many minor injuries suffered during intrastate commerce, the Pennsylvania used such figures as a benchmark for its offers to workers injured in interstate commerce as well.[12]

The Reading Railroad's records for the 1870s reveal a constant stream of petitions from injured workers for assistance, and most received small sums. On November 19, 1874, Elizabeth Russell wrote the company about her brother, George, who had lost a foot some fourteen months earlier, asking "if you would not assist me a little." She got $50 (a bit over a month's average earnings). Two years later, brakeman Frederick Marty, who had permanently injured his arm in the line of duty, pleaded with the company: "I am in want of some help as we have but little to eat and [I] am totally moneyless." He got $25. Wooten also routinely paid small sums to individuals injured trespassing or at grade crossings—no doubt as an investment in avoiding lawsuits. Yet if Wooten was careful with the company's money it was from duty not parsimony, for he also endowed and partially equipped at a cost of $25,000 of his own money a wing of the Reading, Pennsylvania, hospital.[13]

The Reading Railroad was by no means unique, for by 1900 it was "common practice" among railroads to provide payments to some of the injured. In 1897, when W. W. Frisbee, an engineman on the Santa Fe, was killed in a collision, although not liable the company paid his widow $7,000. Most payments were much smaller, however. In 1898 E. P. Delahay wrote Pennsylvania Railroad President A. J. Cassatt requesting financial assistance, claiming that he had gone blind from a fall while working on the railroad. The company thought the blindness a result of cataracts but sent him $25 anyway. Delahay continued to request and receive funds from Cassatt and later President James McCrea over the next decade, and company records reveal dozens of similar stories. Many companies also continued paying the wages of men who were temporarily injured, while some provided employment to the widow or eldest son of a man killed on the line. In the 1880s the Connecticut River Railroad routinely paid the wages of individuals who lost time due to injury, sometimes for several months at a time, while the New York Central's policy was to employ them at half pay. Many companies also provided employment to individuals with permanent but partial disability. Thus while Frederick Marty received only $25 from the Reading, he was also reemployed as a watchman at reduced pay. The Santa Fe typically tried to reemploy injured workers; in 1883 when brakeman Frank Willard lost a leg, the company hired him back in the freight office. Records of the Southern Pacific from the 1880s also reveal efforts to reemploy injured individuals.

The Fitchburg Railroad reported that "when a train employee is injured . . . it is our custom to pay his doctor's bill, to pay in part or in whole for his lost time, and to give him work suited to his condition." The New London Northern, the New York, Providence & Boston, and the New Haven all maintained wages for those temporarily disabled and rehired many employees with permanent partial disabilities. Such policies were compassionate, and they also reduced labor turnover. But more fundamentally, like the more formal settlements, both the policies and their magnitude were indirect consequences of the employers' liability system.[14]

The growth of more formal company benefit programs was stimulated by the expansion of mutual insurance provided by the major railroad unions. The Brotherhood of Locomotive Engineers was the first railroad union to begin such a plan in 1867. These programs arose because of private market failures that made accident or life insurance unavailable or too expensive. While private companies had begun to offer accident insurance in the 1850s, growth was slow, as they were beset by problems of adverse selection and moral hazard, and many companies shunned high-risk occupations such as railroad work. Industrial life insurance appeared after 1875, but these were small policies that covered little more than burial expenses. Thus even if my hypothetical brakeman had wished to buy commercial insurance with his $40 wage premium, he might have found it difficult to do so.

The Locomotive Firemen and Enginemen followed the Engineers in 1876, while the Conductors and Trainmen instituted benefits programs in 1882. Several smaller unions followed later. Membership in all the programs was mandatory for union members who met qualifications, but they covered only death and permanent disability, leaving the decision to cover temporary injuries up to their locals. As union membership grew rapidly during the 1890s, membership in these programs expanded from about 47 percent to 64 percent of those in covered occupations (Table 6.2). In 1900 annual full-time earnings ranged from about $600 for trainmen to $1,100 for engineers and so death benefits for those insured probably averaged about two to three years' income. However, membership in these unions only constituted about 12 percent of railroad employment and so most workers would not have had coverage.[15]

Railroad Benefit Plans

The Baltimore & Ohio Railroad instituted the third form of railroad medical service—the mutual benefit plan—in 1880 at the instigation of an employees' petition. The company's receptiveness to the petition was probably enhanced by memories of the bitter strike of 1877 and a desire to

Table 6.2. **Membership in Union Mutual Benefit Societies, 1892 and 1900**

	Conductors	Enginemen	Firemen	Trainmen	Total***
1892					
Employment	26,042	35,466*	37,747	68,732	167,987
Membership	9,942	18,739*	25,967	24,131	78,779
Average benefit		$2,247	$2195	$1,500	$1,828
1900					
Employment	29,957	42,837	44,130	74,274	191,198
Membership	20,415	26,424	35,801	40,500	123,140
Average benefit		$1,953	$2,199	$1,377	$1,231**

Sources: Employment from ICC, *Statistics of Railways,* various years; membership and benefits from Emory Johnson, "Brotherhood Relief and Insurance of Railway Employees," Bureau of Labor Statistics *Bulletin* 17 (Washington, 1897), and Samuel Lindsey, "Railway Employees in the United States," Bureau of Labor Statistics *Bulletin* 37 (Washington, 1901).

*For 1894.

**For 1908.

***Excludes a small number of members in the Brotherhood of Railway Carmen and Brotherhood of Railway Trackmen.

undermine the union benefit plans. It provided the company a number of other advantages as well. Lawsuits were reduced, while workers largely paid for their own medical care.

On the B&O workers were assessed a monthly fee according to a sliding scale that reflected their income and risk. For example, those operating trains and earning less than $35 a month paid $1.00 a month; other workers with similar incomes, but less risky jobs, paid $0.75. For this payment employees were entitled to both relief features and medical care for both injuries and illnesses. The relief provided a payment of $0.50 a day (about 38% of an average day's pay in 1880) for six months for injuries received while in service, or one year for sickness. Those permanently disabled received $0.25 a day, while fatalities resulted in a lump sum payment of $500. Higher-paid employees who contributed more received correspondingly more. While current employees were given a choice of whether or not to join, participation for new workers was mandatory. The company provided offices, management, and various business services to the association, but most of its income came from the membership.

A staff of seven physicians administered the medical features of the B&O plan. Their primary duties were to examine those applying for benefits and scrutinize medical bills submitted by contract physicians. Beginning in 1884 the B&O also gave medical examinations to all new employees. Members were entitled to free medical care from several hundred physicians along the route of the B&O (or, upon application, from a physician of their choice). Discounts on artificial limbs were also available. The association also contracted with hospitals in major cities. Members received free medical care but paid their own board, which typically ran about $2.50 a week. As the association pointed out, such care and benefits were a far better deal than workers could obtain by buying private accident insurance, but there was a catch: anyone who sued the company over an injury claim automatically forfeited all benefits.[16]

Not surprisingly, several other carriers followed the B&O model and by World War I relief associations covered about 30 percent of all railroad workers (Table 6.1), usually on large carriers. Most other company relief associations were similar to the B&O plan. The Pennsylvania, which had been studying the actuarial aspects of such a program for a decade, finally began a voluntary relief association in 1886. As discussed in chapter 3, one apparent motive was to reduce labor mobility; another was to reduce the lure of unions. Although the company continued its separate system of contract surgeons, the relief association employed doctors as medical examiners who scrutinized claims and performed other functions similar to those on the B&O. When their number was increased in 1892 the company noted that "this increase has . . . contributed to limit the expenditures of the fund by a very much greater amount than the additional cost." In addition, the additional examiners could undertake "frequent visits to disabled members with consequent improvement in the promptness of returning to duty."[17]

The Burlington, which had had a surgical department since 1870, instituted a mutual benefit association in 1889, apparently in response to a bitter strike the year before and the company's dissatisfaction with more informal arrangements. The Reading plan began in 1888. Its motive may have been to reduce the attractions of labor unions. The company hired the Pinkerton detective agency to discover the membership costs and benefit structure of the insurance features provided by railroad labor organizations. While that program was supposedly mandatory only for new employees, one superintendent candidly told his superior, "we will find a way to get them out with out discharging those who persist in not joining." The Reading plan differed from the B&O model as it initially employed no staff physicians, leaving the determination of benefits up to family doctors. Predictably, as the head of the fund noted in 1889, they paid "all benefits claimed," and he warned, "in time such a system will exhaust our funds."[18]

Thus, by the 1890s most of the major carriers had or were developing some form of medical or benefit organization. Many of these organizations soon evolved into large-scale bureaucracies with considerable budgets that provided services for thousands of patients and employment to hundreds of physicians. Despite the rapid growth of organizations that apparently involved contract practice or group hospitalization, they seem to have escaped the wrath of the American Medical Association (AMA). There were several reasons for this. First, most railroad medical organizations employed a chief surgeon, thereby placing control over medical judgments in the hands of physicians, while most doctors employed under contact were usually compensated in the traditional fee-for-service manner. While the fees were lower than those typical in private practice, they yielded few complaints. The lack of AMA opposition to the railroad plans may also have reflected the calculation that they were too big to tangle with: as of World War I, nearly 14,000 physicians (about 10% of the U.S. total) worked full- or part-time for railroad medical organizations.[19]

In addition to the company-sponsored and -managed plans, there were also employee-sponsored relief associations attached to individual carriers. Some carriers also subsidized or otherwise encouraged employees to purchase accident or life insurance. The Illinois Central and Texas Pacific obtained favorable rates from insurance companies for their employees, while in the 1890s, the Alton and the Union Pacific began to subsidize life insurance purchases by employees.[20]

Taken together these various company, union, and other income security programs provided very modest relief to incapacitated workers and their families. A Montana survey of nearly four hundred railroad workers in 1893 provides glimpses of the security some of these various programs provided. Even though the survey oversampled highly paid, unionized trainmen, only 30 percent of workers had any form of insurance against temporary accidents and that averaged $10 a week. Sixty-five percent of workers had some form of death benefit and for that group the average was $3,058, which equaled about 3.3 years' income. Not surprisingly, benefits rose with individuals' incomes and with membership in unions and beneficial societies. Surprisingly, 34 percent of workers also had life insurance, but workers reduced holdings as they got increased union benefits, suggesting that the net effect of union programs on income security was less than their gross impact.[21]

Thus, income security seems to have been elusive. While a few highly paid employees who were brotherhood members might have several benefit policies, others were less fortunate. Only a minority of workers had protection against temporary injuries, except company decisions to maintain wages during the period of recuperation. Insurance against death or catastrophic injury even for the highly paid trainmen rarely covered more than two or three years of wages, and in the Michigan survey cited above, only 60 percent of workers had any insurance at all. Of course, some of these men would benefit from company policies to find work for those permanently but partially disabled. Still, the fate of many injured workers or their families was to receive good medical care only to fall into penury upon recovery. Indeed, such results seem an inevitable outcome of nineteenth-century railroading, for high risks implied expensive insurance that was out of reach for those with low incomes.

The Professionalization and Practice of Railroad Surgery

As occurred in engineering, late-nineteenth-century medicine saw a blossoming medical network of professional societies organized geographically or by specialty, and with so many railroad doctors sharing common experiences, organizations inevitably followed. The first of these were on a company-wide basis. In late 1881, Dr. J. T. Woods, chief surgeon of the Wabash, called a meeting of that company's physicians for January 1882 in Decatur, Illinois. About twenty-two attendees heard two papers and voted to form the Surgical Society of the Wabash. Later that same year, seventeen Pennsylvania Railroad surgeons met, and they too formed a society, as did the surgeons of the B&O, Big Four, Erie, Milwaukee, Santa Fe, and Southern. There were also state societies of railway surgeons in Iowa, Texas, New York, and Ohio. At their 1887 meeting the Pennsylvania society decided to send a circular inviting all railroad surgeons to a national meeting in Chicago in June 1888. About two hundred physicians attended this, the first meeting of the National Association of Railway Surgeons.[22]

The organization grew rapidly, attaining nearly a thousand dues-paying members by January 1891. That year the *Railway Age* began a regular column on railway surgery edited by Dr. R. Harvey Reed, a charter member of the association. In 1894, Reed and some others, having failed to transform the association into a delegatory body, left to form the rival American Academy of Railway Surgeons, whose membership was limited and, the founders hoped, more distinguished. The two organizations co-existed for a decade until 1904, when they merged to form the American Association of Railway Surgeons. In 1894, the Medico Legal Society also formed a department of railway surgery. That same year saw the first issue of *Railway Surgeon,* which had begun as the official organ of the National Association but which also carried news and papers from the American Academy and from local societies. These organizations flourished until about World War I; thereafter they rapidly declined, to be replaced by broader-based national associations. Yet for about two decades, in these journals and at their annual meetings, railway surgeons carved out their professional self-definition, shared medical advice, and debated their relationship to patients and their employers, the railroads.[23]

Railroad doctors realized that their work had to pass the test of business profitability and they needed to advertise that fact to top management. Many of the papers at national meetings and articles in *Railway Surgeon* reflected this theme. In an 1895 survey of railway medical organizations, Warren Bell Outten, chief surgeon of the Missouri Pacific, advanced two main claims. First, a hospital or benefit association "relieves the railway company of paying doctors, board, drugs, funeral and other bills [for employees]." Second, surgical associations, especially those with their own hospitals, would reduce claim costs. This was because good care improved the "mental welfare" of the injured. The goal was to create a caring culture that would reduce the likelihood of lawsuits.

In addition, Outten pointed out that the "union of the hospital department with the claim department . . . enables the accumulation of a truthful history of any personal injury to employees, passengers, and others," thereby preventing "the patient from changing his statement under the scheming advice of shysters." In short, company medical organizations were the key to accurate medical information. Outten claimed the Missouri Pacific settled claims for from $14 to $36 each, which he compared to one carrier where claim costs averaged nearly $600 each. Many other writers made similar points. J. F. Pritchard, chief surgeon of the Milwaukee, thought such a medical organization also improved workers' loyalty. He called a surgical department "as much of a necessity . . . [as] the legal or claim department," arguing that "no road is so poor or so small that it cannot afford the surgeon." For all of these reasons the profession continued to grow and, as Table 6.1 reveals, by World War I virtually all the carriers had some form of medical organization.[24]

The Physician in the Corporate Hierarchy

To perform its proper functions a railway medical system required the proper structure. Outten and those physicians employed by carriers with hospital departments stressed the importance of company hospitals to improve patients' "mental welfare." Physicians from carriers without hospitals disagreed, of course, but all concurred that, as Pritchard put it, "an ideal surgical department . . . should be organized with a competent surgeon as its head," rather than simply having local doctors report to the company superintendent. Pritchard emphasized that "he should be given charge of the department fully," and should be responsible for hiring and firing, contracts with hospitals, and medical examinations.[25]

This concern was one of professional authority and autonomy, for without a chief surgeon nonprofessionals would make medical judgments, as occurred on the Reading. Just as bridge engineers had a professional interest in construing bridge failures as the result of unscientific engineering, so doctors warned companies of the dangers of allowing nonprofessionals to meddle in medical care. To make their case, railway surgeons argued that professionalism would enhance efficiency and be an antidote to opportunistic behavior of contract surgeons and patients. They warned that nonprofessionals might hire incompetent doctors whose improper procedures and unimpressive courtroom performance would cost the company dearly. Writing in *Railway Age* in 1893, R. Harvey Reed warned railroad leaders that "to leave the selection of surgeons to non-professional men; men who are not able to judge the scientific qualifications of surgeons . . . may mean thousands of dollars to the company." When the Lake Shore & Michigan Southern was forced to pay a $32,000 award to a double amputee, Reed attributed this result to their recent decision to dismiss the chief surgeon and rely on local doctors only, thereby returning to "'the age of mollusks' in railway surgery." However, he concluded, "through the educational efforts of the National Association of Railway Surgeons, these wrongs are rapidly becoming righted." By World War I, 86 percent of those carriers with medical organizations employed a chief surgeon.[26]

Thus the ideology of the railway surgical associations held that all railroads should have medical organizations. These should be under the command of a chief surgeon with broad powers and responsible only to

the general manager or superintendent. To the physicians such professionalism promised employment, income, and respect, while to companies it
offered a solution to the problems of contracting, monitoring, and control
that might otherwise plague their medical organizations. Such concerns
seem far removed from the provision of care to the injured, and indeed
they raise a question that excited much discussion among railroad physicians: how could one honorably serve two potentially conflicting masters—patients and profits?

Of course, other professionals faced this question as well. For engineers the answer was that "good engineering" needed to be economic. The
surgeons developed a similar explanation: by serving Science, they would
best serve both patients and profits. R. Harvey Reed bluntly asserted that
"the associations of railway surgeons are wholly scientific." By taking care
of the maimed they "are also serving the best interest of the company," he
announced. Outten saw nothing wrong in providing "complete and perfect" information on injuries to his employer, for thereby he would prevent
fraud. He saw himself as nonpartisan, for he also claimed to advise patients
not to settle claims "if at all uncertain as regards the future." Outten believed "that I am viewed by the employees more as their friend than any
man in the system. All the surgeons are."[27]

As James Mohr has shown, physicians became increasingly embat-
tled as expert witnesses throughout the nineteenth century, and probably
few faced greater conflicts between professional ideals and corporate ob-
ligations than railroad doctors. Testimony in injury lawsuits routinely put
railway surgeons in what Clark Bell, longtime editor of the *Medico-Legal
Journal,* termed a "delicate [professional] position," requiring that they
maintain their objectivity in legal proceedings to which their employer
was a party. Bell condemned the pressure companies sometimes placed
on their doctors to act in effect as claim agents. "Nothing is more calculated to bring the honorable profession of the Railway Surgeon into public
obloquy," he thought. Others agreed. Dr. C. K. Cole, chief surgeon of the
Great Northern, told Bell that there was "too great a tendency of the Railway Surgeon to feel . . . that he must be actively partisan." Cole concluded
that the function of the railway surgeon was "to give clear medical testimony, without prejudice or partiality." Despite such pressures, or perhaps
because of them, Bell was greeted with applause when he told the National
Association that "after an experience of a quarter of a century, I . . . cannot
recall . . . a single instance when the best judgment and opinion of the surgeon has swerved a hair's breadth because of his employment by a railroad
company."[28]

Of course, the claims of impartiality by Bell and Outten were self-
serving. It is impossible to assess how many railroad doctors shaded their
testimony or gave less than "complete and perfect" advice to their patients
but the many physicians who complained of employer pressure suggest
that such events were more common than either Bell or Outten wished to
admit. Alexander Cochran, general solicitor of the Missouri Pacific, de-

scribed the "close, friendly association" that he had with Outten's Medical Association, and it is hard to believe that such associations did not affect physicians' testimony. In addition, the U.S. Railroad Administration survey discovered that about one-fifth of carrier medical organizations reported to the legal or claims department.[29]

The Practice of Railroad Surgery

The primary duty of railroad doctors was to care for the enormous volume of patients that came through their doors. In 1896, the Missouri Pacific treated a total of 29,109 cases. By contrast, seven years later Massachusetts General Hospital saw 36,000 patients. Most patients who saw railroad physicians did not require hospitalization, but even among outpatients many ailments were far from trivial. In 1880 the B&O relief organization treated nearly 2,500 cases out of a total membership of 14,000. It reported them under ninety-three separate injury and illness classifications plus death. The single most common category was "intermittent and remittent fevers and kindred malarial disorders," accounting for 489 total cases. Such a distribution of cases raises the question of why railroad doctors always used the term *surgeon* to refer to themselves and their organizations. Perhaps it reflected the profession's kinship with the Marine Hospital Service or the greater status and pay that surgeons received compared to other specialists. Surgery was also what differentiated railway medicine from most other physicians' practice, for the second most common cause classification on the B&O was "crush[ed] limbs" (184). There were also 85 fractures and 15 amputations. Similarly, of the 788 cases admitted to the Central Pacific hospital in 1882, 158 involved crushed limbs while 22 required amputations.[30]

There are hints that the care a patient received from a railroad medical system might have been quite good. By World War I, when there were about 144 physicians per hundred thousand of population, railway physicians per employee averaged about thirteen times as high (Table 6.1). Companies certainly had strong incentives to provide good medical care to injured passengers, for they were liable for any injury that resulted from unskilled medical care. Liability considerations shaped railroad surgery in other ways. Working through their medical network, surgeons constantly exhorted their brethren to practice "conservative" surgery—implying that amputation should only be undertaken as a last resort. While conservatism was justified by humanitarianism, the president of the Association of Southern Railway Physicians explained another motive: "an empty sleeve or a pair of crutches is a stronger appeal to the jury in a suit for damages than can be made by the most eloquent contingent-fee lawyer."[31]

Dr. Thomas W. Huntington, a graduate of Harvard Medical School, joined the Central Pacific's surgical staff in 1882 and became chief surgeon in 1885. Under his direction the CP hospital established the first antiseptic operating room on the West Coast. Huntington was able to cut fatalities in major surgery from 30 to 40 percent down to 6 to 7 percent. The Gulf,

Colorado & Santa Fe was similarly progressive. Roentgen published his famous findings in December 1895. As Joel Howell has argued, the rapid diffusion of X-ray technology resulted in part from its widespread publicity in the medical literature. The societies of railroad surgeons and their journal were among the groups that spread this and other important developments. In February 1896 an editorial in *Railway Surgeon* on "Professor Roentgen's Discovery" concluded, "the art of shadowgraphy is of immense interest to us as surgeons." In 1897 the Santa Fe's Temple, Texas, hospital purchased an X-ray machine for $525.[32]

Because of their volume of business, railroad medical services were also able to provide patients access to a considerable range of specialists. In 1915 the little Gulf, Colorado & Santa Fe Hospital in Temple boasted a pathologist, cyctoscopist, anesthetist, and eye, ear, nose, and throat specialist. In 1921 the 250-bed Southern Pacific hospital in San Francisco had 115 physicians and surgeons on its staff, including twenty specialists. Exclusive of visiting staff there was one employee for every two patients—as the company pointed out, a ratio that was usually reached only in private, for-profit hospitals. On the visiting staff there were seven physicians and surgeons, one oculist, two aurists, one urologist, one dermatologist, a neurologist, an orthopedic surgeon, and a radiologist. At about the same time, the Santa Fe bragged that its Los Angeles hospital was ranked Class A, certified for intern training, by the American College of Surgeons and employed the "foremost roentgenologist" in the city.[33]

What workers thought of the care they received is unclear. Many historians have argued that the welfare capitalism of the 1920s received little employee support. Certainly the railroad brotherhoods viewed the benefit plans as efforts at union busting, and they opposed the provisions requiring workers to waive their right to sue as a condition for accepting benefits. Many workers also feared and detested physical examinations that might cost them a job. Unions and probably their members also disliked the plans' compulsory features.[34]

Yet worker response could not have been entirely negative. For one thing both the B&O and Santa Fe plans grew out of employee petitions. When the C&O established its hospital plan, representatives of both management and the Brotherhood of Locomotive Engineers traveled to inspect the Wabash medical organization as a model. Moreover, most claims on benefit funds were for illnesses and minor injuries, for which there would have been no legal recourse. As a Burlington executive perceptively noted, this was a contractual right the men must have valued highly, for they were not faced with the choice of begging for a company handout or perhaps suing and potentially losing their job. The Northern Pacific, Southern Pacific, Santa Fe, and other relief plans also involved workers in their management. Perhaps their motive was to fend off threats of unions, for in 1915, after the brotherhoods lobbied the Texas legislature to take over the hospital associations on that state's carriers, Texas railroads quickly added employee representatives to the boards.[35]

The Problem of Fraud

While the primary duty of railroad doctors was coping with casualties, they also had to be on the lookout for inflated bills and faked injuries by individuals trying to defraud the carriers—a function similar to that performed by the carriers' bridge engineers. How common such opportunistic behavior may have been is difficult to ascertain, for railway physicians had an interest in inflating the problem. Its reduction was one of the major benefits they claimed would result from establishing a surgical service. Yet the number of convention papers devoted to the problem, along with much other evidence, suggests that attempted fraud was widespread.

Fraud came in many forms, ranging from slightly inflated bills presented by line physicians, to overstated disabilities, to wholly bogus claims for nonexistent injuries. W. H. Elliott, chief surgeon of the Central Railroad of Georgia, noted that one of his duties was to guard the company's interests whenever an outside surgeon was called in. In 1895, *Railway Surgeon* carried the cautionary tale of the famous Flying Freemans, a family of slip and fall artists that had made a career out of bamboozling doctors and swindling railroads. The case came to light on December 24, 1894, when Mrs. Mary Freeman visited the superintendent of the Rock Island Railroad, claiming that six days earlier, her daughter Fanny had slipped on a banana peel in one of the company's cars and was now permanently paralyzed from the waist down. Mrs. Freeman suggested that a settlement might avoid a lawsuit. Although company doctors—who had stuck pins into her legs—believed the story, claim agent Stuart Wade was suspicious. He put a detective on the case and sent out a circular describing the claim to other area railroads.

Wade discovered that Jennie Freeman, Mary's other daughter, had been injured on the Chicago City Railroad on January 9, 1894. She too was ostensibly paralyzed from the waist down, but had settled with the company for the bargain price of $500. Even more eye-opening was that by June 28, 1894, Jennie had recovered sufficiently to ride the Illinois Central, slip on a banana peel, and become paralyzed all over again. Again, she had gulled the company physician and settled for $200. Only three months later, the deadly banana peel struck again, this time on the Chicago Street Railway, where it felled and paralyzed Mrs. Freeman, who settled for a mere $100.

Concerned that the banana epidemic might prove nationwide, Wade sent his circular to carriers in other cities. Sure enough, he discovered that Boston bananas had felled both Fanny and Jennie—one on a street railway and one on the Boston & Maine Railroad. Thanks to Wade's detective work, the Rock Island soon put the Freemans out of business. But their message to both railroad doctors and management was clear. As *Railway Surgeon* pointed out, the Freemans had duped "learned physicians and sharp witted claim agents." The physician's eternal vigilance seemed to be the price of corporate solvency.[36]

The most common form of dubious claim arose not from wholly

contrived accidents, as with the Freemans, but as a result of real train wrecks that almost invariably yielded allegations of vague, amorphous, or faked injuries. The most celebrated of such claims was "railway spine." First clearly described in 1866 by the famous English physician John Eric Erichsen, railway spine purported to be a syndrome that involved secondary degeneration without any gross lesion to the spine. Erichsen believed that while the lesion might be extremely mild, the disease nevertheless had a physiological origin and he largely dismissed the problem of bogus claims. Later writers, including especially Dr. Herbert Page, thought that railroad spine had psychological origins deriving from the terror of the accident. Page also noted that in such disorders exaggeration was common and need not be consciously faked, thereby blurring the distinction between illness and fraud. Thus the characteristics of train accidents that led the media to amplify risks and the public to dread them also created injuries that might not have resulted from less terrifying experiences.

Whether railway spine was a faked or real injury was hotly debated, but perhaps not surprisingly most railway physicians were skeptical. Dr. Robert Cowan termed it "Erichsen's spook," while R. Harvey Reed referred to it contemptuously as "the 'spook' that haunts every railroad company." Other railroad doctors acknowledged Page's claim that some cases of railroad spine might be a psychological disorder, but they constructed the disease in a manner suggesting that only railroad doctors held its cure. Dr. Arthur Dean Bevan spoke on the topic to the National Association in 1895. Bevan was then a decade into his long career as a distinguished surgeon that would carry him to the presidency of the AMA in 1918. Bevan divided such cases into two categories. The first group included real injuries of the spine that had real lesions, and alleged injuries that were in fact psychoses induced by fright, shock, and medical suggestion. The second group included both wholly bogus claims and errors of diagnosis.[37]

Bevan pointed out that postmortem examinations had never revealed any evidence of spinal injury in the absence of gross lesion. He described his own experience caring for the injured from one wreck. The crash injured a total of 123 individuals. It also yielded a crop of claims for hernia, bunions, blindness, and gray hair. There were in addition twenty-four cases of railway spine supported by physicians. Bevan noted that among the twenty-four, "the cases that did not recover as soon as the scare was over were without exception in the hands of men who regarded them as cases of railroad spine . . . These physicians were more responsible for the condition of the patients than was the accident." By contrast, those under his care soon recovered. One of these had "no evidence of external injury but great fright [and] complained of frightful pains in the back and inability to move the lower limbs." Bevan described his bedside technique: "I picked him up in my arms and put him on his feet and told him he was a fool, that he was simply frightened and not hurt." He had the man walking in ten minutes, he claimed, "and he is most grateful to me for the way I handled him."[38]

Bevan's message was clear. To fellow doctors he asserted that railway spine as a psychological malady could be easily cured with a little common sense by railroad physicians. To employers Bevan demonstrated the need for railway medical departments, suggesting that only company doctors would be properly motivated to "cure" such psychoses.

Preventive Medicine and Public Health

Some railroads implemented physical examinations for employees and other programs to prevent injury or illness to workers or passengers. Given the existence of a medical organization, such programs benefited from economies of scope. They yielded cost savings to medical or benefit programs, reduced accidents and claims, and lowered labor turnover. Yet in matters such as car sanitation railway medical men were often followers, and they employed new understandings of disease transmission to deflect public criticism. While some company medical departments pioneered in first aid work, many physicians felt such programs diminished their professional standing by encouraging amateurs to practice medicine.

The problem of fraudulent claims for preexisting injury or illness led the B&O Relief Association in 1884 to become the first railroad to venture into preventive medicine when it instituted a preemployment physical examination for all employees. As the association explained, the motive was "to strengthen your Relief fund." When manufacturers instituted physicals a generation later in response to compensation insurance, they seem to have been rather superficial. By contrast, because the railroad hospital and benefit plans covered a range of illnesses as well as injuries, their physicals may have been more thorough. On the B&O physicals initially rejected about 7 percent of all applicants. Like the B&O relief plan, both the Santa Fe and Northern Pacific employed physicals not only to screen for injuries but also to exclude individuals with venereal diseases and such chronic diseases as tuberculosis, Bright's disease, and cancer. By World War I, carriers with about 26 percent of employees required physicals for all employees (Table 6.2) and most of them were in mutual benefit societies.[39]

A second motive impelling the railroads to institute physicals was their need to prevent accidents. Such examinations were usually confined to operating employees, and initially tested only for vision and hearing (chapter 3). Most railroad doctors strongly favored such examinations, but their application created a political firestorm for the carriers and legislators and led to sharp challenges to physicians' competence and professionalism.

In both Connecticut and Massachusetts Harvard-educated ophthalmologist B. Joy Jeffries led the campaign to test railroad employees for color blindness. Along with virtually all other physicians, Jeffries advocated tests administered by medical men employing Holmgren's method of matching skeins of colored yarn. When first administered such tests might find 3 to 5 percent of employees to be color blind, many of whom had long, accident-free service records. The result was an outpouring of complaints by the men who, with broad support from the public and often from their supervisors, objected to the "theoretical tests" and urged more "practical" examinations of whether men could distinguish signals. Physicians reacted strongly to the use of what Dr. William Carmalt, who tested railroad workers for Connecticut, termed "flags and lanterns" tests. When that state's Board of Health bowed to public pressure and allowed such examinations, Carmalt penned a hot letter to its superintendent. "I do not care to and will not if I can help it submit to the indignity of having my professional opinion questioned and *contradicted* by an ignorant . . . committee," he announced.[40]

Both the Massachusetts and Ohio laws also allowed examination by nonexperts using nonstandardized examinations, leading Jeffries and others to denounce them. The ophthalmologist Dr. William Thomson, who administered the tests to employees of the Pennsylvania and the Reading Railroads, finally discovered a compromise that suited all parties. Thomson rejected use of Holmgren's test as too time-consuming, but he developed an ingenious modification of it that could be administered by laymen. Those who failed the test were then sent to a physician for reexamination, thereby preserving the principle that only medical men should make medical decisions and satisfying critics such as Jeffries. Thompson also developed considerable skill in demonstrating the utility of his tests to skeptical employees. On the Reading, which had an employees' committee, he was able to show the committee that one engineer who had been able to distinguish a red from a green flag could not distinguish red, green, white, and blue flags even when he held them in his hands. Tests for color blindness gradually spread. In 1902 ninety-three of 128 railroads employed the tests and in 1910 *Railway Age* termed them "well nigh universal."[41]

As the idea spread that companies were responsible for seeing that their employees were physically fit to perform jobs involving public safety, examinations for operating personnel became more widespread and more comprehensive. At the behest of its claim agent, Ralph Richards, the Chicago & North Western began physicals for operating personnel in 1895 in an effort to reduce accidents. The company tested not only for vision and hearing but also for other "defects" such as hernia, loss of fingers, and heart and lung problems that might prevent a worker from doing his duty, and rejected about 13 percent of all applicants. General physicals were instituted on the Burlington sometime before 1897. The exams reduced both claims against the relief fund and the accident rate. By 1900 the ARA had a standard physical examination, and examinations for operating employees were widespread. As Table 6.3 demonstrates, by World War I, many carriers required periodic reexaminations.[42]

The first glimmers of interest in public health date from the 1880s with the efforts of the B&O Relief Association to protect its relief fund. In 1882 the association's secretary, Dr. W. T. Barnard, reported that the medical inspectors were exercising a "rigid sanitary supervision" over the company's properties. Beginning in the fall of 1880 they distributed antimalarial remedies, claiming that "through such precautions large sums have been saved to your treasury and much suffering among members has been averted." In the winter of 1881–82 a smallpox epidemic led the association to administer 12,000 free vaccinations to employees and immediate families and as a result only two members died. The next year Barnard reported that the inspectors focused on the "safety and ventilation of . . . shops, the conditions of water-closets [and] the character of water used for drinking purposes."[43]

Other carriers also began to address matters of sanitation and disease prevention. The Pennsylvania Relief Association also provided free vaccinations to workers and families during outbreaks of smallpox. One writer reported that in 1885, the Northern Pacific medical service attended to sanitation "in a general way," and fifteen years later, when that company instituted a mutual benefit plan, it too included free smallpox vaccinations to employees.[44]

In the 1880s, public health advocates began to focus on passenger car sanitation and ventilation, and a steady stream of papers on such topics began to appear at meetings of the American Public Health Association (APHA), which also formed a committee on car sanitation. The Pennsylvania Railroad seems to have pioneered in the matter of sanitation, issuing its first sanitation circular in 1885 and adopting a disinfectant prepared by its chemist, Charles B. Dudley. A circular issued in 1889 dealt with requirements for cleaning and disinfecting cars and stations as well as care of the water supply. Some railroad doctors agreed on the need for better car ventilation, sanitation, and quarantining of tubercular passengers. According to R. Harvey Reed, fresher air and better ventilation in passenger cars "would save the lives of thousands of passengers annually."[45]

As Nancy Tomes argues, such concerns reflected a marriage of the older sanitary science with the newer germ theory of disease. Railroad cars were notoriously uncomfortable; in the summer they were stiflingly hot, while in winter before the advent of steam heat, they were bitterly cold except for a small radius around the stove, which was hot and stuffy. It was widely believed that such "bad air" was unhealthy as well as unpleasant. Moreover, discovery that germs transmitted illness encouraged physicians to see railroad cars as steam-powered traveling disease vectors. Initial theories of "fomite infection" suggested that hidden dangers might lurk in dusts and furniture. Dr. John N. Hurty, a distinguished public health expert and physician on the Big Four, explained to the APHA that he had seen a consumptive wipe his mouth with the draperies in a railroad coach and he worried about little children who drooled on the upholstery. "Who knows if the diphtheria bacillus is in that spittle?" he wondered. Others worried that the practice of dumping human wastes on the roadbed could transmit disease.[46]

By the 1890s, car sanitation had become a major concern of both railroad physicians and public health officials. In 1896 *Railway Surgeon* devoted an entire issue to the topic, and in 1901 the APHA circularized railroads and state boards of health, urging a variety of sanitary measures. These included better ventilation, the use of smooth surfaces that would be easier to clean, and an end to the practice of flushing toilets along the road, among other things. In response to such pressures the major carriers began to enforce standard rules governing sanitation, and a survey of thirty-one large carriers revealed that most were making modest efforts to improve ventilation and sanitation. As Hurty noted: "unless the railroads do take hold of it [car sanitation] voluntarily . . . the first thing you know they will have some legislation on the subject."[47]

Hurty should have known, for he also served on the Indiana State Board of Health and under his pressure the Big Four had become an early convert to better sanitation. Every six months it fumigated cars with a gas containing formaldehyde and routinely wiped down seats and window ledges in an effort to prevent propagation of diphtheria, and it analyzed drinking water every six months. The Illinois Central also followed similar procedures. In 1899, the Pennsylvania created a division of bacteriological chemistry and began to examine drinking water. That company also disinfected cars, and its physicians inspected company labor camps and the quarters of food service employees.[48]

In the twentieth century as understanding of disease transmission evolved, concern with car sanitation and ventilation diminished. In an important paper to the APHA in 1905, the Pennsylvania's Charles B. Dudley discussed these matters, focusing on tuberculosis. Dudley cited evidence that dried sputum, subject to the action of air and light, was an unlikely source of disease. He also noted that if the tröpschen theory (which held that transmission of tuberculosis resulted from direct contact with the sputum of a diseased person) was accepted, the railroads could not be held responsible for the spread of tuberculosis for it was impossible to prevent consumptives from traveling. Thomas Crowder, chief surgeon for Pullman, also cited scientific literature to downplay worries that railroad cars were an important source of disease transmission. Paradoxically, then, railroad scientists and medical officers were able to employ modern understanding of disease to minimize some concerns with sanitation.[49]

In 1910 a committee of the American Association of Railway Surgeons reported similar views, noting that "the day is past . . . [when disinfecting railway cars] is to be looked upon as a matter of the first importance." Despite such skepticism, the committee stressed "the idea that railway surgeons and Surgical Departments should assume the great duty of promoting the sanitary and hygienic improvement of railway conditions." By World War I carriers with about two-thirds of all railroad employment performed sanitary inspections (Table 6.2).[50]

Beginning about 1913, the carriers began to work closely with the U.S. Public Health Service Committee on Sanitation of Public Conveyances, a group that contained several railroad surgeons, to develop the Standard Railway Sanitary Code. During World War I this work was taken over by the Committee on Health and Medical Relief of the U.S. Railroad

Administration. At least four of the five members of that committee were also railroad surgeons. The code was first published in 1920 and was widely adopted by states during that decade.[51]

As understanding of disease transmission evolved, concern with railway car sanitation increasingly focused on drinking water and ice. Most carriers took on water from any available source and those like the Pennsylvania that paid serious attention to its purity were the exception. In response to the APHA survey in 1899 only two reported using filtered water and only one (the Pennsylvania) performed any testing. In 1914 an investigation by the Minnesota Board of Health condemned thirty-two of the sixty-one sources of railroad water it analyzed.[52]

The most important work of the U.S. Public Health Service in the field of railroad sanitation was to regulate the purity of railroad ice and water supplies. In 1914 the surgeon general issued standards requiring all common carriers to provide water samples to state boards of health for testing. These standards were incorporated into the Standard Sanitary Code, thereby covering intrastate as well as interstate travel. The Norfolk & Western responded to these requirements by establishing a testing laboratory with a bacteriologist and chemist and building purifying plants. By 1923 states had certified about 60 percent of railroad water supplies. One unexpected result of the process was "a coincident improvement in the water supplies of . . . local communities, which in turn serve as an incentive to . . . improve the supplies of other contiguous local [communities]."[53]

As noted, some carriers provided smallpox vaccinations as early as the 1880s. By World War I carriers with nearly 30 percent of all employees required them (Table 6.2). In this century, carriers employing about one-third of all workers began to provide free typhoid vaccinations to potentially exposed employees, while the Rock Island also began a campaign to combat hookworm. By the 1930s the Southern Pacific was giving a TB test to every employee who entered its hospitals and to the families of those found to be infected. Such programs were concentrated in railroads with hospital or beneficial societies, presumably because they reduced costs, but they also benefited the carriers by diminishing labor turnover and perhaps compensation costs. In other industries only a few large employers instituted comparable programs, one of the most comprehensive being that of Tennessee Coal & Iron, a U.S. Steel subsidiary.[54]

Beginning about World War I, carriers employing 28 percent of all workers also became active in antimalarial work (Table 6.3). The most ambitious as well as the best-documented program was that of the Cotton Belt. In 1916 that company's chairman of the board, Ernest Gould, visited its hospital in Texarkana, Texas, and discovered that one-third of the employees admitted had malaria. Like other carriers at the time, the Cotton Belt was experiencing a labor shortage, and as a result the company instituted a sanitary engineering department in July 1917. The company then launched a four-part program that included draining and oiling mosquito breeding grounds, screening of quarters, quinine prophylaxis, and education, draw-

Table 6.3. **Percentage of Railroad Employees with Access to Selected Medical Services, by Type of Program, circa World War I**

	Hospital Association	Mutual Benefit Association	Other Medical Association	Total
First aid training	51	78	63	65
Physical exam (any employees)	90	59	85	78
Physical exam (all employees)	2	41	2	26
Periodic physical (any employees)	64	92	79	79
Smallpox vaccination	52	6	32	29
Typhoid vaccination	55	47	14	34
Malaria prevention*	43	31	19	28
Sanitary inspection	56	82	59	65

Source: See Table 6.1. Figures are percentage of railroad employees working for carriers offering the program, although the program may not be offered to all employees of the carrier.
* Any antimalarial program, such as screening, drainage, and so on.

ing on the aid of the U.S. Public Health Service, local communities, and other businesses. In 1919 it equipped an exhibit car designed to educate workers and others on the cause and prevention of malaria. Workers with malaria were given a quinine treatment and provided blood examination at the company's laboratory. By the mid-1920s it was distributing nearly 75,000 doses of quinine a year. By such methods it cut the hospital malaria admission rate for section workers from 215 to two per thousand.[55]

Employee first aid training began in 1888, when the Lehigh Valley placed kits with written instructions in trains and stations. In 1892 its chief surgeon, W. E. Estes, explained his motives. He had performed numerous amputations on victims of railroad accidents and as a result concluded that surgical shock was not simply a psychological response to the terror of the accident, but rather resulted from loss of blood. Estes reported that his fatality rate after major amputations declined from 7.89 percent before use of first aid to 3.67 percent after its use. At about the same time, Estes privately wrote A. A. McLeod, vice president and general manager of the Reading, describing his results. Estes claimed the policy "has resulted in saving many lives." He noted that at the hospital he saw injury cases from both the Lehigh Valley and the Reading, which had no first aid kits, and "the contrast between the condition of the injured . . . is so great that I am persuaded to appeal to you." The Reading, however, did nothing.[56]

Other carriers were similarly laggard until 1898, when Charles Dickson, a Canadian member of the St John Ambulance Association, gave a paper on "First Aid Classes for Railway Men" to the Association of Railway Surgeons. A committee on first aid that included Dickson was appointed at the next annual meeting and in 1901 the association endorsed a manual on first aid that Dickson had authored and he was asked to begin a first aid department in the *Railway Surgeon.*[57]

Clearly Dickson had struck a responsive chord for at about this time a number of carriers finally began serious first aid activities, and a few even purchased hospital cars. Under the guidance of Chief Surgeon A. F. Jonas, the Union Pacific began first aid work about 1899. Jonas described the program as a form of self-help, or philanthropy. "The surgeon gives freely of his stock of knowledge with the expectation of no other reward than to simplify the subsequent treatment and to shorten the duration of the unfortunate's illness," he explained. As this justification suggests, surgeons who supported first aid saw it as a complement of their own activities in attending to the injured. This vision, in turn, reflected the new understanding of antiseptic and aseptic methods. Commenting on a paper by Charles Dickson, Jonas noted that men invariably provided first aid whether trained or not, and that untrained men would often dress a wound by sticking a cud of tobacco on it and wrapping it in a filthy bandana. "We want to teach these men what not to do and not infect a wound before proper aid can be rendered," he stressed. Other surgeons concurred. In short, the first duty in first aid, as in other medicine, was to do no harm.[58]

In this view such programs enhanced both the stature and value of railway medical associations because they benefited both the injured and the company. Yet if some railway physicians embraced first aid, to many others it was anathema. In the discussion of a paper by Dickson presented to the Association of Railway Surgeons in 1902, W. B. Outten of the Missouri Pacific claimed that "it is very difficult to teach these ignorant men very much about rendering First Aid as it ought to be rendered." Yet the real complaint of those who opposed first aid training was not that the men were uneducable but just the reverse: that rather than being a complement to the surgeon as Jonas imagined, their efforts would be a substitute. Thus, Outten went on to claim that "those who have no hospital department in connection with their railroads are naturally friends of first aid." Another physician, Dr. James H. Ford, worried that "we should not permit them to have the idea enter their heads that they are to displace the local surgeons." Dr. W. A. McCandless also worried about economics. "I believe much of the work entrusted to railway employees in rendering First Aid should be assigned to physicians. They are poor and they need the money for rendering such service," he candidly observed. That such worries were not entirely fantastic was revealed some years later in 1916, when evaluation of the Pennsylvania's first aid program disclosed that the men were doing far more than simply providing first aid and were in some cases even prescribing and administering drugs.[59]

These deep divisions probably help explain the highly uneven character of the early first aid programs. At one end of the spectrum a few carriers such as the Union Pacific provided detailed instruction and "hands on" experience in applying what workers learned, and tried to maintain their skills and interest in various ways. Yet most programs initially offered no training and consisted only of placing a first aid kit on the train. That of the Buffalo, Rochester & Pittsburgh came complete with a manual containing helpful hints on how to stop hemorrhage from an artery. One imagines a harried trainman, in the chaos following a major accident, carefully thumbing through the instructions as his hapless patient bleeds to death. Conditions slowly improved. In 1911, the Red Cross began to offer first aid training to railroad employees and by World War I carriers with about two-thirds of all employees offered first aid instruction, although the programs would later languish (chapter 10).[60]

By World War I, nearly all large railroads boasted some form of medical/ benefit association, and while the railway surgeons' professional organizations declined rapidly in the 1920s, company associations had attained a form they would maintain with only minor changes for a half century. They provided care to their nearly two million employees, about 55 percent of them in hospital or mutual benefit associations, and to passengers and others injured by the new technology. The associations represent an important and neglected example of institutions and organizations that were innovated in part as a response to safety requirements. They emerged from a confluence of high accident rates that characterized early railroading, combined with workers' concerns with security and employers' worries over unions, labor turnover, and accident liability. The programs do not fit easily into the accepted explanation of welfare capitalism. They neither arose out of a crisis in labor relations nor declined as rail unions became more powerful after 1900.

Both the forms the medical associations assumed and the activities they undertook reflected a combination of legal and economic circumstances, internal logic, and the professional aspirations of organized railway surgeons. Hospital associations reflected the need to expand the boundaries of the firm in circumstances where the medical market was as yet undeveloped. In all cases, railroad medical organizations were key sources of information to reduce companies' medical and liability costs and help them control the labor force.

The tensions between physicians' professional aspirations and their relations with the claims department also powerfully shaped railway medicine, and led to much professional agonizing among railroad doctors. They also justified professional demands to place surgeons at the head of medical organizations as well as requests for the latest technology, and they encouraged "conservative" surgery and a peculiar construction of railway spine.

Railroad medical programs pioneered in the private provision of preventive medicine and public health programs. The spread of such programs reflected economic and regulatory incentives and physicians' professional concerns. Employee physicals were among the earliest preventive care programs, and their near universality among operating employees reveals companies' concerns with accident prevention. The carriers with universal physical examinations typically ran hospital associations or benefit plans, and they were also more likely to offer antityphoid or smallpox vaccinations. But companies also valued the private, labor market benefits these programs yielded. As regulatory pressures began to force the carriers to develop expertise in transportation sanitation, railroad doctors used their growing understanding of disease transmission to dampen concern with passenger car sanitation, and guide public policy toward the more serious problem of polluted water. First aid training also became common, encouraged by the spread of ideas on antiseptic treatment. Yet first aid programs also revealed how professional concerns could shape public health programs, for supporters claimed that it enhanced their professional standing while opponents saw them as a potential threat.

Safety Crisis and Safety First, 1900–1920

We do not try very hard to find out whether our trainmen obey their instructions.
—Braman B. Adams, 1902

It is undeniable that collisions . . . could be reduced to an exceedingly small number by the efficient management of block signals.
—Interstate Commerce Commission, 1907

The first two decades of the twentieth century were a time of transition for the railroads. The expansion that had begun in 1828 continued up to about World War I. While track miles grew slowly, the period witnessed a massive reconstruction and upgrading of both roadbed and equipment, increase in traffic density, and rapid expansion of freight and passenger carriage. These developments reduced the social costs of railroad transport, for fatalities per unit of output continued to decline, as they had since the 1880s (Appendix 2). Still, the expansion carried accidents and casualties in the 1907–10 period to levels not seen for nearly two decades, sparking unprecedented public criticism of railroad safety. Congress responded with an avalanche of proposed legislation that increased federal oversight of safety and threatened more far-reaching controls. For a brief period from 1918 to 1920 the federal government ran the railroads, with important consequences for their safety.

The carriers responded to this worsening safety and regulatory environment in both traditional and novel ways. The technical network continued to improve and diffuse safety technology, while the carriers made major investments in track and signaling that improved communication and control. They increasingly coordinated their public position on safety matters, employing the ARA and the newly formed Bureau of Railway Economics and Special Committee on Relations of Railways to Legislation to speak for them. Here as elsewhere centralization in government promoted centralization in the private sector. Of most importance, in a novel response to their agency problems, the carriers also inaugurated organized safety work about 1910. Together these developments resulted in a kind of informal safety compact. Except for especially high-profile dangers, Congress and the ICC would delegate safety matters to the carriers—but on the tacit condition that they would improve safety on their own.

Railway Slaughters, 1900–1910

Despite the carriers' investments in signals, brakes, couplers, and other safety equipment, the economic upswing that began in 1897 sharply increased accident rates. Worker fatality rates jumped one-third, from 0.67 in 1897 to as high as 0.92 per million manhours in 1907 while passenger fatalities shot up about 60 percent, from 14 to 22 per billion passenger miles (Figure 7.1). In 1901 Congress passed the Accident Reports Act, and under its authority the ICC required reports of all train accidents involving injury or loss of $150 or over. The new statistics were to exert a powerful influence on public perception of railroad safety. Derailments, which had steadily risen after 1897, more than doubled from 3,633 in 1902 to 7,432 in 1907, a disturbing number of them from broken rails. One, the Southern *Fast Mail,* which jumped the track and fell 75 feet from a trestle, killing nine crew on September 27, 1903, was immortalized as "The Wreck of the Old '97." In August 1904 a bridge collapse derailed a passenger train near Eden, Colorado, and drowned eighty-eight people, making it the worst American railroad disaster to date.[1]

Collisions also skyrocketed, from 5,042 in 1902 to 8,026 by 1907. While many were minor, between October 1901 and the end of June 1906, the ICC recorded 448 "Class A" collisions, which included all those that killed passengers or resulted in $10,000 in damages. Most of these reflected ancient causes. The New York Central's *Twentieth Century Limited* was perhaps *the* crack passenger train in the country, and it ran on some of the best track. But on June 21, 1905, it wrecked when it ran through an open facing point switch that had no distant signal. Head-on collisions, as usual, reflected

Figure 7.1. **Passenger and Worker Fatality Rates, 1890–1920**

Source: Appendix 2.

some sort of mistake with a train order. Thus Southern Train 12 heading east and Train 15 heading west normally met and passed at Hodges, Tennessee, but on September 24, 1904, a train order scheduled the meet at New Market, 5 miles to the west. The crew of No. 15 forgot the order and ran through New Market to collide with Train 12, killing sixty-three.[2]

More disturbing, however, was the increasing number of collisions that resulted from a failure of the block system, as occurred on January 27, 1903, on the Central Railroad near Westfield, New Jersey. Engineman Davis, young, intelligent, and with a record of six years of good service, was having trouble with his engine and became preoccupied with getting the injector to feed water to the boiler. As a result he failed to see a distant signal, a crossing watchman swinging a lantern, a home signal, and a flagman, and crashed into the rear of a waiting suburban train, killing himself and twenty-two others. Even block signals, it seemed, were bedeviled by agency problems. Perhaps more unsettling than the wreck was the subsequent discovery that the Central had no periodic examinations or tests of enginemen.[3]

The Carriers under Attack

As the death toll mounted after 1900, the carriers' safety record increasingly came under public attack. Such criticism was by no means new, of course, for headlines such as "Another Railway Horror" had sold newspapers for decades. But the breadth of the attack grew with the rising number of popular periodicals, and if anything their vituperation increased. Many writers, it seemed, had entirely lost faith in railroad management, "Who is the Murderer?" wondered the *Washington Post,* as it reviewed railroad casualties for 1902–4. In "Slaughter on Railroads," the *Chicago Daily News* voiced a common theme, bitterly asserting that passengers and trainmen were "being butchered day by day" because railroads "find it cheaper to kill than not to kill." *World's Work* described a "Harvest of Death," while the *Chicago Tribune* compared railroad accidents to the battle of Gettysburg. Many stories contrasted American railroads unfavorably with foreign lines, as did *Scientific American* in "American Railway Slaughters and British Railway Safety."[4]

Thus, while railroads were safer for their passengers and workers in 1900 than a generation earlier, public criticism was far greater. Such stories amplified risk perceptions, but they built on individual experience. In 1902 there were 1,228 passenger trains in accidents serious enough to be reported and probably several thousand more that were unreported. Since the average train carried forty-four passengers, 54,032 people were in reported train accidents that year, and most of them lived to talk about it. Such figures imply that every town in the country with more than 277 people might well have had someone who had been in a train accident within the past five years and could—and no doubt did—talk of its terrors.[5]

In addition, injured individuals seemed less fatalistic, and as Randolph Bergstrom has demonstrated for New York City, lawsuits were increasing. In the South the carriers faced a rising tide of personal injury suits and the companies' success rate was declining. The standards had changed, and the railroads themselves were at least partly to blame. Their critics had been emboldened by the Safety Appliance Act of 1893. Passed in the teeth of carrier opposition, or so they remembered, it had not only sharply improved safety but had benefited the carriers as well. The lessons were clear: the critics knew more about railroading than did the managers, and requiring specific technologies could force safety improvements. More broadly, the powerful demonstration that new technology *could* reduce accidents created the moral imperative that it *should* do so.[6]

Nearly every writer called for laws to require extension of the block system. The *Chicago Tribune* also claimed that overwork of trainmen was a major cause of accidents. Writers such as Frank Haigh Dixon in the *Atlantic Monthly* called for adoption of the "British system of accident investigation." At about this time Edwin A. Moseley, secretary of the ICC and its longtime spokesman, urged the same proposals. The commission thought that railroad safety was out of control, *Railway Age* concluded.[7]

Public figures also began to clamor for greater regulation. In 1904, and in subsequent State of the Union messages, President Theodore Roosevelt called on Congress to legislate block signals, limitation of hours, and accident investigation. Representative John J. Esch of Wisconsin advocated a similar legislative agenda in the *North American Review*. Esch was a member of the House Commerce Committee; he would later write the Transportation Act of 1920 and become an ICC commissioner in 1921. After reviewing the beneficial effects of the Safety Appliance Act, Esch urged hours limits, steel cars, and mandatory block signaling. "Without the most rigid . . . discipline," Esch acknowledged, "the best appliances will not insure absolute safety." Such perfection, he would later argue, could come only with automatic train control (a device that can automatically stop the train if it passes a stop signal).[8]

The railroad brotherhoods, especially the Trainmen and the Locomotive Engineers, joined the clamor for improved safety. They had grown immensely in power and influence since the 1890s, and had been deeply impressed by the impact of the Safety Appliance Act. Yet despite its good results the fatality rate for trainmen was 7.98 per thousand in 1907—one-third higher than in 1896 and nearly equal to the bad old days of 1889. What was needed was some more of that old-time religion, and labor began to press a broad safety agenda that overlapped that of reformers.

The Campaigns for Safety Legislation

Legislation to regulate hours of work had been a goal of the Trainmen since the 1890s. While their concern had more to do with the arduousness of work than its safety, long hours did cause some accidents. In 1902 a collision in North Dakota involved an engineer who had been on duty 23 hours—which he termed "not a very long time." In another, on the Seaboard, on January 20, 1906, a freight crew that had been on duty 25 hours and 32 minutes fell asleep on a sidetrack waiting for another train to pass. When they woke up, thinking it had passed, they pulled out. The ensuing collision killed four men. The issue became more pressing after 1900 with the growth of very heavy, slow freights that led to long, grueling days. Congressional legislation thus represents an early example of a safety rule that in fact had multiple motives.[9]

In 1906 Wisconsin's Senator Robert LaFollette and Representative Esch introduced a law to limit trainmen to 16 hours per day and telegraphers to 9 hours. It was enacted on March 4, 1907—almost fourteen years to the day after the passage of the original safety appliance bill. Although when the bill passed the Telegraphers termed it a great victory, the brotherhoods' enthusiasm was mixed, as the bill limited incomes because it limited hours. But the ICC strongly supported it. Secretary Edwin Moseley presented a list of accidents he claimed had been due to long hours of service, and he dismissed railroad opposition, noting that they had also fought the Safety Appliance Act. Although the railroad press did not oppose the bill, the *Gazette* expressed skepticism that it would do much good. The railroads, however, objected strongly and while their arguments fell on deaf ears they raised issues that were endemic to safety regulation.[10]

Restricting the hours of trainmen necessitated some alternative such as raising train speed or employing more men. Thus the alternatives were not tired versus rested trainmen, they were tired crews versus rested crews with faster trains, or less experienced crews. A similar problem occurred with the law restricting telegraphers' hours. Companies complained that it raised costs because it would require more men and drive up wages, and the result would impede progress of the block system. In fact, its results were complex, for there were a number of good substitutes for the existing approach. The law did indeed shut down an "exceedingly large number" of small telegraph offices as the carriers predicted, thus lengthening manual blocks, and it led some carriers to hire a third man who was "notably deficient in experience." Both alternatives worsened safety. It raised labor costs 50 percent on the Illinois Central manual block system, and so the company abandoned it for the older, more dangerous train order system. But it also encouraged use of automatic blocking and led to rapid growth in the use of telephones, both of which improved safety.

Limiting workers' hours also suffered from two other drawbacks. It focused on only a small part of the safety problem and did nothing to encourage a broader interest in safety. In fact, the safety consequences of hours restriction were extremely modest. The most expansive claims were those by Robert LaFollette, who presented a list of 219 collisions and derailments occurring between July 1, 1901, and September 30, 1906, in which employees had worked more than 15 hours. These yielded ninety-three worker and passenger fatalities, or 1.8 percent of the total—the most that might be prevented if all of these were caused by overwork and if the act were 100 percent effective.[11]

In 1907 the brotherhoods achieved a law requiring that locomotives be equipped with self-cleaning ashpans. The Brotherhood of Locomotive Engineers was the driving force behind state and later federal legislation to require electric arc headlights on locomotives. Unions also campaigned for "full crew" laws that typically required three brakemen to a freight train. Although urged as a safety measure, in reality such laws were efforts to prevent the carriers from achieving the labor-saving consequences of the air brake. Laws to limit train length probably had similar motives. The carriers formed the Bureau of Railway Economics in 1910 to present their side of these matters, which promptly began to generate pamphlets opposing both types of legislation. While twenty states enacted full crew laws by 1915, and one (Arizona) limited train length, the carriers managed to stave off both at the federal level.[12]

In 1908 union pressures finally brought about a federal employers' liability law. It abolished the fellow servant doctrine, which companies had often used to avoid accident liability, and delimited certain other defenses. Although the act received little publicity, it was probably the most important piece of federal safety regulation, for by raising the cost of casualties it helped stimulate a broad and continuing interest by the carriers in reducing all forms of work accidents that affected other aspects of safety.

In 1910 Congress also passed a second Accident Reports Act. This one required reporting of injuries and fatalities (not just train accidents) to the ICC and it finally authorized federal investigation of train accidents—a goal of critics such as Charles Francis Adams and the *Railroad Gazette* since the 1870s. It was an important milestone, for the subsequent investigations would provide the ICC ammunition supporting its later campaigns to improve rails and require automatic train control (chapter 9). That same year carriers also finally paid the price for failing to develop and adhere to strict MCB rules governing a host of details relating to freight car safety appliances when Congress gave the ICC power to govern such matters. The commission promptly proposed detailed, potentially expensive regulations (the ratchet wheel on hand brakes might have no fewer than fourteen teeth, for example). Even worse, the carriers felt, were union-inspired proposals for federal control over railway clearances. One bill would have cost $716 million if it had passed, the railroads estimated, and while federal legislation failed, by 1914, eleven states had regulated clearances either by statute or through their railroad commissions.[13]

The Brotherhood of Locomotive Engineers had pressed for government inspection of locomotives as early as 1904, and the next year New York instituted state inspection. Several bills to regulate locomotive safety were introduced into Congress in 1909 that were both highly detailed and expensive, and the carriers fought them vigorously, working through the Special Committee on Relations of Railways to Legislation. A compromise bill requiring carrier inspection under ICC supervision passed in 1910 and inspections began the next year.[14]

Block signals were far bigger game and this cause was championed not by labor but rather by other reformers, the ICC, and the trade press. The *Railroad Gazette* and its rivals had been urging an extension of the block system since the 1880s, and by 1901 the carriers had blocked about 35,000 miles, or roughly 16 percent of main track. But the failure of the block system to perform as it did in Britain troubled its advocates, for it raised anew the agency problems that had plagued the carriers from the beginning. In the nineteenth century engineers had analyzed the differences between British and American railroad technology, but as the technology converged after 1900 their focus shifted to the labor market. The British engineer H. Raynar Wilson argued that in Britain, since jobs were scarcer, turnover was less and dismissal a greater threat than in the United States. This led British carriers to supply more thorough training and it raised the penalty for disobedience.[15]

Wilson's argument suggested that the agency problems of American railroads were unsolvable, but most railroad men were more optimistic and believed that American workers could learn to behave as safely as their British counterparts. The *Engineering News* explained that British enginemen had been brought up under the block system, which made obedience a fixed habit, while the experience of most American engineers was with the train order system. Disobeying a signal was, therefore, only returning to an older and more familiar system. Moreover, most railroads employed both sorts of systems, which further eroded respect for signals. So did the widespread use of "permissive blocking," which allowed an engineman to proceed after a stop sign as long as his train was "under control." The *Gazette* denounced permissive blocking, claiming that it effectively "trained" engineers to ignore signals and rely instead on flagmen. The common installation of manual block systems without distant signals encouraged enginemen to run to the home signal at full speed, hoping it would be clear.[16]

In the nineteenth century under Matthias Forney and Arthur Wellington, the *Railroad Gazette* had been a powerful and constructive voice for better railroad safety. In a long signed article in the *Gazette* in 1902, editor Braman B. Adams continued that tradition, discussing the problem of collisions under the title "The Superintendent, the Conductor, and the Engineman." Adams developed the themes that would show up later when he served on the Block Signal and Train Control Board, excoriating the carriers for inadequate training of enginemen and especially for their failure to monitor performance. "We do not try very hard to find out whether our trainmen obey their instructions," he claimed, and urged the carriers to hire more inspectors and introduce "surprise checking" to see if train crews obeyed signals. Thus, problems with the block system were not inherent, even when operated permissively; its greater effectiveness in Britain reflected both company procedures and the greater experience of employees in its use.[17]

In 1903 the ICC first proposed mandatory extension of the block system under its supervision. Representative Esch introduced a similar bill, which went nowhere despite the *Railroad Gazette*'s endorsement, but Esch persisted, routinely reintroducing it each year. In 1906, Congress responded to critics' pressures and passed a joint resolution authorizing the commission to investigate block signals and automatic train controls. To no one's surprise, the report, published in 1907, called for mandatory extension of the block system supervised by the ICC. It also encouraged Congress to fund the Block Signal and Train Control Board.[18]

The board, appointed by the ICC with advice from the ARA, could hardly be accused of antirailroad bias. Chaired by Mortimer Cooley, dean of engineering at Michigan, it also included B. B. Adams of the *Gazette;* Azel Adams, signal engineer on the New York Central; and Frank Ewald, consulting engineer to the Illinois Railroad Commission. Between 1908 and 1912 the board issued four annual reports and a final report that ranged widely over the problems of railroad safety and made many useful suggestions. It noted the increasing number of pension plans, which members hoped would improve labor quality and safety. With remarkable prescience it also criticized the absence of any central authority to deal with safety and the problems of legislation that worked in "piecemeal fashion." It urged adoption of a system similar to the British Board of Trade with broad powers to regulate all facets of safety—something that would not occur until the 1970s. But its reports also reflect a gradual shift in reformers' emphasis away from block signals and toward train control.

Although the board advocated mandatory extension of the block system, its skepticism of railroad management matched that of any muckraker. "The primary cause of the lamentable record of collisions in this country . . . is chargeable to railroad officers and their employees," the board pronounced, and it stressed the poor training and lack of discipline, which it thought resulted from high labor turnover and, as British critics claimed, "the conditions of society in America." "The American is not reared with the discipline which becomes a part of the man and governs his actions mechanically," the board intoned, and it went on to note that these problems were "so serious as not to be susceptible to complete cure by any mechanical means, [but] . . . the automatic stop . . . while it might be unnecessary under different social conditions, might be expected to add to safety under conditions which do exist." In another report the board complained that the carriers had not "devoted the same effort toward the development of automatic train control apparatus that has been devoted to the development of interlocking and block signaling devices." A majority of the board members favored "some central authority . . . [that would] deal with the subject of safety in all its phases." Only B. B. Adams, who was as skeptical of the ICC as he was of railroad managers, dissented from this ambitious program. Like a previous Adams, he urged that "publicity often cures obscure evils which can not well be reached by government agencies directly."[19]

Safety First and the Decline in Collisions

The views of the Block Signal Board, like those of British observers such as H. R. Wilson, implied that the carriers' agency problems were insoluble. But most of the railroads' critics disagreed and they were about to be proved right. In 1908, in yet another blistering article, the *Gazette*'s publisher, W. H. Boardman, placed the blame for accidents squarely on management: "the General Manager has the power to stop most of the loss of 1,000 lives in collisions and derailments," he bluntly asserted. Boardman urged better inspection, supervision, and discipline. Two years later the Chicago & North Western decided to try a novel means to achieve these ends. It accepted the proposal from its claim agent Ralph Richards that he implement a campaign he called "Safety First."[20]

In 1910 the railroads killed a staggering 3,383 workers, only 608 of whom died in collisions or derailments. As usual the rest were killed while coupling cars, or connecting air brakes, or working on track, or in a thousand and one other "little accidents," as Richards termed them. The primary focus of Richards's safety activities on the North Western, and of subsequent Safety First campaigns on other carriers, was always to prevent these "little accidents" to workers. But Safety First was also part of a strategy to reduce *all* kinds of accidents. "If the little accidents could be stopped the big ones would take care of themselves thus wiping out the whole accident business on the road," Richards argued. Thus, as the *Railway Age* pointed out, improved worker safety was a means as well as an end. Richards assumed that no appeal to employees to prevent accidents to others would be as effective as an appeal to prevent accidents to themselves, and furthermore, whatever reduced work accidents would reduce other casualties as well.[21]

The *Gazette* termed the North Western's approach "unique," primarily because it included safety committees composed of workmen whose duty it was to discover and report unsafe conditions and practices. This portion of Richards's program was therefore an effort to encourage self-motivated safe behavior because monitoring was difficult. The committees functioned as a voice for operating personnel to communicate dangers across departments and to higher levels and they provided follow-up to ensure a proper response by managers. For example, operating personnel on the New Haven used safety committees to communicate problems with equipment and track to the relevant maintenance people. Workmen's committees were embedded in a safety department that included a system-wide committee composed of the safety supervisor, the general manager, and other top personnel, and division and department committees as well. The safety organization therefore became the focal point to oversee safety work in each operating department.

Douglass North has argued that organizational innovations can reduce the cost of transforming inputs into usable output. George Bradshaw, superintendent of safety on the New York Central, made essentially the same point. "There are so many matters relating solely to safety—matters that do not affect efficiency one way or another—that there is need . . . of a Safety Organization," he claimed. Otherwise safety became simply an "incidental consideration."[22]

The slogan Safety *First* implied that safety was of primary rather than secondary importance. This was both novel and difficult to implement because middle-level managers such as division superintendents and foremen often had a different agenda, which might be termed "Get the Traffic over the Road First." Richards's program was aimed as much at this group as at the men; in modern business terminology, he was trying to reorient the North Western's corporate culture to include safety as well as output as a company goal. One of his mottoes captured the change: "it is better to cause a delay than to cause an accident."

Richards's first work, before he organized safety committees, was to meet with divisional officers, letting them know that he had the backing of top management and selling the idea that time spent on safety would save time reporting accidents and training green men to replace the injured. Another motto made this point: "it takes less time to prevent an accident than it does to make a report of one." On the Pennsylvania, R. H. Newbern, superintendent of the insurance department, also stressed to General Manager S. C. Long "the effect upon operating efficiency" of the company's many injuries. The North Western plan emphasized that safety was a shared responsibility. As a claim agent, Richards was well aware that workers viewed much safety work as a sham largely intended to protect company finances (chapter 4). Accordingly, in a kind of gift-exchange relationship, when safety committees recommended physical improvements that cost money, the North Western implemented them, partly as the price of worker cooperation.[23]

The technical press publicized Richards's activities and Safety First soon spread like wildfire. The enthusiasm reflected new economic realities, for the federal employers' liability act of 1908 was making employee injuries increasingly expensive. Death payments under the new law up to 1910 had averaged about $1,200 (Table 7.1), sharply more than had been common, and other costs were also higher. The spread of safety work reflected the carriers' changed economic circumstances in other ways. The combination of unreliable equipment and rapidly expanding traffic that had induced nineteenth-century railroads to take chances in getting the goods over the road was receding. Richards's slogan "it is better to cause a delay than to cause an accident" might well have fallen upon deaf managerial ears a generation earlier. Facing less difficulty in delivering the goods, managers were more receptive to safety.

But if safety work qualified as a price-induced institutional innovation, political considerations also contributed to its spread among large carriers. In 1912 Frederic Rice of the Burlington and a member of the ARA Committee on Transportation urged the carriers to begin a "campaign for the reduction of accidents." Otherwise, he warned, the carriers would face

Table 7.1. **Worker Injury and Fatality Costs, 1908–1910**

	Average Value of Settlements	Number of Settlements	Average Value of Judgments	Number of Judgments	Average Value, All	Number of Settlements and Judgments
Temporary disability	$70	79,362	$932	251	$73	79,613
Permanent partial disability	$1,296	2,515	$3,515	127	$1,403	2,642
Death	$1,157	5,672	$2,536	276	$1,221	5,948

Source: U.S. 74th Cong., 1st sess., S.D. 68, *Cost of Railroad Employee Accidents, 1932* (Washington, 1935), Table 19.

the extension of "regulation . . . to all the details of operation." The next year the committee tentatively endorsed a North Western-style safety organization. As Charles Francis Adams had expected, public threats finally generated in railroad managers a broad concern with safety. This public commitment, the carriers hoped, would signal that they could be trusted to chart their own course for safety. By 1913, forty-seven carriers with 152,000 miles of road—about 60 percent of the total—had a safety organization like that on the North Western.[24]

Organized safety work complemented other labor policies intended to reduce turnover, improve discipline, and quell unrest. Before 1901 only two carriers had pension plans; between 1901 and 1920, twenty-eight others developed pensions, and by the 1920s they covered about three-fourths of railroad employees. E. P. Ripley of the Santa Fe frankly admitted that the purpose was to maintain employee loyalty in the face of a tight labor market. By diminishing turnover such policies increased firm-specific experience, thereby improving safety. Preemployment physicals and eye tests continued to spread and by 1902 the ARA had a standard physical exam. About 1902 the Queen & Crescent line also introduced stricter employment practices, rejecting men who failed rules examination or who had bad vision, and it reexamined the men periodically. The New York, Chicago & St. Louis tested trainmen at random and several other roads checked to see if men observed signals. The Alton adopted an employment bureau and required signalmen to report failures to observe signals. In 1904 the Erie adopted an employment bureau to centralize hiring and record keeping, which allowed much more careful monitoring of performance. They were followed by the Burlington in 1908, which required physicals for all applicants, and the B&O in 1912, which even required references.[25]

While early Safety First campaigns emphasized workers' voluntary behavior, they contained a subtheme of better discipline, and companies began to increase monitoring and supervision. Many instituted surprise checking, which consisted of setting a trap for the train crew and watching the results. The procedure was later given the more politically palatable term *efficiency testing.* For example, an automatic block signal would be set to stop and the behavior of crew observed. The procedure may have originated with Julius Kruttschnitt, vice president of the Southern Pacific, and by 1907 the Union Pacific and other Harriman lines had adopted it. The SP published results in its *Bulletin,* with names suitably removed. On April 22, 1908, Train 5 exploded a torpedo and stopped, as required, but the flagman was not properly whistled out and the engineer was disciplined. While the brotherhoods opposed the measure, it gradually spread to the North Western, Cincinnati Southern, Big Four, New York Central, New Haven, Lehigh Valley, and Pennsylvania. When W. L. Park left the Harriman lines for the Illinois Central, he introduced the system there.[26]

In response to press stories that stigmatized the carriers as death-traps for passengers, companies also began to publicize their accident investigations. The Pennsylvania made public its findings on the wreck at Mineral Point on February 22, 1907. The Union Pacific and the other Harriman-controlled lines made the most thoroughgoing commitment to candor. In 1907 these companies began to convene boards of inquiry that included employees of the railroad and prestigious outsiders such as merchants, utility executives, retired military officers, and, in one case, the ex-governor of Wyoming, and the company published their findings. In a memo reminiscent of McCallum's work on the Erie in the 1850s, Kruttschnitt emphasized to heads of operating companies "what we desire to insure . . . is to allocate responsibility by name and to publish the findings, that is, THOROUGH INVESTIGATION and PUBLICITY." Results of all investigations were also sent to the general office.

In a talk to the Western Railway Club in 1909, Superintendent W. L. Park, now of the Illinois Central, put the procedures in broader context. He warned of the "growing sentiment that accidents are of too frequent occurrence and unless prevented . . . the strong arm of the government must intervene." To cure the problem required correct diagnosis, he thought, and he admitted that many railroad investigations were simply whitewashes, a result that he feared worsened discipline and corroded respect for authority. Thus, accident publicity, like Safety First, was an effort to encourage safer worker practices, build public trust, and stave off public controls. That the Union Pacific's investigations often exonerated management and blamed worker carelessness was an added publicity bonus.[27]

The Brown system, or record discipline, continued to increase in popularity during these years and by World War I nearly all large carriers used it in some fashion. The Burlington adopted the plan in 1909 to reduce suspensions and therefore the need to hire new men. Around 1914, the Reading and the Long Island also reduced the seasonality of track work to diminish labor turnover, while the latter carrier also raised wages and offered free English classes to its mostly Italian workforce. The primary motives were to improve efficiency and cut costs by stabilizing the force, but safety must also have benefited from the reduction in new hires.[28]

As noted, effective safety work required the carriers to demonstrate organizational commitment, which implied the proper signals from top management. Following the North Western example, companies invested in safety equipment and held mid-level managers accountable. In 1913 the Pennsylvania reported that in two years it had spent $500,000 to comply with safety committee recommendations. In 1915, Robert Lovett was chairman of the executive committee for the Harriman lines, and he kept a careful watch on company safety. After one Kansas wreck, he sharply questioned A. L. Mohler, president of the Union Pacific: "are you satisfied that your Kansas Division had not been asleep much of the time?" he asked.[29]

Block Signals and Better Cars

After 1900, rising traffic density and collisions encouraged major extensions in block signaling. Blocking sharply improved companies' ability to communicate with and potentially control trainmen, while Safety First, surprise checking, and other labor policies helped ensure that agency problems would not nullify its good effects. As electric motors improved and wages of telegraphers rose, automatic signals became increasingly attractive. The spread of central power-generating systems that rapidly cheapened alternating current also encouraged both the shift to automatic signals and a movement away from older battery-powered DC installations. Companies also discovered a range of other benefits from automatic signals. They increased capacity, caught broken rails, decreased overtime, and reduced the need for stops and sidings, which resulted in a substantial return on investment. For such reasons mileage of nonautomatic signals slowly declined after 1915 while total miles of block signaling rose from about 16 percent of main track in 1901 to 30 percent in 1909. By 1922, nearly 132,000 miles, or about 45 percent of all main track, was blocked and 21 percent was automatic.

Such figures understate its importance because they ignore traffic density. Just as market forces allocated steel rail and better car heating to high-density carriers, so lines with high traffic density were most likely to install blocking and presumably they placed it on their high-density routes. Although data on route density are unavailable, I have weighted each carrier's passenger miles by its proportion of track that was blocked to generate an estimate of the proportion of passenger miles traveled under the block system for selected years. The result (Figure 7.2) reveals that by 1909, blocking protected 43 percent of passenger miles (versus 30% of main track) while by 1922 the figure had risen to 71 percent (versus 45% of track) while 33 percent of passenger miles were governed by automatic block signals. Even these calculations must be an undercount, for the percent of passenger miles block signaled on a carrier surely exceeded its percent of track block signaled.[30]

The carriers also improved signaling hardware and practice. Traditionally a white light at night meant clear, while green meant caution. Such colors were not failsafe, for a broken stop signal would show white for clear. After such a false clear caused a wreck on the New Haven in 1898, companies rapidly shifted to yellow for caution and green for clear. Begin-

Figure 7.2. **Track and Passenger Miles Protected by Block Signaling, 1900–1939**

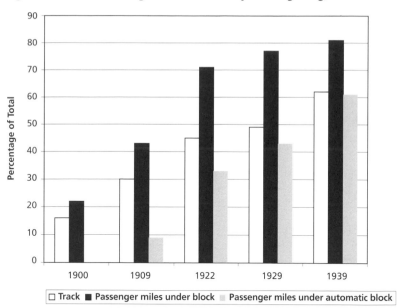

☐ Track ■ Passenger miles under block ▨ Passenger miles under automatic block

ning about 1905, research at Corning Glass into the optics of signal lenses by Dr. William Churchill and others resulted in standardized colors and improved light transmission, and after 1915 signal lights began to replace the semaphore. They were more visible in poor light while various combinations of color and position could be employed to communicate more information such as speed control. Companies also modified practices to make the block system safer. In 1903 the ARA modified rules allowing trains to proceed governed by signals only. While the motive had been efficiency, supporters also claimed that by preventing the misreading or forgetting of orders the change would do away with a fruitful source of collisions. In 1910 the Union Pacific led the way in getting rid of the flagman.[31]

Beginning in about 1902 experimentation with steel passenger cars also proceeded rapidly. The Pennsylvania placed the first major order—for two hundred cars—in 1907. By the end of 1911 about 9 percent of all passenger equipment was steel or steel frame, but these accounted for about 90 percent of new construction. By 1916 about 31 percent of all passenger equipment was steel or steel frame, and of course these protected a disproportionate number of passenger miles.[32]

Experience in the form of numerous wrecks soon led to many modifications in design. Early work had focused on constructing a massive underframe that would withstand the shock of collision. But this led to problems that were starkly revealed by a collision on the Lackawanna on July 4, 1912, near Corning, New York, and another in December that year on the

Milwaukee at Odessa, Minnesota. In the first wreck a steel car demolished wooden equipment, while at Odessa a steel Pullman was telescoped by a dining car with a steel underframe. A year later a derailment on the Oregon Railroad & Navigation Co. near Lakeview, Washington, on May 12, 1913, yielded a similar result. Such accidents led Pullman and other builders to develop collapsible, shock-absorbing end frames (nearly sixty years before the idea was introduced into automobile design) and steel frames to retrofit wooden cars. The Pennsylvania's cars, which had been designed with a stronger superstructure, revealed their value in a high-speed collision at Tyrone, Pennsylvania, on July 30, 1913. Although many were injured only one person was killed because the collision energy was absorbed in crushing the car ends.[33]

Coincident with this shift to steel, casualties per passenger train in collisions and derailments declined even as the number of passengers per train began to rise. Although some of the decline may simply reflect the effects of inflation on the reporting of minor accidents, casualties per accident had declined in the nineteenth century as well (chapter 2), and steel cars helped continue the trend. A statistical analysis suggests that by 1916, this shift to steel may have reduced fatalities per derailment or collision by as much as 43 percent and 12 percent, respectively, although rising train speed may have offset some of this decrease.[34]

After 1900 the carriers also continued the upgrading of air brakes on both passenger and freight equipment that had been going on since 1872. About 1906 Westinghouse released a modified freight brake with a new K-type triple valve. The new brake valve released air from the train pipe at each car, thereby cutting in half the time it took to apply brakes in a service application on long freights. It also featured more uniform release that reduced train shocks and hence collisions from break-in-twos. "It has been applied in large numbers . . . during the past two years," *Railway Age* reported in 1908. While Westinghouse initiated most of these improvements, railroad engineers also began to anticipate future needs. In 1909, at the behest of the Pennsylvania, New York Central, and other large carriers then building steel cars, the MCB and representatives of Westinghouse began to study brake requirements for very large cars weighing over 130,000 pounds. As a result Westinghouse introduced new "PC" brakes that employed much higher train pipe pressures, two brake cylinders (one for service and both for emergency), and two air reservoirs per car. The new brake featured quick release and quick recharge, while dual air reservoirs allowed an emergency stop even after a service operation. Tests on the Lake Shore demonstrated that the new equipment stopped a train traveling 60 miles per hour in about one-third less distance than the high-speed brake. In 1912 the Reading and the Central of New Jersey also began to employ clasp brakes (with two shoes per wheel) on passenger equipment.[35]

The combined effect of improved signals, safer equipment, and the new labor policies was a sharp decline in collisions. In 1907 the number of collisions and associated casualty rates for workers and passengers reached a cyclical peak. Thereafter they all fell steadily except for a brief upsurge during World War I. The number of passenger trains in collision fell 50 percent between 1902–4 and 1914–16 even as passenger train miles rose about 45 percent. On individual carriers the results were sometimes even more spectacular. The rapid growth of block signaling on the Union Pacific from almost nothing in 1905 to over 40 percent of all track in 1908 no doubt reduced collisions. But that company's record reveals that *all* forms of train accident per million locomotive miles fell in half in 1905–6—a result the company attributed to accident publicity and surprise checking. On the Illinois Central train accidents fell one-quarter between 1910 and 1911 even as output rose. Company Vice President W. L. Park attributed the improvement to modifications in record discipline and the introduction of surprise checking and accident publicity. On the Burlington train accidents per million locomotive miles fell from 30.8 in 1906 to 14.5 in 1913—a fact it attributed to better discipline, the block system, and improved equipment. Even the ICC was impressed—if only temporarily. In 1912 it noted that the safety committees held "great possibilities . . . [for] accident prevention."[36]

Regulatory Skirmishes

Even as the carriers were improving safety, their critics were busily proposing new regulations, and so the railroads mounted a propaganda offensive to publicize their good works. In 1908 the Union Pacific began to feature the safety of its block system in passenger advertising, billing itself as the "safe route." The Lackawanna's corporate symbol Miss Phoebe Snow, whose dress of white advertised the cleanliness of the "road of anthracite," began to hymn its safety too.

> Miss Snow draws near the cab to cheer
> The level headed engineer
> Whose watchful sight makes safe her flight
> Upon the Road of Anthracite.

By 1909 the Pennsylvania had hired Ivy Lee as a publicity agent and was planting news stories about its safety. About 1909 the carriers formed the Special Committee on Relations of Railways to Legislation. A loose association of railroad executives, the committee coordinated the carriers' response to proposed legislation and gathered statistics intended to show progress on safety issues. In 1910 the Bureau of Railway Economics also began to present the carriers' side of public issues. In 1912, the *New York American* published a long uncritical interview with Julius Kruttschnitt of the Harriman lines under the headline "Our Railways Are Safest."[37]

The carriers pointed to the decline in collisions after 1907 and marshaled the evidence on their investments in steel cars and the block system to stave off expensive new regulations. The year 1911 saw the first of many bills that would require *all* passenger cars be made of steel. Some focused on new cars only; others would have required replacement of the entire stock within four years, or ten years, or left the matter up to the ICC. But by 1913 when hearings were finally held, the ICC had concluded that the carriers were buying steel cars about as fast as possible and so it no longer pressed for legislation.[38]

The carriers persuasively argued that legislation was either unnecessary or undesirable. Requiring new cars to be made of steel was irrelevant, they claimed, for statistics gathered by the Special Committee demonstrated that in 1913, 97 percent of all new cars were made of steel. To replace all 47,000 wooden passenger cars with steel in four years exceeded the capacity of the car builders. Equally important, it would be expensive, costing about $615 million, the railroads argued. An impressive parade of small carriers claimed that they had an almost infinite number of projects that would make better use of the money. Even a ten-year phase-in would be expensive because passenger cars were very long lived and many carriers had stocked up during the recent boom. The argument impressed even such longtime critics of railroad safety as Representative Esch. He quoted *Railway Age* that steel cars "might absorb so much money . . . as to force neglect of other improvements more important in the interest of safety," and noted that only about 3 percent of all railroad fatalities were passengers.[39]

The carriers also successfully fended off legislation to require the block system. They publicized their investments in signaling while the Special Committee marshaled cost data that led Congress to shy away from legislation. The committee claimed that one bill to require block signals on all passenger lines would have a capital cost of $465 million. In addition, as the reports of the Block Signal Board reflected, the carriers' critics were losing faith in the ability of block signals to deliver safety. In 1910, when the Accident Reports Act finally gave the ICC legal authority to investigate accidents, it began to focus on those collisions that reflected lapses in the block system or other forms of human error.

On July 4, 1912, a Lackawanna engineman ran a stop signal in the fog at Corning, New York, killing thirty-nine, and ten days later a Burlington engineer missed a signal at Western Springs, Illinois, killing thirteen. In its annual report that year the ICC excoriated the carriers for lax discipline and obsolescent rules, warning that the block system "by no means insures immunity from collisions." In 1913 three collisions on the New Haven vividly reinforced the point; they resulted from an antiquated signal system and an organizational structure that even the general manager described as "disorganized . . . bad . . . [and] out of gear." The ICC noted that of the seventy-six train accidents it investigated in 1913, fifty-six resulted from dereliction of duty. Of course, such findings were not the results of a random sample but were rather a reflection of the commission's choice of which accidents to investigate. The report indicted the habitual violation of rules by trainmen and the carriers' lack of supervision over safety matters. It called for federal legislation to standardize rules and ensure proper instruction. The alternative, it suggested, was automatic train control.[40]

Despite the ICC's veiled threats, on the eve of World War I, the carriers' Safety First and related activities had successfully fended off most federal safety regulations. By then the contest between reformers and railroads had resulted in a draw. Congress and the ICC would formally regulate especially high-visibility risks, but most matters, such as rules, equipment, track, and roadbed, would be left up to the carriers, subject to strictures from the authorities, and only if they showed progress in safety. This was the sort of regulation by exhortation that had enamored Adams and other nineteenth-century critics. It was a fragile contract; the brotherhoods routinely pressed for stricter controls while the carriers did not always deliver on safety promises. Nor had they imagined that ICC economic regulations would eventually make it increasingly difficult to do so.

The Emergence of Derailments, 1902–1915

Derailments, long the curse of American carriers, had steadily declined in the late nineteenth century, as chapter 2 detailed. But the economic expansion that began in 1897 reversed this trend, bringing an increase in derailments and casualties in every year from 1902 to 1907, and contributing to public clamor over railroad safety. Thereafter the number of derailments steadily increased, in 1908 surpassing collisions, which were on the decline, and thoroughly alarming the ICC. In 1902 there had been 1,609 derailments from equipment defects, most of them involving freights; by 1920 the total had risen by nearly 600 percent, reaching 11,172. Roadway-related derailments rose an astonishing 855 percent, from 577 to 5,508 during the same period. Yet after 1908 casualty rates from derailments fell, albeit somewhat unevenly.

The likely explanations for this decoupling of derailments and casualties lie in both the real forces affecting train derailments and the ICC's reporting rules. As chapter 2 pointed out, most derailments were very minor affairs, usually involving freights. In the nineteenth century these came by the tens of thousands; they caused few injuries and little damage, and were ignored in the *Railroad Gazette*'s annual compilation. ICC definitions, which excluded any train accident that yielded no casualties and damage less than $150, were also intended to exclude these minor affairs, but price inflation increasingly caught them up. The result provides a good example of how faulty statistical gathering can help create a public issue. I have employed a statistical analysis to estimate the impact of inflation on reporting of derailments and then adjusted the data accordingly. The adjustment (Figure 7.3), although no doubt only approximate, suggests that in the aggregate, derailments adjusted for both inflation and car-miles declined slightly from about 1905 on. The reduction in casualties from derailments came about primarily for two interrelated reasons. The increasingly scientific orientation of the railroad technological community continued to improve the reliability of technology, while economic changes such as the growth of traffic and changes in cost induced a host of investments in better track and roadbed.[41]

Figure 7.3. **Derailments, Adjusted for Inflation and Car Miles, 1902–1920**

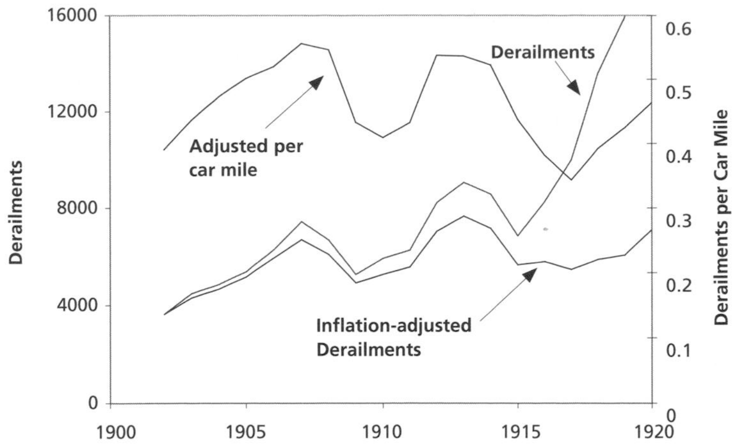

Source: Appendix 2.

Improving Equipment: The Cast Iron Wheel

As use of 50-ton freight cars spread after 1900 and train speeds rose, wheel failures increased sharply, becoming "the subject of some apprehension," as the *Railway Review* observed. About the same time the *Gazette* also noted "rumors afloat" that wheels were performing poorly. In 1904 the "growing frequency of broken flanges" led the Pennsylvania to launch a major investigation of wheel failures that largely confirmed what others were claiming. The breaks were most frequent under heavily loaded hopper cars. The weight caused truck bolsters (the beam on which the car body sits) to sag and led center plates (which allow the truck to turn under the car) to crack, both of which prevented the truck from curving. The company also admitted that men controlled trains by setting hand brakes on only a few cars; not surprisingly, therefore, most problems were discovered at the foot of heavy grades.[42]

Since many failures were broken flanges, in 1904 the MCB responded by increasing flange size, and they raised wheel taper from 1:38 to 1:20 to make wheels track away from the rims. The Pennsylvania implemented these changes and promptly reported a marked reduction in wheel failures. In 1898, Pullman had begun to experiment with antifriction (ball) side bearings, with good results. A report on their use on the Duluth, Missabe & Iron Range to the MCB in 1902 revealed that they reduced both rail and wheel wear and the dangers of derailments from sharp flanges, and their use gradually spread. Flange lubrication yielded similar benefits and its use spread as well.[43]

In 1909, in order to present a united front, the wheel makers incorporated as the Association of Manufacturers of Chilled Car Wheels. It proved to be an important organizational innovation, for the association soon began an advertising campaign to combat inroads from steel wheels and launched an effort to improve MCB specifications. While it may seem strange for a trade association to demand stricter standards for its products, each seller's quality influenced the sales of all. That year at the association's urging, the MCB adopted larger 725-pound wheels for 50-ton cars, and three years later, also at the behest of the wheel makers, it modified its recommended wheel sizes to account for brake pressure. This had been a particular problem for 30-ton refrigerator cars. Because they had a higher ratio of empty to loaded weight, their higher brake pressure overheated and broke the wheel employed on standard 30-ton cars. In later years the association even inspected members' plants to ensure that they maintained standards. One change clearly intended to reduce the incentive freight car interchange created to buy bad wheels was to have them marked with the name of the purchaser. Such identification, it was hoped, would prevent makers from simply reselling a lot that had failed tests.[44]

Finally, both the MCB and the manufacturers developed long-term arrangements with major engineering schools to study wheel design. These contracts reflected a growing awareness of the need for a more thorough understanding of wheel metallurgy. In 1908 the *Railroad Gazette* asked rhetorically "who knows . . . the internal stresses of the wheel? We do not know one single item of what is going on inside that metal," the editor claimed. As with rails (see below), this interest in more systematic research was encouraged by threats of competition and regulation. Steel wheels were declining in price, and some companies began to put them under 50-ton cars. In 1907, after a broken wheel on a Norfolk & Western freight caused a derailment that also wrecked a passenger train, the *Railway Age* called for steel wheels on freights.

Regulatory pressures also encouraged wheel improvements. The Block Signal Board looked into wheel-manufacturing methods, and beginning in 1911 the ICC regularly urged an investigation of the material in car wheels and their stresses in service. As a result of these concerns, in 1914, the MCB arranged for engineering studies at Purdue University to measure stresses in wheels induced by braking. About the same time, the Association of Manufacturers Chilled Car Wheels began a similar arrangement with the University of Illinois and persuaded the Bureau of Standards to take up the problem as well. Collectively this research resulted in numerous design improvements.[45]

These changes contained wheel failures. Although reported wheel failures roughly doubled from 1902 through 1918, about half the increase simply reflected the growth in traffic, for the number of cars (and therefore wheels) rose about 55 percent. Moreover, inflation-adjusted damage per accident fell about 20 percent while both fatalities and injuries also declined, suggesting that the remainder of the increase resulted from changes in reporting. This continued ability of American carriers to use cast iron wheels well into the twentieth century reveals much about the interrelationships between technological progress and increasing safety. Few items are as prosaic as a cast iron wheel, and while the basic form of the wheel of 1920 appeared little changed from that of 1850, it was in fact a far better product, allowing greater productivity and greater safety. The improvements had typically been an outcome of safety problems brought about by the drive for larger cars. Initially they were mostly the result of cut-and-dry empiricism, but in the twentieth century the process of change became far more scientific.

Track and Roadbed

As locomotive weight steadily increased in the early twentieth century, loads on engine driving axles grew to 30 tons or more. Passenger and freight equipment grew in size as steel replaced wood, and as traffic density increased the heavy equipment sharply increased track stresses. Such trains multiplied risks of derailments, for they might spread rails, especially on corners or where ties were weak; or they might break a rail or compress the roadbed, as apparently occurred to the *Twentieth Century Limited,* which derailed near Hyde Park, New York, on March 3, 1912, injuring seventy-two. These developments disturbed technical people; one maintenance engineer complained that track was becoming increasingly overstressed and in 1913 *Railway Age* called for an investigation of track stresses. The next year the American Railway Engineering Association (AREA) formed a committee, headed by Professor Arthur N. Talbot of the University of Illinois, that began a comprehensive investigation of track stresses that was to bear fruit in the 1920s (chapter 9).[46]

The carriers also responded to these challenges by upgrading main line roadbed and track. In 1915 one writer noted the gradual shift from unballasted line to the use of sand, cinder, or especially crushed stone, which had fallen in price due to advances in crushing machinery. Lines with heavy traffic increased the depth of ballast—to 18 inches, in the case of the Pennsylvania. They also broadened roadbeds from 14 or 16 to 18 or 20 feet. Rising hardwood prices and improved methods led to a boom in tie preservation. By 1913, one-fifth of all ties were treated and more plants were in the works. With ties less likely to rot, mechanical destruction became the constraint on their life. A preserved tie might last fifteen to twenty years and outlive several rails but could be destroyed if cut by rails without tie plates or destroyed by spikes. Investments in ties, therefore, induced increased use of tie plates and one survey in 1901 indicated that their use was spreading, while companies also experimented with screw spikes. Longer tie life yielded safety benefits because it reduced dangerous track work. For example, the Lake View derailment on the Oregon Railroad & Navigation Company on May 12, 1913, that killed four and injured seven people resulted because workmen were replacing ties and had not tamped the track.[47]

While all these improvements led to a safer, more substantial track structure, as always, the carriers only undertook those justified by traffic density. Many marginal carriers—most of them in the South and the West—

failed to upgrade or even maintain track, especially on branches. A report on one section of the Fort Worth & Denver City noted that "the track is rough, in bad surface, and a large percentage of the cross ties need renewing." With a few exceptions, "no ballasting has been done." ICC accident investigations unearthed a disturbing number of wrecks that resulted from dreadful track. On the New Orleans & Northeastern on May 6, 1912, a passenger train went off some bad track, killing five. A year later, when the ICC investigated a wreck on the Central of Georgia, it found 56-pound rails laid in 1883, many of which had broken and been patched, and track it termed in "deplorable condition."[48]

Rail Problems

Despite improving track and roadbed, the early years of the twentieth century brought an epidemic of broken rails. "Since 1899 every year . . . has brought a worse rail than the previous one," grumbled W. H. Cushing, the Pennsylvania's chief engineer of maintenance of way, in 1905. On the company's lines east of Pittsburgh, the number of rails removed due to defects rose from 769 in 1902 to 1,332 in 1906. Other companies experienced similar problems, and both accidents and casualties from broken rails rose sharply after 1902, even after adjustment for inflation and car miles. The carriers claimed these failures resulted from "bad steel." Early specifications typically kept phosphorus below 0.08 percent, but low-phosphorus ores were becoming scarce. Shortly after 1900—right after U.S. Steel was formed, as some writers observed—and at a time of sharply increasing demand for rails, the steel makers began to refuse the carriers' specifications and phosphorus rose to 0.1 percent. British observers felt this was far too high, especially in high-carbon steel. The rail makers also finally dropped their five-year warranty on rails, while the rising demand for steel led them to skimp on the amount of steel discarded from the head, leading to piped and segregated "A" rails (the top rail in the ingot).

No carrier escaped the deluge of bad rails. In 1905 the B&O found that 22 percent of one lot of rails developed split or crushed heads within one year. In 1906 an especially bad lot of rails from Cambria Steel provoked A. C. Shand, the Pennsylvania's new chief engineer, to write Cambria's President Powell Stackhouse, urging him to increase discard. When Stackhouse agreed—but only if the price of rails were raised $2.00 a ton—Shand tried to turn the issue away from economics back to quality. "I cannot think it possible that you would be willing to have your company get the name of furnishing rails . . . that did not answer the reasonable requirements of wear or safety," he threatened.[49]

Manufacturers blamed rail failures on rising train weight and speed, and claimed that rails were simply too skimpy for the loads they bore, a conclusion others shared (see below). Thus derailments resulted not from bad steel but carrier penny-pinching. The makers also stressed the role of poorly counterbalanced locomotives and flat wheels. Studies revealed that the impact of a flat wheel was proportionate to the wheel load and the square of speed. *Railway Age* told of one bad wheel on a fast passenger train that broke 200 85-pound rails in 14 miles. The steel men also claimed (correctly) that the standard ASCE section was poorly proportioned for new, larger rails, with too much metal in the head relative to the web and base. This reduced its girder strength and also led to differential cooling and required it to be rolled at too high a temperature, while the thin base was fracture-prone.[50]

During these years several groups in the technological network tried to develop better industry-wide rail standards. First out of the blocks was the newly formed American Society of Testing Materials (ASTM), whose rail committee largely reflected manufacturers' interests. In 1901 it proposed specifications calling for 0.1 percent phosphorus and a drop test once in every five heats. This protocol was less strict than that which many companies had employed a generation before. Thus while the wheel manufacturers who faced interindustry competition had an interest in strict standards, the steel makers who were not threatened by substitute materials did not. Robert Hunt was head of a company that provided rail inspection services to the carriers and had been a respected figure in rail manufacture since the 1880s. Hunt thought the specifications a joke. "You might as well have it [a drop test] once or twice a day," he observed. The British journal *Engineering* referred to the standards as "a tale of unrestricted commercialism on the part of manufacturers" and a "surrender to the steel interests." Thereafter little was heard from the ASTM. The AREA had been formed in 1899 to reflect the increasingly technical nature of track and roadwork, and in 1904–5 it proposed specifications calling for testing an "A" rail from an ingot in each heat. It also included a clause developed by the Pennsylvania to limit the amount of shrinkage that was an effort to prevent rolling at too high a heat. Although these specifications would go through almost continuous modification, they were the first industry-wide step to enforce quality control on rail makers.[51]

At this point, rail failures shouldered collisions off the front page. In 1907 a report of the New York Railroad Commission showed that from January through March there had been 1,331 rail failures on that state's roads in 1905, 816 in 1906, and 3,134 in 1907. The *New York Times* headlined the report "Amazing Increase in Broken Rails," and followed up with several articles adopting the railroads' perspective and blaming the problem on the rail producers' greed. "The manufacturers squirt molten steel upon a railroad right of way, take their pay at $28 a ton, and expect their careless product to serve its purpose," the editor raged. He called for remission of the tariff on rails. *McClures* and *Scientific American* also featured the problem, and hued to the carriers' position. *Outlook* favored "government supervision over every department of transportation," and called for "national action."[52]

Government supervision was the last thing either the carriers or the rail makers wanted. The result was a flurry of public finger pointing as each side tried to pin the tale of greed on the other. The carriers, happy no doubt to have so appealing a villain to blame, concentrated their fire on The Steel Corporation (U.S. Steel), which outdid even the railroads in public disfavor. In a representative piece, the *Railroad Gazette* denounced the "criminal willingness of the Steel Corporation . . . to manufacture rails that cost human life." *Iron Age* responded for the makers, blaming the breaks on "severer service" and suggesting that the carriers were too stingy to buy rails heavy enough to be safe.[53]

While the trade press was fighting the paper war, individual companies and the AREA began to tighten rail quality control. In 1908, the Reading required that all rails be labeled by position in the ingot—and promptly discovered that the "A" rail accounted for about 65 percent of all failures. About the same time, the Great Northern hired an independent company to monitor rail making at Lackawanna Steel and reported a "very decided improvement." The New York Central experimented with adding titanium to the ingot to reduce segregation and claimed the result was more homogeneous metal. The Harriman lines, complaining that U.S. Steel would not make rails with less than 0.1 percent phosphorous, shifted their business to Tennessee Coal and Iron, which could make open hearth rails. Steel companies experimented with ingot size and shape, deseaming processes, and a host of other matters. In early 1907 the Pennsylvania set up a rail committee that began systematic collection of failure statistics by manufacturer, which it distributed to them, and it experimented continuously with rail specifications, adding a nick-and-break test. If the "A" rail survived the drop test it would be nicked and broken and if it showed defects all "A" rails in the heat would be rejected.[54]

The ARA also became involved, developing in concert with the rail makers two new, heavier, rail sections with more metal in the web and base that could be rolled at lower temperatures. It then handed matters back to the AREA, which joined with the Rail Manufacturers' Technical Committee to study ways to reduce failure. In 1908, the AREA began to collect the first industry-wide statistics on rail failure. Just how useful such monitoring could be had been demonstrated by the Harriman lines in 1907. The Harriman data computed failure rates controlling for rail weight and section per 100 miles of track for three-year periods. Just as the railroads claimed, the newer, heavier rail had the highest failure rates.[55]

The AREA also initiated systematic research on the causes of rail failures, employing as its engineer of tests Max Wickhorst, who had been similarly employed on the Burlington. Funding came from the ARA—its first significant venture into cooperative research—while the steel companies furnished use of their plants. By 1912–13, Wickhorst and others had completed several dozen studies and important conclusions began to emerge. The shape of the section proved less important than production methods and steel quality. As Wickhorst dryly observed, the section could not cure bad metal. Research also revealed that segregation worsened with ingot size, but as the New York Central had discovered, titanium and other additives such as aluminum could reduce it.[56]

More companies also initiated process inspection as well as product testing. In 1912 Robert W. Hunt began to offer special inspections, guaranteeing that his men would monitor every stage during rail making. By the end of the year Hunt claimed that thirty-three carriers with 122,000 miles of track (about one-third of the U.S. total) had bought his service, and by 1917 the proportion was up to 60 percent. The result, he claimed, was a 30 percent decline in failures. Improved product testing came more slowly, however. Some rail men had wanted to apply the nick and break test to the top rail in every ingot, rather than simply to the drop-tested rail, but manufacturers opposed it. Instead, the carriers increased the frequency of the drop test, and some eventually applied it to rails from three ingots per heat.[57]

About this time rail failure again became a public issue. Accidents from broken rails, which had declined for several years from their peak in 1907, rose from a low of 196 in 1909 to 249 in 1911. In response the National Association of Railway Commissioners began an investigation with an eye to "the advisability of Government inspection and legislation." The Indiana Railroad Commission also investigated rail failures, while the *Scientific American* continued to denounce manufacturers, referring darkly to a "ring of steel makers who make bad rails."[58]

Up to 1911 the carriers had been winning the publicity battle over rail failures. But on August 25 of that year, Lehigh Valley Train No. 25 struck a broken rail near a bridge outside of Manchester, New York. Nine cars pitched into the stream below, killing twenty-nine and injuring sixty-two passengers. James Howard, an engineer physicist at the Bureau of Standards, conducted the subsequent ICC investigation, one of the first under the new Accident Report Act of 1910. Howard was a metallurgist who had been engineer of tests at the Watertown Arsenal. Much of his work there had involved testing steel for the railroads and he was well known and widely respected by the technological community. Howard reported that the broken rail on the Lehigh Valley was not a result of bad steel—"the mills are making the steel they are required to make under the specifications," he claimed. Instead, the cause was an internal transverse fissure, a result of metal fatigue from the rolling action of high wheel loads, and he concluded "the remedy does not consist in increasing the weight of the rail but in diminishing the intensity of the wheel pressure." As the *Gazette* quickly pointed out, the effect of this bombshell was "to throw the responsibility of rail fractures on current railway specifications."[59]

The steel companies chose this moment for a preemptive publicity strike. A congressional committee was just then slow-roasting U.S. Steel. The company's president James Farrell took the offensive when questioned about rail failures. Following the trail Howard had blazed, Farrell testified that rail failures had worsened in the 1890s after the carriers had begun to demand higher-carbon steel. This made harder rails and extended their life in the face of rising wheel loads, Farrell noted. And—he emphasized—it was cheaper than a heavier section. Railroad greed, it turned out, was the problem after all. A. W. Gibbs, chair of the Pennsylvania's rail committee, understood the strategy: "It is evident that we will be put on the defensive in the event of any rail breakages," he warned. In fact, while the point about rail weight was well taken, carbon specifications had not increased since the 1880s. Just how sensitive a public issue rail failures were is revealed by a letter from the Union Pacific's Chairman of the Board Robert Lovett, to its president, A. L. Mohler. The company's rail requisition forms, Lovett noted, sometimes used the term *poor condition* to justify replacement. He worried that "a jury might conclude that it means dangerous or unsafe," and urged Lovett to change the justification.[60]

With the carriers on the defensive, the rail makers now called for a truce. On February 15, 1912, U.S. Steel Chairman of the Board Judge Elbert Gary gaveled a meeting of top steel and railroad men. Gary scolded both sides, and he warned that unless rail failures were reduced "we shall find ourselves in the hands of commissions of some sort like they have in England . . . which shall lay down for us formulas in accordance with which we shall be obliged to manufacture our rails."

The steel men had chosen an opportune moment, for three months later in May, and then again in August, the ICC released reports of two derailments from broken rails. The first, on the Great Northern, killed five and injured eighteen people, while the second, on the Wabash, resulted in eighty-seven injuries and two deaths. In both cases the offending rail had been rolled by Illinois Steel, a U.S. Steel subsidiary, and the steel was seamy and highly segregated—a result of manufacturing defects.[61]

These two disasters marked the crest of this flood of rail failures. Broken rails resulted in 363 accidents and fifty-two deaths in 1912, an all-time high that provoked the ICC to warn, "the danger of serious derailment is ever present." But in November the National Association of Railway Commissioners delivered a more optimistic review, concluding that the rail situation was improving. *Iron Age* took its measure with the headline "No Government Inspection of Steel Rails." A year later the Indiana Railroad Commission agreed that "all is being done that could be practically done." The carriers agreed that rail quality was getting better. In late 1911 the Harriman lines claimed there had been a "marked improvement" in its rails, and other observers agreed.[62]

The trend toward heavier rail sections also reduced breakages, just as had occurred a generation earlier. Thus, when the Pittsburgh & Lake Erie introduced new 2-8-2 Mikado locomotives with 75,000 axle loads in 1916, rail failures erupted and only declined when it shifted from 100- to 115-pound rail. New locomotives also became less destructive. Work at the Pennsylvania's testing plant about 1910 led the company to design much lighter moving parts and valve gear into its Atlantic-type locomotives, thereby allowing both heavier wheel loads and less pounding from im-

Figure 7.4. **Rail Failures, Five-Year Rates, Adjusted for Track and Traffic, 1908-1939**

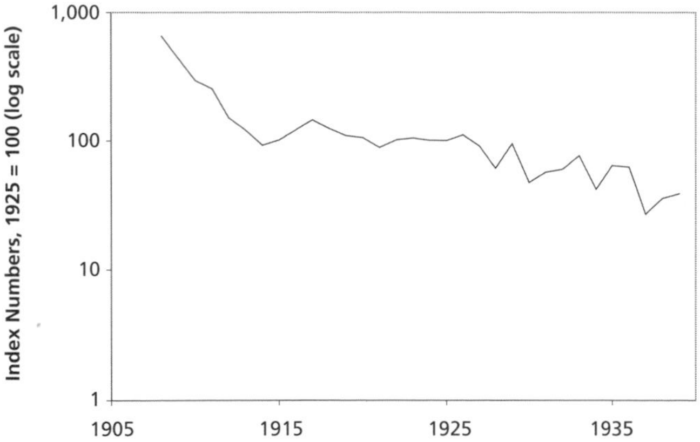

Source: Appendix 2.

perfect counterbalance. At about the same time the shift to vanadium steel in driving rods also reduced rail failure by diminishing the hammer blow delivered by the driving wheels. The full benefits from both innovations appeared only gradually, however, as the locomotive stock turned over. In 1913 the AREA published the first five-year rail failure rates for rail made in 1908 and 1909. An index of these data (Figure 7.4) adjusted for train and track miles fell from 650 to 150 by 1912. By 1920 the failure rate was down to 105, and it continued to decline throughout the interwar years. Both accidents and casualties gradually declined. The rail outlook seemed rosy indeed, except for the troubling problem of transverse fissures that would take another generation to unravel.[63]

The AREA's figures suggest that only about 30 percent of the improvement resulted from the shift to open hearth steel, which was lower in phosphorous, while the remainder resulted from better testing, monitoring and quality control, additives, and manufacturing modifications. As with the improvement in the cast iron wheel, the decline in rail failures is a good example of the large cumulative improvements that spring from a well of small causes. The episode also provides another example of the workings of the railroad technological network. It reveals the increasing ability of engineers to understand, monitor, and control product quality. It also suggests the role of economic incentives in shaping institutional behavior. Much of the push to improve wheels had come from the makers, who felt threatened by regulation and the competition from steel wheels. But the driving force for better rails was clearly the carriers—quite possibly because they were more directly affected by regulatory threats while the steel makers faced little competition.[64]

Thus, the reduction in casualties from derailments that began after 1907 reflected the continuous improvement in track, rail, and roadbed that grew out of the railroad technological community's increased skill in puzzle solving. Yet, as with the decline in collisions, many other forces also made their contribution. Steel cars reduced casualties in derailments perhaps even more effectively than in collisions (note 34). In 1902–4, 1,450 passenger train derailments killed 184 people; in 1914–16, 1,787 derailments killed only 148. The *Railway Age* attributed the relatively low death toll in a 1912 derailment on the Pennsylvania at Warrior Ridge to the safety of steel cars. Companies such as the Lehigh Valley that installed telephones at sidings and crossings discovered that they helped prevent derailments, as men could report dangerous conditions. Finally, the decline of derailment dangers also reflected the carriers' new labor policies. Carriers began to have freight crews inspect trains before departure and some required maintenance forces to scan passing trains looking for problems.[65]

The Course of Little Accidents

Accidents that didn't involve collisions or derailments received very little public attention, although the problem of grade crossings continued to concern state regulators. Rather, the railroad brotherhoods and the railroads took the lead in these areas, with mixed success. Casualty rates to employees and passengers from accidents other than collisions and derailments continued their long-term decline, as did casualties due to trespassing, but despite major efforts by both the carriers and regulators, grade crossing accidents worsened.

Little Accidents to Workers

The most visible safety achievement of the brotherhoods was the Locomotive Inspection Act of 1910. The act had been contentious; that it was implemented so smoothly resulted from the skills of the Chief Inspector Frank McManamy, who guided the bureau from 1913 to 1918. McManamy had risen through the ranks on the Père Marquette; he then joined the civil service to become an ICC safety appliance inspector and finally chief inspector of locomotives. Although an ex-union man, in his public pronouncements to railroad groups McManamy stressed cooperation with the carriers and that the act would benefit them, and he ensured that his inspectors, who were also union men, behaved with a high standard of professionalism. On lines with already well-developed procedures the law prevented the usual practice of waiving inspection when the supply of locomotives was tight, and on small carriers it greatly improved inspections. The tiny Grand Rapids & Indiana, for example, so seldom removed washout plugs that they had to be cut out and that company also reported that it routinely ran locomotives in violation of the regulations. From 1905 to 1909, the fatality rate from boiler explosions averaged 0.17 per thousand trainmen, or about 2 percent of all fatalities. In 1912–15, after legislation, the rate averaged 0.09—about a 47 percent decline, and in the three years 1918–20 it averaged 0.04.[66]

Yet some of this reduction was a byproduct of technical progress that had been steadily making locomotives more productive and safer for decades (chapter 4). Flexible staybolts came into widespread use after 1900, reducing the likelihood of boiler explosions, while water gauge glasses were improved, making them less likely to burst. About 1911 employees of the Santa Fe developed the Jacobs-Shupert firebox, which was made in sections, doing away with the need for staybolts. In a series of widely publicized tests by Professor W.F.M. Goss of Purdue, the Jacobs-Shupert firebox survived water levels that blew up the standard firebox. It was widely used on the Santa Fe, but less so elsewhere due to expense.[67]

Many other aspects of locomotive operation and construction were also being improved. As locomotives became larger and more expensive and trainmen's wages increased, the cost of downtime rose steadily. By 1903 *Railway Age* reported that many carriers were building water-softening plants, which resulted in less downtime and danger from burned-out flues and fireboxes. When the Union Pacific shifted to better water, repair costs per engine mile fell 37 percent. Water treatment on the Santa Fe began about the same time under the direction of its chief chemist. The company reported dramatic reductions in leaky flues and cracked boiler sheets as a result. By 1906 the Rock Island and many other carriers were shifting from Stephenson link motion to Walschaert valve gear, which both reduced maintenance costs and engine failures from broken valve motion. The shift to chrome vanadium steel in axles, frames, pistons, rods, and crank pins also improved trainmen's safety even as it saved weight and preserved track. Records of the Lake Shore, the Michigan Central, and New York Central showed an 80 percent reduction in the failure rate of main and side rods after the shift to such material in 1910.[68]

Thus, federal inspection, while important, was but one of a host of forces contributing to better locomotive safety. Moreover, boiler explosions, long hours, ash pans, tight clearances, and freight car appurtenances all together accounted for no more than 10 percent of all fatalities to trainmen in 1907–8. In the nineteenth century the application of air brakes and automatic couplers had yielded dramatic reductions in work accidents, but such options no longer existed. Dedicated safety regulations therefore could never make a real dent in employee casualties. Hence, the most important piece of railroad safety legislation to emerge during these years was the Employers' Liability Act, for it helped spark a generalized interest in accident reduction.

Roads with an active safety program achieved remarkable results. From 1912 through 1915 the Norfolk & Western reduced worker fatalities 49 percent even though traffic grew by 42 percent. In 1918 the North Western calculated that in the seven and a half years ending in 1917 it had implemented over 27,000 safety suggestions and reduced worker fatalities 39 percent compared to the previous seven and a half years, while injuries were down one-quarter. The company also reported reductions in injuries and fatalities to passengers and others as a result of its program. All told, safety work had prevented nearly 19,000 casualties and saved the company about $2.7 million in injury costs. The New York Central and some other carriers experienced similar results.[69]

Yet the early safety movement was uneven, for some lines did little or nothing, and by 1915 the *Railroad Trainman* was dismissing it as "simply another way of throwing dust over the accident record." After 1917, when the U.S. Railroad Administration (USRA) took over the carriers, it sharply expanded safety work, instituting a safety section within the division of operation and setting up regional safety offices as well. The section sent out brochures urging the carriers' federal managers to develop safety departments modeled on the North Western. As a result, lines such as the Delaware & Hudson and Northwestern Pacific began safety work for the first time. On the B&O, federal control modified the safety organizations to include regional agents whose full-time job was safety work and a general safety committee whose membership included the top company officers.[70]

In late 1918 some regions began "no accident weeks," and by 1919 safety work was in full swing. During October 18–31, 1919, the USRA conducted a National Railroad Accident Prevention Drive with a propaganda blitz that resulted in 154 fewer worker fatalities (a 50% decline) and 3,051 fewer injuries (a 21% decline) than during the same period in 1918. These gains were not costless. For the year the carriers reported 1,789 safety meetings that included 7,231 officers and 15,460 men as well as inspections that took 20,356 manhours. They corrected over 16,000 unsafe conditions and 5,300 unsafe practices. The payoff to these practices (and the decline in business) was that for the year worker fatalities fell about 38 percent compared to 1918, while fatalities of passengers and others declined as well.

By 1920, while train miles were only slightly less than in 1907, fatality rates to trainmen from causes *other than collisions and derailments* were down from about 5.5 to 2.7 per thousand. Over the same period fatality rates to all railroad workers from such little accidents also fell in half. Some of these gains reflected the laws discussed above, but most resulted from continued improvements in technology and organized safety work. All the railroad unions enthusiastically endorsed the safety programs of USRA and expressed the hope they would continue. When the carriers returned to private control, the ARA demonstrated its commitment to the new safety compact, adopting both the structure and accident campaigns of the USRA.[71]

Trespassers and Passengers

During the nineteenth century trespassing on railroads was like the weather: state commissions and railroads talked about it but no one did anything about it. In the early twentieth century, as the toll of casualties mounted, the railroads and the technical press began to publicize the problem to try to force state action. Their motives were the usual mixture of humanitarian concerns, economics, and politics. As noted, accidents to trespassers were not costless to the carriers, and as their numbers increased so did the motivation to act. The Safety First campaign that was just beginning heightened concerns with all forms of accidents. On the Chicago & North Western Ralph Richards was primarily interested in protecting employees, but he also pointed out the number of others the company killed. A campaign against casualties to trespassers generated good publicity at a time when the carriers' safety record was under fire from every quarter and it allowed them to turn the tables upon their critics. For if trespassing was the problem, the cause—refreshingly enough—was government rather than railroad incompetence.[72]

In 1911, the Pennsylvania reported that it had killed 7,997 trespassers and injured another 7,838 in the ten years after 1900. In the worst year, 1907, the toll came to 916 killed. When the company began a safety campaign in 1908, fatalities dropped to 585, the fewest in the decade. The Central, the Long Island, the New Haven, and other carriers also began campaigns to cut down on trespassing. The companies tried to induce courts and legislatures to crack down by demonstrating that the victims were mostly solid citizens, not tramps and hobos. In 1912, Frank Whiting, general claim agent for the New York Central, published results of an investigation of 1,000 trespassers killed the previous year on the Central. Whiting found that only 93 were intoxicated at the time. Most were local people, and only 50 could be identified as hobos simply traveling through. There were 268 laborers, 81 shopkeepers and mechanics, 44 farm hands, and a miscellany of other occupations as well as 70 school children. The Association of Railway Claim Agents published a larger survey of over 10,000 trespassing fatalities in 1914 that yielded similar findings. The problem, as Whiting summarized his findings, was not a matter of tramps and hobos, "but with trespassers, who in many instances are regularly employed, well-to-do and respected citizens." He emphasized that prevention was "wholly within the hands of the local authorities."[73]

This was a congenial refrain and several other carriers joined in the chorus. In 1912 the presidents of both the Burlington and the Frisco wrote open letters to the governors of the states in which the companies operated. They pointed out how much the carriers spent to prevent the few accidents to passengers and blamed the states for failing to prevent the far more numerous accidents to trespassers. The *Railway Age* also jumped in, reporting that the New York Central had plastered signs on its right of way and stepped up arrests, but found that getting convictions was difficult. Out of 5,400 individuals arrested for riding cars during 1912, only about 1,500 were jailed. The *Age* also surveyed the Chicago General Managers Association, which reported information on 4,785 arrests made for trespassing on member lines between June and August 1912, nearly all of them by railroad police. Of these, only about 2,200 received any punishment at all and many of these cases involved not just trespassing but also petty larceny, throwing stones at trains, and drunkenness. In Chicago, where the railroads were in the process of spending $150 million to elevate tracks and eliminate dangerous grade crossings, individuals routinely trespassed on

Figure 7.5. **Fatalities of Trespassers and at Grade Crossings, 1890–1920**

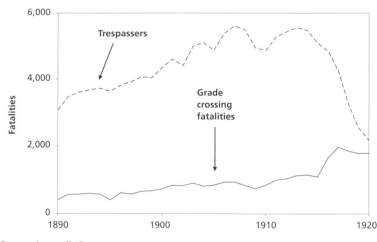

Source: Appendix 2.

the elevations but were rarely punished. "In no other civilized country," the *Age* fumed, "can there be found public officials whose hearts are so tender that they are pained more by having men arrested than by having them killed."[74]

In fact, after rising for decades, casualties to trespassers had already peaked in 1907, when 5,612 individuals were killed. Thereafter the numbers fluctuated in a narrow range up to World War I, and then plunged in half by 1920 (Figure 7.5). This great decline in mortality owed something to the carriers' safety campaigns, but it probably owed more to the spread of cheaper, more convenient means of transportation. Around 1900, the rapid growth of electric interurbans, which offered lower fares and faster service than steam railroads, generated traffic from individuals who would otherwise have walked the track, while rising real wages and the availability of affordable automobiles encouraged riding rather than track walking (chapter 9).

Some of these same forces also improved passenger safety. As discussed in chapter 4, the decline in accidents to passengers at stations and from getting on and off cars or falling from cars made a major contribution to improved safety after 1890. In 1918 the AREA rather belatedly concerned itself with station design. It found that many eastern carriers had separated traffic at high-density, double-track stations. Many others employed gates, or at least painted "dead lines" to keep passengers away from trains. They also improved lighting and built wider platforms. The continuing increase in the length of passenger journeys also improved safety. From 1900 on, the railroads began to lose short-haul passengers first to interurbans and then to automobiles, and as a result, the average journey jumped from about 28 miles in 1900 to 37 in 1920, or by one-third. Other things being the same, the result was to raise passenger miles relative to trips to and from the station, and exits and entrances to trains. Thus, longer journeys should have reduced accident rates from these causes by somewhat less than one-third. In addition, the passengers that abandoned steam railroads must have been those who placed a high value on their time and would, therefore, probably have been most likely to have run risks jumping on and off trains or cutting across tracks at stations.[75]

The Grade Crossing and the Automobile

Efforts of some states and cities to prevent grade crossing accidents by abolishing crossings during the nineteenth century saved many lives but were inadequate to stem the tide of crossing accidents. From 1890 to 1900 fatalities at crossings to nontrespassers rose from 301 to 559, or by 86 percent, as both population and train miles increased, and they rose another 27 percent to 710 by 1910. Thereafter, even as train mileage stagnated, such accidents exploded, rising 138 percent from 1910 to 1920 (Figure 7.5). The automobile age had arrived.

Before the automobile, grade crossing accidents had been predominantly an urban problem, involving horses, wagons, and pedestrians and might be largely addressed by crossing elimination. After 1910, however, matters changed rapidly and radically. Automobiles vastly increased mobility, and dangerous rural crossings that had experienced little traffic in 1900 suddenly became a much more serious problem as automobile traffic expanded. Moreover, as motorists traveled farther from home they experienced unknown dangerous crossings. The speed and noise of early cars also increased dangers—especially where visibility was obstructed. Where crossings were at the top of a hill, drivers of early underpowered vehicles had to either get a running start or risk stalling out on the track, and both situations were dangerous. By 1914–15 one large state reported that over half of all crossing accidents involved automobiles.[76]

In 1915 the Committee on Grade Crossings of the National Association of Railway Commissioners concluded that the advent of the automobile called for a "careful reconsideration of our present methods of crossing protection." While the committee immediately thereafter reaffirmed its faith in the eventual elimination of all crossings, it increasingly focused on the problem of protection rather than elimination. The carriers especially drew back from the implications of elimination. In 1914 the Pennsylvania noted that it had 5,442 crossings on its lines east of Pittsburgh. While their total abolition was out of the question, guarding could also be expensive. That company concluded that to install and operate a disk signal would cost $110 a year on a single-track crossing, or about $600,000 a year if applied to all crossings. Yet accidents could sometimes be expensive too. The Delaware & Hudson records for this period reveal a number of crossing accident claims the company was forced to settle for $10,000 to $15,000 each. In addition, the prospect of colliding with a car or truck sharpened the railroads' interest in crossing safety, for it was far more likely to wreck a train than was hitting a horse and buggy. In 1915 representatives from the ARA and the railway commissioners met with representatives of the American Automobile Association to try to discover common ground. They agreed that the automobile had made insufficient the usual warning sign placed next to the tracks. Indeed, tests by the Pennsylvania demonstrated that cars with modestly worn brakes traveling at 35 miles per hour took 200 to 300 feet to stop. By the time a motorist could see the sign it was too late. What was needed, all agreed, was uniform caution signs placed 300 to 500 feet before the crossing. The National Association made a resolution to this effect and by 1917 eleven states had such a requirement.[77]

Warning motorists was the easy part; getting them to stop for a train was a different matter. The problem, no doubt, reflected both the rising cost of motorists' time and the general tendency of individuals to ignore routine, low-probability risks. Drivers entered crossings millions of times a year, yet there were only about a thousand fatalities. Such accidents, one could imagine, would happen to other, poorer drivers. Although as of 1917, only about 10 percent of crossings were protected, most of the rest had signs. Yet few drivers heeded any warning. The B&O observed drivers at a crossing at Uniontown, Pennsylvania, on September 12, 1914. Out of 729 automobiles, only 28 heeded the stop sign and only 224 slowed down. At about the same time the Southern Pacific observed 17,021 drivers at a number of crossings; it too found that most neither stopped nor looked in both directions.[78]

More heavily traveled crossings had bells to warn motorists, but the automobile encouraged a shift to visual signals such as wig-wags or flashing lights, for audible warnings were hard to hear in a noisy car with the windows up. Gates operated by a flagman were thought the most effective protection (they were not; see chapter 9), but sometimes these also failed. In a two-year period the Southern Pacific reported that 525 motorists simply crashed through. One draconian solution adopted by the Long Island but apparently not widely employed was to use gates the size of telephone poles. No doubt these were effective, but were probably hard to raise and lower. The Long Island also applied Safety First techniques to crossing dangers, erecting immense 500-square-foot warning signs at some crossings. Other carriers placed "Stop, Look, Listen" articles in popular magazines. The Southern Pacific publicized crossing dangers and claimed that it reduced the number of individuals who ignored warnings. On the whole, however, little was accomplished and crossing accidents continued to increase.[79]

The first two decades of the twentieth century represented a great divide for American railroads as three major upheavals in their environment reshaped the course of safety. These were the increase in regulation, the shift from construction to reconstruction, and the advent of the automobile. The upsurge in accidents that began in 1897 resulted in an avalanche of proposed safety regulations and rising accident costs. In response, the carriers innovated safety departments. Safety organizations symbolized a corporate and indeed an industry commitment to reduce all forms of accident.

The commitment reflected an informal safety contract between the industry and its overseers. Regulation by exhortation would leave the railroads largely free of formal safety rules—but only if they did improve safety. For the first time, the carriers began a concerted attack on little accidents to workers, a policy they hoped would also diminish other forms of accidents. Regulatory threats and increasing traffic density led to the rapid diffusion of the block system. Better wheels and rails reflected the increasing application of engineering analysis by the carriers and university researchers and their increasing ability to monitor and control physical processes. These improvements were made under pressure in varying degrees by regulatory threats, and yielded safety and productivity gains. Better safety and productivity diminished the social costs of railroading, which fell in half between 1900 and 1920 (Appendix 2), the former by reducing accidents and accident rates and the latter by increasing output per worker.

After 1907, the headlong expansion of track, traffic, and employment that had characterized every decade from the 1830s on was replaced by more modest growth. Slower growth improved safety directly, for fewer new men were needed and fewer new and dangerous lines opened for traffic. Slower growth in passenger traffic also reflected the development of competition from electric interurbans and then automobiles. By 1920 the automobile was helping to tame the trespassing problem, but it worsened the dangers of grade crossings, ultimately transforming them into an issue of public safety.

Lobbying for Regulation
Transporting Hazardous Substances, 1903–1930

We don't give a car of explosives as much care as we do a car of horses.
— Charles B. Dudley, Pennsylvania Railroad, 1905

In all important countries . . . [regulation of explosives] is a work of the government and it should have been so in the United States. But it is now being conducted entirely by the railroad companies.
— James McCrea, Pennsylvania Railroad, 1908

Pressure from the public and the brotherhoods was the driving force behind the wave of proposed safety regulations that swept over the railroads in the two decades before World War I. The carriers' response usually ranged from indifference to implacable hostility. But in the transportation of explosives and other hazards, the push for regulation came not from critics but from the carriers themselves. The story has its roots in the nineteenth century, but the move for federal regulation began in 1903. On November 1 of that year, a freight car loaded with dynamite exploded in the yards of the Pennsylvania Railroad in Crestline, Ohio. The cause of the blast was unclear; probably it was the result of rough switching. The explosion injured scores of people, two of them seriously, set fire to about five hundred freight cars, derailed locomotives hundreds of yards away, and blew a crater 40 feet deep. It also set in motion a sequence of events that led to sweeping organizational and institutional innovations in the shipping of hazardous substances.[1]

Led by the Pennsylvania Railroad, the American Railway Association (ARA) developed rules governing the shipping of explosives, to be enforced by the newly formed ARA Bureau of Explosives. The ARA also lobbied successfully for federal controls, and in 1908 Congress granted the ICC power to regulate the transportation of explosives. The commission promptly adopted ARA rules and left enforcement to the Bureau of Explosives. These procedures remained in force until the 1960s, and they sharply reduced accidents despite rapid growth in shipments of dangerous cargo.

The regulation of explosives does not fit easily in most of the accepted stories about why businesses pursue regulation. It was not an effort to achieve market power. Nor did the carriers support regulation of explosives because they thought it inevitable or because it provided competitive advantage. The dominant motive for regulating explosives was the carriers' desire to internalize and reduce social costs. In some respects the process resembles the associationalism of Herbert Hoover during the 1920s. Hoover hoped that private institutions and expertise could perform public functions but without the drawbacks associated with government bureaucracies. A decade before Hoover's efforts, the railroads evolved an associationalist solution to the difficulties of regulating the transport of hazardous materials. The carriers created the Bureau of Explosives—an industry-wide private organization that both rationalized and administered rules for shipping dangerous goods, and provided the scientific and technical expertise to improve practices. But while Hoover's associationalism downplayed coercion, the carriers' need to prevent "free riders" led them to support legislation giving the ICC power to transform the bureau's rules into federal regulations.[2]

Regulatory Origins, 1883–1903

In the years following the Civil War the explosives industry grew, well, explosively, reflecting the expansion of mining and heavy construction. In the quarter century before 1905, the value of output increased five times. From 1899 to 1904 dynamite production rose 53 percent—from about 43,000 to 65,000 tons—a bit more than half of which was used in railroad and other construction, with the remainder going to coal and hard rock mining. Black powder production more than doubled between 1899 and 1904, increasing from 46,000 to 108,000 tons, and 80 percent of it was used to mine coal.[3]

Inevitably, some shipments blew up. In 1866 several explosions, including one involving nitroglycerine at the Wells Fargo building in San Francisco that killed fourteen people, led to a federal law that year that forbade shipment of explosives on passenger vessels and otherwise governed their transportation. But the law failed to specify an effective enforcement mechanism and was therefore widely ignored. In 1871 a freight car full of dynamite on the Boston & Albany Railroad went off in Worcester, Massachusetts, doing about $50,000 of damage but—miraculously—killing no one. Seven years later in Michigan a carload of nitroglycerine blew up, killing seven. In 1883 an explosion of 400 kegs of "giant powder" on a freight train in Kentucky killed three people. On May 19, 1888, on the Santa Fe, a runaway car containing 17,000 pounds of giant powder collided with a car of naphtha, destroying all twenty houses in the tiny town of Fountain, Colorado, and killing three people. By the twentieth century, one expert for the railroads calculated that on any given day there were about 3,300 freight cars carrying explosives, and accidents such as that at Crestline had therefore become common. Indeed, almost precisely one month after Crestline, on December 2, 1903, another carload of dynamite went up, killing one person and injuring thirty-three, and doing about $25,000 of property damage.[4]

Problems on the Pennsylvania

By the end of the nineteenth century executives of the Pennsylvania Railroad had begun to call the company "the standard railroad of the world." It was the largest American carrier, accounting for over 8 percent of the nation's freight ton miles and nearly 6 percent of passenger miles in 1900. As Alfred Chandler has shown, the company pioneered in modern management. As we have seen (chapter 2), it also pioneered in the application of technical expertise to business problems, setting up the first chemical laboratory about 1875.[5]

The Pennsylvania had long grappled with the dangers of shipping explosives. Until 1883 the company had refused to accept them. In that year the superintendent of the Repauno chemical works urged the railroad to reconsider this longstanding policy. The company turned for assistance to chemist Charles B. Dudley, director of its Altoona research laboratory, whom the reader has met before (chapter 6). In 1883, Dudley visited Repauno, where he got a demonstration on dynamite safety. As he later recounted the story, the company superintendent sent a boy with a box of 50 percent nitroglycerine dynamite to the top of a 30-foot tower. He was to drop the box on a pile of rocks, as Dudley and the superintendent stood close by and watched. Although the superintendent assured him that "there is no danger," and indeed none of the dynamite went off, Dudley took refuge behind a pile of rails.[6]

These events led him to formulate in that year the company's first regulations for transporting explosives. One rule that provides a microcosm of the problems companies faced concerned the placing of signs warning "Explosives" on freight cars. They were to be 4.5 feet above the floor, the reason being that dynamite was rarely packed that high and hunters often used the signs for target practice. The company's rules were widely publicized and formed the basis for most of the regulations adopted by other large carriers. Despite the existence of such rules, the Pennsylvania began to be plagued by explosions. In Ohio alone, company records reveal disasters at Forest in 1887, Kings Mills in 1890, and Lawrence in 1893 that collectively killed twelve and injured twenty-four people and did about $200,000 of property damage.[7]

In 1895, these events led management again to focus on ways to improve the safety of transporting explosives. Just as Americans looked to Great Britain for models of regulation and accident investigation, Dudley also reviewed British experience for guidance, as that country had regulated explosives since 1875. He corresponded with the British inspectors and obtained copies of their annual reports. British experience, along with his own investigations of the physical properties of black powder containers, led him to develop company regulations similar to those in force in Britain. The Pennsylvania also met several times with manufacturers, but nothing concrete seems to have been accomplished. As one company report summarized matters, "the high explosives people thought high explosives safer to transport than black powder . . . [and] the black powder people thought that the great trouble was due to the carelessness of the railroad employees."[8]

Expansion of the chemical and petroleum industries forced the carriers to contend with the transport of a host of other dangerous products in addition to explosives. Breakage of the glass jars (carboys) in which nitric acid was shipped was a constant source of fires. On May 12, 1902, a switching crew punctured a tank car of naphtha in the Pennsylvania's Sheridan yard, just outside of Pittsburgh, resulting in the sort of complex, closely coupled disaster Charles Perrow has called a "normal" or system accident. A switch light ignited the naphtha, which in turn set fire to a number of oil cars, two of which blew up; still more cars ruptured, sending a stream of oil into the sewer, which also exploded. Twenty-nine people were killed and ninety-four injured.[9]

The explosion at Crestline was therefore one in a long line of disasters that were raising the cost of transporting explosives and other dangerous articles, but it proved to be the last straw, finally provoking the carriers to craft an institutional solution. Other railroads were also beset by these dangers but the Pennsylvania was uniquely sensitive to the problem. As noted, it was the largest carrier in the country and a 1905 survey revealed that it had forty-eight manufacturers of explosives located along its lines—more than any other railroad. In addition, the Pennsylvania was a major trunk line that received in interchange large numbers of cars that also contained hazardous materials. Men with scientific training dominated the Pennsylvania and, thanks in part to Dudley, management was alert to the dangers of such substances. The company's size also gave it considerable influence in the ARA and Congress. As a result, the Pennsylvania sought both a political and an industry-wide solution to the problems of transporting hazardous cargo. Thus while Steven Usselman has claimed that the Pennsylvania symbolized the triumph of engineering over innovation, in this instance that company's management proved highly innovative. This decision, and the ability of the company to enlist the ARA, was also shaped by the nature of private liability for damages resulting from explosion, from agency problems and transaction costs, and from competitive pressures.[10]

The safe transport of explosives has some of the qualities of a public good like clean air—for example, no one can be excluded from the benefits. Yet private companies will provide public goods when they can capture some of the benefits. Such was the case with explosives transport. When a carload of explosives blew up, determining the actual cause of the disaster was nearly always impossible and liability for the damage usually fell on the carrier, thereby internalizing to the railroad the benefits from handling explosives with care. The difficulty, from the point of view of the Pennsylvania and other large carriers, was that manufacturing and packaging also influenced the dangers of transportation. Yet manufacturers would be unlikely to face liability from a transportation disaster and therefore behaved like free riders. Similarly, feeder lines might accept explosives in dangerous condition, or place them in poorly constructed cars, transport them for only a few miles, and then hand them off to one of the trunk lines. Thus the cost shifting associated with explosions occurred not between business and bystanders but among businesses.[11]

Of course, the Pennsylvania might have dealt with these unilaterally. It could have inspected and refused cars in interchange, but that would have been an expensive nightmare. As the Pennsylvania's 1895 report observed, it could have refused to accept black powder that was not packed in double-jacketed containers or accepted them only at a premium rate but did not do so, "as of course it would drive traffic to other lines." Thus regulation of hazardous substances faced the same sort of network problems that had beset the introduction of automatic couplers and air brakes two decades earlier. Competitive rivalry was worsened by agency problems, for while the Pennsylvania had developed detailed rules governing shipments of explosives, the company freely acknowledged that they were routinely violated by its own agents to get business. Stricter enforcement might also lead to "surreptitious shipping of powder." In short, individual private action was likely to be expensive and ineffective, and so the Pennsylvania turned to regulation to enforce a cheaper, more efficient alternative.[12]

Institutional Innovation in Shipment of Hazardous Cargo, 1904–1908

In March 1904, five months after Crestline, the House and Senate commerce committees held hearings on a bill drafted by Pennsylvania First Vice President (and soon to be President) James McCrea to regulate interstate shipments of explosives. While McCrea seems to have been the driving entrepreneurial force in back of explosives regulation, the bill reflected Dudley's admiration of British practice, for like that system it would have authorized the ICC to appoint government inspectors of explosives, who would be given access to manufacturers' plants. McCrea and Dudley, the only two witnesses who testified on the bill, explained the need for federal control as they saw it.[13]

McCrea informed the committee that most states had little or nothing in the way of regulations governing shipment of explosives. Pennsylvania, he pointed out, only prohibited their shipment in passenger cars. "The public does not understand to what extent they are living over a powder mine," McCrea warned the committee. He explained the need to cover packing and loading by the manufacturer because the railroads' "inspection after the car is loaded cannot be thorough." McCrea claimed that "the railroads themselves have stringent and detailed regulations [but] . . . if we make them too strict, the public is open to the danger of smuggling." He also noted that "nothing we [the Pennsylvania] can do will govern all the other railroads in the country."[14]

Dudley provided the necessary scientific reinforcements to McCrea's arguments, playing on the mysterious, frightening qualities of dynamite. He told the committee that he had once taken a small quantity of nitroglycerine mixed with wood pulp and hit it with a hammer without exploding it. But on the second hit, the substance always exploded, he claimed. Dudley also noted that poorly manufactured dynamite might contain excess acid, which made it much more unstable—a result that could not be discerned from after-the-fact inspection of the product. Dynamite also needed to be packed on its side or it might leak nitroglycerin. Dudley had visited the plant that made the dynamite that blew up at Crestline. The company employed a nondrying absorbent, failed to use antacid, and placed the product in thin boxes, he discovered. All these problems were magnified by freight car interchange. "It is in connection with this reception of material from connecting lines that we feel it is essential to have legislation," Dudley explained. In short, McCrea and Dudley were asking Congress to direct the ICC to prevent certain forms of competition in transporting explosives. No single firm could enforce such rules due to the impossibility of inspecting factories and the prohibitive costs of inspecting freight cars and detecting smuggling, while competition would ensure that efforts to do so would drive business elsewhere.[15]

Despite the eloquence of McCrea and Dudley, the bill went nowhere and, faced with certain defeat, McCrea withdrew it. The Pennsylvania had not bothered to enlist the support of other railroads, and the bill predictably stirred up the makers of explosives, who viewed it as unconstitutional. DuPont by then had emerged as the dominant producer. Like the Pennsylvania it was a management pioneer, and while company executives were coming to see safety as an aspect of efficiency, they opposed giving government inspectors access to their plants. As one DuPont representative later remarked, "we opposed it [McCrea's bill] as best we could." Recognizing the need for allies, McCrea then proceeded to work through the ARA and to negotiate with manufacturers before the next political assault.[16]

The American Railway Association Becomes Involved

At its April 5, 1905, meeting, at McCrea's request, the ARA appointed a committee to consider the advisability of adopting "uniform regulations governing the transportation of explosives *and other dangerous articles*," with McCrea as its chair (emphasis added). As the title indicates, while most of the focus was on explosives, from the outset the committee intended to cover a broad range of hazardous materials. Probably this was Dudley's doing. In response to disasters such as that at Sheridan, the Pennsylvania had already developed rules governing shipment of acids and flammables. As Dudley later argued, it was impossible to regulate explosives without also regulating substances that might detonate them.[17]

A scant month later, on May 11, events gave the committee's task a new urgency. A sixty-eight-car freight—of which only the front half was air-braked—moving at about 6 miles per hour in the Pennsylvania's Harrisburg yards, made an emergency stop to avoid a train ahead. The rear half of the train collided with the front, buckling a light car and derailing others, which pitched over onto another track. An incoming passenger train plowed into the boxcars, setting off a carload of dynamite. The blast killed twenty-three and injured 110 people and cost the Pennsylvania about $643,000 in damage and compensation.[18]

As the *Engineering News* dryly remarked, the blast "started public discussion" on how best to ship explosives. The *New York Times* reported the accident with the front-page headline "Train Hit Dynamite; 163 Reported Dead." The next day the paper reduced the carnage to "20 Dead, 100 Hurt in Dynamite Wreck," but it was still page 1 news. Further stories followed, as did two editorials, one of which accused the carriers of having killed regulatory efforts (an error that it never bothered to correct), and called for "enactments which will better safeguard the traveling public." Yet despite the carriers' fears and the *Times'* call for regulation, the transportation of explosives never became a public issue. Other major newspapers failed to join the *Times* while popular magazines and the railroad unions also ignored the matter, probably because explosives caused comparatively few accidents and these groups preferred to concentrate on other aspects of railroad safety.[19]

The Harrisburg blast did precipitate a major review of policy at the Pennsylvania, however. On May 19, at a meeting chaired by W. W. Atterbury, then general manager and later president, the company's top officials assessed what needed to be done. Transportation officials noted that explosives originated from at least ninety-nine points on the system and that they routinely shipped seven hundred to eight hundred cars a month. As noted, the company's regulations were modeled on those of Great Britain, and they had most recently been revised in 1899. Officially the Pennsylvania had announced that the Harrisburg disaster "was an accident that could not be avoided. Every precaution known to train operation was in effect, and . . . there was not a single violation." Despite such brave claims, Atterbury urged the group to review company procedures carefully, for "our action . . . will . . . largely govern the practice of the country."[20]

In fact, company rules were far from perfect; officials concluded that the freight at Harrisburg had broken because of the practice of combining both heavy and light cars in the same train. Nor should explosives be shipped in trains with fewer than two-thirds of the cars air-braked. But all agreed that the more intractable problem was enforcement of its own rules. The company agent had not inspected the car before shipment. Nor could careful behavior be enforced on trainmen. Atterbury suggested one solution to that problem: "if that car of dynamite is right next to the engine and cab, you are not going to run into much." As Dudley summarized the whole problem, despite the company's complex rules, "we don't give a car of explosives as much care as we do a car of horses."[21]

The difficulties were those McCrea had described to Congress a few months before: agents were too busy to inspect or too hungry for business to condemn shipments in dangerous cars. Stricter regulations would lead to manufacture at the point of use (i.e., at a mine or construction site), or lose freight to the B&O or the Reading, or—worst of all—result in smuggling. Dudley claimed that for years the New York Central did not officially accept explosives but hauled them anyway, labeled as sugar, and the Lackawanna did the same thing.

The Harrisburg accident was, therefore, a nearly perfect example of a technological accident that reflected both complexity and close coupling of events. An emergency stop, a partly air-braked train, train makeup, and the unregulated shipment of explosives combined with traffic density and bad luck to create disaster. While such accidents may have been normal, they were not inevitable, and as Atterbury closed the meeting there was general agreement to increase the proportion of air-braked cars and to tighten enforcement of existing regulations. As a result, the Pennsylvania revised its requirements and added an inspector of explosives who soon confirmed that the company's rules were indeed largely honored in the breach and set about to remedy that deficiency. Little or nothing could be done about cars accepted in interchange, however. That was one of the issues for the ARA.[22]

Meanwhile McCrea's committee was making haste, spurred on perhaps by the specter that Harrisburg might arouse Congress and lead to regulation it could not control. In June the committee surveyed individual railroads as well as states and every city with over a hundred thousand inhabitants, requesting their rules for transporting explosives. McCrea had also learned from his previous mistakes and the committee requested suggestions from the explosives manufacturers and met with them to hash out differences. In an effort to educate ARA members on the need for regulation, the committee also called upon Charles B. Dudley for a technical report on the issues involved and requested him to consult with outside experts as well.

Dudley promptly enlisted Professor Charles Munroe, an explosives expert and professor of chemistry at George Washington University; Henry S. Drinker, a mining engineer and president of Lehigh University; and Dr. Charles McKenna, a consulting chemist and also an expert on explosives. Dudley presented their report to the ARA on October 25, 1905. It was substantially the same as the testimony he had given to Congress the previous year—suggesting that the outside experts were enlisted largely for their prestige—and pointedly noted the number of explosives manufacturers located on the routes of each carrier in the ARA.[23]

At the same meeting McCrea explained that he had appealed to the secretary of war and ultimately to President Theodore Roosevelt himself to obtain the services of Major Beverly W. Dunn as an expert consultant. The choice of a military man again probably reflected Dudley's enthusiasm for British procedures, for British inspectors of explosives were drawn from the Royal Artillery. Born in Louisiana in 1860, Dunn had graduated from West Point in 1883 and in 1890 joined the Ordnance Department, where he invented both a detonating device and an explosive intended for armor-piercing ammunition ("Dunnite").[24]

The committee's survey of existing regulations disclosed a hodge-podge that ranged from carriers with no special rules governing transport of explosives to the detailed regulations of the Pennsylvania and a few other trunk lines. In combination with Dudley's report, it made a pow-erful case for standardization, and the committee drew up a set of rules that were virtually identical with those of the Pennsylvania. At its October 25 meeting, the ARA adopted the rules unanimously. In effect, they com-mitted members not to compete for explosives traffic in ways that might compromise safety, thereby solving both the interchange problem and that of competitive rivalry—in theory at least. However, as the Pennsylvania's experience demonstrated, rules did not enforce themselves, and so at the same meeting the ARA also directed the committee to develop plans for a bureau of inspection to enforce the rules. Finally, the committee was to work for a law forbidding misrepresentation of explosives offered for ship-ment.[25]

In March 1906, the committee reported to the ARA that 185 carriers with 124,000 miles of road had adopted the rules and that it had submitted two bills to Congress. In January it had met with representatives of various general managers' associations with whom it formulated a constitution and bylaws for what was called "The Bureau for the Safe Transportation of Explosives and Other Dangerous Articles." The committee offered four justifications for such a bureau. First, in its absence each railroad would need to develop expertise on explosives and this was clearly inefficient. Second, common enforcement was necessary to prevent companies from modifying rules in order to obtain traffic. Third, the bureau would "be-come the repository of information, and study will be made of the subject with consequent progress . . . in the safety of transportation. If there is no central authority, such studies . . . will not be practicable." In short, the bureau would enforce the rules of the safety cartel while making it both efficient and progressive. Finally, the committee warned that "unless the railroads do something of this kind themselves . . . they will be liable at any moment to become subject to regulation by legislation which will be much more onerous, without any gain in efficiency." Apparently persuaded, the ARA adopted a constitution and bylaws for such a bureau and directed its organization.[26]

Private Rules Become Public Regulations

On June 10, 1907, the ARA Bureau for the Safe Transportation of Explosives and Other Dangerous Articles opened to enforce the ARA rules on the transportation of explosives. While neither the bureau's activities nor the ARA rules had any legal foundation, that was soon to change. In 1906, the ARA had submitted a bill to Congress that made it a crime to mislabel explosives and required each carrier to develop its own regulations for explosives and other dangerous substances. It also set up the ICC as umpire to adjudicate differences. Who wrote the bill is unknown, but it reflected the carriers' generally favorable experience with the regulation of air brakes where ICC rules requiring minimum numbers of workable brakes achieved what the ARA had been unable to do alone. The bill was submitted too late for passage; it was reintroduced in 1907 and subsequently modified in three important ways. The carriers had discovered the federal law of 1866. It required dynamite to be packed in metal boxes, which was unsafe, and carried a fine of $1,000 for each violation, half of which was to go to the informant. The law had never been followed, but the carriers were understandably worried that it could lead to mischief and so the new bill explicitly repealed previous legislation.[27]

Of greater importance was the narrowed scope of the bill. Although it retained the phrase "other dangerous articles" in its title, for reasons I have been unable to discover, the bill gave the ICC power to regulate the transport only of explosives.[28] The bill directed the ICC to promulgate such regulations within ninety days of the act's passage. It also stated that the commission might modify these regulations at any time either on its own initiative or in response to petition from an outside party. The railroads supported the bill, although they were disappointed that it covered only explosives. McCrea, now president of the Pennsylvania, concluded in a letter to Dudley that "half a loaf is much better than no loaf at all." The Pennsylvania also lobbied strongly for the bill, and McCrea wrote personal letters to Representative William P. Hepburn and Senators Stephen Elkins, John Kean, and Joseph B. Fouraker. Other carriers also lobbied for the bill. The hearings were brief, lasting only two days. With both the ARA and the manufacturers of explosives solidly in support, and no one else interested, it passed easily and become Public Law 174 on May 30, 1908.[29]

As noted, while the title of the act referred to "other dangerous articles," it authorized the ICC to regulate only shipment of explosives. Just prior to the bill's passage, in April 1908, the bureau adopted rules governing shipment of acids and flammables. These were to take effect on October 15, 1908, the same date the ICC promulgated its regulations for shipping explosives. The commission, with only three months to develop a code of regulations and with no expertise in explosives, met with the ARA and adopted its rules, as no doubt had been intended. Although the law did not directly grant the commission authority beyond the regulation of explosives, both it and the bureau argued that the latter's rules covering acids and flammables had the same legal standing as the federal regulations governing transport of explosives. Echoing Dudley, they claimed that explosives could not be regulated without also regulating shipment of other hazardous materials, which soon came to include, in addition to acids and flammables, compressed gasses and poisons. This awkward situation was not rectified until 1921, when Congress modified the law to include other dangerous cargo.[30]

The new procedures represented a corporatist middle ground between the old laissez-faire approach that left rules up to each carrier and the Pennsylvania's initial efforts to achieve a British-type centralization of authority in the ICC. The carriers, working through the ARA Bureau of Explosives, would take the lead in self-regulation while the commission would supply the legal muscle to uphold private regulation. The success of the approach therefore depended crucially on the ability of the bureau to develop and enforce its new rules.

Developing Regulatory Procedures, 1907–1912

When the bureau opened for business in June 1907, it had a chemical laboratory, sixteen inspectors, Charles B. Dudley as its president, and Major Beverly W. Dunn as chief inspector. The carriers realized that obtaining the cooperation of manufacturers was vital to the bureau's success, and to do so they needed a chief inspector who was both expert and perceived as impartial. Dunn was a superb choice—he remained in the job until 1935—but he proved difficult to obtain. Dudley organized a lobbying effort that included personal appeals from various important railroaders to President Roosevelt and Secretary of War Taft. In a letter that reveals much about business-government relations at the turn of the century, McCrea lectured Taft that it was his duty to deploy Dunn to the railroads. "In all important countries ... [regulation of explosives] is a work of the government and it should have been so in the United States," he instructed the secretary. "But it is now being conducted entirely by the railroad companies and the least that the Government can do is give us the benefit of Col. Dunn's experience."[31]

Selling the Bureau to the Railroads

The bureau's initial membership consisted of seventy-eight carriers with about 130,000 miles of track, or about 40 percent of the total. Such important lines as the B&O, the Milwaukee, the Great Northern, and the Northern Pacific had not joined. The laggards were probably responding to a mixture of motives. The B&O had adopted regulations similar to the Pennsylvania's and probably saw no need for anything more, while the Milwaukee had only two manufacturers of explosives on its lines and the Great Northern and Northern Pacific had none. The bureau's constitution granted it considerable powers. It was authorized, for example, to stop dangerous shipments, but despite (or perhaps because of) his military background, Dunn realized that his power to command was more apparent than real. The bureau was a voluntary association—a safety cartel designed to limit some forms of competition for shipments of explosives—and like many other cartels, it might well fall apart unless he could demonstrate that it yielded net benefits to members. Nor with so few inspectors could the bureau have enforced its regulations, even if Dunn had been so inclined. Finally, Dunn was familiar with the Pennsylvania's enforcement difficulties, and he had few illusions that left to their own devices companies would enforce the new rules. For all these reasons, his initial strategy was to encourage voluntary compliance, primarily by informing and educating key railroad personnel both on what the new rules required and why they were necessary, charting a course of action that Frank McManamy followed two years later at the Bureau of Locomotive Inspection (chapter 7).[32]

Dunn attacked on several fronts. Inspectors were ordered to provide a general evaluation of procedures on individual carriers that was then given to the general manager. Dunn himself met with representatives of both the railroads and manufacturers. The small makers of explosives were suspicious that the bureau was an agent of the DuPonts, and at one meeting they took the "opportunity to express freely their objections to all proposed restrictions." To allay suspicions Dunn promised no major changes in rules without prior consultation, and he urged all the explosives manufacturers to form a conference committee. In 1908 this relationship was formalized as the bureau developed a class of associate members that included many manufacturers.[33]

Both the inspectors and Dunn also gave scores of speeches each year to individual companies, railroad clubs, chambers of commerce, and state regulatory commissions. In his reports and talks, Dunn justified the bureau's work with numerous examples, his goal being to preserve and expand its membership and to encourage voluntary compliance with its rules. He emphasized that compliance by the carriers would require more than simply distributing the new rules to employees. Dunn described an inspection during which the station agent was asked if he had a copy of the regulations. "The agent was not quite sure. It depended on whether this particular circular was printed on paper soft enough to use in cleaning lamp chimneys. If so it had probably been expended for this purpose. If not it might be in a box where he kept a lot of circulars he had never found time to read."[34]

Dunn illustrated what could happen to carriers who failed to follow the bureau's guidelines by describing a disaster on the Michigan Central, near Essex, Ontario, in August 1907. That company had handled a car containing 5,000 pounds of dynamite so roughly that the boxes (which were improperly stayed) broke loose and landed on end. This led them to leak nitroglycerine, which the conductor—oblivious to the dangers—playfully used to wash another employee's face. The nitroglycerine also dripped from a hole in the floor onto a wheel, and several workers later recalled hearing small explosions every time the car was moved and the wheel made contact with the rail. Apparently after the car had stood for a while, a pool of nitroglycerine built up and when the car moved the explosion set off its contents. Property damage amounted to $200,000; the company's liability came to about $50,000 and it was forced to plead guilty to criminal negligence and fined $15,000. The bureau had not yet inspected the Michigan Central, and Dunn simply stated that "enforcement of the regulations would have prevented this unfortunate accident."[35]

The bureau's annual reports also served as educational devices. In the first report covering about half of 1907, bureau inspectors found that 47 out of 141 factories it visited had serious safety violations, and 59 of 134 storage magazines (used by manufactures to house inventories of explosives) contained leaking dynamite. Preliminary figures also recorded 17 accidents with explosives that killed 31 and injured 78 people while destroying $500,000 of property. These were the first such accident statistics ever published. As the data became more complete they again served as a guide for future regulations—as had the earlier ICC statistics on coupler injuries and train accidents. In 1908 the bureau reported that it had condemned as unsafe 4,852 boxes of high explosive and 531 kegs of black powder; the next year the numbers were, respectively, 10,069 boxes and 1,468 kegs.[36]

The reports also provided concrete examples of both violations and accidents that were intended to justify the bureau's activities. Dunn reported finding one car loaded with 10,000 pounds of black blasting powder. "Kegs had not been properly stayed and had shifted, breaking open some of the kegs and spilling powder on the car floor." Not all such stories had a happy ending. In 1909 a carload of 10,000 pounds of black powder being transported in open mine cars by the Oriskany Ore and Iron Company blew up, killing four men and injuring two others. The carrier was not a member of the bureau, Dunn's report emphasized, and the incident illustrated the dangers of failing to follow bureau and ICC procedures.[37]

Technological Change in the Shipping of Hazards

The bureau soon became an agent for discovering as well as disseminating new knowledge. The chemical laboratory immediately began testing dynamite for exudation and it developed an improved wrapping. In 1909, after the ARA expanded its regulations to cover compressed gasses and liquids, the laboratory also began to test acetylene tanks and fireworks that were subject to spontaneous ignition. The bureau also drew upon the larger railroad technological network, prevailing upon the Pennsylvania Railroad to study freight car impact forces at its Altoona laboratories so that package design and staying could be put on a more scientific basis. Not until 1919 would the bureau have its own engineering facility.[38]

Inspection also yielded new knowledge that led to safer practices. Those who had agitated for the bureau focused on dangerous manufacturing and shipping practices, but they had not thoroughly appreciated the dangers of storage magazines. Bureau inspectors soon discovered there were thousands of such magazines, which were dangerous because many contained old, leaky dynamite, and also because they were often located within a few feet of the tracks. In 1909, the bureau, along with the conference of explosives manufacturers, studied other countries' regulations governing the location of magazines and in 1910 came up with its own American schedule that related the amount of explosive stored to a minimum distance from surrounding buildings.[39]

Such activities help explain why an increasing number of railroads chose to join the bureau. Membership rose steadily and by 1912 totaled 300 carriers with 252,130 miles (68%) of track. Thus, while the bureau's work had some qualities of a public good, more and more carriers chose not to ride free. There were two important benefits that accrued only to members. First, the bureau's inspections and advice were valuable in reducing risks—even for those carriers that only transshipped hazardous material. Second, the bureau was a quasi-governmental organization; it not only enforced regulations, its members also helped shape their content. Membership, in short, was the ticket to political influence. Probably for similar reasons many explosives manufacturers became associate members in 1909 and could, in Dunn's phrase, "share in the direction and support" of the work. In addition, the bureau and its members were not above a little arm twisting to encourage outsiders to join. Few eastern carriers could have wished to antagonize the Pennsylvania, and that company's records reveal that President McCrea personally appealed to the presidents of the Lehigh Valley and the Richmond, Fredericksburg & Potomac to join. Dunn also employed the ICC and the courts to raise the benefits of membership. In 1912 he sent a circular letter in which he quoted ICC Commissioner E. E. Clark on the importance of uniform instruction of employees and cited a court case that bound intrastate carriers by the ICC's regulations. Sixty-two small roads promptly applied to join the bureau.[40]

Just as the logic of regulating explosives had led the railroads to extend coverage to other hazardous materials, so regulation of rail freight inevitably led to inclusion of other methods of transport. As early as 1908 Dunn began to worry that the rules of express companies (which operated as freight forwarders) differed from those of the ARA. Many conferences led to harmonization and in 1910 the bureau allowed express companies to become members. Steamship lines proved more recalcitrant, for in 1909 Dunn reported that they were not complying. When the railroads threatened not to accept shipments that failed to comply they had a change of heart and in 1910 eleven major lines joined the bureau.[41]

Thus within about five years of its foundation the bureau included virtually all the carriers, and Dunn had formulated the basic structure and procedures that he and his successors would employ for the next half century. The bureau's accident statistics, inspections, and chemical investigations yielded information on successful regulations and potential hazards. Armed with these data, the bureau could direct its own research to discover solutions and coordinate with manufacturers, shippers, and their associations to develop safer practices and to propose to the ICC regulatory modifications to reflect new technology and newly discovered dangers. Although the Pennsylvania had once proposed a British model of government regulation, it now preferred the more flexible, railroad-driven corporatist solution. As early as 1911 McCrea had concluded that while the bureau cost the Pennsylvania $7,000 a year in dues, "I would rather pay this than have the matter put under government control."[42]

Regulating the Transport of Dangerous Substances, 1908–1935

As accident and inspection data and reports from the chemistry department came in, a clearer picture of risks began to emerge. The dangers of magazines were dealt with and a conference of manufacturers finally yielded modifications intended to make black powder kegs less likely to leak. In 1914 the bureau met with match manufacturers to establish specifications for corrugated boxes. Cardboard was gradually driving wood boxes out of the transportation of matches and the shift was responsible for a disproportionate number of fires. The new rules yielded improvements but cardboard remained comparatively risky.[43]

Improving Safety

The laboratory routinely tested new explosives, condemning those it found prone to spontaneous ignition or otherwise unstable. The combined impact of careful inspection and new technology was remarkable. Although production (and therefore presumably shipments) of explosives continued to grow, accidents declined sharply, averaging fifteen to twenty a year in the 1920s, while casualties also dropped precipitously relative to their prewar levels (Table 8.1).[44]

But if the dangerous shipping of explosives was quickly brought under control, such was not the case with other hazards. The extension of regulation to cover first acids and flammables, and later other hazardous substances enormously increased the scope and complexity of the bureau's activities. Moreover, production of explosives was growing slowly and so, presumably, were shipments. The same could not be said of many of the "other dangerous substances," for the early twentieth century saw an extraordinarily rapid expansion of the chemical and petroleum industries. Between 1909 and 1919 sulfuric acid production doubled from 2.8 million to 5.6 million tons and then rose another 50 percent to 8.5 million tons by 1929. Production of compressed gasses also expanded. Anhydrous ammonia production more than tripled during 1909–27 while that of oxygen increased from 4 million cubic feet in 1909 to 3.1 billion by 1929. Most important of all was the booming demand for gasoline, which drove production from 540 million gallons to 17.5 *billion* between 1909 and 1929.[45]

Table 8.1. **Accidents and Injuries from Transportation of Explosives and Other Dangerous Substances, 1907–1930**

	Explosives		Other Dangerous Articles	
Year	Accidents	Injuries	Accident	Injuries
1907	79	132	—	—
1908	22	79	—	—
1909	12	13	—	—
1910	16	3	385	34
1911	10	6	254	32
1912	9	6	326	23
1913	11	4	479	38
1914	11	7	508	124
1915	11	6	609	635
1916	17	112	1,095	63
1917	5	1	1,223	66
1918	11	5	1,342	119
1919	10	4	1,724	78
1920	11	13	2,057	63
1921	19	25	1,627	67
1922	8	1	1,373	65
1923	16	38	1,445	57
1924	16	4	1,402	16
1925	28	0	1,703	75
1926	20	9	1,494	30
1927	4	0	1,340	39
1928	12	0	1,398	124
1929	20	1	1,246	34
1930	15	2	1,145	31

Source: ARA Bureau of Explosives, *Annual Report of the Chief Inspector,* various years, Appendix I.
*Includes fatalities.

Regulation Begets More Regulation

The extension of bureau activities to cover these hazards did not begin smoothly, for while the regulations governing shipment of explosives had been developed in concert with manufacturers, the rules for other hazards had not been so effectively sold. Prior to their announcement Dunn had met with the Manufacturing Chemists and Wholesale Druggists and he had their official endorsement, but for many members of these groups and other trade associations the regulations came as a nasty surprise. As Dunn noted, they "prescribed practices that differed radically from those followed previously by railroads and shippers," and the result was "ignorance, misconception, and opposition." In their confusion, some shippers simply labeled everything; in early 1909, Dunn reported that 25 to 30 percent of all boxcars were being placarded with warning labels, and a few station agents reported that 80 to 90 percent of cars were placarded.[46]

The bureau responded in three ways, the first being educational. It increased the number of inspectors from sixteen in 1907 to twenty-six in 1910, and in hundreds of talks and meetings with railway agents, shippers, and representatives of trade groups, Dunn and his inspectors explained the new rules. Again, the annual reports were used to educate shippers with case studies of the dangers from poor packing in breezy format under the titles of "Bureau Briefs" and "Explosive Questions and Answers." Dunn proved willing to relax those rules that seemed burdensome. In July 1909, the bureau reduced the flash point for flammables from 100 to 80 degrees F (but not 70 degrees as the manufacturers wished); it specified that small quantities of material need not be labeled, and exempted some articles. Such changes cut the proportion of placarded boxcars in half. A decade later, in 1919, when the bureau proposed "Rules Governing the Location of Loading Racks and Unloading Points" for gasoline that would have placed them no nearer than 80 feet from the tracks, a storm of protest erupted from the oil companies. In a series of nine meetings with industry representatives throughout the year Dunn agreed to significant modifications of these and other provisions of the new rules.[47]

But the bureau could also play rough. In 1908, when one drug manufacturer proposed to label everything in an effort to nullify the rules he was told that such mislabeling would be interpreted as a violation of both the ICC regulations and the Food and Drug Act. One railroad that failed to comply was told that its shipments would be refused at junction points. Several years later one shipper who persisted in loading flammables into unapproved barrels was embargoed.[48]

Despite such efforts, enforcement remained a problem, and Dunn repeatedly returned to the topic in his annual reports. In 1913 he complained that many railways seemingly left matters entirely up to bureau inspectors and he urged the carriers and shippers to designate someone whose job it was to see that regulations were obeyed. Finally, in 1919 under the authority of the U.S. Railroad Administration the bureau published a circular and organization chart suggesting the lines of authority, duties, and responsibilities for dealing with hazardous substances. The following year Dunn reported that the "plan has been quite generally adopted . . . [and its] good effect . . . is shown by the reports of our inspectors."[49]

As with explosives, the bureau also worked with shippers and manufacturers to improve the safety of containers used to ship a host of other dangerous substances. In 1914 it published standard tests for gas cylinders. As early as 1910 reports of accidents confirmed the dangers of shipping acids. Glass carboys were a particular problem: poor cushioning led to breakage, while many carboys were defective and others through repeated use had developed leaky stoppers and necks that could not be sealed. Despite meeting with manufacturers, little was accomplished and shipments of acid continued to cause a disproportionate number of accidents. In response to these difficulties, in 1919 the bureau established its own test and specification department, employing two mechanical engineers. The department was generally charged with inspecting and testing approved containers, and the ARA soon gave it the mandate to develop standard containers for all kinds of products, whether hazardous or not.[50]

One of the department's first tasks was to develop standardized tests for carboy containers as well as a method to discover the shocks they experienced during transport. In 1921 its engineers reported that they had developed a machine to grind the neck of carboys to improve their seal, as well as a better stopper and gasket and a machine to test for weak carboys. As Table 8.2 reveals, these and other changes resulted in a dramatic drop in the hazards from transporting acids, especially those associated with breaking carboys.[51]

The bureau's accident data demonstrated that while transport of acids and corrosives was a leading cause of accidents, flammable liquids caused far more damage. For example, in a typical year (1914), while there were more than twice as many accidents attributed to acids and corrosives as flammables (192 versus 80), the latter accounted for $152,000 of damage compared to only $5,000 for acids and corrosives. As if to emphasize the

Table 8.2. **Accidents and Damage from Transporting Acids, 1915–1929**

	All Containers			Carboys	
Year	Accidents	Injuries	Damage $	Accidents	Damage $
1915	257	26	175,487	—	—
1916	560	35	69,963	—	—
1917	690	27	277,077	—	—
1918	675	29	137,982	—	—
1919	812	32	59,136	—	—
1920	1,060	15	90,181	913	38,455
1921	801	19	34,759	744	22,635
1922	667	37	31,188	613	21,493
1923	652	7	28,162	536	10,943
1924	483	19	14,731	359	8,356
1925	553	19	20,519	376	11,203
1926	405	13	19,302	277	18,244
1927	419	5	17,713	290	8,259
1928	333	13	9,725	237	3,870
1929	335	13	11,463	204	5,655

Source: ARA Bureau of Explosives, *Annual Report of the Chief Inspector 1930* (New York, 1931).

problem, one of the most spectacular and deadly accidents in the history of railroading occurred the next year, on September 27, in the Santa Fe yards in Ardmore, Oklahoma. A carload of casinghead gasoline exploded, demolishing a substantial fraction of the town, killing forty-one people and injuring 458, and costing the railroad about $593,000 in claims and damages. Casinghead gasoline is highly volatile and the shipper of the Ardmore tank car had left insufficient space to allow for expansion. But the principal cause of the disaster was the removal of the dome cap by an employee, which led to an eruption of gasoline that was ignited by a switch lamp, some 215 feet away.[52]

The Ardmore disaster revealed not only a new kind of hazard, but also that the bureau needed to broaden its focus from the packaging of individual shipments to include freight car design. It led the bureau to propose special tank cars for highly volatile liquids, and in 1916 the bureau reported that it had detailed a special inspector to ensure that ARA specifications for tank car construction were followed. Dunn also noted that special investigation of accidents demonstrated that existing specifications for tank cars were part of the problem, and the bureau began to look at issues associated with safety valves, dome closures, and outlet valves until war forced curtailment of the investigation.[53]

Dunn returned to these problems in successive reports. He noted that bureau educational work had led to improved handling of tank cars by railroad personnel, especially during wrecks and derailments, while cooperation with the various oil industry groups had improved car maintenance and loading practices. But design problems remained. The most serious difficulty involved bottom outlet valves, which often leaked and were prone to break off in an accident. Safety valves also leaked at too low a pressure, and manhole covers leaked and could be removed with the contents under pressure, leading to disaster, as Ardmore had impressively demonstrated. To remedy these defects, in 1919 the ARA delegated a special Tank Car Committee of its mechanical section with representatives from the bureau and the American Petroleum Institute to develop better designs. In the mid-1930s the Tank Car Committee was also the driving force that led the railroads to replace the old-style arch bar car trucks with safer, cast steel designs (chapter 9).[54]

In 1927, in yet another example of associationalism, the ARA committee employed an engineer on loan from the National Bureau of Standards to evaluate proposed tank car modifications. That same year Dunn reported progress. Manhole covers were modified so they need not be opened for loading and unloading. A positive sealing bottom outlet valve was developed, as was a new, more leakproof safety valve. On July 1 of that year, the ICC required all of them on new construction. Each of these devices was in turn further improved in the light of evidence gleaned from accident statistics, and eventually these modifications also became standard on all tank cars. The bureau's data soon demonstrated the payoff to these changes. In 1931 Dunn reported that during the past six years there had been twenty-five accidents involving the new discharge valves that had cost an average of $25 each, whereas 189 accidents from the old-style valves had cost an average of $4,200 each.[55]

On February 1, 1935, Beverly Dunn, who had been ailing for several months, finally stepped down as head of the Bureau of Explosives, and on May 10, 1936, he died. As advisor to the ARA in 1905–6, and during his twenty-eight years as head of the bureau, Dunn played a leading role as the railroads invented the modern procedures for shipping hazardous substances. His combination of expertise and independence eased the transition of all parties to the new regulatory regime and his role, like that of Dudley's, reflects the importance that key individuals have sometimes played in shaping business regulation. The bureau's pioneering developments provided a model when government controls were extended beyond the railroads, for the ICC later applied its procedures to highway transportation in the 1930s while the Coast Guard and Civil Aeronautics Board adopted them for marine and air transport as well.[56]

This associationalist system of controlling shipments of hazardous substances developed out of a particular confluence of historical events. It represented an early federal application of social regulation; yet unlike the associationalism of the 1920s it was less a reflection of ideas and ideology than of stark necessities. By 1904, led by the Pennsylvania, the railroads recognized the increasing dangers from shipping the products of the modern chemical industry. Individual company responses to these dangers were inadequate in the face of competitive pressures and so institutional innovations became necessary. With manufacturers opposing a statist solution, associationalism was the outcome. Thus arose the rather strange partnership between the Bureau of Explosives and the ICC. The bureau investigated accidents, educated and cajoled carriers and shippers, enforced its own rules and ICC regulations, and proposed modifications of both technology and regulations that the ICC then adopted.[57]

Regulating explosives required a grasp of both science and technology, and this system located the driving force for change in the private sector at a time when the government—as represented by the ICC—had little technical expertise. It encouraged the carriers and major shippers to develop safety expertise and led to improvements in the technology of hazardous materials transport. The experience underlines the historically contingent nature of such concepts as "normal accidents," for the new procedures contained disasters even as shipments of hazards rose. Because regulations developed out of private bargaining they generated little opposition. As a result, while many other associationalist schemes withered with the passing of the 1920s, the system of regulating shipments of dangerous materials remained in place largely unchanged until the 1960s.

That decade saw an increasing number of highly publicized accidents involving hazardous materials that heightened public fears of technological catastrophe (chapter 10). These resulted not from a failure of the bureau but rather from the upsurge in derailments that reflected the railroads' long-term deteriorating finances. But this time the carriers could produce no Charles B. Dudley or James McCrea, and the new dangers led to a wholesale federalization of railroad safety in 1970.[58]

(LEFT) In August 1853, a wreck at Valley Falls, Rhode Island, on the Providence & Worcester Railroad killed fourteen people. The accident resulted from the conductor's failure to follow the company timetable. (George Eastman House)

(BELOW) Twenty-three persons died in this derailment on the Camden & Amboy Railroad in late August 1855, at Burlington, New Jersey. The train struck a carriage while backing up at a high rate of speed so as to reach a siding before another train passed on the same track. (Library of Congress)

(ABOVE) On December 18, 1867, the rear car of the Lake Shore & Michigan Southern Railway Express jumped the track while crossing a bridge near Angola, New York, killing forty-two passengers. Equipment failure at high speed plagued early railroading in the United States. (*Harpers Weekly,* Jan. 11, 1868)

(RIGHT) "Death Rides the Rails." In the late nineteenth century, Harpers Weekly and other periodicals luridly depicted the dangers of railroad travel. (*Harpers Weekly,* Sept. 23, 1865)

(LEFT) In a mishap that probably took place in the 1880s, this Denver & Rio Grande train derailed after striking a cow. For good reason the projection on the locomotive came to be called the "cowcatcher." (Colorado Historical Society)

(BELOW) Stubb switches that broke the main track, along with bad track by itself, made nineteenth-century American trains derailment-prone. (B&O Railroad Museum)

(TOP) Bad track, the curse of all early lines, provoked the Pennsylvania Railroad to establish performance contests. This inspection train carried company officers, who awarded prizes to workers for the best-maintained sections. (*Railroad Gazette,* Feb. 2, 1877)

(MIDDLE) In August 1890, when a careless Old Colony Railroad employee left a jack on the tracks, this derailment took place near Quincy, Massachusetts, killing twenty-three travelers. Faulty safety rules proved the bane of early track maintenance. (Massachusetts Board of Railroad Commissioners, *Annual Report,* 1890)

(BOTTOM) The first air-braked train on the Pennsylvania, proudly photographed at the end of the nineteenth century. (*Locomotive Engineering,* Feb. 1900)

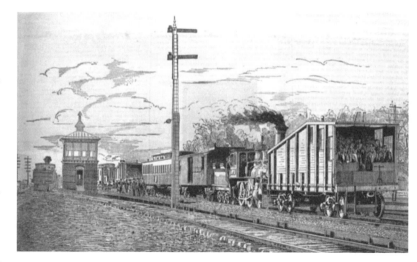

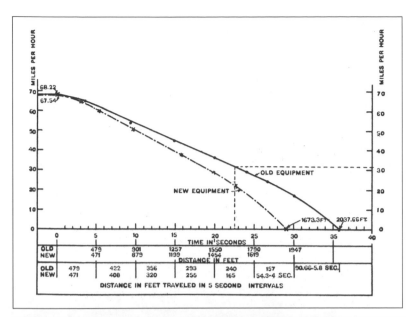

(TOP) Early airbrakes dramatically improved safety. Tests demonstrated that later improvements, which included high-speed brakes, largely offset increasing train speed and weight. (W. V. Turner, *Developments in Air Brakes for Railroads* [Westinghouse, 1909])

(MIDDLE) While this slow-speed accident near Silver Creek, New York, in the early fall of 1886 did little damage to the Nickel Plate locomotives, it nonetheless—due to poor car braking—telescoped several passenger cars, killing twenty riders. (*Locomotive Engineering,* June 1893)

(BOTTOM) "Another Sacrifice to the Devouring Car Stove." Train accidents often upset these heating devices, setting fire to wooden cars and causing much collateral loss of life. (*Harpers Weekly,* Feb. 19, 1887)

(TOP) "Winter and the Deadly Car Stove Routed by the Steam Car Heating Forces." By the 1890s, pressure from state regulators and railroad journals led the carriers to scrap the old car stove and install steam heat in passenger cars. (*Railway Master Mechanic,* Jan. 1890)

(BOTTOM) Boiler explosions like this one on the B&O in June 1886 resulted from flaws like thin boiler iron—and poor maintenance. Such accidents killed many trainmen. (*Locomotive Engineer's Journal,* Dec. 1931)

(LEFT) A typically dangerous downtown rail-street intersection. (*Century Magazine,* May 1898)

(BELOW) Many more people died at rail stations and in yards than in train wrecks; for many years, railroads in the United States allowed passengers, bystanders, and even playing children free or nearly free access to the track. Under threat of legal liability, companies grudgingly put up fences and built urban bridges and underpasses. (*Outlook,* Aug. 1913)

(RIGHT) Poling freight cars allowed a locomotive to switch cars on another track. The procedure notoriously traded worker safety for yard productivity. (*Railroad Gazette,* Aug. 18, 1893)

(BELOW LEFT) "A Tell-Tale That Tells" in this instance turned out to be a hollow staybolt that leaked steam if broken. New technologies gradually diminished the risk of boiler explosion. (*Locomotive Engineering,* Aug. 1899)

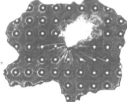

LOCOMOTIVE ENGINEERING.

A Tell-Tale

You can't Neglect the Warning Given by a Broken Stay-Bolt—if it's a "Falls Hollow." It won't be Happy 'till You fix it.

Absolutely Safe. No waiting till there are many broken and then —trouble. Every Stay-Bolt is a Detective. Ask for Circular.

Falls Hollow Stay-Bolt Co. Cuyhoga Falls, Ohio.

That Tells.

(LEFT) Problems with automatic couplers in the 1890s included their bewildering variety but also the extra expense of weeding out those of poor quality. Only slowly did manufacturers adopt testing equipment like this machine. (*Railway Review,* July 7, 1894)

(ABOVE AND OPPOSITE PAGE) In the late nineteenth century depressed labor markets (along with high passenger fares) led thousands of unemployed men to become hobos, many of whom died while hopping freight cars. Hobo trespassing, and mortality, reached a peak in the early twentieth century. (*Century Magazine,* Apr. 1893)

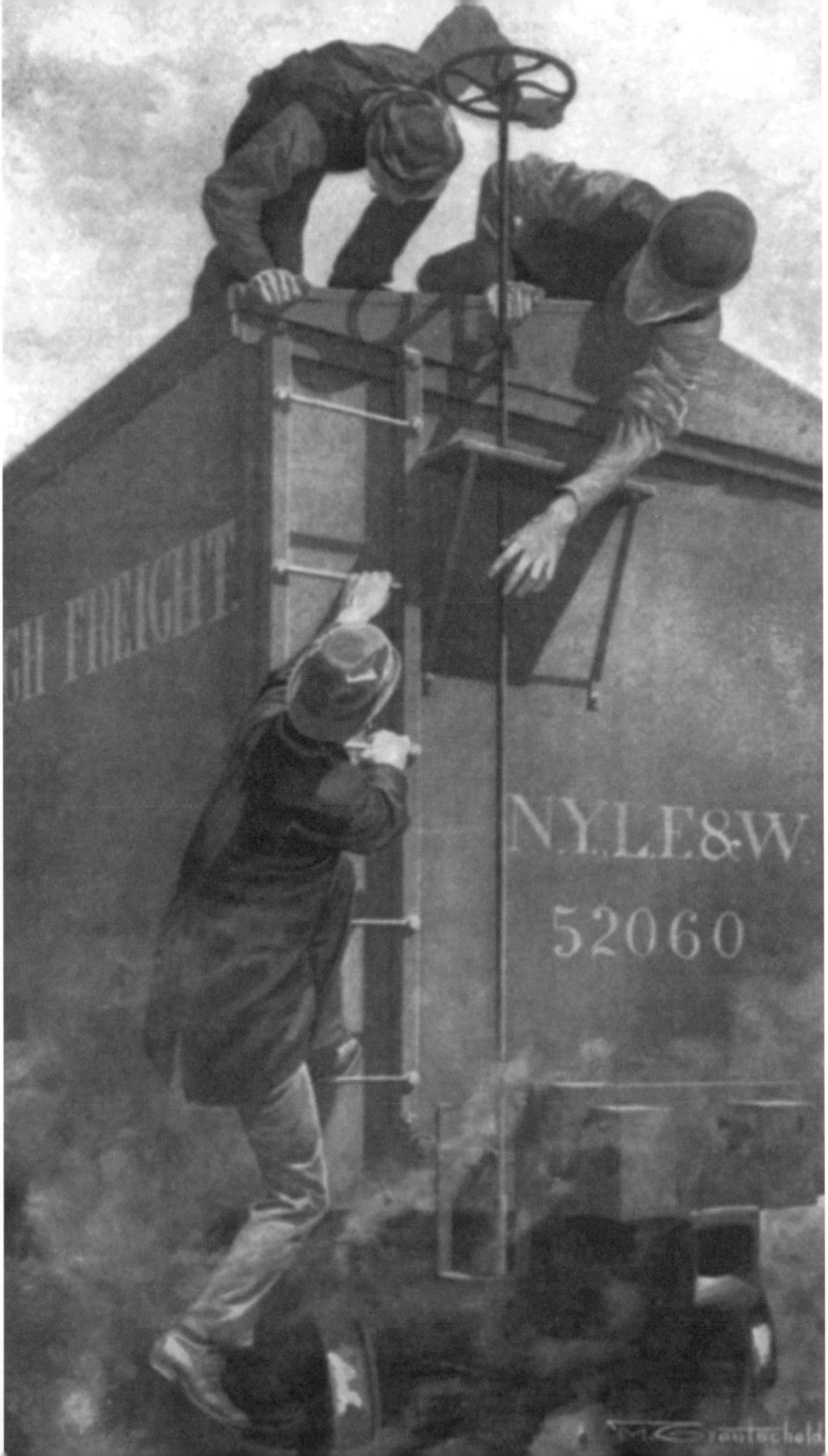
GH FREIGHT
N.Y.L.E&W.
52060

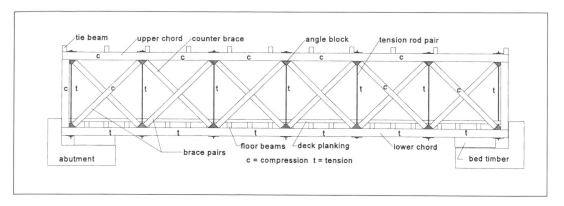

tie beam   upper chord   counter brace   angle block   tension rod pair

c   c   c   c   c

c | t   c | t   t   t   t | c   t   c | t

t   t   t   t   t   t

abutment   brace pairs   floor beams   deck planking   lower chord   bed timber

c = compression t = tension

(TOP) Wooden Howe Truss bridges, relatively cheap and quick to erect, were built by the thousands, but they were high-maintenance structures and routinely failed from fire, rot, and overloading. (Joseph Nelson, *Spanning Time* [New England Press, 1997])

(MIDDLE) Near Ashtabula, Ohio, in late 1876 a downed bridge—supposedly due to a broken car axle—killed eighty rail passengers. Poor design remained a central problem of bridge failures, as American engineers saw matters. (*Frank Leslie's Weekly,* Jan. 20, 1877)

(BOTTOM) This wooden-truss bridge on the Brattle-boro & Whitehall Railroad, badly stressed for some time, finally collapsed in mid-August 1886. (Windham Co., Vermont Historical Society)

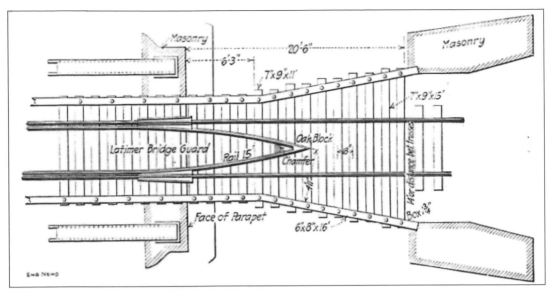

(ABOVE) The Childs-Latimer guarding and rerailing system—this one on the New York, Ontario & Western Railroad—could prevent most bridge knockdowns, and professionals championed its use. (*Engineering News*, Jan. 28, 1888)

(LEFT) "Marks' Patent Artificial Limbs, with Rubber Hands and Feet, Are natural in action, noiseless in motion and the most durable in construction." Advertisements in railroad journals took a matter-of-fact approach to the risks of working around trains. Companies responded to commonplace injuries by establishing complex medical and beneficial associations and in some cases by hiring rehabilitated workers. (*Railway Review*, Jan. 6, 1893)

(TOP) Although locomotives had become much safer by the late nineteenth century, boiler explosions still continued to kill many trainmen. In 1911 the Brotherhood of Locomotive Engineers successfully pressed for federal locomotive inspection. (ICC, Bureau of Locomotive Inspection, *Annual Report*, 1912)

(MIDDLE) A derailment at Lakeview, Washington, on the Oregon & Washington Railroad killed five people in May 1913. Steel cars may have saved many lives, but early designs failed to provide the security the industry and advocates of steel promised. (Union Pacific Museum)

(BOTTOM) Early automatic block signals, operated by weights and electromagnets, were unreliable, expensive, and accordingly rare. (*American Railway Signaling*, 1953)

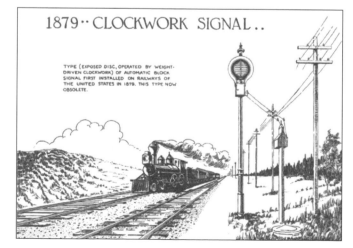

1879 ·· CLOCKWORK SIGNAL ..

TYPE (EXPOSED DISC, OPERATED BY WEIGHT-DRIVEN CLOCKWORK) OF AUTOMATIC BLOCK SIGNAL FIRST INSTALLED ON RAILWAYS OF THE UNITED STATES IN 1879. THIS TYPE NOW OBSOLETE.

(TOP) By 1908 rising traffic density and improving technology led Union Pacific to install many miles of automatic block signals and advertise its devotion to safety in illustrated brochures. (Union Pacific Railroad, *Making Travel Safe* [Omaha, 1911])

(MIDDLE) In 1908 new American Railway Association rules and ICC regulations required the proper packing and prominent labeling of freight cars carrying explosives and inflammable materials. (ICC, *Regulations for the Transport of Explosives*)

(BOTTOM) This wreck near Victoria, Mississippi, on the St. Louis & San Francisco Railroad in October 1925 owed to a transverse fissure in a rail. Twenty-one died. Transverse fissures continued to bedevil carriers and puzzle engineers until the late 1930s, when scientific breakthroughs established their causes. (ICC, *Report of the Bureau of Safety*)

INFLAMMABLE PLACARD

(TOP) Research by the carriers and steel companies led to improved rails and to the Sperry rail-flaw detector car, an early application of non-destructive testing. Electromagnetism detected hidden transverse fissures. (Original source unknown, author's collection)

(MIDDLE) "Flirting with the Undertaker." The issue of grade crossing accidents as the railroads and trainmen interpreted them. (*American City*, Feb. 1927; courtesy of the Brotherhood of Locomotive Engineers, *Engineer's Journal*)

(BOTTOM) This rear collision on the Long Island Railroad near Richmond Hill, New York, on November 22, 1950, killed seventy-nine passengers. It resulted when a trainman ran a signal; it revealed flaws in the ICC requirements for automatic train control. (Courtesy, Richmond Hill Historical Society)

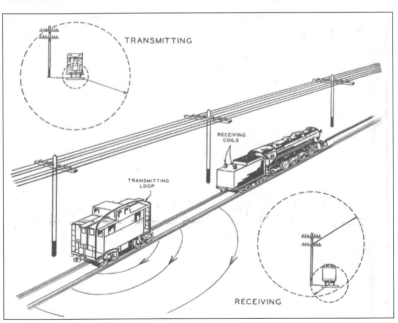

(TOP) After World War II worker motivation continued to dominate industry safety efforts. In 1954 crewmen with the worst safety record on the Southern Pacific's Rio Grande division had to tow this caboose as a reminder of their failings. (*Modern Railroads*, Feb. 1955)

(MIDDLE) New technologies such as the Servo infrared hotbox detector (circled) helped control derailments in the 1950s. (*Modern Railroads*, Mar. 1958)

(BOTTOM) Inductive train radios allowed voice communication within a moving train and with stations. Timely warnings of track obstructions and dragging equipment increased both productivity and safety. (*Railway Signaling*, Nov. 1944)

TRANSMITTING

RECEIVING COILS

TRANSMITTING LOOP

RECEIVING

(TOP) Postwar safety technologies included ultrasound, nondestructive flaw detectors that could find rail cracks hidden under splice bars as well as internal flaws in wheels and axles. (*Railway Engineering and Maintenance*, June 1948)

(BOTTOM) A detector for broken wheel flanges in action. Passing over the detector, the wheel flange completed an electronic circuit that would break if the flange were chipped or broken. (*Railway Signaling*, September 1955)

9

Private Enterprise and Public Regulation

Safety between the Wars, 1922–1939

*It is of course recognized that improvements in the roadbed [and], bridges . . . must
be made . . . but at the same time there must be progress in methods of protecting
trains [from collisions].*

—Interstate Commerce Commission, 1925

*The basic principle which should govern all expenditures . . . is . . . to secure the
greatest possible return per dollar spent.*

—ICC Commissioner Frank McManamy in dissent, 1925

Normalcy for the railroads did not arrive immediately with the end of
World War I. The year 1919 brought an inflationary bubble, while the U.S.
Railroad Administration returned the carriers to private hands in March
1920. Comparative stability finally returned in 1922, but the good times
lasted only about eight years until 1929 ushered in a decade that for the
railroads can only be termed catastrophic. Yet these tumultuous conditions had far less impact on safety than the longer-term forces that had
been reducing accidents for decades.

By the early 1920s passenger traffic was declining while freight faced
new competition from trucks, waterways, and pipelines. Employers' liability and workers' compensation laws had ballooned injury costs while wage
pressures by the brotherhoods along with ICC rate regulation contributed
to the profit squeeze. The carriers responded in predictable ways. Reducing
all types of accidents became a corporate goal that was institutionalized in
both the ARA and individual safety departments. Especially in the 1920s,
many lines made major investments to improve safety and efficiency. In
addition, the carriers and their suppliers expanded and institutionalized
research efforts, leading to a host of important but incremental technological improvements. In accord with the informal safety compact that had
emerged, the ICC intervened sparingly in these events, but it did mandate
a large-scale experiment with train control and helped force major improvements in rail and brake technology.

As a result of these forces, the social costs of railroading (measured as fatalities per unit of output) continued to decline. Safety for workers and passengers improved steadily throughout the interwar years, and so railroad safety largely disappeared from popular view, becoming instead the province of regulators, unions, and managers. The automobile also powerfully shaped safety, but in contradictory ways, reducing trespassing but enormously magnifying crossing accidents and transforming them into a public and ultimately a federal matter. The carriers' deteriorating finances also brought into the open a conflict between economic and safety regulation that would grow in the postwar decades.

Safety Organizations and Personnel Management

The USRA had given safety work a powerful boost and legitimized it to the brotherhoods. The unions had largely dismissed the carriers' earlier efforts as a public relations smoke screen and as H. R. Lake, trainmaster on the Santa Fe, later recalled, their refusal to cooperate had doomed early safety work on his line. When the carriers returned to private control they determined not to make this mistake again. In 1920 Richard Aishton became president of the ARA; he was committed to better safety, having been general manager on the North Western when Ralph Richards began his campaign. In mid-1921 the ARA took over the safety section from the USRA, while Isaac Hale, the Santa Fe's chief of safety, met repeatedly with the brotherhoods, gaining the support of the Trainmen and some of the other unions as well.

Yet while safety men pursued labor's blessing they rarely invited the unions into their councils. Although President William G. Lee of the Trainmen sometimes spoke at safety conventions, and the brotherhoods were often represented on local safety committees and at national meetings, neither the Safety Section of the ARA nor the National Safety Council included union representation. The failure to include labor was a major blunder, for labor went its own way on safety, continuing to propose expensive and often irrelevant measures.[1]

A second shortcoming was that the ARA Safety Section remained largely a hortatory body. It never developed the expertise or authority of the carriers' other technical associations such as the Bureau of Explosives or the AREA, and it rarely coordinated with these groups—failures that prevented it from pressing effectively for industry-wide reforms and narrowed its focus to rules and behavior. Inclusion of labor might have usefully broadened the focus of safety workers. Because labor tended to be overly enamored with technological fixes their representation might have led to a more balanced attack on accidents. For example, the Brotherhood of Railroad Trainmen, not the carriers' safety men, led the charge for better handbrakes, although with uneven success, and the issue would arise again in the 1950s.[2]

Individual companies' continued interest in safety was impelled by the steady rise in injury costs throughout the 1920s and 1930s. Employee

Table 9.1. **Cost of Fatalities, Selected Carriers, 1908–1942**

Year	Group	Employees			Others		
		Claims $	Suits $	Total $	Claims $	Suits $	Total $
1908–1910	All carriers	1,557	2,536	1,221	—	—	—
1916	D&H	2,469	2,650	2,559	—	—	—
1924	PRR	3,612	6,459	4,546	1,444	2,657	1,933
1925	PRR	4,063	5,614	4,402	2,554	4,301	2,904
1926	PRR	4,014	5,126	4,219	2,697	2,648	2,682
1927	PRR	4,435	6,895	4,944	1,376	3,218	2,094
1932	All carriers	4,157	5,854	4,561	—	—	—
1936	D&H	—	—	5,032	—	—	—
1942	D&H	—	—	11,404	—	—	—

Sources: All carriers from "Cost of Railroad Employee Accidents, 1932,"74th Cong., 1st sess., S.D. 68 (Washington, 1935), Tables 7 and 18. Delaware & Hudson from various boxes, series 2, Delaware & Hudson Collections, New York State Archives. Pennsylvania Railroad from "Payments for Injuries to Persons," Pennsylvania Railroad Insurance Department, box 635, PRRC, HML.

fatalities that had cost $1,200 around 1910 cost the Delaware & Hudson $2,559 in 1916 (Table 9.1). By the mid-1920s the Pennsylvania was paying between $4,000 and $5,000 for worker fatalities while all other fatalities, including those of passengers, averaged somewhat less. These figures reflect the liberalization of liability rules that began in 1908, but workmen's compensation laws may also have increased injury costs. The Pennsylvania employed compensation laws as the basis for settlement even for injuries in interstate commerce where the law did not apply. While public policies often focused narrowly on specific types of accidents or technologies, rising accident costs encouraged the carriers to take a broad approach to safety. The result was a kind of "organizational learning" that yielded a decline in all major types of accidents to all groups.[3]

Reducing the Little Accidents to Workers

In 1922 railroad workers experienced a fatality rate of about 0.39 per million manhours—roughly half its typical prewar value. Despite this progress, a staggering 1,587 men were killed that year, and as usual about 84 percent of them died in "little" accidents. The first decade of safety work had concentrated on persuading workers to follow rules voluntarily, and the procedures often resembled a revival meeting where earnest safety workers preached the gospel to a backsliding congregation, and companies justified their faith by good works. This focus had led the USRA to stage a nationwide safety campaign in 1919, and in 1923 the ARA Safety Section continued the contests, setting a goal of a 30 percent reduction in employee casualties by 1930. Many companies also staged accident reduction contests, pitting groups of workers against one another to see who could achieve the largest reduction in casualties.

Yet during the 1920s the railroad man's legendary freedom that had made his trade both romantic and dangerous came under assault as companies increasingly shifted safety procedures away from exhortation to rules, monitoring, and discipline. In 1927 *Railway Age* concluded, "ballyhoo methods of accident prevention, if not already in the discard, are rapidly headed in that direction." Earlier Progressive and wartime concerns had heightened interest in efficiency, and companies began to see accidents as symptoms of inefficiency and as management failures. The Norfolk & Western computed accidents per dollar of payroll and used the resulting figures as a measure of the efficiency of its safety committees. "Safety is largely a matter of instruction and supervision," was how the Pennsylvania put it in 1927. "A preventable accident is a record of failure . . . [by] the employee, his supervisory officer or both." It was "a form of inefficiency," the company concluded. In 1929 the company issued a new code of rules—there were ninety-seven governing the conduct of maintenance of way employees—and fired men who habitually violated rules. Companies began to require that shop employees wear goggles, and they sent home those who arrived for work unequipped. The Chicago & North Western enforced rules forbidding men to ride on locomotive footboards. It also made foremen responsible for accidents—and watched rates fall. The Wabash encouraged workers to report unsafe conditions, while the Union Pacific and other carriers undertook increasingly minute investigations of all accidents.[4]

The shift from exhortation to discipline and supervision paralleled similar changes that were occurring in safety work in other industries. But railroad safety work remained closely associated with personnel management and the claims department, probably because of the continued relevance of employers' liability, while in manufacturing safety had a stronger engineering orientation. As a result railroad safety organizations were more likely to recommend organizational and behavioral solutions to safety problems, and less likely to be involved in equipment design and purchase. As with the failure to include labor and develop stronger industry-wide institutions, this weakened safety work, especially at a time of rapid technological change, making it reactive rather than proactive.

Safety workers had always emphasized the importance of employee selection, and the continued spread of personnel departments and pension and benefit plans reinforced their efforts. As with accidents, companies began to see turnover as a symptom of waste; they also began to keep records, and to advocate interviews, physicals, and references as a condition of employment. The Rock Island studied why track workers quit and tried to modify conditions accordingly. By 1927 thirty-seven carriers had formal pension plans covering about 1.2 million employees and another fifty-one made informal arrangements. All together by that date about 82 percent of all railroad workers were in pension plans. The motive, clearly, was labor control. Virtually all computed benefits were based on years of service and none were vested—which surely must have reduced turnover. For similar reasons in the early 1920s, "one railroad after another" began to investigate group life and accident insurance, and the Delaware & Hudson, Lehigh Valley, Lackawanna, and other carriers instituted various programs.[5]

Such efforts, along with the decline in railroad employment, sharply reduced turnover on the Delaware & Hudson, which dropped from about 45 percent in 1926 to less than 2 percent in the early 1930s. A survey of seven large carriers also revealed quits and dismissals of 1.55 percent in 1933. The reduction in turnover allowed companies to be more choosy about whom they hired and retained, and also to invest money in training. In 1926 H. G. Hassler of the Pennsylvania reported that in the company's shops "at present our labor turnover is very small which gives us an opportunity to select more desirable men." Better discipline also arrived in the wake of labor stability. E. B. Hall of the North Western stressed the importance of low turnover in improving selection and allowing better training. In 1930 a representative of the Lackawanna noted that labor turnover "has almost been reduced to a vanishing point," a fact he thought had greatly contributed to the decline in accidents. Low turnover allowed the Reading to institute a policy that no inexperienced yard employees could ride cars at night. One Pennsylvania yard where employees ranged in experience from ten to thirty-five years ran 534 days without any damage or injuries.[6]

Safety work was reinforced by a host of new technologies that both increased productivity and reduced accidents as a byproduct of their normal operation. Almost certainly improvements in medical technology contributed to the gains. Although accident severity information dates only from 1934 and shows little change during these years (Appendix 2), permanent nonfatal injuries rose relative to fatalities—probably because medical workers were able to save the lives of amputees and others who would have died in earlier days.[7]

The dangers from locomotives continued their steady decline, not only as a result of federal inspection but also due to a host of incremental technological improvements. Faced with rising wages and increasing competition from trucks, the carriers began to value faster, more powerful, and more reliable motive power. As a result, after 1923 they scrapped large numbers of older engines, replacing them with fewer, newer machines. Newer power was more likely to have automatic fire doors and other safety devices, while staybolts came to be made of tougher alloy steels. Use of roller bearings allowed redesign of reciprocating parts to make them lighter. By the early 1930s about 40 percent of locomotives had power reverse gears, which were less dangerous than the old "Johnson bar" reverse, and after a petition by the Brotherhood of Locomotive Engineers that resulted in ICC hearings and ultimately a Supreme Court case, they became universal. The continued spread of thermic siphons was but one of a host of small changes that improved both locomotive safety and efficiency. Although these devices were hard to maintain they improved fuel efficiency, and by carrying a supply of water to the crown sheet, reduced boiler explosions. Introduced about 1920, they soon proved their utility and by 1930 ran on ten thousand locomotives, or about one-sixth of the total.[8]

By the mid-1920s the Canadian Pacific and Canadian National boiler design ensured that failure would be gradual rather than catastrophic. Boiler blowout plugs, which worked like a fuse and were intended to allow an overheated boiler to fail safely, had never worked well, but about 1932 the Santa Fe developed an improved blowout plug that virtually ended boiler explosions. Boiler water softening continued to spread. By 1933 the carriers treated about 40 percent of all boiler water while research into the problem of embrittlement of boiler steel led to use of tests and water additives to prevent it. As a result of these changes, efficiency improved and both locomotive defects and accidents plummeted from about 1923 to the early 1930s.[9]

Yet the experience with siphons and blowout plugs also provides a good example of the failure of safety and regulatory organizations to focus on engineering design. Siphons could have been retrofitted on most locomotives for they would have paid their way in addition to saving lives, while better blowout plugs might also have saved many lives. Yet the safety men largely ignored both devices. In the early 1920s the ICC Bureau of Locomotive Inspection had discovered that water gauges on many locomotives gave dangerously misleading readings and it campaigned effectively for change. But it was never again so proactive and it too failed to pressure the carriers to install siphons, better blowout plugs, low-water alarms, or other safety devices.[10]

Many other investments in new technology that were intended to boost productivity also improved safety. Bridges designed with walkways were not only safer, they were easier to inspect. In the late 1920s companies introduced electromechanical car retarders in freight yards, and floodlights made their debut about the same time. Car retarders controlled freight car movement in gravity yards without the need for a brakeman to ride the car. They are a good example of a wage-induced technological improvement that saved labor and reduced risks as well, for they sharply increased productivity and nearly abolished personal injuries. On the Indiana Harbor Belt Line retarders reduced monthly injury costs from $2,263 to $55.[11]

The above improvements required no special procedures to improve safety, but the impact of some new technologies was contingent upon the development of proper procedures and training. An important function of safety organizations was to ensure that these new production methods did not compromise safety but their record in this regard was mixed. The best examples derive from the revolution in maintenance-of-way work that began in the 1920s. Traditionally such work had involved large gangs of common labor with little heavy equipment; turnover was enormous and the work was dangerous. Not only were there many injuries from hand tools and heavy lifting, but passing trains routinely killed men. In addition, work trains often followed "smoke orders"—which is to say, no orders—

sometimes with terrible results. In one instance on September 10, 1918, a Burlington work train running without orders or flag protection collided with a passenger train near Birdsell, Nebraska, killing eleven.[12]

World War I had precipitated a labor shortage that continued into the early 1920s as a result of the cessation of immigration. The need to economize on track labor, in combination with the introduction of the internal combustion engine and portable power equipment, began to change maintenance procedures. Companies introduced an array of power-driven adzes, tie tampers, and other small equipment, along with motor cars and off-track cranes and crawlers. The new equipment also allowed the carriers to reorganize work crews, employing fewer but better-trained workers who worked longer sections. Experiments with year-round maintenance spread—both to reduce turnover and to employ the expensive equipment.[13]

The impact of these changes reflected the uneven effectiveness of carrier safety organizations. With proper guarding, rules, and supervision, the new capital-intensive procedures might increase safety. Off-track equipment was less likely to foul track in front of an oncoming train, while smaller gangs of experienced men were easier to supervise. Yet machinery was inherently dangerous and in the absence of hazard assessment, learning came the hard way. Thus, on the Burlington proper guarding appeared only after a man lost an arm in the unguarded friction clutch on a pile driver. When companies introduced mechanized spike-pulling equipment, they discovered that it often broke off spikes, which might then be thrown by adzing machines, with deadly results. Such hard-won wisdom suggested not only the need to guard the adzes but to involve safety men in a hazard assessment from the outset.[14]

The use of track motor cars raised productivity and allowed supervisors to cover more territory. But early models without a clutch had to be jump-started, which was dangerous, while front wheel drive made some cars derailment-prone. Cars on track without flag protection caused a spate of accidents. In one case, on April 29, 1927, a freight on the New York, Chicago, and St. Louis ran into a track repairers' car, killing three; in another, on July 21, 1927, a Chicago, Great Western train hit a self-propelled mower, killing five. In response, some lines promulgated strict rules abolishing smoke orders and requiring lookouts and slow orders to provide protection against trains for crews and track cars. The North Western required motor cars on double-track lines to get off the track when freights passed. The Southern Pacific wired a bell into the track circuit to warn men of impending danger. Yet many companies' procedures were ineffective and motor car safety remained weak.[15]

Whatever the net impact of mechanization during these years, the rise in liability costs and the carriers' broad commitment to Safety First procedures contributed to a sharp decline in fatality rates for all major classes of employees (Figure 9.1; Table 9.2). Moreover, the commitment to worker safety spilled over and generated improvements in passenger safety as well. Statistical analysis reveals that the companies most successful in

Figure 9.1. **Passenger and Worker Fatality Rates, 1922–1939**

Source: Appendix 2.

Table 9.2. **Employee Fatality Rates, by Broad Occupational Group, 1922–1926 and 1935–1939 (per million manhours)**

Group	1922–1926	1935–1939
Executive, professional, and clerical	0.061	0.035
Maintenance of way	0.462	0.253
Maintenance of equipment	0.176	0.121
Transportation (other than train, engine, and yard)	0.109	0.091
Transportation (yardmasters, etc.)	0.376	0.158
Transportation (train and engine)	0.860	0.610
Total	0.338	0.225

Sources: ICC, *Accident Bulletin,* various years, author's calculations.

reducing accidents to workers were also most successful in improving passenger safety—just as Ralph Richards had predicted.[16]

As the above analysis suggests, the reduction in accident rates did not result because as a group, companies traded output for safety during these years. Instead, successful safety work was largely a matter of increasing organizational capabilities and of investment and technological change that raised both efficiency and safety. Had accident prevention been simply a matter of allocating more resources to safety, then, given total resources, comparatively safe carriers would have been those that allocated resources away from production to accident prevention. Such procedures imply a negative relationship between safety and worker productivity, yet Figure 9.2 shows little relationship among carriers for 1922–23. Instead, the wide variation in safety/productivity packages suggests differences in company efficiency. Similarly, the improvement in safety from 1922–23 to 1938–39 reflected both organizational learning and investment in better technology. However, the greater scatter for the latter period suggests the unsurprising possibility that learning may have been easier within than between com-

Figure 9.2. **Safety and Productivity, 1922–1923 and 1938–1939**

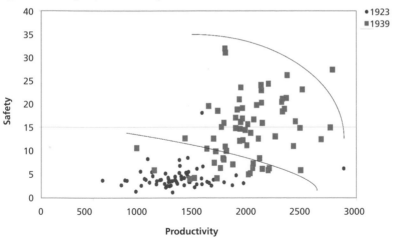

Note: Safety is (1/injury rate) x 100. Productivity is constant dollar total revenue per thousand manhours.

panies. Such progress eventually impressed a skeptical ICC, but not before it mandated an expensive attempt to stamp out collisions with automatic train control.[17]

Controlling Collisions

As chapter 7 detailed, during the first decade of the twentieth century railroad critics had been nearly unanimous in blaming collisions on the carriers' sloppy management and primitive signaling. The trade press remained optimistic that companies could improve management and discipline sufficiently so that better signals would improve safety. But the ICC and many other observers took a more pessimistic view, and as use of the block system grew, they increasingly focused on its failure. What was needed, it seemed, was a foolproof remedy for in-discipline.[18]

A spate of disasters that came in the wake of the wartime traffic boom reinforced this view. At Amherst, Ohio, on March 29, 1916, a New York Central train running too fast in a fog passed a signal and caused a wreck that killed twenty-six. On February 27, 1917, at Mount Union, Pennsylvania, the engineer ran full steam past a distant signal, killing twenty. More wrecks came in 1918, and in 1919 disaster struck early. On January 12, on the New York Central at South Byron, New York, an engineer ran an automatic block signal and collided with the rear of the preceding train that had suffered an engine failure, killing twenty. The list continued into 1920 with three collisions, and 1921, when, on February 27, an engineer on the New York Central missed a signal causing a wreck at Porter, Indiana, that killed thirty-seven.[19]

These and other collisions occurred on some of the best-managed carriers in the country. They often involved crack, main line passenger trains, and all resulted from mistakes or errors in judgment that involved signaling. Most American carriers used "permissive" blocking with the manual block system. That is, trains were allowed to enter an occupied block but told to run "under control." With automatic blocking some carriers also allowed enginemen to proceed "under control" after stopping at the entrance to an occupied block. Because automatic blocks were short, often only one or 2 miles long, enginemen usually controlled their speed, but manual blocks were 10 or 15 miles long, due to the need to economize on labor, and enginemen were more likely to violate rules. Under such circumstances the manual block system was, as the *Railway Age* remarked, "little better than a farce." But even automatic signals were not a panacea, as several disasters revealed. Reviewing the accident at Porter, Indiana, the editor observed that "there is good reason to believe that an automatic train control or train stop would have prevented this accident as well as a large number of others such as [those] at . . . Amherst . . . Mount Union, and South Byron." Perhaps, the journal suggested, it was "time to supplement the signal system with train control on lines [with] . . . high speed schedules."[20]

Section 26 of the Transportation Act of 1920, written by Representative John J. Esch of Wisconsin, authorized the ICC to require automatic train control or "other safety devices," which presumably meant block signals. The next year Esch became an ICC commissioner. For decades he had urged mandatory extension of block signals, as had other safety advocates. Yet the wartime disasters had finally destroyed reformers' confidence in the block system. On January 10, 1922, the commission issued Tentative Order 13,413, requiring all carriers with revenues over $25 million to install train control on at least one passenger division. The distinction between the order of the commission and the views of the *Railway Age* reflected a fundamental disagreement on how best to improve railroad safety. While the *Age* saw train control as a potential complement to blocking on a few high-density lines, the editor thought that individual safety investments were best left to railroad management. The commission, by contrast, viewed train control as a superior substitute for blocking and proposed to require all large carriers to install it, irrespective of their traffic density or signaling.[21]

Constructing Train Control

The idea of an automatic train stop to prevent accidents is simple enough: it is some form of device that can apply the brakes without the aid of the engineman, usually in an emergency. The idea had beguiled inventors since the 1850s and several carriers began serious experimentation in the 1890s. The Boston Elevated made the first permanent installation in 1901. The Block Signal Board began to investigate train control devices in 1908, and in 1913 the ICC picked up where it left off. As early as 1912, the *Railway Age* had urged the carriers to develop train control but opposed mandating its use. The reasons were economic: with the carriers pinched by rate regulation, the editor wondered where the funds would come from. And in all events, there were better uses for the money, he concluded. In 1919 the Committee on Automatic Train Control of the USRA, which was dominated by the carriers, echoed similar views. Using ICC figures it demonstrated that such devices might prevent only about 6 percent of all fatalities to nontrespassers. "The first step," the committee concluded, "is not . . . the adoption of some form of train control device, but the adoption of the block system itself."[22]

While all forms of automatic train control involved a marriage of the air brake to the track circuit, by about 1920 inventors had evolved several different approaches. A simplified categorization is as follows.

Character of Control	Class of Device	Type of Device
Intermittent	Contact	Mechanical
Intermittent	Contact	Electrical
Intermittent	Non-contact	Induction
Continuous	Non-contact	Induction

The simplest were intermittent mechanical contact train stops. On these a roadside ramp connected to the track circuit would be placed at full braking distance before the stop signal. Should the signal be at stop and the engineman pass the ramp, it would energize a contact on the locomotive to apply the brakes unless the device was equipped with a forestaller and the engineman employed it to override the system. These mechanical or electrical devices were sensitive to the elements, while differences in clearances among companies made locomotives noninterchangeable, and so companies preferred induction-based systems. More sophisticated devices included automatic speed controls as well as stops and could be designed to reduce speed should an engineman pass a distant signal too fast. The most complex forms were continuous; that is, they constantly monitored the train, not just at specific points on the line, and could slow or stop it at any time.[23]

The Chicago & Eastern Illinois had installed an experimental train stop in 1911, followed by the C&O in 1916 and Rock Island in 1919. Several interurbans had also installed short sections of automatic stops. By 1920 these devices were no longer experimental. That year, speaking for the ICC, the chief of its Bureau of Safety W. P. Borland informed a group of engineers that "the next logical step in their development is their installation and use in actual service." He brandished a list of sixty-eight collisions that killed 375 people from 1911 to 1920 that he claimed train control could have prevented.[24]

As its order made clear, the ICC saw train control simply as a technical fix for a safety problem, and for that reason it ruled out devices that could be forestalled, thereby requiring that any train running a distant block signal that showed a restrictive aspect be stopped. By contrast, the carriers evaluated train control as both an economic and a safety investment. Previous technologies that improved safety, ranging from steel rails to automatic couplers, better brakes, and block signals, were installed because they also increased productivity, thereby earning at least their cost of capital. Similarly, to the carriers automatic train control had to compete for scarce funds with other investments that could improve safety and productivity. As *Railway Age* repeatedly emphasized, train control had to be paid for. Moreover, the simple automatic stop promised no productivity gain, and if applied without a forestaller it would raise costs (for stopping a train was expensive) and reduce average speeds and therefore productivity. More sophisticated devices that involved continuous speed control did not suffer from these flaws and would even increase capacity. But these truly were experimental, for no carrier had yet installed such a system.[25]

The hearings on Order 13,413 revealed the width of the gulf that separated the commission and the carriers. To make train stops more economic the carriers requested that the commission allow use of a forestaller. But the commission's focus was on a technological fix for in-discipline: "this factor of human judgment is the factor which an automatic train control device is designed to eliminate," it lectured. This ruling shifted the railroads' interest to more sophisticated devices, which they argued were experimental. The commission, continuing to focus on a simple train stop, rejected the claim. Similarly, the carriers emphasized the costs of more sophisticated speed control devices, an argument the commission of course found exaggerated, as it emphasized the use of simple stops.

The railroads' most powerful argument echoed the views of the trade press and the USRA that train control was not a substitute for block signals but a marginal supplement to them. Many of the carriers had little or no mileage in block signaling, and improved signals or other investments such as grade crossing elimination or extra tracking might yield more safety per dollar. Indeed, as late as 1926 the Wheeling & Lake Erie had no automatic block signals at all, yet it ran seventy-two trains a day on some divisions. To this the commission blandly and irrelevantly replied that "all of these are unquestionably desirable [but] automatic train control will still be a necessary safety measure."[26]

The carriers' response to the commission's order was to develop the more sophisticated speed control devices. The Pennsylvania, for example, began to install continuous speed control, complete with cab signals. Cab signals were a European invention; they included either lights or audible signals that indicated in the cab the position of the block signal, thus greatly reducing the likelihood that an engineman would miss a signal. Developing the new technology took time, and on January 14, 1924, irritated by the lack of progress, the commission slapped the original forty-nine carriers with additional installations and applied its order to forty-two smaller roads as well. One, the Western Maryland, ran only four trains a day at an average speed of 22 miles per hour over the specified territory. At the hearings the carriers trotted out the same arguments and were met with many of the same replies. C. L. Bardo of the New Haven urged that grade crossing accidents were a worse problem than collisions and in a priceless response Commissioner Aitcheson wondered if it wasn't important to stop both sorts of accident. But the railroads won a victory of sorts when the commission rescinded the order for the second group of carriers and reinstated use of the forestaller. The Pennsylvania immediately shifted from speed control to a continuous stop with forestaller and many other carriers also changed to simpler devices.[27]

Even more significant was the dissent by Commissioner McManamy, who was not on the commission when the original order was issued. McManamy, the reader will recall, had been an engineman on the Père Marquette and eventually became the ICC's chief inspector of locomotives. More than any other commissioner he was intimately familiar with the dangers of railroading. The carriers had convinced him that there were better safety investments than train control for the money.[28]

The 1924 order marked the high tide of this episode of train control. The next year, dissenting from a ruling in which Esch and other members of the commission refused to exempt the Great Northern from its order, McManamy took issue with the commission's approach. He accused fellow commissioners of having abandoned the goal of the Transportation Act, which was "not the installation of any particular device . . . but the protection of life and property." Rather than a blanket order, McManamy favored requiring specific safety devices fitted to the needs of specific carriers.[29]

In July 1927 the commission announced a general investigation of both train control and block signaling. This was a far more open-ended inquiry than the earlier "show cause" orders, and it reflected a change in the commission's operating procedures. The ICC had reorganized its workload, and train control was now the province of Division 6, that included not only Joseph Eastman, a stalwart on train control, but McManamy and Richard Taylor, who had been a railroad man. The carriers realized that a strong presentation emphasizing their commitment to safety might forestall a new order. At the hearings in 1928 they no longer argued that train control was experimental and concentrated instead on the miniscule number of accidents it might prevent, claiming that the record demonstrated that voluntarism worked and that they could make safety investments if left to their own devices. At least provisionally the ICC accepted the carriers' proposed safety contract. McManamy and Taylor concurred that a "far greater measure of safety" might be achieved through such investments, and refused to order an extension of train control.[30]

By 1930, under the first two orders, at a cost of $26 million, the carriers had installed train control on 15,208 miles of track (5% of the total). But about 15 percent of locomotives were equipped for train control, which is a better index of passenger miles covered. In the 1930s, as passenger traffic declined, some carriers successfully petitioned to remove the apparatus. The Pennsylvania substituted cab signals for train control on some divisions, as did the Union Pacific, while the commission allowed the Boston & Maine and Great Northern to scrap train control entirely.[31]

The episode raised a host of issues that remain contentious to this day. The fundamental policy issues were how much safety was worth, and what was the best way to achieve it—by voluntarism or specific standards. Finally, there was the question *Railway Age* asked that would grow in importance as the carriers' financial position crumbled in the postwar years: how could they afford such safety investments given the commission's control over railroad profitability?

Initially the ICC failed to address these matters. Dominated by lawyers and still blinded by the glare of the Safety Appliance Act of 1893, it preferred to see train control as a technological end rather than a choice to be balanced against competing means. It ignored entirely any connection between safety and rate regulation. Its focus ensured that the original order was poorly crafted as it covered some lines with low traffic density and few signals, and failed to specify the location of the installation. A statistical analysis of carriers' collision rates in 1929 also finds that train control had no discernable impact, which is not surprising, for Borland's list of collisions noted above amounted to only about 0.47 percent of all rear-end and head-on collisions during the period. Yet the commission's orders were not a complete loss, for they spurred the development of more complex forms of train control and cab signals. Indeed, A. H. Rudd, signal engineer on the Pennsylvania and no friend of the ICC, claimed that "we would never have had the cab signal if it had not been for that order."[32]

Better Brakes for Bigger Freights

While the ICC was mandating train control it was also pressuring the carriers to improve freight car brakes. Yet the development and diffusion of what came to be called the AB brake reflected not only regulatory pressure but also competitive rivalries and nearly a decade of research and development. Since 1905 the standard freight air brake had been the Westinghouse K-type (chapter 7). As the reader may recall, when the engineman reduced air pressure in the brake pipe a triple valve emitted air from a car reservoir to the brake cylinder until pipe and reservoir pressure were equalized, and it reduced train pipe pressure locally, thereby speeding up operation. The brake had several major flaws, however. Because air pressure came from a reservoir on each car, the train would have inadequate pressure for an emergency after a full service stop. This problem was magnified as trains became longer and heavier, and by World War I, as freights of a hundred cars and more became common, the K-brake was becoming obsolete.

In addition, the K-brake required careful maintenance, which it rarely got. In 1919 a general discussion of brakes at the ARA Mechanical Division disclosed the usual complaints; one company reported that it tested 175 cars received in interchange and only nineteen met MCB requirements for brakes. Later discussions revealed improper repairs resulting in braking ratios that ranged from 10 to 400 percent of the ARA standard on the two trucks of the same car. If dirty or leaky or improperly adjusted, the brakes might not work, or perhaps provide an unexpected emergency stop. Either mishap was highly dangerous and could lead to a runaway or perhaps buckle the train. Despite such difficulties brake failures were not a major safety problem; ICC data implicate them in only about thirty to forty collisions a year. They were more likely to result in derailments but in neither case were brake problems a fruitful cause of casualties.[33]

As early as 1910 the Block Signal Board had investigated an alternative to the Westinghouse brake that by World War I was being test-marketed by the Automatic Straight Air Brake Company. Like the Westinghouse, it used a cylinder and car reservoir. However, in a service application it applied air to the cylinder from the brake pipe. Hence a leaky cylinder or excessive piston travel were not major problems; they simply required more air from the air pump. A service application still left the auxiliary reservoir available for emergency. The latter trait made it especially attractive to heavy coal haulers such as the Virginian, which successfully tested it on a 100-car 7,600-ton coal train in July 1918. The Norfolk & Western, Denver & Salt Lake, and Chicago & Eastern Illinois also experimented with the new brake, and, acting on a petition from its maker, the ICC opened hearings on power brakes in early 1922, basing its authority on the Transportation Act of 1920.[34]

The hearings lasted off and on until 1924. They revealed that some carriers were continuing to use brakemen to control freights on heavy grades. The ICC, which was aware of the process and had seemingly condoned it in 1907, now reversed field. In an apparent effort to pressure the carriers to develop better brakes it brought suit against the Pennsylvania and other carriers. In 1925, when a runaway freight killed two Pennsylvania trainmen, the commission's report on the accident contained a stinging rebuke to that carrier's management, terming it "regrettable [that] serious accidents . . . [are] the only means [of correcting] unsafe practices of longstanding." Finally, in 1927, nearly thirty years after the Safety Appliance Act went into effect, the Pennsylvania stopped using brakemen to help control freights.[35]

Enginemen also testified at the hearings that they often controlled long trains with the engine brake only, as they feared to use the train brake. The commission concluded that major improvements in maintenance and design were needed. While the carriers expressed little interest in improving their brakes, neither they nor the trade press contested the commission's findings. The ARA issued new standards for brake inspection in early 1925 and by the next year the commission's Bureau of Safety reported that "air brake conditions throughout the country are the most satisfactory ever known."

The ARA also appointed Harley Johnson as director of research; he used the organization's Purdue facilities to begin developing a brake that met the commission's requirements. Johnson was a Purdue graduate in mechanical engineering and had worked on a number of lines. Interestingly enough, he came highly recommended by Westinghouse. That company and the Automatic Straight Air Brake Company submitted brakes for tests and the whole process, including laboratory and road testing, lasted until 1931. Ironically, the Straight Air Brake, which had precipitated the quest for improvements, fell by the wayside. The new brake, Westinghouse type AB, represented a major advance over its older cousin. It was designed for easier inspection and maintenance, was faster than the K-brake, provided protection against unwanted emergency operation, and allowed an emergency stop after a service application. In 1934, Westinghouse developed an automatic empty and load feature that further improved control of long trains. The ARA made the new brake standard and required it on all new freight equipment after 1933.[36]

The development and diffusion of the AB brake demonstrates the ICC and voluntarism at their best. The experience was far less contentious than train control in part because the commission's pressure to develop a new brake was comparatively inexpensive to the carriers, for most of the development cost was borne by Westinghouse, which wished to preserve its market. Second, the ARA mandated the new brakes only on new freight cars, which made them comparatively painless. Moreover, in pressing for better brakes the commission was clearly doing the ARA's business, for the brake yielded an economic as well as a safety payoff, as it allowed longer, faster freights and helped prevent wheel failures from overheating. One widely circulated study estimated that the new brake cost about $718 per car but would save $186 a year, or about a 26 percent return on investment, *if widely adopted*. But that was the rub; cars that might be gone two-thirds of the time would earn only $0.33 \times 0.26 = 8.6$ percent. Once again freight car interchange with fixed per diem rates impeded the diffusion of better technology as did the depression-induced shortage of funds. By 1940 only one-fifth of cars carried the AB brake.[37]

The Decline in Collisions

Between 1922 and 1929 passenger fatality rates from collisions virtually disappeared, falling from about two per billion passenger miles to 0.04, and they remained low throughout the 1930s. Employee casualties in collisions followed a similar pattern. The AB brake played little or no role in these gains; not only were brakes responsible for few collisions, there were too few cars with AB brakes to make much difference. Nor, as noted, was automatic train control an important contributor to better safety.

The decline in collisions resulted in part because the carriers extended the block system to new lines and replaced long, permissively operated, manual block systems with automatic signals. Signal investment, which had averaged about $5 million in 1922, rose to $32 million by 1926 and remained high for the remainder of the decade. The share of main track with automatic signals rose from 21 percent in 1922 to about 28 percent in 1929 to 33 percent in 1939. Of greater importance, passenger miles traveled under the automatic block system increased from (at least) 34 percent in 1922 to (at least) 61 percent in 1939 (Figure 7.2).

Table 9.3. **Collision Rates, Selected Carriers with No Change in Block Signaling, 1922 and 1929**

Company	1922			1929		
	Percent Signaled	Percent Auto Signaled	Collision Rate	Percent Signaled	Percent Auto Signaled	Collision Rate
Bessemer & Lake Erie	97.2	0	10	98.1	0	4.03
Chicago & Erie	100	98	3.26	100	97	3.33
Delaware, Lackawanna & Western	94	94	3.31	91	90	2.47
Richmond, Fredericksburg & Potomac	100	100	7.11	100	100	6.29
Chicago & Alton	75	62	5.33	83	60	2.31
Minneapolis, Saint Paul & Sault Ste. Marie	44	0	1.64	34	0.1	0.97
Reading	83	71	5.88	87	72	4.56
Virginian	97.2	0	10	98.1	0	4.03
Average Collision Rate	—	—	5.25	—	—	3.31

Sources: ICC, *Statistics of Block Signals* (Washington, 1922 and 1929), and *Accident Bulletin* (Washington, 1922 and 1929).

Notes: Percent Signaled is passenger track with manual or automatic block signaling; Percent Auto Signaled is passenger track with automatic block signaling. Collision Rate is collisions per million locomotive miles. Weighting carriers' collision rates by locomotive miles changes the averages to 4.42 in 1922 and 3.36 in 1929.

Yet the spread of block signaling was not the whole story, for as Table 9.3 shows, collisions declined sharply even on lines that made little or no change in signaling. Statistical evidence from a broader sample confirms this conclusion. Controlling for traffic density and signal use, by 1929 carriers had reduced collision rates 74 percent from their 1922 levels. Such findings reveal that, like air brakes and automatic couplers before them, the effectiveness of block signals was contingent upon a host of operating practices and minor improvements.

Some of the decline reflected organizational innovation and learning in the form of surprise checking, Safety First, and other labor policies. In addition, a host of minor modifications in equipment, signal design, and operating practices continued to improve the efficiency of the block system. Companies replaced banjo or semaphore signals with colored lights that had no moving parts to fail and were much more visible in poor light and fog, while by 1939 about 8 percent of locomotives were equipped with cab signals. Tragedies such as that at Amherst, Ohio, where poor visibility caused the engineer to miss a signal therefore became less common. Some lines also scrapped both train orders and the flagman, running by centralized traffic control of signals, but this was largely a post–World War II development (chapter 10).

The diffusion of steel passenger cars, which reduced casualties in both collisions and derailments in the prewar years, continued in the 1920s and 1930s. Steel or steel frame passenger coaches rose from about 42 percent of the total in 1922 to 97 percent in 1939 and since they were used exclusively on high-density lines they carried an even larger share of passenger miles. Companies also began to install roller bearings on passenger equipment, and in the 1930s safety glass made an appearance on the streamliners. Steel cars on the New Haven had antitelescoping bulkheads and a rear collision on February 13, 1934, yielded no fatalities. By contrast a similar accident at comparable speed on the Erie, which did not use such bulkheads, killed thirteen people. However, statistical analysis suggests that while the growing use of steel cars would by itself have substantially reduced casualties per passenger train in derailment or collision, their impact was offset by other forces and the probability of being killed given that one was in a collision or derailment shows little change during the interwar years.[38]

Investment, Technology, and the Decline in Derailments

In 1922 the ICC reported 13,155 derailments—an astonishing number, but down from the all-time high of 22,477 only two years earlier. Of course, most of these involved freights and most were not very serious, for wartime inflation had ballooned the number of reportable accidents (chapter 7). Thereafter, derailments fell steadily—to 9,871 in 1929 and 3,224 in 1939. This decline was not simply the result of falling prices in the early 1930s, and while some of it reflected declining car miles, rates per freight and passenger car mile declined as well, and so did casualty rates to both employees and passengers. The decline in derailments and casualties reflected many of the same long-term forces that were reducing collisions and other kinds of accident during these years. In particular, these involved the development and diffusion of safer technology along with major investments in roadbed during the 1920s.

The introduction of safer technology was largely an outcome of the drive for efficiency that spurred research and development. In 1926 *Railway Age* described the process. Technological change was "revolutionizing transportation," the journal noted, and the editor highlighted the importance of technological sequences and competition in motivating the improvements. Sometimes "changes in transportation conditions create a demand for new kind[s] of equipment . . . [Alternatively,] new kinds of equipment . . . brought out by the manufacturers really create the demand from the railroads," he thought. The *Age* also stressed the importance of supplier and buyer rivalries in encouraging innovation. "It seems very probable that the revolutionary progress . . . would not be occurring except for the division of the initiative between those who specialize in railroading and those who specialize in manufacturing," the journal concluded.[39]

Yet cooperation was as important as competition. Since the 1890s the MCB and the Master Mechanics had become increasingly involved in standard setting, testing, and applied research, and had formed close ties with Purdue and the University of Illinois. When the USRA took over railroading in 1917 it folded these societies into the ARA Mechanical Division, and after the war the association itself became much more active in supporting and coordinating their equipment research and in using inter-change rules to require use of the new technology. In addition, the division fielded a corps of inspectors who checked on company repair practices and helped maintain ARA standards. By the 1930s the carriers felt the need for further centralization to cope with New Deal policies, and in 1934 the ARA and several other organizations became the Association of American Railways (AAR), which for the first time had a division of research.

While the AREA remained independent, it continued its decades-long research into rail and track, which was also partly funded by the ARA and then AAR. The Pennsylvania and other major carriers also continued research and testing and were joined by some others such as the Cincinnati Southern and the Rio Grande. While large suppliers such as Westinghouse maintained their own research, trade associations composed of smaller companies began to support scientific and engineering investigations. These included the Association of Manufacturers of Chilled Car Wheels, the Association of Coupler Manufacturers, the Steel Wheel Manufacturers, and the Draft Gear Manufacturers.[40]

Building a Better Freight Car

Most derailments involved freight trains and the freight car was a major culprit. The share of steel or steel frame freight cars rose from 66 percent in 1922 to 96 percent in 1939. This reduced derailments because on wooden cars the truck bolster would sag on heavily loaded equipment, thereby put-ting weight on side bearings and making cornering difficult. The need to upgrade service to compete with trucks also encouraged better mainte-nance to prevent delays and derailments. In 1929 the Pennsylvania built a special inspection pit in its Altoona yards to check for defects in eastbound freight cars. In one sixty-four-day period the inspector discovered 503 de-fects that ordinary inspection missed—103 of which might have caused accidents. Other companies similarly improved inspections.[41]

In 1920 nearly every freight car rode on the arch bar truck. Consist-ing of plates and springs and bolsters, and held together by bolts, it was a maintenance nightmare and could not be kept square. Occasionally such equipment might also affect passenger safety, as occurred when an arch bar on a tender broke on the Pennsylvania on February 15, 1912, causing a derailment at Warrior Ridge that killed five. While manufacturers had pro-duced steel side frame trucks since the nineteenth century, their cost along with freight car interchange had precluded widespread adoption. But in the 1920s rising car weight and increasing concern with accidents—especially those involving tank cars (chapter 8)—finally spelled the demise of the arch bar truck. The ARA made the cast steel truck recommended practice in 1923 and required it on all new cars built after 1928. At about this time, the arch bar was still causing an average of twenty-five wrecks a month on the Pennsylvania. The Rock Island also complained of increasing arch bar failures. Replacing trucks on every car in the country was a Herculean task, while the expense of new trucks in Depression times slowed their diffusion. Finally, in 1939, the AAR banned the arch bar in interchange.[42]

In the late 1920s, manufacturers such as the T. H. Symington Company began to investigate truck dynamics, and they discovered that the interaction of truck springs and uneven track could cause cars to rock sufficiently at high speeds to derail. As a result companies offered improved springs that dampened oscillations. The new trucks also ran on better wheels. In 1916 the Association of Manufacturers of Chilled Car Wheels contracted with the University of Illinois to study stresses in wheels. The association also set up its own research laboratory, developed standards for foundry practice, monitored members' plants to ensure compliance, and required companies to make their test results public. These procedures led to a major redesign in car wheels and to important improvements in their metallurgy. Drawing on the Illinois research, in about 1926 the Griffin Wheel Company produced a single-plate wheel which, because it cooled more uniformly, proved far stronger than the older design. The association laboratory also studied annealing and found that proper heat soaking yielded a marked increase in wheel strength.[43]

These and many other changes sharply improved equipment reliability. They reduced maintenance requirements and they made a major contribution to the decline in derailments during the interwar years. In 1922–24 the railroads averaged about 24 billion freight car miles, and trucks, wheels, and axles caused an average of 4,763 derailments a year, or about one third of the total. In 1937–39 freight car miles averaged 22 billion and these same causes led to 1,121 derailments, which was 29 percent of the total. Over the same period fatalities from these causes fell from twenty-four to seven.

Reengineering Track and Rail

In the beginnings of American railroading, as the first chapters of this book have detailed, the carriers built a track structure that, while flimsy by European standards, was well suited to American conditions. In the first decades of the twentieth century as labor costs rose, density increased, and cars and locomotives grew heavier, major carriers started to construct a much more substantial, European-style, main line track structure. In 1914, the AREA had begun a long-term study of stresses in track under the guidance of Arthur N. Talbot. Talbot was an 1881 graduate of the University of Illinois who had returned to teach after a spate of railroad engineering. He was a distinguished professor and his work with the AREA symbolized the increasingly scientific character of railroad engineering and its reliance upon university resources.

Talbot's reports were the first large-scale scientific study of track stresses and by the 1920s they began to influence both track and equipment design. The stress on rail and track reflected both load and span, and Talbot's work revealed the need for heavier ballast and more careful fitting of plates, rails, and ties, and demonstrated that canting rail inward could reduce stress—a practice that companies initiated in the mid-1920s. Experiments on 2-10-2 Santa Fe locomotives also revealed very high stress from the two-wheel trailing axle because of its great distance from the drivers. As a result, when Lima introduced a new experimental locomotive it was designed as a 2-8-4, which promised lower track stress than had the older 2-8-2. Talbot's work also reinforced the trend toward use of much lighter alloy steels in locomotive reciprocating parts that had begun about 1910. By the 1920s, probably most newly built equipment was designed with the new metal that was not only less likely to break but also put less stress on the track.[44]

As late as World War I old pin-connected truss bridges were still occasionally collapsing. But as better technology reduced the cost of earth-moving, steel, and concrete, the carriers replaced the truss with embankments or girders or reinforced concrete bridges. Except for washouts bridge accidents virtually ceased. As research demonstrated the value of a good roadbed in reducing maintenance requirements, rising labor costs led the carriers to undertake what *Railway Age* described as a "record breaking . . . program of track betterment." Preserved ties, which saved labor and wood, rose from about 44 percent of those newly installed in 1920 to 79 percent in 1930, while tie plates and heavier ballast became much more common. Much of this upgrading involved southern and western carriers. The Cotton Belt, which was thoroughly rundown by the end of World War I, made heavy investments in road and track as did the Texas & Pacific. When the Rio Grande upgraded plant and equipment after 1924, *Railway Age* claimed that "one of the most important results has been the large reduction made in the number of train accidents."[45]

In the tradition of Ashbel Welch and Arthur Wellington, companies also studied the "economical selection of rail." One analysis by the Kansas City Southern presented to the AREA in 1930 claimed net operation and maintenance savings as rail weight increased up to about 130 pounds per yard. In response to such studies companies increased the weight of new rails, which slowly raised the average weight of rail in track from 82 pounds to the yard in 1920 to 94 in 1939. The AREA also adopted new sections, which were heavier and deeper, thereby providing greater stiffness and girder strength. Rail lengths first of 39 feet and then of 78 feet also reduced the number of dangerous joints. Because rising train weight and speed caused rails to creep, anticreepers were widely installed. Their utility was revealed in the breach on September 24, 1941, when creeping rails buckled, causing a derailment on the Norfolk Southern that killed the engineer and a brakeman. Metallurgical developments even improved the lowly fishplate bolt, as heat-treating raised its elastic limit. Combined with heavier rail that was less flexible, stronger bolts yielded stronger, safer joints.[46]

These developments finally ended a century in which European railroads built a far more substantial permanent way than did American lines. An AREA study disclosed that American carriers now laid heavier rail and employed more ties per mile of track than did those in Europe, not only because of heavier traffic but also because the rails and ties were far cheaper relative to labor cost in the United States. In early 1928 the *Railway Age* concluded that "the roadway and structures are now in the best condition they have ever been."[47]

The spread of automatic block signals discussed above also helped reduce derailments. From 1920 to 1924 signals on the Northern Pacific detected 2,719 rail breaks along with hundreds of defective or open switches, which the ICC listed under miscellaneous causes of derailments. From 1931 to 1934 the Rio Grande found 256 broken rails and 79 open switches in one 615-mile stretch of block signals. In 1923 the Northern Pacific also became the first carrier to connect the signal system to remote sensors that monitored track conditions, and other companies soon followed. Thomas Kendreck, a yardmaster on the Seaboard, recalled that wiring fences that controlled rock slides to the signal system on that line "saved lots and lots of lives and bad wrecks." When the Pennsylvania wired its signals to fences they caught eight slides that fouled the track during 1933 and 1934. The Southern Pacific also wired the block system to fire and washout detectors on bridges and snow sheds.[48]

Despite these changes, the safety of rails remained a problem. While the carriers had forced major improvements in rail manufacturing during the prewar years, the transverse fissures, first discovered in 1911 by James Howard in the wreck of the Lehigh Valley (chapter 7), continued to puzzle metallurgists and threaten the carriers with further regulation. In 1917 and 1918 several more wrecks turned up the heat. The worst of these occurred on April 15, 1918, when a troop train on the Long Island Railroad derailed near Islip, New York, killing three soldiers and injuring thirty-six—the result of a transverse fissure in a rail that had been laid in 1898. Speaking for the ICC, Howard called for "concerted action," and in its annual report for 1917, the commission's Bureau of Safety asked Congress to grant it the power to investigate the railroads' physical conditions and operating practices.[49]

The first major discovery came in 1919 as a result of investigations by F. M. Waring, the Pennsylvania's engineer of tests, and his assistant, E. K. Hofammann. In early 1916, the company had removed all the rails in one heat containing five transverse fissures, shipping them to the manufacturer, Illinois Steel, for study. There Waring and steel company engineers found hints of small cracks in the heads of some of the rails. When the steel company balked at publicizing the findings, the Pennsylvania shipped the rails to its own Altoona, Pennsylvania, laboratory. Using traditional methods of polishing and etching steel with mild acids, Waring and Hofammann were initially unable to discover defects. They then tried deep etching with acid and found minute "shatter cracks" in the rails that had transverse fissures. Further study revealed that shatter cracks also existed in new rail. Researchers at the Bureau of Standards soon confirmed these findings. About the same time, work at the Bureau of Standards also determined the similarity of transverse fissures to the "snowflakes" that had been discovered during World War I in gun forgings.[50]

These were clearly important results, as nearly all researchers immediately recognized, and they seemed to vindicate the railroads' position that the origins of transverse fissures, like earlier forms of rail failure, could be traced to poor steel. The solution, it appeared, lay with the steel men rather than the carriers after all. Only James Howard dismissed the new findings. He pointed out that no causal relation between the shatter cracks and fissures had been found, and he noted that rail without such cracks also sometimes displayed transverse fissures. Howard's dismissal of shatter cracks went beyond the skepticism that new discoveries might reasonably warrant, suggesting that he had become wedded to his wheel-load explanation. The ICC soon followed Howard's lead. In 1923 it issued an official report written by Howard that provided details on some eight thousand transverse fissures. The report demonstrated that most had been found on the wear side of the rail and had come from track subject to high-density traffic, but it said nothing of Waring's new discoveries.[51]

That same year, the Seaboard Railroad wrote Commerce Secretary Herbert Hoover, urging a joint investigation of transverse fissures. This was exactly the sort of business-government partnership the secretary had long been advocating, and he referred the matter to the Bureau of Standards, which had been studying transverse fissures for nearly a decade. In June 1923, the bureau proposed a joint research effort with the AREA Committee on Rail, the steel producers, and the ICC. Although the AREA was reluctant, thinking that little new would come of the arrangement, it agreed to participate, contributing the statistics on transverse fissures it had already decided to collect.[52]

Initially nothing happened. The AREA lacked the funds to set up a large-scale experiment to measure rail stresses in track, and the discovery of shatter cracks had convinced the carriers that transverse fissures were largely a manufacturing problem anyway. In January 1925 the ICC's Bureau of Safety, impatient at the lack of progress, called a conference of railroad and steel company technical people to discuss a Santa Fe rail that contained fifty-three fissures in 33 feet. The meeting reviewed present knowledge and derived a set of questions that were sent out to the major carriers and resulted in a second, larger meeting in May, which again reviewed the evidence. In what appeared to be an important shift in its thinking, the Bureau of Safety, in addition to its usual exhortation to the carriers to investigate track stresses, now urged the steel makers to investigate internal cooling strains of rail.[53]

At about the same time European researchers conducted metallurgical investigations that would support the carriers' position. British engineer C. Peter Sandberg's firm of consulting engineers had developed a method of hardening rails, called the sorbitic process, that involved quenching them at high temperatures, and Sandberg worried that the process might also produce shatter cracks. Technological complementarities sometimes

shape technical change, and the development of the sorbitic process focused Sandberg's research on the problem of shatter cracks and helped
suggest a solution.

Sandberg knew that the ductility of steel varied sharply with its temperature and reached a low point at around 500 degrees C. Sandberg also
knew from earlier work that as steel is cooled, "a uniform rate of contraction occurs above the critical range [700 to 750 degrees C for steel of rail
composition], a sudden expansion during the critical range, and a steady
contraction from this temperature to atmospheric." He reasoned that
stresses built up when the interior and exterior of the rail head cooled
through this range at different rates, thereby causing shatter cracks, and
that the most dangerous range would be when the steel was least ductile—
roughly between 400 and 750 degrees C. He therefore designed an oven to
control cooling of rails, which was built at the Cargo Fleet Iron and Steel
Works in 1928, and applied for a patent the same year.[54]

The late 1920s also witnessed a breakthrough in the methods of detecting transverse fissures. The first practical device was invented by Elmer
Sperry, who learned of transverse fissures while employed by the Santa Fe
to design a track recorder car to measure roadbed alignment. Sperry began
work on his detector about 1923, and in 1926 he brought the invention to
the attention of the AREA. Sperry obtained three rails that had already developed fissures, and using his device, found a number of other fissures—
including eleven in one rail. He then called in the AREA rail committee
and as they watched, he broke the rails to reveal the flaws with, as he put it,
"100 percent results." The committee was duly impressed, and it persuaded
the ARA to underwrite the costs of perfecting Sperry's contrivance.

Sperry's detector marked the first significant use of scientifically
based nondestructive testing on railroad materials. The device employed
two brushes per rail, which transmitted a current of several thousand am-
peres at about a 0.5 volt, setting up a magnetic field around the rail. After
some experimentation, he devised a pickup that relied on the fact that any
internal discontinuity in the rail would reverse the flow of current, which
could be detected by a set of coils suspended over the rail. The coils were
attached to a transformer and a secondary circuit connected to a pen and
recording device. Sperry also rigged a paint gun that automatically sprayed
the rail at the suspect spot. The entire apparatus was set in a test car that
traveled at 6 to 8 miles per hour. The device could detect incipient fissures
that were as small as 2 percent of the rail head and it yielded no false fissures. In one startling instance, the machine indicated a fissure and when
the rail was removed, James Howard tapped it several times on the indicated spot with a ball peen hammer. The rail broke.[55]

The carriers accepted Sperry's detector on October 2, 1928, and
promptly placed two in service. By 1933 ten were in use and they had tested
135,000 miles of track. The Pennsylvania even installed one in a plant
where it reconditioned used rails. Although the detector was a dramatic
success from the outset, further minor innovations greatly improved its ac-

curacy. Sperry discovered the need to "pre-energize" the rail to ensure the proper polarity of fissures. Soon, the improved detector cars found nearly twice as many fissures per mile as the initial ones had, and the number of detected fissures skyrocketed. In addition, the procedure could find split and crushed heads and other defects as well. The initial development had cost the railroads $2,000, which may have been the best safety investment they ever made, but the invention also led to a nasty patent squabble as Sperry refused to sell his improved device and would rent only at monopoly prices, eventually leading the carriers to invent a detector of their own to escape his patents.[56]

In 1927, even as Sperry was developing his detector car, the AREA Rail Committee and the Rail Manufacturers Technical Committee finally began the joint investigations that had been promised since 1923. The agreement reflected a growing expert consensus that shatter cracks and high wheel loads were complementary, not alternative, explanations for transverse fissures. In the face of findings by Waring, and pressed by the ICC, the steel makers could no longer deny the need to investigate steel quality. The carriers could not ignore their own and ICC data demonstrating that transverse fissures—unlike shatter cracks—were never found in new rails and nearly always occurred on the wear side. Writing in 1926, John B. Emerson, the AREA's new engineer of tests, articulated the new consensus, asserting that "both mill and track must join in pleading guilty."[57]

The railroads' decision to initiate joint research was motivated by worries over both safety and regulation. With disastrous wrecks there was always the chance that the ICC might obtain more control over safety decisions. Once the carriers began to collect statistics on fissures in the mid-1920s, they discovered a large and gradually worsening problem, even as other kinds of rail failure were receding. Thus, by the late 1920s, the problem of transverse fissures had evolved from being a minor matter into a dangerous and politically charged safety problem. As if to underline the need for research, a sprinkling of spectacular derailments continued to hit the news. October 27, 1925, brought a wreck on the Frisco near Victoria, Mississippi, that killed twenty-one people and injured seventy-five. On June 9, 1928, a transverse fissure led to a derailment on the Missouri Pacific that injured fifty-one, and on December 1, 1929, a fissured rail on the Pennsylvania caused a wreck that killed nine people and injured thirty-one.[58]

In 1931 the rail committee contracted with the Engineering Experiment Station at the University of Illinois. As has been seen, the AREA had long-term ties to the university. Its Professor Herbert F. Moore was a colleague of Talbot and an expert on metal fatigue. The carriers contracted with him for a five-year study to determine the causes of and solution for transverse fissures, with the costs of $50,000 a year to be borne equally by the rail makers and the ARA.

Four years later, in March 1934, when Moore delivered his first progress report, the sponsors must have thought they had gotten value for the money. The researchers picked up from the work of Waring, determining that shatter cracks in rail reduced its fatigue strength about 50 percent, to between 27,000 and 32,000 psi. Employing a rail-testing machine, Moore's team had been able to develop transverse fissures in such rails with loads as low as 40,000 pounds. Confirming Howard's views, Moore stressed that the wheel load rather than the bending moment was the key factor, and hence heavier rail was no solution. "The picture of an internal fissure starting from intensified shearing stresses at a shatter crack seems consistent and convincing," Moore concluded. Following up on earlier work, the researchers tested ordinary rails against those that had been subject to controlled cooling and found that the latter process sharply reduced the incidence of shatter cracks.

Finally, as James Howard had been urging since 1911, the Illinois researchers employed methods developed by Talbot to measure actual wheel loads from two locations on the B&O and Pennsylvania. They discovered that the dynamic augment resulting from flat wheels and poor counterbalancing of locomotives was "very noticeable," and that about one out of 1,000 (later estimated to be one out of 500) wheel loads exceeded the critical threshold of 40,000 pounds. Such findings suggest an explanation for why transverse fissures began to occur when they did. As wheel loads and train speed increased after 1900, they resulted in dynamic loads that increasingly exceeded 40,000 pounds. As traffic density increased, disaster eventually ensued.[59]

The Illinois findings provided strong support for the experiments already in progress to reduce shatter cracks by controlled cooling of rails. Yet even as research demonstrated the efficacy of Sandberg's process, other investigations revealed that he had arrived at it for the wrong reasons. In 1935, German researchers attempting to discover the origins of flaky steel were unable to prevent the flakes by retarding cooling in the 700 to 400 degrees C range, but were able to do so when they continued the process down to nearly 200 degrees C. Yet this was far below the range within which thermal stresses should have occurred. The presence of hydrogen had long been known to make steel brittle, and research indicated that when hydrogen was present in sufficient quantities, rapid cooling could raise its solution pressure above the tensile strength of the metal, resulting in a flake or crack. Moore and other researchers confirmed the finding and were able to produce shatter cracks by treating steel with hydrogen.[60]

British rail makers had employed the Sandberg process for cooling since 1928. In 1929 Sandberg had come to United States, where he interested Bethlehem in the process. After much experimentation in applying the process to the higher-carbon American rails, Bethlehem began production in March 1935. As other manufacturers adopted similar methods, controlled cool rail rapidly supplanted all other processes. By 1939, it comprised 91 percent of all new track miles, which represents a breathtakingly rapid rate of diffusion. As use of rail manufactured by controlled cooling technique supplanted older rail in the early 1940s, transverse fissures finally started to decline. As so often happens the solution to one problem ameliorated another one as well, for controlled cooling also reduced other forms of failure.[61]

Little Accidents to Passengers and Trespassers

While passenger fatality rates fell sharply during the interwar years (Figure 9.1), only about 36 percent of the improvement resulted from the decline in train accidents. The biggest gains resulted from reductions in little accidents at stations that reflected changes in the composition of traffic.

Even before World War I, the electric interurbans and then automobiles began to cut into the carriers' short-haul passenger traffic. In the early 1920s the decline in traffic turned into a rout. In 1922 the railroads hauled about 967 million passengers on an average journey of 37 miles, resulting in 35 billion passenger miles. By 1939 only 450 million passengers took the train; their journey now averaged 50 miles, for a total of 23 billion passenger miles. Individual carriers confirmed that the automobile and the bus were supplanting the train as the dominant means of short-distance rural travel. New bus service over the 23 miles between Norfolk and Suffolk, Virginia, reduced ticket sales on the Norfolk & Western from 110,000 in 1916 to 29,000 in 1923. On the B&O, passengers hauled the 6 miles between Berkeley Springs and Hancock, Maryland, fell from 4,254 in June 1916 to 875 during the same month of 1924. Overall that carrier hauled 38 percent fewer passengers in 1923 than in 1916, and other lines told the same story.[62]

These changes in the type of passenger traffic improved safety in several ways. First, the length of haul rose about 43 percent during 1922–39, which as earlier discussed induced a similar decline in accidents getting off and on trains and at stations. In fact, such accidents fell about 73 percent relative to passenger miles during these years, for the trend to longer journeys was reinforced by a shift of traffic from comparatively dangerous rural stations to better-designed urban and suburban structures. In addition, the shift to long-haul traffic moved passengers into steel cars with vestibules and gates, all of which improved their safety. Ironically, then, the automobile improved the railroads' safety record as it destroyed their business. And, of course while the shift of travel from train to car improved train safety it worsened overall transportation safety. For example, in 1926, 23,400 Americans died in automobile accidents. There were 128.6 billion automobile passenger miles traveled that year, and so the fatality rate was 182 per billion passenger miles, compared with 4 on the railroads. Thus every billion passenger miles the railroads lost to the automobile took 178 lives. Twentieth-century Americans appeared to prefer speed and convenience to safety—just like their nineteenth-century ancestors.[63]

Riding the Rods and Walking the Tracks

Trespassing on tracks and trains had been deeply embedded in the structure of nineteenth-century working-class life, and the predictable result was that American railroads killed trespassers by the thousands. As earlier demonstrated, however, around World War I rising real wages and better transportation began to undermine this world. Some hobos who had once ridden the rails became "rubber tramps" who hitched rides or drove beat-up Model Ts. Along with agricultural mechanization that reduced the demand for migrant farm work, these forces steadily reduced trespassing casualties, but after 1929 a decade of Depression turned back the calendar to an earlier era and deaths of trespassers rose sharply.[64]

As unemployment rose, men and boys and a few women and girls took to the road. While many hitchhiked they soon became too seedy-looking to attract a ride and had to shift to the rails. "In '30 and '31 you'd see freight trains, you'd see hundreds of kids, young kids, lots of 'em just wandering all over the country looking for jobs," one man remembered. Between September 1, 1931, and April 30, 1932, the Southern Pacific ejected 416,915 trespassers, many of them boys. Not all of them got to their destination; ICC data confirm that the largest increase in fatalities after adult hobos was among individuals ages fourteen to twenty-one.[65]

Figure 9.3 presents trespassers killed on trains and on the tracks from 1926 through 1940. As can be seen, fatalities to track walkers continued to fall in the 1930s probably because these were local people and their numbers were only modestly affected by economic circumstances. Yet the Depression briefly reversed the long-term decline in fatalities on trains as eroding wages and the rise in unemployment led thousands to steal a ride in search of work. But by the late 1930s, as wages and automobile registrations rose and unemployment waned, the long-term trend again emerged and trespassing fatalities plunged with the advent of wartime prosperity.[66]

Figure 9.3. **The Impact of Depression on Trespassing Fatalities, 1926–1940**

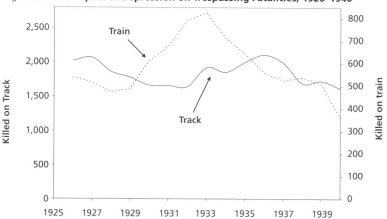

Source: ICC, Accident Bulletin, various years.

New Ideas and New Policies for Grade Crossings

The problem of accidents to trespassers did not simply go away; with the advent of the automobile it metamorphosed, so to speak, into the grade crossing problem. The explosive growth in motor vehicle travel in the context of large numbers of poorly guarded crossings and dense rail traffic touched off a sharp rise in crossing accidents that began before World War I and continued throughout the 1920s. By 1921 grade crossing accidents were killing or injuring about 6,500 people a year. Two years later, on *one day* in August 1923, the carriers killed thirty-six individuals at railroad crossings. One eastbound express on the Pennsylvania killed four people in Highland, Illinois, and several hours later hit another car in Indiana, killing nine more.

In addition, while collisions with pedestrians or horses rarely did much damage to the train, collision with a car or truck might derail a passenger train and generate a bumper crop of liability claims. In May 1922, the Lehigh Valley's *Black Diamond Express* hit a car, killing three train passengers and injuring three others. Such accidents rose steadily; in 1929 crossing accidents accounted for 174 train derailments and five fatalities to passengers and employees. When the first of the streamliners—the Burlington *Zephyr*—appeared in 1934, much of its weight was placed in the nose to withstand crossing collisions. That company's Ralph Budd happily described it as a "steel sheathed bullet." Another company dropped the height of the locomotive pilot 4 inches, preventing derailments when it struck vehicles. "On the other hand," a spokesman cheerfully observed, "several accidents have occurred at grade crossings where the highway vehicle has been sheared in two and thrown clear of the track."[67]

Moreover, because the automobile increased both the volume of travel and range of individual journeys, it enormously multiplied the number of potentially dangerous crossings. While the ARA continued efforts to standardize crossing signs and signals, its main approach to crossing safety involved publicity. The carriers' experience with Safety First campaigns led them to try the technique on motorists, and the ARA Safety Section began its first Careful Crossing campaign from June through October 1922. Persuading drivers to "cross crossings carefully" turned out to be more difficult than exhorting workers to follow rules, however, for accidents were higher in 1922 than in the previous year. Yet some of these campaigns clearly had an impact. The Northwestern Pacific wrote letters to drivers seen violating crossing regulations, urging them to obey the law and sometimes company representatives even made house calls on the guilty parties. As a result its accidents at California crossings declined during 1922–26 even as the statewide totals for all carriers increased.[68]

At least as important to the carriers, such campaigns defined the crossing problem in ways they found congenial. The carriers had two main goals: first, to shape perceptions of crossing accidents so that public rather than railroad funding would seem appropriate, and second, to establish priorities for removal and guarding so that not all crossings would be candidates for abolition. Should accidents come to be seen simply as the result of dangerous crossings, pressure might build for their abolition or for much more widespread guarding. *Railway Age* worried that drivers, who were also voters, would push through massively expensive programs to guard or separate rail-highway crossings at the carriers' expense. Shifting the focus to careless, stupid, or drunken motorists suggested that stricter laws, better driver testing, and public funding might be more appropriate—all of which were conveniently costless to the carriers.[69]

Thus in 1927 H. A. Rowe, chairman of the ARA Safety Section committee on grade crossing accidents, reported that "the Safety Section is entitled to feel gratified with the results accomplished." By this he did not mean that accidents had declined, for 1927 was only marginally better than 1926, but rather that "there has been an undeniably lessened . . . hostility to the railroad in crossing mishaps [and] there has been recognition of the automobile driver's responsibility for practically all such accidents." To support their claims the carriers emphasized that a large fraction of crossing accidents involved drivers who ran into trains. One motorist, the *Railway Age* reported, hit the fifty-fifth car of a freight train and another rammed a caboose. In 1928 the ICC reported that nearly 1,300 of the 5,752 grade crossing accidents that year involved a vehicle running into a train. Even terminology could be important: railroad lobbyists urged Congress not to think of "railroad grade crossings" but rather "highway-railroad crossings."[70]

The popular press largely accepted the railroads' position that the solution to crossing accidents was better driving. *Literary Digest* often reprinted articles from *Railway Age* such as "Engineer's View of the Crossing Fool," that gave the carriers' perspective, while the *Saturday Evening Post* chipped in with "Gray Hairs for Casey Jones: Automobile Drivers and Grade Crossings." Only *American City* pushed aggressively to eliminate crossings, and it acknowledged a government responsibility in the matter.[71]

In the 1930s highway engineers began to challenge the driver carelessness explanation, just as safety engineers had repudiated employee carelessness as a cause of work injuries a generation earlier. The Federal Aid to Highways legislation, which began in 1916, not only initiated construction of a nationwide road system, it encouraged states to create highway departments and centralized concern over both roads and grade crossings. Traffic engineers staffed these institutions; by training they looked for physical causes for events and so they found insufficient the assumption that drivers were careless. ICC statistics supported their efforts, for the commission published accident data by time of day and engineers pointed out that a disproportionate number of motorists who ran into trains did so at night—suggesting that bad lighting rather than alcohol was the likely culprit. Engineers also began to estimate the value of time lost stopping at crossings and they discovered—unsurprisingly—that motorists were more likely to run a signal the longer they had to wait. Thus when work trains activated signals for long periods, such "crying wolf" encouraged motorists to ignore them. Yet such reasoning might lead to conclusions the railroads found congenial, for if crossing accidents were the outcome of a complex of causes, they were not exclusively a railroad problem, and state and federal agencies, and even the American Automobile Association, started to urge public funding.[72]

Beginning with California in 1917, a number of states inaugurated ambitious programs to catalogue and guard or eliminate grade crossings. In 1926 New York authorized a $300 million bond issue for the purpose, with the railroads to pay half of any elimination and localities one-quarter, but progress was slow as both the towns and carriers balked at the expense. In the 1930s the AAR successfully campaigned to reduce the carriers' share of costs in crossing elimination. In 1935 the railroads finally gained some legal support for their position when the U.S. Supreme Court struck down a law that would have made the Nashville, Chattanooga & St. Louis pay half the costs for an underpass on a road designed to take business away from the carrier. "It is the railroad which now requires protection from dangers incident to motor transportation," wrote Justice Brandeis for the majority.[73]

In 1938 New York modified its crossing law and the state assumed the entire cost. Maryland reduced carrier funding to 25 percent of the cost of crossing elimination on state roads and Indiana placed it at 20 percent. By 1948, Ohio, New Jersey, and West Virginia had all reduced carrier funding to 15 percent or less while Indiana, Michigan, and Pennsylvania passed laws shouldering some or all of the cost of crossing protection as well.[74]

As the states assumed a greater share of the cost of guarding or elimi-nating crossings they immediately faced a problem of priorities. With over 230,000 public crossings in the country, and the number increasing each year, which should be chosen for elimination or guarding and how should it be done? When funding had largely been left to the carriers and munici-palities each had checked the other—indeed, they had sometimes done so too well. But with state and later federal funding the "incentive to econ-omy on the part of municipalities has been removed," New York's Railroad Commission ruefully observed as it faced a blizzard of gold-plated pro-posed eliminations from municipalities that also included new highways, stone-faced bridges, sidewalks, landscaping, and recreational facilities.[75]

But even without gold-plating, the problem remained of how to maximize the benefits for a given expenditure of resources. It was, of course, no different from the issue of whether automatic train control or crossing protection yielded the most benefits per dollar. As we have seen, the carriers had been solving such problems since the dawn of railroad-ing, and they have since become a standard in economic and public policy analysis. But their formal analysis was comparatively new to policy makers in the 1920s. The problem was especially important because crossings var-ied enormously in their dangers. In California in the late 1920s and early 1930s about 72 percent of all crossings had no accidents during a seven-year period while 13 percent had more than one. Yet these 13 percent of crossings had 85 percent of accidents during this period. New York and no doubt other states found a similar concentration of dangerous crossings.[76]

The carriers had long advocated that crossings should be ranked for elimination or guarding according to hazard, and in the 1920s, state highway engineers developed formal tools to allocate funds. Initially many programs simply used "judgment" to rate crossings but in 1934 when Illinois funded a program for crossing protection, Warren Henry, an engineer for the state's commerce commission, developed the first mathematical formula for hazard assessment. It was widely mimicked and modified, and in 1941, engineer-economists at the Bureau of Public Roads summarized several of the models and proposed their own hazard index derived from an analysis of accidents over a five-year period at 3,563 crossings.[77]

$$A = 1.28[H^{0.170} \times T^{0.151}]/P^{0.171} + K$$

Here A is the probable number of accidents in a five-year period; H and T are daily vehicle and train traffic. The term P measures the effectiveness of the means of protection (signs, gates, signals, etc.), and was derived from accident and traffic experience of such equipment at other crossings, while K was a correction factor. Employing such a formula, the authors argued, was necessary to achieve the goal of maximizing the hazard elimination for a given expenditure. By the late 1940s eighteen states used this or similar formulas and an additional fourteen relied on expert engineering assessment. These formulas and judgments combined with nonsafety benefits such as the value of time-focused public funding on high-payoff crossings. Just as improvements in science and engineering-economic analysis contributed to the decline in train accidents, so they also began to contribute to crossing safety.[78]

A New Deal for Grade Crossings

Until the 1930s grade crossing accidents were largely the concern of states, localities, and railroads. Large-scale federal participation started with the New Deal. In 1933–34 about $34 million was spent on crossing elimination under the National Industrial Recovery Act, while both the Works Progress Administration and Public Works Administration spent about $5 million on crossing elimination up through 1936. The Emergency Relief Appropriation (ERA) Act of 1935 appropriated $200 million for crossing elimination, and the federal aid highway system appropriated $100 million in 1938–39 and another $90 million for 1940–43.[79]

Unlike much federal aid to states these grants were unconditional rather than matching. They were apportioned not politically, as some New Deal spending may have been, but rather on the basis of population, railroad mileage, and highway mileage qualifying for federal aid. Such an allocation raises two obvious questions: first, how closely did this process reflect safety needs, and second, did federal funds supplement or supplant state and local spending?[80]

The answer to the first question is that the allocation of funds matched accidents only loosely. While accidents account for about 60 percent of the variance in the allocation of ERA funds among the states, the appropriation discriminated against high accident states. The average number of crossing accidents in a state in the late 1930s was about eighty-two per year; a state with twice that many received only about 69 percent more funds. The result was to allow relatively safe states to become even safer.[81]

The funding was a boon to the carriers. By 1940 federal funding had eliminated or reconstructed 181 crossings on the Pennsylvania at a total cost of $27.5 million, of which the railroad paid only $1.5 million, or 5.6 percent. A statistical analysis suggests that compared to the years 1926–35, there were an additional 350 crossing separations a year from 1936 to 1942, or a total of 2,450. Thus it does not look as if states and railroads simply offset the New Deal funds by reducing their own efforts. Since federal spending amounted to about $440 million and may have led to the reduction of an additional 2,450 crossings as well as some guarding, the average cost was $163,265. While this was high for the period, probably federal money went to relatively expensive projects.[82]

In fact, grade crossing accidents peaked about 1928, when 2,568 individuals were killed. Both fatalities and injuries declined unevenly throughout the 1930s. Overall, between 1925 and 1941 fatalities declined about 12 percent. This occurred despite the 172 percent increase in vehicle miles over the same period, which statistical analysis suggests would by itself have increased fatalities about 34 percent. Increasing road traffic was in part offset by the decline in train miles, which fell about 18 percent from 1925 to 1941, generating a roughly 21 percent decline in fatalities. While part of the decline in train miles reflected the Depression-induced eclipse of passenger traffic, it also resulted from railroad efficiency gains that increased freight train size about 27 percent over this same period. Absent such an improvement, train miles would have fallen hardly at all. Such figures suggest how perverse train limit laws were, for such legislation could have crippled efficiency gains, thereby worsening crossing fatalities.[83]

In addition, however, controlling for train and vehicle traffic, crossing fatalities declined about 2.3 percent a year, reflecting a combination of behavioral and physical changes. Companies such as the Erie required engineers to sound the whistle until the engine arrived at the crossing and it employed spot checks to monitor compliance. Safety campaigns also contributed to the decline, while the *Railway Age* believed that as motorists became more accustomed to stop-and-go traffic signals they were more likely to obey crossing warnings. But surely much of the gain must have reflected the abolition or guarding of the most dangerous crossings.[84]

Automobiles probably contributed to the reduction in grade crossings, for by increasing travel speed they reduced the need for multiple crossings. A striking feature of the campaign against crossing accidents was how much was accomplished by guarding and removing a relatively few crossings. The total number of crossings in the United States peaked at 242,809 in

1929. By 1941 that number had declined to 230,285. Over this period a bit more than 25,000 crossings were removed, or 10 percent of the 1929 total, but about 13,000 were created, leaving the net decline of 12,524, or about 5 percent of the 1929 total. Of those crossings removed, only about 3,413 were the result of grade separation; the rest reflected declines in railroad mileage, road straightening, and similar events. The entire decline was in unprotected crossings as those with some form of protection actually rose from about 12 to 14 percent of the total, while the typical means of protection increasingly shifted to the familiar flashing lights. In short, guarding an additional 2 percent of crossings and removing about 5 percent of them reduced crossing accidents even in the face of sharply rising automobile traffic.

On the eve of World War II American railroads were far safer than they had been only two decades earlier. Regulatory threats and rising liability costs induced the carriers and the ARA to develop for the first time an industry-wide commitment to reduce not simply collisions and derailments, but all types of accidents. In response, Congress ignored railroad safety while the ICC, after its train control order, largely allowed the experiment with voluntarism to proceed, although it intervened forcefully and effectively to press for better brakes and rails. The threat of intermodal competition and rising labor costs encouraged technological change and investment, and a host of improvements in rails, ties, work equipment, cars, and locomotives transformed railroading. Better rail, brakes, and signals increased productivity and safety while the impact of other forms of mechanization was often contingent upon successful safety work. Other institutional innovations in the form of closer ties to trade associations and universities also yielded both safety and productivity.

Worker and passenger safety also benefited from the long-term forces that were reducing labor turnover and traffic density. Similarly, the eclipse of trespassing casualties except for a Depression-era spike reflected Americans' rising affluence. Thus the automobile, while it generated an explosion of casualties at railroad crossings, probably improved railroad safety on balance, because it yielded an even larger decline in trespassing accidents.

As accidents receded, their political economy evolved. While improving safety during the nineteenth century had made the public impatient with accidents, now it contributed to public indifference. Increasingly safety was becoming the province of specialists and interest groups. Yet for all the enormous improvement, on the eve of war all was not well with railroad safety. Companies were entering a period of rapid mechanization, yet they had failed to integrate the role of safety organizations into equipment specifications and purchasing. As *Railway Age* had begun to note, eroding profits were diminishing the funds for future safety investment. In the postwar years, these forces would again make safety a public issue and finally destroy the safety compact.

10
Safety in War and Decline, 1940–1965

Safety, like every other worth-while thing, must be bought and paid for. It is not cheap . . . there are no bargains.
— Frank Cizek, Superintendent of Safety, Delaware, Lackawanna & Western, 1943

The [Interstate Commerce] commission persistently prove[s] it is not a qualified judge of [the railroads' safety] by preventing them from making enough earnings to provide for their total needs.
—*Railway Age*, 1947

When war began in Europe in September 1939, American railroads were still recovering from a decade of depression. Passenger and freight traffic remained far below their 1929 levels, as did employment and manhours. Military demands soon put an end to excess capacity, however, and by 1943 the carriers were straining to meet the colossal increase in traffic. Predictably, the boom swamped the longer-term forces that had been improving safety, and accident rates rose from their Depression-era lows, but the increase was in fact quite modest. Two decades of research and investment in new technologies and procedures that enhanced both efficiency and safety now paid off as the railroads carried the arsenal of democracy to war.

With the end of war in 1945, secular trends reasserted themselves. The railroad technological network generated a host of improvements, many of which had war or prewar origins, and which yielded safety and productivity gains. As a result, most forms of safety steadily improved down to the mid-1950s. But the decline in traffic that had begun in the 1920s also resumed. By the 1950s less-than-carload freight was increasingly shifting to trucks, while passenger traffic was collapsing as automobiles and airplanes stole the carriers' market. In addition, union work rules and wage pressures raised costs while stultifying state and federal taxes and regulations contributed to railroad woes. Returns on investment sank to abysmal levels and by the late 1950s many eastern and Midwestern lines were in serious financial trouble. Deteriorating finances curbed both investment and safety work. Partly as a result while passenger risks fell unevenly, employee safety worsened for nearly two decades after 1957 and freight derailments rose, becoming for the first time a public issue. The erosion of safety progress undermined the existing regulatory bargain; in 1966 Congress transferred control of railroad safety from the ICC to the newly formed Federal Railroad Administration and four years later federalized control over nearly all aspects of rail safety.

Carrying the Arsenal of Democracy, 1939–1945

The expansion of railroad output from 1939 to the wartime peak in 1944 is without any historical parallel. Passenger miles quadrupled, rising from about 22 to nearly 96 billion while freight haulage more than doubled, from 335 to nearly 741 billion ton miles. To accomplish these Herculean feats the carriers increased manhours about 70 percent even as the Selective Service skimmed off many long-time employees. New entrants to the railroad labor force jumped from 239,000 workers in 1939 to 820,000 in 1942. A tight labor market also sharply raised turnover. By 1943 the carriers were hiring women and high school boys, yet labor shortages remained widespread. The Delaware & Hudson, which had seen resignations and dismissals all but vanish in the 1930s, lost nearly one-quarter of its force in 1944.[1]

Trading Safety for Victory

As wartime demands raised traffic density from Depression-era lows, and employee experience and supervision declined, worker safety deteriorated, with fatality rates rising one-third from 1939 to their wartime peak in 1942. Indeed, the wartime years represent one of the few times when we can directly observe safety being traded for increased output. Derailments, which had averaged around 4,000 a year in the 1930s, rose to 9,379 in 1945. Accidents from transverse fissures jumped, as the carriers' detection programs were unable to keep up with traffic growth. While most derailments involved freights, passenger fatality rates from derailments, which had been zero in 1935, ballooned to 1.87 per billion passenger miles in 1943—a level not seen in decades. The Atlantic Coast Line (ACL) seemed particularly derailment-prone. In 1941 two passenger train derailments caused serious injuries, and on December 16, 1943, near Buie, North Carolina, a transverse fissure spilled one passenger train onto the track in front of another, killing seventy-seven people. Less than a year later, on August 4, 1944, another broken rail killed forty-seven on the ACL near Stockton, Georgia. The Pennsylvania's turn came on September 6, 1943, when a hot box broke an axle, causing a wreck near Philadelphia that killed seventy-nine.[2]

Collisions also rose sharply, from an average of about 1,500 a year in the late 1930s to a peak of 4,989 in 1943. Again, while most such wrecks involved freights, passenger casualty rates briefly hit levels characteristic of the early 1920s. On June 14, 1942, when the Seaboard's *Silver Meteor* recalled its flagman too soon, the following freight ran an approach signal and the resulting wreck killed nine. The worst such wreck—a rear collision on the Southern Pacific, near Bagley, Utah, on December 31, 1944—oc-curred when the engineer ran a block signal and killed fifty people. Failure to protect a train stopped for a hot box yielded a rear collision on the Great Northern on August 9, 1945, that took thirty-four lives.[3]

As these examples suggest, much of this upswing in accidents was due to employee negligence. In early 1942, *Railway Age* began to worry about the accident record and it warned that because new hires were in-creasing faster than superintendents, monitoring was getting thinner. Later that year the journal observed that seventy of the previous one hundred accidents studied by the ICC reflected poor supervision or training. That same year one superintendent reported that because of new hires and re-tirements he had one and sometimes two green men on every train. One solution was better safeguards. After the Signal Inspection Act of 1937 clari-fied its authority, the ICC finally began to require block signals on espe-cially dangerous stretches of road. In October 1942, after a rear collision on the Denver & Salt Lake, the commission ordered blocking on a section that had experienced five collisions, five derailments, and fifteen rockslide accidents in the previous two years.[4]

As usual while most casualties from collisions and derailments in-volved workers, most worker casualties resulted from other causes. Even in 1942 at their wartime peak, train wrecks accounted for only about 17 percent of worker fatalities. As always, the little accidents were far more deadly, and here again, the problem lay in deteriorating experience of workers and supervisors. Locomotive accidents rose sharply. Many re-sulted from defects as the carriers strained to keep power on the road, but the most deadly—boiler explosions—also reflected inexperienced crews and inadequate training. In the ethos of railroading, locomotive failures on the road reflected badly on the engineman, and crews would go to great lengths to make a station, even in the face of a low-water warning on an en-gine. Yet such behavior also reflected poor training, for few trainmen had been instructed that when a waterglass went dry there might be only 500 gallons over the boiler crownsheet. Since a large locomotive could evapo-rate 80,000 pounds of water an hour, a crew might have 5 minutes before disaster.[5]

Nor did inexperienced track supervisors always appreciate that mechanized work required proper procedures and supervision. On August 4, 1942, a section gang on the Wabash employing noisy power tampers and without a lookout failed to hear an oncoming freight. Ten men were killed. A little more than a year later an almost identical accident on the Big Four killed nine more men.[6] While there is no evidence on the extent of experi-ence dilution during these years, records of the Railroad Retirement Board indicate that the expansion of employment reduced workers' median age by five full years from 1939 to 1942. Statistical analysis suggests that each year of age reduced trainmen's fatality rates by about 0.26 points. If so, labor market changes during these years may have accounted for about a 19 percent rise in fatality rates.[7]

Yet wartime conditions reduced the incidence of some kinds of accidents. Although railroad passenger travel exploded, automobiles remained the mode of choice for short trips. Accordingly, the average railroad passenger journey jumped from 50 to 104 miles between 1939 and 1944, and so station accidents and those from getting off and on trains declined relative to passenger miles. Fatalities from trespassing also resumed their long-term decline, a reflection of falling unemployment and rising incomes, while grade crossing fatalities also continued modestly downward—in good part due to gasoline rationing.

Hence the striking fact about railroad safety during World War II was not how much but how little it deteriorated. As noted, worker fatality rates jumped sharply from 1939 to 1942, but thereafter they trended downward while passenger fatality rates failed to rise at all. *Railway Age* was fond of comparing railroad safety in World War II with that of World War I, and the comparison is instructive. Figures 10.1 and 10.2 show fatality rates for passengers and workers for World War I and II. Had the risks of the first war obtained during the second, 5,478 more workers and 2,097 more passengers would have been killed from 1940 through 1945. The total of 7,575 fatalities would have exceeded the number of Americans killed on Iwo Jima—one of the bloodiest battles of the Pacific war. Similarly, had fatalities of others per locomotive mile remained at World War I rates, an additional 7,193 individuals would have died trespassing or at grade crossings.

Both the institutionalization of safety work and the carriers' investments in improved technology yielded wartime dividends. As supervision and experience declined, many carriers responded with better training for new workers. The Erie expanded its program that taught supervisors how to teach, employing a "tell-show-discuss-repeat-use" approach. High school boys employed as brakemen and signalmen or in track work were given special instruction. The Milwaukee employed a handbrake instruc-

Figure 10.1. **Passenger Fatality Rates, World War I and World War II**

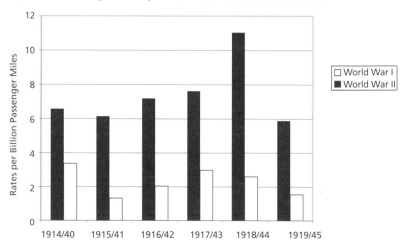

Source: ICC, *Accident Bulletin,* various years.

Figure 10.2. **Worker Fatality Rates, World War I and World War II**

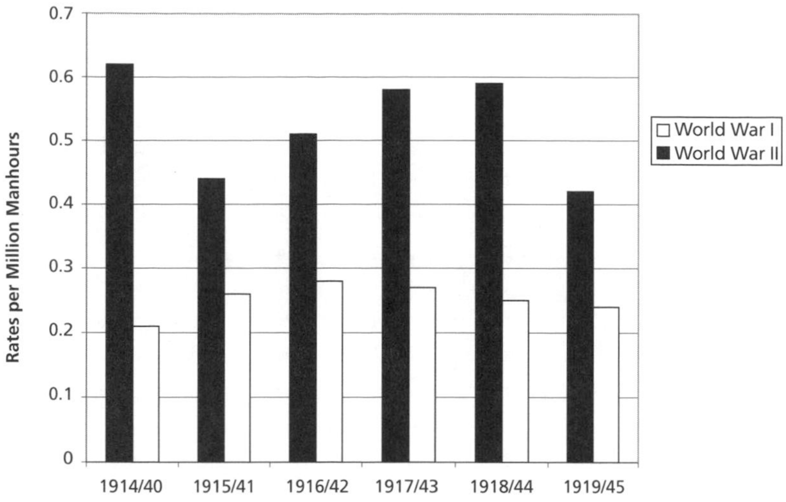

Source: ICC, *Accident Bulletin,* various years.

tion car and it provided special training in the dangers of low water for its engine drivers. The Rio Grande used a replica of a locomotive to teach newly minted enginemen about the dangers of low water while the Milwaukee and Lehigh Valley developed an instruction book for the same purpose.[8]

The newer, more efficient technologies introduced in the 1920s and 1930s improved safety because they allowed the carriers to increase passenger and ton mileage without commensurate increases in train miles. For example, treating locomotive boiler water reduced road breakdowns and hence contributed to gains in average speed as well as safety. Similarly, the "superpower" steam locomotives and diesels allowed greater speed and increased train length. Larger freight cars contributed to the increase in tons per train, and by hauling goods in fewer cars as longer trains, they reduced the chance of derailments as well. As a result, in 1944 the railroads carried 2.21 times as many passenger miles and 1.81 times as many ton miles as in 1918, even though passenger *train* miles were only 0.88 and freight *train* miles 1.13 times the earlier levels. They achieved such results by increasing passengers per train to 2.51 and tons per train to 1.61 times 1918 levels by increasing car loading and train length.[9]

Overall the carriers' wartime safety experience illustrates the interplay of cyclical and secular forces in work safety, as is illustrated in Figure 10.3. The great wartime traffic boom forced the carriers to make a short-term trade-off of safety for enhanced output. But the long-term improvements in railroad efficiency and safety organizations yielded greater output and far better safety than had been possible only a generation earlier.

Figure 10.3. **The Changing Trade-off between Safety and Output, 1918 and 1944**

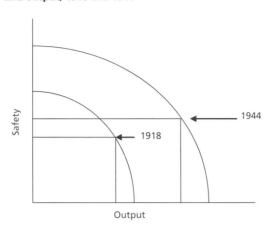

Postwar Safety: Environment and Structures

For about a decade after World War II, most aspects of railroad safety improved; thereafter, the picture becomes more mixed. In part these uneven results reflected the conflicting incentives generated by the carriers' economic circumstances. The rising cost of passenger and worker fatalities encouraged safety organizations to campaign against negligence and lax practices that caused accidents. The relative rise in labor costs also encouraged the generation of and investment in a host of new technologies with a strong labor and accident saving bias. As a result, labor productivity grew rapidly, while most aspects of safety also improved down through the mid-1950s. Yet the increasing squeeze on carriers' net earnings finally undermined their ability to supply safety, leading to increased derailments and raising work accidents. In addition, rising freight car weight may also have worsened derailment risks.[10]

Safety Institutions in the Postwar Years

The dominant institutions shaping railroad safety after 1945 remained the railroad technological network, the AAR and individual carrier safety organizations, the technical press, and the ICC. In the past, each had contributed to the remarkable improvements in rail safely. But individually and collectively they failed to respond to new challenges.

By the 1930s the railroad technological network had become institutionalized in individual carriers and engineering organizations and their relations with universities and suppliers. In 1948, after decades of debate, the AAR finally set up a central research laboratory on the campus of the Illinois Institute of Technology. Centralization allowed the AAR to focus research on important new technologies common to all roads, while that organization also demonstrated a new willingness to speed up diffusion of innovations by mandating them in interchange. In addition, new companies entered the railroad supply network and new technologies yielded sharp improvements in both safety and productivity (see below).

Yet there were also major failings in railroad research. Commenting on the AAR facility, *Railway Age* observed it "is definitely not of the scope contemplated by the proponents of a centralized research agency." Others complained that its budget was "pitifully small"—only $600,000 by 1957. Nor did individual carriers fund much research, for railroad R&D remained a tiny fraction of revenue. Railroad research also remained narrowly focused on engineering and testing. There was no safety research section and the carriers were slow to discover the value of operations research for problems such as freight car productivity. Such an analysis might have encouraged greater automation, reducing the need for dangerous yard work and improving safety as well as production.[11]

The carriers' safety organizations changed little from the form they had assumed in the early 1930s. The AAR Safety Section remained focused on supervision and behavior, largely leaving engineering to other branches of the association, and the same was true of safety departments on individual carriers. When the Safety Section proposed to employ a full-time safety engineer in 1948, the AAR refused to fund the position. With a few exceptions, safety workers failed to press for needed industry-wide improvements and they rarely functioned as an organized force for better safety. The national meetings of safetymen also became stale; one looks largely in vain for analysis or even awareness of how the new machines and practices sweeping through railroading were affecting safety. The Safety Section did develop uniform codes of safety. Unfortunately, many of the rules resulted from experience in the form of accidents rather than from a proactive hazard assessment.

The absence of formal labor representatives at high levels also remained a problem. As noted, their inclusion might have broadened the focus of safety organizations and perhaps have encouraged the unions to employ safety experts who could have channeled their efforts in more constructive ways. With the decline in railroad employment, the unions increasingly equated safety with job maintenance, and they remained overly enamored with technological fixes for safety problems. In 1958 the brotherhoods achieved legislation giving the ICC power over brake maintenance, the apparent goal being the prevention of changes that might allow longer freights, and they continued to press for clearance legislation, both of which were overly narrow and addressed minor sources of risk. The brotherhoods also continued the campaign for full crew and train limit laws, and managed to persuade some states to apply full crew laws to diesels, which wasted resources and worsened safety.[12]

Finally, neither the technical press nor the ICC performed well as informed critics during these years. Voluntarism as practiced by Charles Francis Adams assumed expert public oversight, and for decades, railroad safety had benefited from outside criticism from the ICC and the technical press. At major journals a parade of forceful editors, including Henry

Varnum Poor, Matthias Forney, Arthur Wellington, B. B. Adams, W. H. Boardman, and others had been a powerful voice for a host of improvements in equipment and practices. But in the postwar years, concerns with regulation and crumbling finances largely drove safety from the pages of the trade publications, thereby removing an important source of informed criticism.

The ICC might also have been a much more effective force for better safety. In part, the commission's failure reflected the laws that it administered, which tended to be technology-specific and therefore subject to obsolescence. The ashpan law, to take the most extreme example, had little relevance in a world of diesels. When the commission occasionally campaigned for broader laws, modifying locomotive inspection or governing train operation, for example, Congress rebuffed it.[13]

Yet the commission was not simply a captive of bad laws. It had usually construed regulations narrowly and failed to use its own resources imaginatively. As noted, important improvements in boiler design such as thermic siphons, better fuse plugs, and boilers that failed safely could have sharply reduced deaths from boiler explosions long before the diesel had they been more widely used. The Bureau of Locomotive Inspection could have called conferences, publicizing the new designs, and urged and monitored their adoption, but it did not. Similarly, the Bureau of Safety seems to have been obsessed with minutiae. It investigated accidents and inspected freight cars, largely confining itself to counting defects, but it never focused on the need for better safety training or publicized especially effective company safety organizations. Indeed, the commission admitted to Congress it knew nothing of such programs. It could have provided detailed studies of the benefits and dangers of new technologies such as nonspin hand brakes, or pressed for better safety training, all of which was routine practice at the Bureau of Mines. In its youth the commission had been an independent and sometimes constructive force for safety; in the 1890s it had helped marshal the campaign for safety appliances and after 1900 had campaigned to extend them. In the 1930s it effectively pressed for the AB brake and better rails, but by the 1950s in old age it had largely lapsed into passivity.[14]

Little Accidents to Passengers, Trespassers, and Workers

Public interest in railroad safety in the postwar years was, as always, dominated by concern with passenger train accidents, but as usual it was the less visible little accidents that were of greater importance. Fatality rates to passengers declined unevenly (Figure 10.4) largely due to causes other than train accidents, which reflected the continuing evolution in passenger traffic. By the 1960s rail traffic consisted of either commuters, who accounted for about one-fifth of passenger miles, or long-haul travelers, whose average journey of 135 miles sharply exceeded that of the 1930s. As has been noted, such changes resulted in less station exposure, and they suggest that

Figure 10.4. **Passenger and Worker Fatality Rates, 1940–1965**

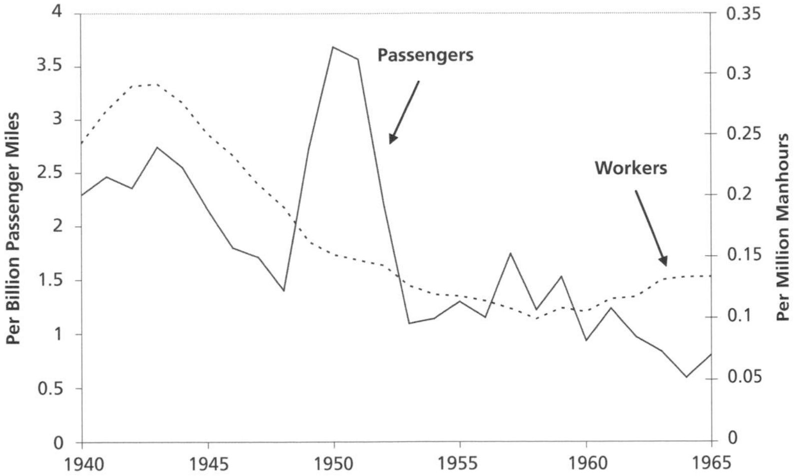

Source: Appendix 2. Data are three-year moving averages.

traffic shifted to larger, safer stations. In the 1922–39 period there were 1.58 passenger fatalities not involving train accidents for every fatality from trains; in the 1946–56 period the ratio fell to 0.61. Had it not fallen, passenger safety would have shown no improvement.

Trespassing fatalities continued their long-term decline into the 1960s, falling 60 percent to 634 from 1,592 between 1945 and 1965. Both the ARA Safety Section and the Railroad Section of the National Safety Council had committees on trespassing. Along with individual carriers such as the B&O, Santa Fe, Cotton Belt, Illinois Central, and New York Central, they conducted antitrespassing campaigns in schools. Rising real incomes, auto-mobility, and the decline in train miles continued to reduce casualties in the postwar years.[15]

Worker safety improved for a decade after World War II, with fatality rates falling in half from about 0.23 to 0.1 per million manhours between 1947 and 1957 and severity rates declining sharply over the same period; thereafter both measures rose (Figure 10.4). No single cause or small set of causes can account for the postwar safety improvements, for as Table 10.1 reveals, they accrued broadly to all major groups of employees. Nor did the substantial shifts in employment greatly affect safety. Had the broad occupational distribution remained unchanged between 1935–39 and 1953–57, fatality rates would have declined by just about the same amount. Similarly, the deterioration of safety after 1957 was also general, and it was not simply a reflection of the boom of the 1960s, for the rise in risks continued well into the 1970s. While maintenance-of-way workers experienced the largest increase in risks after 1957, safety would still have deteriorated even if their fatality rate had remained at the 1953–57 level. Nor did changes in the distribution of work among occupations have much effect on the rise in fatality rates after the 1950s.[16]

Table 10.1. **Employee Fatality Rates, by Broad Occupational Group, 1935–1939 to 1971–1975**

Group	1935–1939	1953–1957	1961–1965	1971–1975
Executive, professional, and clerical	0.035	0.012	0.022	0.024
Maintenance of way	0.253	0.157	0.228	0.174
Maintenance of equipment	0.121	0.069	0.068	0.077
Transportation (other than train, engine, and yard)	0.091	0.037	0.032	0.037
Transportation (yardmasters, etc.)	0.158	0.074	0.055	0.042
Transportation (train and engine)	0.610	0.237	0.208	0.222
Total	0.225	0.108	0.111	0.119

Sources: ICC, *Accident Bulletin,* various years; author's calculations.

Technology, Organization, and Work Safety

The need on the part of carrier safety departments to justify themselves created a continuing force for improved safety in the postwar period. Sharply rising personal injury costs also supported their efforts. The 1939 amendments to the Federal Employers' Liability Law did away with the assumption of risk defense and sharply limited contributory negligence. Thus, in 1952, when Walter Kintz, a machinist on the Lehigh & Hudson River, lost an eye while chipping steel, even though he had failed to wear safety glasses the company settled with him for $11,124 (about three years' earnings). On the Pennsylvania the average cost of worker death claims rose 48 percent from $11,900 to $17,700 between 1945 and 1951. While the rise was less than the rate of inflation, with railroad rates regulated the nominal increase must have encouraged accident saving on the part of the carriers.[17]

On some lines worker safety committees continued to play a vital role in identifying and correcting job hazards. In the mid-1950s the Barstow, California, shops of the Santa Fe had active worker safety committees that not only discussed rules but also recommended a host of safety improvements. As a result the company constructed permanent scaffolds and platforms to be used instead of ladders for diesel work, thereby reducing risks from falls. One of the duties of its safety supervisors was also to seek out hazard information from employees. The ACL centralized its safety work in 1952 but retained worker committees. The Frisco's master mechanic claimed that the "keynote of the program's success is its emphasis on safety suggestions from . . . employees." About 1946 the B&O Chicago Terminal revitalized its safety program. In addition to rules and discipline the company also emphasized employee selection and training, accident investigation, and the need for worker safety committees to identify and report hazardous conditions. Injury rates declined from nearly twelve per million manhours in 1946 to less than four in 1955. In 1948, the Milwaukee trained employees in a handbrake instruction car, and saw falls from cars nearly disappear.[18]

Because safety work emphasized employee behavior many companies tried to enlist family and local union support. The Katy, the Rock Island, and the Frisco all held family safety rallies or developed ladies' safety councils. The Frisco's safety committees included union men, "for the organizations are solidly behind this work." Union representatives also often participated in national meetings of the AAR Safety Section. John Banks, a committeeman on the Southern Pacific in 1948, reported to the AAR that labor representatives were included in every accident investigation. "Labor put over safety," he concluded. Records of one B&O Trainmen's local in the 1950s reveal a close relationship between the union representative and the superintendent that effectively identified and reduced job hazards. Howard Banta, chair of Trainmen's Local 1000, routinely wrote B&O superintendent R. C. Diamond notifying him of job hazards and Diamond routinely responded with a "Dear Howard" letter thanking him for the interest and agreeing to fix the problem.[19]

Institutionalized safety work also smoothed the transition to diesels. While diesel locomotives were inherently safer than steam, they also presented new risks and so their full impact on safety was contingent on the adoption of a host of new rules and procedures. As has been discussed, locomotive safety improved right after the founding of the Bureau of Locomotive Inspection in 1911 and another sharp reduction in accidents began in 1923. The third major improvement in locomotive safety started about 1947, coinciding with the wholesale scrapping of steam power for the diesel.

Diesels affected work safety in many and complex ways. They revolutionized shop work and did away with coaling, water towers, and ash cleanout. On the road, dynamic braking—which resulted from running the electric motors as generators on a downhill grade—improved train control. The overall result of such changes was probably better safety. On the other hand, because diesels were quieter than steam they may have worsened risks to maintenance-of-way work and they yielded higher wheel loads than had steam locomotives. The net impact of such changes is unknowable. However, for many years the ICC data on locomotive accidents and injuries disclose how the diesel directly affected the safety of trainmen's lives. Paradoxically, while diesels were far safer for trainmen than steam had been, their triumph only marginally improved trainmen's safety.[20]

The ICC reported accidents to steam locomotives under fifty-five separate categories plus "miscellaneous." Of course the most deadly of all accidents, for which there was no counterpart on a diesel, was a boiler explosion. From 1945 to 1954 sixty out of a total of ninety-three fatalities on steam locomotives resulted from this cause. It was hardly a coincidence that in 1955, with the sun setting on the age of steam, the ICC's Bureau of Locomotive Inspection announced "for the first time in . . . 44 years, a full fiscal year has elapsed without the occurrence of a steam locomotive boiler explosion."[21]

Table 10.2. **Accidents and Fatalities, Steam and Nonsteam Locomotives, 1945–1954**

	Steam	Nonsteam
Accidents	2,307	524
Locomotive-years	300,118	153,274
Worker fatalities	103	9
Accidents per thousand locomotives	7.69	3.42
Fatalities per thousand accidents	45	17

Source: ICC Bureau of Locomotive Inspection, *Annual Report of the Chief Inspector,* various years.

The bureau first took official note of "locomotives other than steam" in 1927. Reports separated by type of motive power continue throughout the period of dieselization until 1954, by which time the steam locomotive was all but extinct. Table 10.2 presents data on worker fatalities and accidents by type of locomotive from 1945 to 1954 and as can be seen nonsteam (by this time mostly diesel) locomotives maintained a modest but significantly lower accident rating throughout the postwar years. When there were accidents, diesels tended to be considerably less lethal than their steam counterparts; the typical steam locomotive accident was abut 2.6 times as likely to result in a worker fatality.

In contrast with the bewildering number of dangers on a steam locomotive, the bureau listed accidents to nonsteam locomotives under just nine categories plus "miscellaneous." Diesel oil caused dermatitis, and it made floors slippery, raising risks from falls. Diesels were also subject to fires and short circuits, and crank case explosions, while starters and generators might also cause trouble. For example, the bureau reported that on September 24, 1950, the flash from a grounded shunt regulator injured one man on the Santa Fe. In December that year on the North Western a crankcase exploded due to an overheated bearing and it too injured a man.

As use of diesels spread, companies developed rules, procedures, and equipment to cope with the risks. A Union Pacific superintendent of safety invented an uncoupler that reduced dangers in switching, and later marketed a similar device. Because of electrical dangers companies forbid shop men to wear jewelry or metal frame glasses when working on the new locomotives, while tight-fitting clothing was also required to avoid the many revolving parts. Sometimes the very safety of diesels caused problems. They did away with cinders, which had been a major source of eye injuries, but the ACL reported that its trainmen responded by wearing safety glasses less frequently, and so the company had to step up monitoring and enforcement of rules.

While the above actions were simply intended to improve work safety, other safety procedures were motivated by the desire to keep an expensive piece of capital equipment operational. Because diesel oil is not particularly flammable, both manufacturers and carriers initially overlooked fire danger. Early diesels carried only one tiny carbon tetrachloride extinguisher. After a rash of fires proved this inadequate, companies shifted to larger foam or CO_2 extinguishers. They employed spectrographic analysis of lubricants to search for traces of worn parts or overheating that might presage a crankcase explosion, and installed protection devices that would shut down the engine at high crankcase pressures.[22]

Thus it seems clear that the shift from steam locomotives to diesels, in combination with modifications in safety rules and procedures, sharply reduced accidents per locomotive and diminished their severity by perhaps 60 percent. Yet these declines yielded very modest reductions in worker accident rates. This was partly because diesels reduced exposure: two steam locomotives required two crews while a four-unit diesel required only one. In addition, by the 1930s fatality rates to trainmen from steam locomotives were already very low—in the range of 0.04 to 0.06 per thousand, while injury rates were usually less than one per thousand. Even large percentage decreases in small accident rates will generate small safety gains.[23]

Unfortunately, we cannot directly measure the risks of working with nonsteam locomotives because the division of workers or worker hours between types of locomotives is unavailable. An alternative approach is to compare fatality rates to trainmen from locomotive accidents for two periods, between 1927–30, when nearly all locomotives were steam, and 1950–54, when over half were nonsteam. Fatality rates declined by about two-thirds between these periods, from 0.059 to 0.022 per thousand trainmen but the absolute decline was only 0.037 per thousand, and probably not all of this could be attributed to dieselization. Over the same period fatality rates of trainmen from all causes fell from roughly 1.7 to 0.7 per thousand. Hence the greater safety of diesels accounted for less than 3.7 percent of trainmen's safety gains. Diesels were a revolutionary technology that changed nearly everything about railroading, but their direct impact on trainmen's safety was modest indeed.[24]

Safety concerns sometimes encouraged the adoption of new technologies. The Lackawanna employed photoelectric circuits in its shops to prevent collisions between gantry cranes on the same track. Railroad repair shops began to use colored paints not only to improve visibility and therefore productivity, but also to signify danger, employing diagonal black and yellow stripes to identify moving equipment. Improved shop lighting reduced defective parts and improved safety.[25]

On the Pennsylvania Railroad, safety was a significant, although probably secondary motive for the gradual diffusion of car retarders. In 1946, the Pennsylvania operated fifty-seven classification yards, and only four used retarders. That year the company's claims department began a program to reduce personal injuries, which the war had ballooned. It assembled data showing the safety gains and cost savings from use of car retarders in an effort to spur their use. The company calculated that installing retarders in its thirty-four rider classification yards would have cut personal injuries by 70 percent, from 24.8 to 7.6 per million cars moved, saving about $1.2 million in annual injury costs. Injuries declined because retarders raised productivity, thereby reducing labor requirements and because they reduced the risks of yard work. They were, therefore, both labor and accident saving. As one executive put it, "I know of no place where we can get hold of our injuries to employees quicker than by the installation of retarders as they are, at the present time, the most valuable capital expenditure facility to decrease the number of employees exposed to hazardous occupations." Combined with the reduction in labor costs, such savings encouraged the spread of retarders. By the 1950s they were becoming increasingly automated, with all switches and signals controlled from a central tower. But lack of funds limited their diffusion. In 1955 the AREA reported only eighty-four retarder yards in operation while yard automation also spread slowly.[26]

Another instance in which safety shaped the technological trajectory reveals the strengths and weaknesses of railroad safety organizations. Despite decades of federal regulation and union efforts, handbrakes on freight cars continued to cause accidents, accounting for thirty-two deaths in 1946–48. In 1948, in a rare foray into engineering matters, the AAR safety men contacted the mechanical division to try to improve brake safety. By this time a bewildering variety of brakes with very different properties were in use. Some had a vertical wheel, some horizontal. Some were power, some not, and required a club to tighten them. Some might spin when released, and throw a man from the top of the car. The goals of the safety men were standardization and—because a disproportionate number of accidents resulted from use of old equipment—the universal use of newer, nonspin brakes. In 1950, both safety and mechanical representatives met with suppliers, and all agreed to standardize. But the AAR postponed universal nonspin brakes, and no more than a handful of individual carriers adopted them during these years. It was a lost opportunity; had the safety men combined forces with the brotherhoods, and had the trade press and the ICC supported them, far more might have been accomplished.[27]

A few carriers integrated safety concerns into equipment design and purchase. In the early 1950s the Norfolk & Western safety department investigated safety devices for new machinery. Somewhat later, the safety department of the Southern Pacific ensured that the company's new cabooses had safety glass, rounded inside corners, nonskid paint, and outside handrails painted white for better visibility. In 1957 the Pennsylvania's chief engineer, L. A. Villella, reported that his company involved the safety department in new equipment purchases. By the 1950s the Lackawanna studied new work equipment for hidden dangers. It sometimes required additional safety features and occasionally rejected unsafe equipment.[28]

The Erosion of Work Safety

Despite such examples of successful safety work, employee fatality rates deteriorated in the years after 1957 (Figure 10.4), and injury severity also worsened (Appendix 2). The problem reflected broad changes in the demand for railroad services as well as organizational and financial difficulties. The dangers of freight work had always exceeded those of passenger transportation and the evolution of railroads from passenger and freight lines to all freight therefore worsened risks. The shift to freight work had been going on during the years when safety was improving, but it accelerated in the postwar years. A statistical exercise illustrates the impact of these changes. Had traffic not shifted away from passenger toward freight train miles between 1957 and 1965, worker fatality rates would have been about 25 percent lower in the latter year.[29]

Still, even controlling for changes in the nature of railroading, safety worsened during these years as the AAR and company programs failed to meet new challenges. On a number of individual lines, the emphasis on management responsibility led to a parallel deemphasis on safety devices and worker involvement, leading some companies to abandon safety committees. Both problems were evident on the Pennsylvania by the mid-1940s. On May 7, 1946, a steam hammer threw a 177-pound piece of cylinder head, striking its operator John Kunsman on the head and killing him. A hard hat might not have saved Kunsman's life, but the machine could and should have been guarded. While the company still had worker safety committees, they seem to have been largely moribund by this time, reducing a valuable source of information on hazards. The company paid dearly for this failure when on May 27, 1947, an employee became permanently paralyzed after falling 20 feet through an unguarded opening at the North Pennsylvania station—an obvious hazard that should have been identified and remedied. The company thereupon hastily erected a cheap guard, but a court awarded the man $250,000 (about 101 times annual earnings) and a judge upheld the verdict.[30]

While the Pennsylvania required its safety department to review purchasing decisions, most companies did not. Compared to safety work in manufacturing, which became proactive, emphasizing hazard assessment and engineering design during these years, on the railroads it remained reactive and focused more on supervision and rules. One survey in 1962 found that only 5 to 10 percent of carriers routinely involved the safety department in new purchases and construction. The result was sometimes dangerous equipment that had to be modified on the job or by a later purchase. Moreover, the flood of new machines for track work and materials handling that had begun in the 1920s increased after World War II as the shift to a 40-hour week in 1949 raised the payoff to mechanization.

This rapid transition to ever more mechanized track work affected worker safety in complex ways. The carriers shifted to smaller, more specialized track gangs that could be better supervised. Mechanization also reduced injuries from materials handling and from using hand tools, and some of the latter could be deadly. In 1950, C. E. Hightower of the Denver & Rio Grande told of a flying tool chip that cut a maintenance worker's throat, killing him. New technologies and work practices could also make use of hand tools safer: as a result of the accident the Rio Grande began Magnaflux inspections of tools (see below) and required chippers to wear safety goggles.[31]

Yet mechanization of maintenance coincided with a ballooning of accidents after 1957 (Table 10.1), probably because most companies failed to integrate safety into purchasing and to develop proactive rules. These were not new problems (chapter 9), but they probably worsened with the burst of new machines after World War II and the deterioration of company safety programs (see below). In 1952, J. P. Hiltz, maintenance-of-way engineer on the Lackawanna, admitted that the new equipment had caught his company's safety work "flat footed," and injury rates rose. In response, the company studied machine design and operation, tightened inspections, and improved worker training. In addition, use of personal protective equipment was highly inconsistent. While most shops required safety glasses for dangerous work, enforcement was sometimes lax, as the experience of Walter Kintz, recounted above, suggests. Use of hard hats among maintenance-of-way workers remained uneven and wearing safety shoes was usually voluntary. By the 1950s, most companies had also abandoned first aid training.[32]

Being struck by a train remained the worst risk from mechanized track work. Charles Hill of the New York Central remembered that when that company introduced track tampers "we had a lot of fatalities." Most companies had learned that watchmen needed whistles to overcome equipment noise. As use of "walkie-talkie" radios became widespread, the Detroit, Toledo & Ironton and some other lines required foremen to obtain written permission on a company form to occupy track. They were also to put out special signs that trains might not pass until the foreman informed the engineer by radio that the track was clear.[33]

Track cars and trains were natural enemies, and in a contest the train always won. William Doble was both a signalman on the New York Central and an official of the Signalmen's Union from the 1930s through the 1950s. Men complained to him that the company installed too few set-offs to allow a worker to remove the car when a train appeared. Cars were also too heavy for one man to set off quickly, yet not heavy enough to trip crossing signals, leading to many accidents with automobiles. The Missouri, Kansas & Texas and some other lines installed indicators so that motor car operators could tell if there was a train in the block. But many companies' procedures for operating track cars remained hopelessly lax; car operators were sometimes given a train lineup but told that it could change at any time. Doble complained that some train dispatchers would not provide car operators with a written train lineup for fear of liability in case of an accident. In 1954, at the request of the ICC, the AAR Safety Section belatedly looked into the problem and it issued a code for proper use. Procedures remained inadequate, however, and the rail unions made several abortive efforts to induce Congress to legislate on the matter.[34]

By the mid-1950s eroding profitability compounded these problems. As noted, the decline had many causes but the main culprit must have been oppressive federal rate regulation, for since deregulation railroad profitability has improved. The railroads' nominal yield on assets during these years ranged between 2 and 3 percent, which was not always enough to offset inflation. This evaporation of net income reduced the payoff to safety/productivity packages; it also reduced internal funds and raised the cost and reduced the availability of external funding as well. ICC studies for these years invariably point to large amounts of "deferred maintenance." Inflation-adjusted expenditures for maintenance of way per track and ton mile, and for maintenance of equipment per ton mile, both declined sharply. Freight cars were aging while track and equipment were not being upgraded or were actually deteriorating, and matters worsened after 1965.[35]

A 1962 survey of maintenance-of-way supervisors found that twenty-one of forty thought deferred maintenance was "piling up." On some lines car inspectors were let go and the number of bad order cars skyrocketed. Workers reported cluttered yards and roadbed that was choked with weeds and brush, causing treacherous footing and poor visibility. One anonymous worker on the Pennsylvania complained to Vice President J. M. Symes that there was "too much junk" in the yards and "when someone gets hurt by tripping or falling the boss is always ready to put the blame on us." About the same time another Pennsylvania worker informed Symes he had resigned as a shop safety chairman. "Any recommendation for the elimination of a safety hazard usually ended up in the waste basket," he explained, and proceeded to list examples. It was a devastating indictment. Safety First had always involved an implicit contract: the carriers would spend the money and the men would do their part, but now finances were leading the lines to welsh on the contract.

Financial constraints also slowed the diffusion of safer technologies. Edwin Mansfield's study of new technologies including car retarders, diesels, and centralized traffic control (CTC) demonstrated that the speed of their diffusion reflected their profitability, but that the carriers were slow to adopt even these profitable innovations—perhaps because of financial constraints. Indeed, railroad journals reported that even wheel bearing lubricating pads could be too expensive. Similarly, the slow spread of automated switching and unit trains, both of which would have reduced dangerous yard work, resulted from weak profits and regulatory constraints.[36]

Thus rising worker accidents did not result because the carriers as a group traded safety for output, which would imply a positive relationship between accidents and profitability. It reflected instead a broader organizational collapse on many carriers that saw both profits and safety decline. For such reasons greater profitability would have improved safety. On some unprofitable carriers, such as the Boston & Maine, and the Reading,

Figure 10.5. **Worker Fatality Rates, Three Weak and Three Strong Roads, 1950–1965**

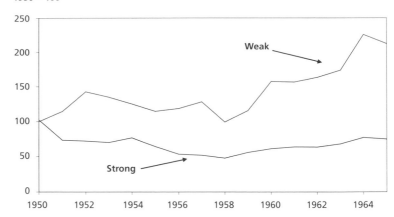

Index Numbers,
  1950 = 100

Note: Weak roads are B&M, Reading, and North Western; strong roads are Norfolk & Western, Union Pacific and Southern Pacific. Data are three-year moving averages.

worker safety dramatically worsened as their financial pictures darkened in the early 1960s (Figure 10.5). Statistical analysis for the 1959–65 period supports this conclusion. Less profitable carriers were indeed more dangerous for their workers. In 1964, a large sample of carriers earned an average return on assets of 3 percent; a 2-standard deviation increase in the return to 7 percent would have reduced casualty rates about 29 percent (Appendix 2).[37]

The New Focus of Crossing Safety

In the postwar years, crossing safety gradually evolved from a private to a public function as states assumed a larger share of funding. Partly for this reason the focus shifted from an almost exclusive emphasis on grade separation to use of automatic guards. In the 1930s, federal and state funding had concentrated research on crossing hazard assessment. Such studies suggested that not all crossings merited abolition; for some guarding was a better solution. But which crossings should be eliminated, which should be guarded, and with what sort of guards?

   Where crossings were previously unguarded, the payoff to guarding came in reduced accident claims, which heightened the carriers' interest in determining the comparative effectiveness of differing forms of guarding. Warren Hedley, a civil engineer on the Wabash, made the first serious attempt at such an evaluation. Hedley used that company's data to evaluate differing forms of crossing protection under similar traffic conditions. His findings—later substantially confirmed by other researchers—revealed that watchmen were surprisingly ineffective while the combination of flashing lights and gates yielded dramatic increases in safety. Yet it was rising wages more than safety concerns that led the carriers to substitute mechanized guards for watchmen. When the North Western replaced men with automatic gates at nine crossings in Wheaton, Illinois, in the late 1950s the annual wage saving yielded a 35 percent return on investment. Thus substituting automatic guards for watchmen was a labor-saving investment that reduced costs and accidents.[38]

The statistics on grade crossings reflected these trends in thinking during the postwar years. Grade separations, which had averaged 300 to 500 a year in the late 1930s, declined to 25 to 50 in the postwar years before rising to around 100 a year in the late 1950s with the construction of the interstates. The big change came in crossing protection, as reflected in Table 10.3. Crossings with some protection other than a simple crossbuck that warned "Railroad Crossing" rose from about 15 to 21 percent of the total from 1945 to 1965. The effectiveness of protection increased as well. Early crossing guards had included bells and wigwags, but by the 1950s companies had largely settled on flashing lights with or without gates—a decision that was reinforced by Hedley's findings. To prevent signals from crying wolf, a number of lines began to install crossing signals with complex controls that provided a constant warning time irrespective of train speed or that could be disengaged if work or switching trains were in the circuit. Companies also responded to the need for better night visibility at crossings. The ACL, the Great Northern, and other lines lettered freight and passenger cars and warning signs with "Scotchlite," which was far more reflective than white paint.[39]

I have mapped "protection factors" derived from Hedley's work onto the ICC's list of crossings for selected years and the results are in Table 10.3. The reader should take such calculations with a dose of salt, for they require a good deal of guesswork. What they suggest was surely true, however; while the number of crossings declined only modestly, the efficiency

Table 10.3. **Trends in Crossing Protection, Selected Years, 1925–1965**

	1925	1939	1945	1965	Percent Change, 1945–1965
Total crossings	233,633	231,104	226,153	215,961	−4.5
Percent protected*	11.7	13.7	14.7	20.5	+39
Average protection factor (protected crossings only)*	1.26	1.38	1.57	2.21	+41
Average protection factor (all crossings)	1.04	1.05	1.08	1.25	+16

Source: Computed by mapping protection factors derived from Warren Hedley, "Second Report on the achievement of Grade Crossing Protection," AREA *Proceedings* 53 (1952): 985–1003, Table 17, onto ICC data on crossing protection.

*Protection other than fixed signs.

of crossing protection, at specially protected crossings and at all crossings, rose sharply after 1945.[40]

Numerous real-world examples breathe life into such findings. The Illinois Commerce Commission recorded the results of 3 large-scale protection projects that closed 26 crossings and upgraded 62 others from warning signs to automatic gates and flashing lights. Accidents had killed 77 people and injured 84 in the 10 years before the work; in the subsequent decade the numbers dropped to 2 and 17. In another instance, the Grand Trunk put gates and flashing lights on 23 previously unprotected crossings. Prior to the work they had killed 72 and injured 95 individuals in 297 crossing-years; afterward the numbers dropped to 5 killed and 2 injured in 255 crossing-years.[41]

Guarding and crossing removal were complemented by the publicity campaigns of the AAR and the National Safety Council (NSC) that emphasized the need for drivers to cross carefully. The B&O and the ACL also conducted their own publicity campaigns. They placed announcements in local papers, and monitored crossings for drivers who ran signals, sending them warning letters. Such campaigns might reduce accidents; they also maintained the focus on driver error, which encouraged public funding and sympathetic juries. Thus in 1946 when the carriers helped fund an NSC campaign the *Railway Age* enthused that this would emphasize that crossing accidents were "a *traffic*, not a *railroad*, problem." Still, states and localities were slow to see guarding as a public responsibility. Throughout the 1950s, at least, funding continued to come primarily from the railroads, although by the end of the decade the trend was to share costs.[42]

While these safety activities were reinforced by a decline in train miles, automobile travel exploded. A statistical analysis of the 1941–65 period shows that by itself rising automobile travel would have increased crossing fatalities by about 4.3 percent a year while declining train mileage and crossings reduced fatalities by about 2.9 percent a year. Thus changes in travel alone would have *raised* crossing fatalities 1.4 percent a year. The guarding of some crossings and the elimination of others, as well as other forces captured by a trend, more than offset travel changes and so crossing fatalities declined roughly one percent a year. As a result, while vehicle miles traveled rose by 250 percent between 1945 and 1965—from 250 to 887 billion—and while total highway fatalities increased to 49,000 from 28,000 over the same period, deaths at railroad crossings fell about 20 percent, from 1,900 to 1,533.[43]

The Course of Train Accidents

War-induced inflation sharply increased the number of reportable train accidents. In response, in 1948 the ICC raised its reporting threshold to $250, and then in a series of steps it increased the threshold to $350 in 1957 and $750 the next year, where it remained until 1975. The net effect of these reporting changes was first to overcompensate for inflation and reduce

Figure 10.6. **Derailments, Adjusted for Inflation, Reporting, and Car Miles, 1940–1965**

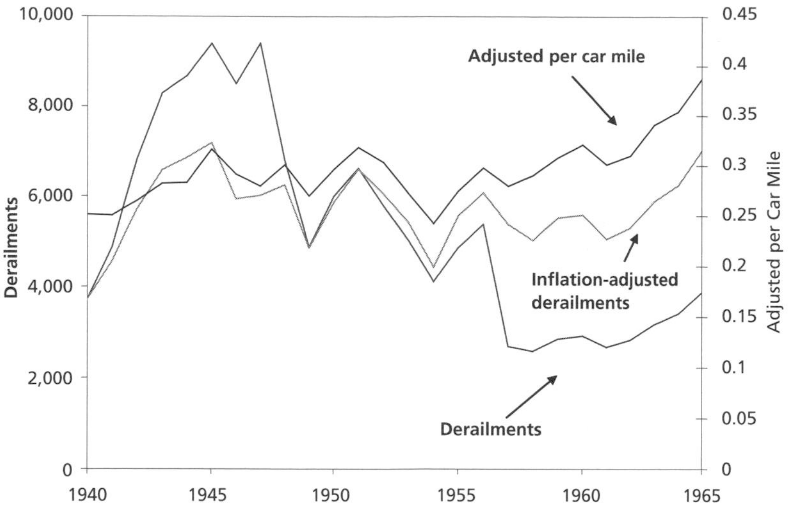

Source: Appendix 2.

the number of reportable accidents—which probably dropped out a large number of minor freight derailments and yard collisions—and then to undercompensate after 1957. Correcting for these effects and expressing the results relative to train or car miles reveals very little change in collisions between 1948 and 1965, while derailments rose well into the 1970s (Figure 10.6 and Appendix 2).[44]

Casualties are less likely to be significantly affected by such reporting changes, however, and for passengers casualty rates from train accidents show little change from the very low levels that obtained in the late 1930s. There was, however, much variability, for the nature of train accidents also changed during these years. The collapse of short-haul passenger traffic raised average train size sharply from about fifty passengers in the early 1930s to double that by 1960. Thus, while accidents per passenger mile declined after World War II, the number of casualties per accident rose from the 1930s on, resulting in a disproportionate increase in the number of disasters—just as had occurred a century earlier in the 1850s. Indeed, some of the worst wrecks in the history of American railroading occurred soon after World War II. Because of its size and extensive passenger operations, many of these occurred on the Pennsylvania. Thus on February 18, 1947, the engineer of that company's *Red Arrow* took a curve too fast and derailed near Gallitzin, Pennsylvania, killing twenty-four. On September 11, 1950, the same company's *Spirit of St. Louis,* with a sixty-eight-year-old engineer at the throttle, plowed into the rear of a troop train near Coshocton, Ohio, killing thirty-three.[45]

Of course, most collisions and derailments involved freights, and most of the resulting casualties were trainmen and here the record looks better. Fatality rates from collisions and derailments for employees dropped from roughly 0.02 per million manhours during the 1930s to half that level by the mid-1960s. For trainmen who were in the front lines at such accidents, rates dropped from 0.29 per thousand workers in 1939 to 0.10 in 1965.

Containing Collisions

Major disasters, of course, alarmed the ICC, which blamed them on higher speeds and the increasing use of lightweight equipment. In 1947, in a sweeping proposal reminiscent of the train control order of 1922, it mandated cab signals or train control where passenger speeds exceeded 80 miles per hour, and block signals on all lines where freight speeds exceeded 50 and passenger speeds 60 miles per hour. The proposal was similar to one that the Railway Labor Executives Association had recommended. The carriers argued that speed was a poor measure of risk but to no avail. The two parts affected respectively 18,600 and 27,200 miles of track, all of which was to be equipped by the end of 1952. Full compliance, the carriers estimated, would cost about $158 million. Once again *Railway Age* argued that the carriers had to earn a competitive return on investment to fund such unprofitable safety ventures. Since the commission prevented the carriers from earning their cost of capital, it forfeited the right to judge safety requirements, the editor fumed. By now this had become a tired claim and the commission predictably ignored it.[46]

The order's effects are unclear. The proportion of road or main track under the block system remained unchanged at about 50 percent throughout this period. While the North Western, the Union Pacific, and some others increased their train control territory, the amount of track under train control or cab signals rose by only about 4,600 miles during 1947–53. Many carriers decided to avoid compliance costs by reducing speeds—an option chosen by the Pennsylvania, the Union Pacific, the Milwaukee, the Santa Fe, the Burlington, the Seaboard, and others. As automobiles continued to reduce demand for rural, short-haul service, the railroads scrapped some 33,000 miles of passenger track from 1947 to 1953. By reducing speed and raising costs, the order thus contributed in a small way to the shift of passengers from rails to automobiles—a far more dangerous means of travel.[47]

Companies had begun to install CTC in the early 1930s, but rising postwar labor costs spurred its spread and some 26,000 additional miles were equipped from 1945 to 1965, bringing the total up to one-third of all road. CTC was in fact a marriage of block signals with interlocking. It might be installed on single or double track, and on the latter companies usually ran trains both ways on both tracks. On a typical installation the train dispatcher, watching a track diagram that showed the position of signals and trains, might control interlocked signals and switches to sidings, while the main line was governed by automatic block signals operated permissively. Such a system could control meeting points, run following trains around slower preceding trains, and generally speed up traffic. Predictably, therefore, companies applied CTC to high-density single-track lines to increase capacity without the need for double tracking. It also allowed the removal of many sidings and signals and so it might save both capital and labor. Where CTC was installed on a line without signals, or substituted for permissively operated manual blocking, it greatly increased safety as well. The carriers also upgraded signaling, scrapping semaphores for position lights and introducing three- and four-aspect signals for high-speed trains that showed track conditions several blocks ahead.

Collisions from break-in-twos were diminished by the AB freight brake. Its use had spread slowly and in 1945 only about half of all cars were equipped. Exasperated, the ICC mandated its use on all cars that year, which was probably its most constructive safety order in the postwar years.[48]

Improved passenger cars helped offset the effects of rising train speed on risks from collisions and derailments. Although speeds increased, lighter weight decreased momentum and raised the ratio of strength to inertia, thus reducing the likelihood of telescoping. In 1954 the AAR laboratory even began compression tests on entire cars using equipment donated by the Pennsylvania. New passenger equipment increasingly employed electro-pneumatic brakes (first developed about 1910), which operated more quickly, and they employed both speed-pressure controls to raise brake pressure at top speeds and "decelostats" to prevent wheel sliding. Disc brakes also came in the late 1930s and they helped reduce wheel failures. Cars employed tightlock couplers and pinned trucks to keep the train in line during a derailment, and the Southern Pacific even installed truck guides shaped like an inverted U to fit over the rail and hold cars in line.[49]

While these improvements contributed to passenger safety they contained serious omissions. Signaling policy reflected a combination of the ICC regulations, which used accidents or speed as a measure of dangers, or the carriers' approach, which installed CTC when justified by costs and productivity. What was missing was a more complex risk assessment that might have improved signaling on especially dangerous lines before an accident occurred. The need was compellingly illustrated on the Long Island Railroad, a Pennsylvania subsidiary. The Long Island was almost wholly a commuter road. The company operated most of its low-speed but high-density lines using an automatic block system, and on February 17, 1950, a motorman ran a signal and ignored a cab buzzer to plunge head-on into another train at 25 miles per hour, killing thirty-one. Nine months later on the same line, on November 22, another motorman ran the signal near Richmond Hill, Long Island, plowing into the rear of another train at 30 miles per hour, killing seventy-nine people.

The Richmond Hill wreck briefly catapulted railroad safety into the public spotlight again. The *New York Times* editorialized on "A Night of Horror," while to the *Daily Mirror* it was "Bloody Murder." The Pennsylvania had controlled the Long Island since 1901, and the *Daily News* used the tragedy to revive an old charge that the parent company had bled the Long Island, thereby preventing it from installing proper signals. The wreck precipitated a blizzard of bills filed in Albany, although their spon-

sors thoughtfully confided to the Pennsylvania that these were only window dressing for constituents. The ICC and the New York Public Service
Commission investigated the accident, and even Congress held hearings.
The editor of *Railway Age* again took the occasion to point out that ICC
economic regulation and safety were antithetical, and he mailed copies to
the House Commerce Committee investigating the accident. The immediate outcome of all this activity was that the Long Island agreed to install a
$6 million system of automatic train control, including speed control—a
result that should have come decades earlier.[50]

A second flaw in both government regulations and company rules
was their failure to deal adequately with human shortcomings. While train
control might prevent an engineer from running a signal, many systems
would not prevent excessive speed, as occurred on another Pennsylvania
commuter line, at Woodbridge, New Jersey, on February 6, 1951, where a
train going 50 miles per hour over temporary track derailed, killing eighty-
four people. Passenger fatalities cost the Pennsylvania nearly $49,000 each
(about fifteen years' earnings), and this string of disasters led the company
to appoint a special committee to assess its safety procedures for passenger
trains. The report turned up a number of unsafe practices and conditions.
Concerned that aging engineers might lose their skills, the report recommended much more rigid compliance with physical examinations. The
committee also recommended three-aspect block signals in high-speed
territory and use of speed recorders on locomotives. (In fact, the company chose to follow ICC recommendations and install speed control—a
sophisticated version of train control—on 1,100 locomotives instead.) By
the mid-1950s a few companies such as the New York Central were using
speed recorders system-wide, and their absence on many lines reflected a
major gap in railroad safety procedures, for such checks might have helped
prevent engineer negligence. Given the carriers' interest in both discipline
and equipment monitoring devices, it was a strange omission.[51]

The Rise in Derailments

The increase in inflation-adjusted derailments relative to car miles in the
postwar years (Appendix 2) occurred despite the continued spread of technologies introduced in the 1930s as well as a host of postwar developments.
Their diffusion was spotty, in good part for financial reasons. In addition,
low profitability also retarded track and equipment maintenance. In combination with the rapid rise in freight car weight, these developments steadily
worsened risks of freight derailment and helped transform railroad safety
once again into a public issue.

Because virtually all new rail had been manufactured using controlled
cooling since 1940, its share of track miles rose steadily for years thereafter,
contributing to a steady decline in failures. Sperry also improved the use
and marketing of its rail tester in the postwar years. It taught operators to
study rail burns from slipping wheels, as these might conceal a crack. The

company developed statistics showing how repeated testing sharply raised the proportion of failures discovered by tests instead of by accident. In response some companies tested main track several times a year. By 1958 Sperry claimed to have tested over 3.4 million miles of track and found 1.6 million defective rails. Detectors owned by the AAR and individual companies no doubt had similar experiences. In response AREA data on rail failures adjusted for ton miles fell steadily throughout the 1950s.[52]

Both the Sperry detector car and controlled cooling reflected the carriers' increased interest in research that began in the 1930s. This newfound receptiveness came about just as a host of new suppliers were developing techniques that sharply improved the ability to monitor and control track and equipment, and that revolutionized train communication. Communicating with or along a moving train had bedeviled railroading since the beginning, resulting in a complex of signs, signals, and whistles. At the end of World War II voice communication finally became practical. Although designed to expedite traffic, it proved valuable in preventing collisions and derailments by allowing much better communication and closer monitoring of men and equipment. The carriers had experimented with various forms of train radio before World War I, but high costs, inadequate equipment, and lack of available frequencies from the FCC hindered its use. Military developments during World War II greatly improved high-frequency equipment, and by about 1944, Bendix, Farnsworth, and other new suppliers began to offer high-frequency (around 500 mc) train radio to the carriers. The systems allowed communication from locomotive to caboose as well as with yards and stations.

After the FCC allocated a portion of the electromagnetic band to train radio, and much experimentation with FM and AM systems and various frequencies, the systems spread rapidly. Meanwhile, Union Switch and Signal had developed an induction system of communication that grew out of its earlier work on cab signals and the need at the time to avoid FCC regulation. This system operated at low frequencies (around 100 kc) and because it operated at very low power, it needed no license. It induced a signal in either rail or telegraph wires that could be picked up elsewhere on the train or at a station or yard. It too found favor, being adopted by the Pennsylvania, among other carriers. Between 1948 and 1959 the number of miles of road operated with train radio rose from a few hundred to 118,000, or about 54 percent of the total. Train radio improved safety in several ways. In yard work, companies noted that it reduced dangers of missed signals. On the road, a stalled train might emit a signal to warn a following engineman of the danger. The Duluth, Missabe & Ironton reported a number of incidents in which a conductor on one train spotted dragging equipment on a passing train or a broken rail in another track and was able to warn the other train in time to stop. The Bessemer & Lake Erie and Pennsylvania both reported instances where conductors notified the engineer of hot boxes and sticking brakes without the need to induce a dangerous emergency stop that might have broken the train.[53]

Radio was one of several ways the carriers improved their ability to monitor and control men and equipment during these years. Others involved the increasing use of remote sensing devices, some of which also employed radio. In the 1930s, companies had developed rock slide and dragging equipment detectors, and other remote sensors connected to block signals (chapter 9). After World War II, as rising labor costs reduced the number of section gangs and track walkers, use of these electronic substitutes for the human eye slowly increased. One survey showed that dragging equipment detectors in use had risen from thirteen tracks in 1936 to 616 in 1948 and that they had been tripped 4,600 times that year. In addition, some companies installed fire, flood, and bridge alignment detectors, high load detectors, and seismographs and snow detectors. Most of these were simply some form of trip that was wired into existing signals. But the Santa Fe installed one flood detector up a dry wash some miles from the track that, when activated, sent a radio signal of an impending flood to the dispatcher, which gave him about 6 hours to divert traffic.[54]

In addition to these communication devices, the increasing use of nondestructive testing greatly improved monitoring and control of track and equipment. Sperry's rail flaw detector was among the first of these, but it would not catch bolt-hole cracks between joint bars, and by the late 1940s the decline in other forms of failure focused attention on them. In response, in about 1950 Sperry and a new supplier, Branson Instruments, in cooperation with the Pennsylvania Railroad independently developed ultrasonic testers that changed tone at a crack. By 1952 both were coming into wider use and discovering thousands of dangerous defects. About 1955, Sperry also developed a portable ultrasound device that could test car axles in place. The C&O, which had experienced a string of freight wrecks from broken axle journals in 1952, cooperated with Sperry to develop the new detector while the Southern, working with Curtiss-Wright, developed a similar device. Several years later Sperry modified its detector to require less skill in operation. The C&O then purchased a fleet of them, and so did the Southern. Wheel and axle manufacturers also routinely employed ultrasound for quality control.[55]

Another new source of information was the use of photo-elastic methods to study stress concentrations in models of rails and joint bars. Arthur Talbot employed such methods in research for the AAR in the 1940s and so did individual carriers and suppliers. The fruits appeared in 1946, when the AREA introduced three new rail designs that reduced localized stresses in the upper web and fillet. Later research in the mid-1950s by Colorado Fuel & Iron in cooperation with the Rio Grande yielded still stronger designs.[56]

The economic size of rails had long interested the carriers, and a series of studies proved the need to embody the improved designs in heavier rail, which was disproportionately stiffer and would save on rising labor costs by reducing maintenance requirements. In 1950 the Pennsylvania found that the shift from 133- to 150-pound rail reduced annual maintenance costs by $1,400 per mile, 90 percent of which was labor costs. By the early 1960s the B&O and C&O were using operations research to determine optimal rail size. Traditional stress repetition (S-N) studies by the Pennsylvania found that rail failures at joints were increased by air pollution-induced corrosion, for the pitting of the metal yielded stress concentrations that reduced its fatigue strength. A simple solution was to grease the rails at the joint. But getting rid of joints was even better, for they were not only dangerous but maintenance-intensive as well, and in the 1930s rising labor costs had begun a shift to longer rails. For similar reasons companies experimented with continuously welded rail, but a host of problems prevented its widespread use until the mid-1950s.[57]

Other forms of nondestructive testing spread after the war. While the motives were primarily to improve safety, as *Railway Age* pointed out, they also extended the useful life of parts and equipment by removing much of the guesswork from the determination of useful service life. Some carriers had employed X-ray testing for boiler and tank car welds since the early 1930s but its wider use began after World War II. In 1949 the Pennsylvania bought a portable machine that it used in purchasing and to qualify welders, and in 1952 the Southern Pacific began to employ cobalt to X-ray castings, while the Rio Grande used radioactive wrist pins to trace engine wear in diesels.[58]

Far more important was the gradual spread of Magnaflux and similar forms of testing in the postwar years. First employed on the railroads in the mid-1930s, one such test passed a heavy electric current through an axle or joint bar, causing a dusting of magnetic powder to "find" any tiny cracks in the metal. The Union Pacific used a similar process that employed a dye penetrant first employed to test jet aircraft turbines. Brittle paints were used to find cracks under stress. By 1950 many carriers tested diesel axles, as well as crankshafts, pistons, and many other locomotive and car parts. One survey found defects in one percent of all car axles tested; some could be turned down on a lathe but two-thirds of them were scrapped. With 1.7 million freight cars in service, the survey implied that testing might remove about 86,000 dangerously defective axles a year. By the mid-1950s the Pennsylvania enhanced the value of these new techniques with efficiency studies of derailments that revealed large savings from a Magnaflux check of center plates and truck parts during routine air brake inspections.[59]

Car axles had also been the subject of a ten-year research project by Timkin in cooperation with the AAR that began in 1937. The company's interest in roller bearings had led it to study axle failure, and the research, which employed photo-elastic studies of axle models, led to improved designs. But while some axle failures reflected poor design or metal fatigue, more resulted from the perennial problem of hot boxes. As noted, during World War II they had been responsible for some spectacular accidents, but to the carriers they were mainly a source of delay and, therefore, expense. As freight train speed and car weight rose, they became increasingly common. In addition, the improved reliability of diesels and the decline in other sources of delay and derailments focused attention on hot boxes. In 1946 they accounted for 4 percent of derailments, and 13 percent of those due to equipment failures; a decade later the proportion had risen to 14 percent of derailments and 30 percent of those from equipment. Hot boxes could always be reduced by more intensive maintenance involving stricter inspection and use of better oils and waste, but such methods were increasingly expensive as labor costs rose. A few companies such as the ACL and C&O switched to roller bearings on freights but these too were costly, and of course the economics of interchange limited them to special cars that tended to stay on the home road. As a representative of the ACL remarked, "to equip other cars would be entirely charity on our part."[60]

Finally in the early 1950s several technologies arrived that promised to contain the hot box problem. A number of companies began to market lubricating pads with brand names such as "Plypack," "Redipak," and "Cell-O-Pak." Maintenance costs and axle failures pushed the Pennsylvania, the Norfolk & Western, and other large carriers to experiment with these and other pads while the AAR laboratory tested them as well. Their effectiveness was contingent upon adequate service and inspection, but properly employed they sharply reduced hot boxes. As companies substituted them for the old-style waste packing, lubrication improved and "waste grab" (which wrapped the waste around the axle and yielded a hot box) declined. In 1955 *Railway Age* hopefully concluded that waste packing was "on the way out," but two years later another journal was still grumbling that lack of cash led some companies to stick with loose waste or to employ cheap, inadequate pads. Despite foot-dragging by cash-strapped roads the AAR mandated use of lubricating pads in 1960—only a decade after their introduction.[61]

In the 1950s, hot boxes also yielded to a new detection device. There had been many attempts to invent such a detector. One, developed by the Southern Pacific, automatically set the brake. Another, a product of the New York Central, would give off smoke and a foul smell when it overheated, but AAR tests revealed that all yielded too many false alarms to be economic. The Rock Island experimented with an electronic detector developed by Farnsworth Corporation. Finally, in 1955, at the request of the C&O, which had seen infrared detectors employed to monitor bearing temperatures in industry, the Servo Corporation developed a workable trackside indicator that used an infrared detector. Tests on the C&O, the Pennsylvania, and other lines demonstrated its efficacy. Companies placed the detectors at strategic points and wired them into stations, which would receive a printout of the offending car and journal. The Seaboard even employed a detector that would broadcast a warning in English to both the train crew and the dispatcher. Soon General Electric, Union Switch and Signal, and others entered the market. By 1959 some twenty-three carriers employed the new equipment, and use spread slowly thereafter.[62]

The postwar years also saw older suppliers innovate new or sharply improved products. The AB brake reduced derailment as well as collision risks, and in 1948 Westinghouse introduced a more effective empty and load brake. As usual the problem was a matter of both safety and economics. As the ratio of loaded to empty weight of cars rose steadily, brake pressure that was low enough to prevent wheel sliding on empty cars was too low for those fully loaded. Thus the empty and load brake allowed the carriers to obtain the full benefits of large cars without raising risks of longer stopping distance.[63]

While the alliance between the AAR and the manufacturers' association continued to improve the lowly cast iron freight car wheel throughout the 1940s, the shift to 100-ton cars finally reached the limits of cast iron. Aware that the future seemed bleak, its makers redeployed their expertise to cast steel. The Southern Wheel Company began research into cast steel as early as 1941 and field tests in 1947. The new wheels, which had the chemical composition of die steel and were highly resistant to heat, finally made their debut in 1954. Demonstrating a new willingness to force the diffusion of improved technology, the AAR banned iron wheels on new cars in 1958 and phased them out entirely a decade later. Wheel safety also benefited from the development of improved monitoring and detection devices. Ultrasound was employed to detect internal defects, and in the early 1950s broken wheel flange detectors first made an appearance. Installed in a yard, the detectors employed "fingers" that would complete a circuit if the flange was broken, sounding an alarm in a tower. By early 1958 eight carriers were installing the new detectors.[64]

Yet despite the development of these new technologies, derailments rose. Part of the problem was that deteriorating finances slowed the diffusion of the new technologies that might have reduced derailments. The case of lubricating pads has been noted, but installation of other safety equipment was painfully slow. Only eighteen carriers installed hot box detectors in 1959 and ten more did so in 1960. For wheel detectors the numbers were five and three, respectively. The number of wrecks from burned-off axles suggested the need for far more careful inspection and more intensive use of hot box detectors. By the mid-1960s regulators were also calling for a stepped-up program of rail flaw detection.[65]

Financial problems also led many carriers to skimp on maintenance of track and equipment. By 1957 the Pennsylvania was reporting that deferred maintenance amounted to $249 million, the equivalent of nearly four years of net income. On that company one worker complained that "we have tracks that the ties are gone from 8 to 12 feet and no spikes for 18 to 20 feet. The joints are so bad that when two cars set side by side . . . they all but touch each other at the roof." Like a contagion, poorly maintained track and equipment might spread derailments even to sound lines through freight car interchange and because companies sometimes ran trains on each other's tracks.[66]

Deteriorating maintenance interacted with problems that resulted from the rapid shift to much larger freight cars. In the early 1960s, the spur of competitive pressures led to a wholesale increase in freight car size as well as rising train speed and length. Just as occurred in the 1880s (chapter 2), these changes were not effectively coordinated with track and road-bed and so, like a ghost from the past, a host of problems with train/track dynamics reappeared. Long trains that included empty and loaded cars might jackknife when the brakes were applied. Problems with cornering and oscillation of cars began to cause derailments. As coal and tank car weight rose to 125 tons, wheel loads exceeded 30,000 pounds. In 1959 engineers recommended standards that would have reduced loads per inch of wheel diameter, but they were not adopted. The heavy loads yielded dangerous bending stresses on light rails and shelling on heavier rails that might precipitate a detail fracture. Rail failure rates, which had steadily declined since 1908, began to rise after 1957 (Table A2.8), doubling between that year and 1965, while accidents from broken rails also skyrocketed.[67]

The impact of rising car size on derailments may well have been conditional upon maintenance levels but neither contribution can be measured statistically. Profitability, however, is subject to statistical test. Previous work has found a weak inverse relationship between company profitability and all train accidents during the 1960s. My own research for 1959–65 confirms that in most years, after controlling for car miles, company derailments were increased by low profitability. In 1962, for example, carriers in the sample earned an average return on assets of 3 percent. Had their profitability increased 2 standard deviations to 7 percent, derailments would have declined about 37 percent (Appendix 2). On individual lines derailments skyrocketed between 1959 and 1965 (Figure 10.7). The wheezy old Boston & Maine saw derailments per car mile rise 60 percent while the Reading experienced a similar increase. There was a 35 percent increase on the Northwestern and a 46 percent rise on the Milwaukee. These findings suggest that rising derailments resulted not from companies that traded

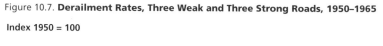

Figure 10.7. **Derailment Rates, Three Weak and Three Strong Roads, 1950–1965**

Index 1950 = 100

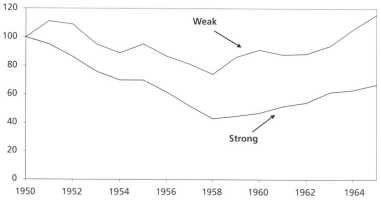

Note: Weak roads are B&M, Reading, and North Western; strong roads are Norfolk & Western, Union Pacific and Southern Pacific. Data are three-year moving averages.

safety for output but rather from market pressures that reduced their ability to deliver either one. Even collisions might be increased by deteriorating maintenance: on July 24, 1963, a side collision on the Pennsylvania that killed two and injured twenty-seven people resulted because the rails were too rusty to complete the track circuit. Railroad journals had warned for decades that the carriers' poverty would eventually undermine safety, and in the 1960s the wolf finally came.[68]

A Change of Regimes, 1965–1969

The rise in derailments occurred as the carriers were shifting to larger cars and longer trains, and while the proportion of hazardous substances that they carried showed no increase, the volume grew with traffic. This was a deadly mix that made freight train derailments seem a major threat to public safety and finally destroyed the old safety compact. In September 1956 the *New York Times* carried the story of a derailed Pennsylvania Railroad tank car full of cleaning fluid that exploded in Greenfield, Indiana, injuring four firemen. Three years later, on January 24, 1959, a Missouri Pacific tank car derailed and exploded in Monroe, Louisiana, killing five. On June 28 that year a tank car of butane derailed on a trestle near Meldrim, Georgia, and blew up, killing twenty-two swimmers and picnickers below. The Meldrim wreck must have caused Arthur Wellington to stir in his grave. The tragedy resulted from a buckled rail on a trestle, and seventy-five years after Wellington had inaugurated the campaign for guard rails, the Seaboard's trestle lacked an inside guard, which the ICC concluded would have prevented the tragedy. About a week later a bad journal caused a fiery wreck of ten cars of alcohol in Marion, North Carolina, that required evacuation of nearby houses.[69]

While there are no statistics of hazardous spills from this period, it seems likely that they rose with traffic. Yet once again the public response was not simply to risk, but rather to risk perception, for such accidents, although far less lethal to the public than grade crossing risks, were far more frightening. Derailments that might include poisonous gas or radioactive materials, or yield conflagrations like that at Meldrim, were especially terrifying, and newspaper stories probably amplified public worries. Times had changed since 1908, when derailments of dangerous cargo could be left to such private organizations as the Bureau of Explosives. The 1960s were years of rapidly expanding federal power over industry, marked by the establishment of the Environmental Protection Agency and the Occupational Safety and Health Administration, as well as countless laws that extended the federal reach. Railroad safety, with its limited public role, seemed dangerously out of step.

By 1965 Congress was fed up with ICC safely regulation. In that year the House Committee on Government Operations held hearings and delivered a report that sharply criticized the commission. The next year Congress transferred rail safely to the newly created Federal Railroad Administration, which soon discovered that 95 percent of all train accidents could not be addressed under existing federal law.

In 1968, in congressional hearings, Senator Harley Staggers termed railroad accidents "a most serious and grave situation." The next year the Senate again held hearings on railroad safety with Vance Hartke in the chair. For years, Hartke noted, safe railroad transportation had been largely left to railroad management and labor "because no one outside of the industry was directly threatened, [and] this myopia has allowed an extremely dangerous situation to develop." Hartke worried that "the railroads today are transporting extremely flammable, explosive . . . and poisonous substances . . . throughout the Nation's metropolitan areas." The committee listed thirty-nine major accidents that required evacuation of population between 1964 and 1969. Rising freight derailments now seemed a major public safety issue, and in 1970 Congress passed the Federal Railroad Safety Act, finally establishing federal control over all forms of railroad safety.[70]

While World War II saw the carriers trade away safety for military victory, it also witnessed the payoff in safety and productivity from the previous decades of technical and institutional innovation. The twenty years after World War II brought sharp changes in railroad safety and in public concerns and conceptions. In some areas progress was continual. Trespassing fatalities steadily declined. That crossing safety was a public good requiring public funding became increasingly widely accepted. As a result the goal of separating all grade crossings quietly disappeared as railroad and highway engineers focused instead on choosing which crossings to guard and how to guard them. In response to this more cost-effective approach, crossing accidents declined continuously. Passenger safety improved, although much of the gain reflected changes in traffic, and as passengers increasingly fled the railroads, its relevance declined.

Yet the period also witnessed worsening train accidents and the end of a century of improvement in work safety. The cessation of progress resulted in part from the carriers' shift to freight traffic and from a rush to larger freight equipment that ignored problems of train/track dynamics. But there were deeper problems as well. Eroding profits undermined safety programs, constrained the spread of safer technologies, and caused widespread deterioration of track and equipment. The railroads' own shortcomings compounded these difficulties. Safety work became formulaic while neither the technical press nor the ICC adequately addressed the carriers' lapses. Derailments that involved hazardous chemicals rose at a time when public worries over technological hazards were increasing, thus changing railroad accidents from an industry concern to a public safety issue and bringing to a close the long experiment with voluntarism.

Conclusion

The Political Economy of Railroad Safety, 1830–1965

In 1935 Joseph Eastman delivered a speech claiming that the carriers' safety improvements had nearly always been forced down their throats by government pressures. "Practically every safety device required by government . . . and bitterly resisted [by the carriers] . . . has ultimately [benefited them]," Eastman pronounced. This was too much for the editor of the *Railway Age*. "It may be worthwhile to recall that the initiative in inventing, developing and promoting every . . . [safety device] has been taken by private enterprise," he sputtered. In fact, as previous chapters have demonstrated, railroad safety was the outcome of a political economy more complex than either of these combatants wished to admit, and its complexity yields insights into why the modern world has become so much safer. Down to perhaps the 1870s, the *Age* was essentially correct: safety was largely an affair of private parties, as mediated by the common law of liability. The regime of private enterprise had built what were by European standards extremely unsafe railroads, but it was continually in the process of making them safer.[1]

As I have argued, more productive and safer railroading was the outcome of a technological network that helped to innovate, evaluate, and diffuse improvements in both "things" such as track and equipment, and institutions and organizations, and that responded to market incentives. All forms of train accident were expensive and passenger injuries made them more so, but better technology and business organization reduced worker and passenger risks from the 1840s on. The political scientist Aaron Wildavsky summarized such a dynamic, noting that danger begets safety and that "richer is safer." Railroad history suggests some explanations for why this is so; it is in good part because market forces press the development of technology, organizations, and institutions that not only enhance wealth but respond to risks as well.[2]

In a few instances the new technology raised productivity but worsened safety—the link and pin coupler probably increased worker risks and the spread of freight car interchange after the Civil War may have done so as well—but most were both labor- and accident-saving. Examples include automatic grade crossing gates, heavier rail, remote sensors, and mechanized track equipment. Alfred Chandler has argued that capital and human capital were complements in the large corporation, and in railroads at least, technological change was embodied not only in capital but also in institutions and organizations that enhanced safety. Internal labor markets, track inspection contests, surprise checking, the Bureau of Explosives, rail inspection, company health and benefit plans, and Safety First were among the new institutions and organizations that bettered safety. Most of these improvements were, in economists' terminology, "induced." That is, they were a response to the economic and political context of railroading. Sometimes the mechanism was rising labor costs, as with the mechanization of maintenance and the diffusion of automatic signals. But accident costs as well as broader labor market concerns and—as Eastman claimed—political pressures shaped this evolution as well.

Most new technology worked by increasing corporate control over man, nature, and organizations. Washouts, snowstorms, track subsidence, fire, and fog were all natural hazards, and railroading became steadily safer as it became less "natural." Control over human behavior was even more important; block signals and later CTC reduced train accidents while the diffusion of the Standard Code of train rules, materials testing, surprise checking, and training in use of hand brakes also diminished risks. Institutional innovations such as the Bureau of Explosives or MCB standards dealt with the need for a system-wide approach to agency problems and technology.

In some instances, the accident-reducing effects of these investments were all but inevitable—car retarders provide an example—but in most cases the full impact on safety was contingent upon follow-on improvements in organizations and institutions, and sometimes upon business decisions on where to allocate the new technology. Companies had to develop new rules and training to obtain the full benefit of air brakes and automatic couplers, and for years block signals were a great disappointment until companies instructed train crews and developed surprise checking to institutionalize obedience. Mechanization of track work sometimes improved safety, but without proper organizational and institutional adaptations it could worsen risks. The impact of better roadbed, steel rails, steam heat for cars, block signals, crossing guards, and many other investments was contingent upon their effective allocation, and as demonstrated, economic forces pressed their deployment so as to maximize safety.

A broader view of this process suggests that accidents and casualties declined roughly in proportion to the railroads' increasing technical and organizational complexity. Commuter traffic had become extremely "close-coupled" by the 1890s, to borrow terminology from Charles Perrow, while in that decade and in the 1960s, the complex causation associated with the train/track dynamics of long freights yielded accidents in surprising ways. While as Perrow stressed, the immediate result was "normal accidents," the longer-term outcome has been that the social costs of railroading steadily declined. In the 1890s, about seven men died as the result of each million dollars of transportation services, but these costs fell steadily and by the 1950s had declined about 90 percent.

While I have emphasized that many accident-reducing technologies were induced by a variety of mechanisms internal to the carriers, railroad safety also benefited from a host of improvements that arose outside the industry. Good examples include developments in electrical equipment and electronics that the carriers and their suppliers applied to train control, CTC, nondestructive tests, and train radio. Trespassing casualties declined only with the advent of rising incomes, cheap cars, and better roads, while the sophisticated analysis of grade crossing hazards came out of railroad and state highway department efforts. The carriers and their suppliers benefited from the development of university engineering departments that were themselves under strong entrepreneurial pressures. The programs at MIT and especially Purdue and the University of Illinois provided vital expertise to the carriers and regulators on a range of safety problems.[3]

But if this largely private regime delivered better safety, it did so in ways that the public quickly found unsatisfactory. In 1835, Massachusetts required its few and tiny railroads to post warning signs at grade crossings, marking the first government foray into railroad safety. Moreover, safety supply creates its own demand, and the carriers' very successes at reducing accidents proved that they could do more, and so a complex political economy evolved. As states groped for solutions, two issues emerged: what should be regulated and how should it be done? In the 1870s the Massachusetts railroad commissioners led by Charles Francis Adams formalized what would be the dominant answers for nearly a century.

Adams's focus was quite narrow, emphasizing train accidents to passengers, and in this his concerns differed little from those of the carriers themselves. Both regulators and carriers were responding not to risks, but to risk perceptions. Train accidents to passengers seemed dreadful and the public press often amplified their horror. But for a very long time, the ideology of individualism at least partly shielded many dangers from public scrutiny. Workers assumed the normal risks of the job, so the courts usually found. If a man was fool enough to run for a train or walk the tracks, his injury was his own fault, or so courts, regulators, and legislators typically concluded. Massachusetts's early crossing law required railroads to warn of trains, but a properly warned individual was responsible for his or her own safety.

If passenger safety was the proper end of public policy, the means, in Adams's vision, was the expert commission that might use public pressure to press the carriers for "voluntary" improvements in passenger train safety. The veiled threat of regulation would thus provide the incentive for improvement, correcting the failings of private enterprise. Without the need for a lumbering bureaucracy, regulation would be fast and expert. It was a vision with considerable appeal and it was forcefully advocated by a parade of distinguished engineer-editors at major railroad journals that included Forney, Wellington, and Boardman, among others.

An alternative to Adams's voluntarism was specific safety regulations. In the 1880s states began to regulate passenger car heating, bridge safety, and a variety of train and track appurtenances intended to reduce worker risks. Social scientists have proposed a host of explanations for the rise of regulation during these years, many of which focus on company efforts to shape policies to their own ends. However, most of the emphasis has been on economic regulation and none of the theories seem to fit the early rise in state safety regulation very well. There is little evidence that the carriers supported it or that this growth in regulation reflected decreasing effectiveness of either Adams's approach or the private liability system. Passenger accidents had always been expensive and several careful studies suggest that employers' liability costs for work accidents may have been rising. What was changing was historical experience with railroads.[4]

The very expansion of railroading helped to broaden the focus of public concern. Early carriers had been too poor to do much about grade crossings. But by the 1890s they were rich and powerful, and busy crossings in towns were taking a fearful toll. The carriers could now afford to shoulder the burden of grade crossing elimination. Such views encouraged state regulations and led to sympathetic jurors. Reformers and regulators began to look at other forms of accident in a different light as well. Accident statistics gave a focus to these worries while technological competence undermined the old idea that accidents were simply the result of carelessness. Modern readers have heard claims such as "if we can go to the moon, surely we can collect the garbage." If the carriers could span the continent, surely they could make a safer locomotive or coupler. Why couldn't railroad stations be made safer, as was done in Europe?

Engineers like Matthias Forney and Arthur Mellen Wellington were impatient with incompetence and offended by bad designs; they were positioned to know in intimate detail the danger of railroad work, and to do something about it. The improvement in bridge design illustrates the emergence of a self-conscious safety ethic as engineering professionalized. Accident statistics so shocked Henry Cabot Lodge that when better couplers appeared, he concluded that it was immoral not to use them. Thus wealth led to safety in part because technological competence created a safety imperative. Regulation was path-dependent as well; apparently successful safety regulations became the justification for further twentieth-century regulation. At their best such specific regulations might accomplish much. But they could never substitute for the proper managerial incentives. So the major improvements in safety during these years remained an affair of private enterprise.

By about 1900 the focus of railroad safety included a broader public that comprised passengers and workers, but other citizens only as they legally crossed the tracks. The private regime was increasingly subjected to public and labor pressures for better safety, and in the decades prior to World War I the pressure increased. The period yielded effective rate regulation and many specific safety rules—an unstable mixture that would finally explode a half century later. These years saw liability legislation that sharply raised the cost of work injuries and threats of far more extensive controls. The carriers had already begun a massive upgrading of track and equipment to deal with surging traffic. They responded to these threats as Adams had imagined they would, with a host of additional investments in steel cars and block signals and with organizational innovations such as Safety First to deal with their pervasive agency problems. To anyone who would listen, the carriers claimed that these commitments signaled that they could be trusted to chart their own safety course.

Died-in-the-wool Progressives never accepted the carriers' claims. Similarly, states pressed for an expanded program of crossing protection, while labor continued to clamor for stricter rules. But in practice the carriers carried the day. Except for legislation that gave the ICC power to authorize train control, the carriers fought off most of the prewar safety legislation and by the 1920s what I have termed a tacit safety compact had evolved. Congress and the ICC would leave safety largely up to the carriers, but only if they held up their end of the bargain, which was to reduce accidents. It was in fact simply the old voluntarism of Charles Francis Adams on a national scale, and for a very long time it delivered the goods.

Such a construction of events stresses the continuity in safety from the nineteenth into the twentieth century, and indeed the carriers' Safety First commitment was a change in means more than in ends. As I have shown, accident costs and political pressures for better safety had been increasing for some time and formal safety organizations and rules were but one of the railroads' safety-enhancing improvements that ranged from steel cars and block signals to pension plans and automatic crossing gates. Not surprisingly, therefore, the trends in worker and passenger fatality rates show steady decline with no sharp breaks.

There were major failures as well as successes in the railroad safety record. Conceived and raised under a regime of employers' liability, the carriers' safety organizations overemphasized behavioral rather than physical causes of accidents. The railroads were slow to develop an industry-wide research organization. And when they did, its research was underfunded and misdirected. Nor did the ICC usually provide the sort of expert prodding that Adams had envisaged, while by the 1950s the railroad press had evolved from critic to cheerleader. As deteriorating profits began to increase derailments and worker casualty rates after 1957, there was no one to sound the alarm.[5]

Yet if the worsening of safety during the 1960s is striking, that is only in comparison with what had earlier been achieved. That long-term record seems to suggest two important truths. First, formal safety regulation (as opposed to informal pressures) had only marginal effects on safety.

As has been stressed, few safety appliances could exert a decisive impact on accident rates. Moreover, regulators responded to risk perceptions, not risk, and so while safety laws have sometimes reduced risks in a cost-effective manner, they have sometimes squandered funds on marginal or trivial problems. It is hard to imagine that railroad managers could not have used resources to improve safety more effectively than did the ICC's two experiments with train control. A second and related point is that long-term safety gains have never resulted from narrowly conceived "safety appliances" or because the carriers allocated resources from producing transportation to reducing accidents. To companies, accidents might signal inefficiencies that when removed could add to profits. And they focused invention as well, leading to long-term improvements in physical capital, institutions, and organizations. Thus the history of railroad safety teaches that it comes wrapped in a package that might yield more or less safety or productivity, depending upon managers' choices. But without healthy, profitable railroads to deliver the package, better safety would never have arrived at all.

Appendix 1
Nineteenth-Century Railroad Accident and Casualty Statistics

Prior to the establishment of the ICC, no nationwide data on American railroad accidents, casualties (fatalities and injuries), or casualty rates exist. In Britain, casualty figures date from the 1840s but casualty rates are similarly nonexistent. This appendix describes and presents some of the data and calculations employed in chapter 1 to derive fatality rates on American and British railroads and it presents a summary of the train accident data that underlie chapters 2–6.

American State Data, 1846–1900

Almost every state with a railroad commission collected some data on railroad casualties, but much of it is of little value. Sometimes casualties are for the state only, while employment or passenger miles are for the entire line. No state defined what it meant by an injury and so I have computed only fatality rates. The American data (Table A1.2) underlying Figures 1.2 through 1.4 derive from railroad reports of the various states. They are consistent but fatalities are probably underreported, especially for workers and "others." The information on other (neither passenger nor worker) fatalities is for Massachusetts only. For passengers the following states collected usable data: Massachusetts from 1846 on; New York, 1852 on; Minnesota, 1873 on; Wisconsin, 1876 on; Illinois, 1888 on; Iowa and Ohio, 1897 on. For workers the following states are included: Massachusetts from 1846 on; Ohio from 1868 on; Connecticut and Michigan from 1874 on; Illinois and Iowa from 1878 on; Minnesota from 1883 on.

For antebellum years the data reflect Massachusetts and New York only. Table A1.1 compares the passenger miles and workers in my sample for select years with Fishlow's national estimates. As can be seen, my samples range from 10 to 30 percent of the relevant population, but even for the years following the Civil War they omit all southern and western states. Probably for this reason, they underestimate risks nationwide (chapter 1 and below).

British Railroad Fatalities and Rates, 1847–1900

These data are all for the United Kingdom (England, Wales, Scotland, and Ireland) and all derive from reports to the Board of Trade or calculations based on these

Table A1.1. **Railroad Output and Input, Selected Measures, 1839–1880**

Year	Passenger Miles (billions)		Passenger Train Miles (millions)	Employment (thousands)	
	Total	**Sample**	**Total**	**Total**	**Sample**
1839	0.09	—	1,731*	5	—
1849	0.47	0.15	8,545	18	3.6
1859	1.9	0.56	32,203	35	6.0
1870	4.1	1.34	—	230	35.9
1880	5.7	2.35	—	416	159.1

Sources: Albert Fishlow, "Productivity and Technological Change in the Railroad Sector, 1840–1910," in *Output, Employment and Productivity in the United States after 1800* (New York: National Bureau of Economic Research, 1966), 583–646; author's calculations.

* Derived from passenger mile data employing passengers per train calculated to be 52 in 1839 from Franz Anton Ritter von Gerstner, *Early American Railroads*, ed. Frederick Gamst (Stanford: Stanford University Press, 1997), 55 in 1849 from Massachusetts railroads, and 59 in 1859 from New York railroads.

reports. The statistics on "other" (i.e., neither passenger nor worker) fatalities are straightforward and will not be discussed here. The Board of Trade published estimates of employees from 1847 to 1860, and then from 1874 to 1901 with gaps.[1] For the post-1873 period I filled in the gaps by interpolation. The following equation estimated for 1852–60 and 1874–1901 relating observed employment (E) to train miles ($TRNMI$), which begin only in 1852, allowed estimation of employment from 1861 to 1873.[2]

$$\text{Log}(E) = -2.976 + 0.791\text{Log}(TRNMI) + 0.011\,Trend$$
$$(458.52) \qquad\qquad (11.46)$$
$$R^2 = 0.99,\ N = 37,\ D.W. = 1.56,\ Rho = 0.777$$

The board also published estimates of employee fatalities for each year from 1840 on. These are defective in several ways. For the early years they contain some fatalities for construction workers as well as railroad employees. But simple observation of the data suggests that there was very substantial underreporting prior to 1872, when a new law was implemented, tightening reporting requirements. Accordingly, I estimated the effect of reporting changes on the employee fatality rate (EFR) as follows:

$$(EFR) = 1.88 - 0.058\,Trend + 2.03\,Report$$
$$(5.72) \qquad (9.11)$$
$$R^2 = 0.77,\ N = 52,\ D.W. = 2.32$$

Report is a dummy that is zero prior to 1872 and one from 1872 on. Despite such findings, I have not corrected the original data (Table A1.3) for two reasons. First, U.S. figures are also subject to underreporting and, second, the size of the *Report* coefficient depends upon the functional form of the above equation.

From 1840 on the board also required companies to report passenger fatalities and passenger journeys, but passenger miles are available only through 1859.

Table A1.2. **Passenger and Worker Fatality Rates, American State Data, 1846–1900**

Year	Workers	Worker Fatalities	Worker Fatality Rate*	Passenger Miles (thousands)	Passenger Fatalities	Passenger Fatality Rate*
1846	2,000	2	1.00	81,250	1	12.31
1847	2,706	11	4.07	101,329	9	88.82
1848	3,294	18	5.46	128,076	6	46.85
1849	3,601	27	7.50	147,151	7	47.57
1850	4,582	22	4.80	155,926	7	44.89
1851	4,955	24	4.84	147,598	3	20.33
1852	5,306	21	3.96	515,827	29	56.22
1853	6,765	20	3.47	730,215	34	46.56
1854	6,176	33	5.34	686,177	19	27.69
1855	6,710	27	4.02	613,687	21	34.22
1856	6,357	23	3.62	202,447	10	49.40
1857	1,390	5	3.60	480,103	10	20.83
1858	5,766	10	1.73	543,805	27	49.65
1859	5,950	9	1.57	561,559	15	26.71
1860	6,498	11	1.69	581,708	23	39.54
1861	6,453	11	1.70	510,618	14	27.42
1862	6,471	12	1.85	527,234	31	58.80
1863	6,061	23	3.79	665,312	22	33.07
1864	7,475	29	3.88	911,412	106	116.30
1865	8,021	29	3.62	1,054,242	49	46.48
1866	8,966	34	3.79	985,509	29	29.43
1867	10,256	41	2.84	1,011,267	29	28.68
1868	30,763	31	1.01	1,036,309	90	86.85
1869	34,819	37	3.04	1,120,864	37	33.01
1870	35,869	44	3.15	1,340,707	41	30.58
1871	37,098	40	2.16	1,408,788	77	54.66
1872	43,063	74	3.41	1,561,306	35	22.42
1873	49,615	57	3.00	1,821,847	35	19.21
1874	70,031	37	1.98	2,049,406	26	12.69
1875	66,863	27	2.41	1,915,034	31	16.19
1876	64,754	29	2.13	2,050,302	28	13.66
1877	65,749	36	2.10	1,960,795	17	8.67
1878	115,866	34	2.08	1,955,869	37	18.92
1879	129,406	42	1.94	2,015,711	52	25.80
1880	159,056	54	2.94	2,351,848	35	14.88
1881	191,259	96	2.87	1,307,059	22	16.83
1882	169,946	87	3.55	3,054,468	42	13.75
1883	188,322	106	3.24	3,287,604	67	20.38

Continued

Table A1.2. Continued

Year	Workers	Worker Fatalities	Worker Fatality Rate*	Passenger Miles (thousands)	Passenger Fatalities	Passenger Fatality Rate*
1884	185,378	76	2.66	3,286,253	45	13.69
1885	138,593	56	1.97	3,411,547	38	11.14
1886	245,636	101	1.53	3,718,772	79	21.24
1887	217,901	111	1.37	4,103,406	97	23.64
1888	289,169	122	1.93	5,151,502	163	31.64
1889	283,306	99	1.56	4,954,009	66	13.32
1890	368,634	113	2.08	5,389,551	82	15.21
1891	397,284	124	1.99	5,911,343	119	20.13
1892	449,528	143	1.90	6,412,962	95	14.81
1893	474,186	198	2.07	7,147,792	106	14.83
1894	474,186	93	1.21	8,382,454	143	17.06
1895	417,359	111	1.03	6,685,736	44	6.58
1896	338,483	140	1.29	7,032,733	45	6.40
1897	347,756	97	1.27	7,358,544	87	11.82
1898	387,292	370	2.01	7,951,658	56	7.04
1899	402,761	85	1.06	8,083,400	67	8.29
1900	366,742	103	1.68	8,531,448	45	5.27

Source: See text.

* Fatality rates are per thousand workers and per billion passenger miles.

These are all usefully summarized in Neison in the years up through 1852.[3] I employed these data only from 1847 through 1859, as the earlier figures derive from unreliably small samples. From 1860 to 1870, Hawke[4] has calculated passenger miles for England and Wales. I calculated the average length of journey that these imply, assumed that it held for the United Kingdom, and multiplied it by passenger journeys for the United Kingdom to get passenger miles. Dorsey estimated passenger miles for 1884.[5] I employed his figure and used it to estimate the average length of journey for that year. I then interpolated the average length of journey from 1870 to 1884. For 1885–1900 I assumed that average journey length remained constant. Again, there is evidence of underreporting before 1872. I estimated its impact on passenger fatality rates (PFR) using the following equation:

$$(PFR) = 30.06 - 0.643\,Trend + 20.56\,Report$$
$$(5.84) \qquad (5.98)$$
$$R^2 = 0.40,\ N = 54,\ D.W. = 1.99$$

where *Report* is a dummy that is zero prior to 1872 and one from 1872 on. Although for reasons discussed above I have chosen not to correct the original data, the size of the correction suggests that British railroads were considerably more dangerous than contemporaries imagined (Table A1.4).

Table A1.3. **Employee Fatality Rates, United Kingdom, 1847–1900**

Year	Employees	Employee Fatalities	Fatality Rate
1847	47,218	122	2.58*
1848	52,688	138	2.62
1849	55,968	129	2.30
1850	59,974	129	2.15
1851	63,563	113	1.78
1852	67,601	123	1.82
1853	80,409	159	1.98
1854	90,409	112	1.24
1855	97,958	125	1.28
1856	102,117	142	1.39
1857	107,660	93	0.86
1858	109,329	131	1.20
1859	116,270	117	1.01
1860	127,450	121	0.95
1861	125,587	128	1.02
1862	129,780	109	0.84
1863	139,360	98	0.70
1864	152,868	103	0.67
1865	164,375	122	0.74
1866	169,442	100	0.59
1867	222,310	105	0.47
1868	206,191	83	0.40
1869	189,213	151	0.80
1870	202,434	115	0.57
1871	214,259	213	0.99
1872	224,984	632	2.81
1873	233,753	773	3.31
1874	250,000	788	3.15
1875	255,000	765	3.00
1876	260,000	673	2.59
1877	265,000	642	2.42
1878	270,000	544	2.01
1879	280,000	452	1.61
1880	290,000	546	1.88
1881	300,000	521	1.74
1882	315,000	553	1.76
1883	330,000	554	1.68
1884	346,426	546	1.58
1885	353,666	451	1.28
1886	360,706	425	1.18

Continued

Table A1.3. Continued

Year	Employees	Employee Fatalities	Fatality Rate
1887	367,746	422	1.15
1888	374,786	396	1.06
1889	381,626	435	1.14
1890	396,540	499	1.26
1891	409,454	549	1.34
1892	423,368	534	1.26
1893	437,282	460	1.05
1894	451,196	479	1.06
1895	465,112	442	0.95
1896	488,112	447	0.92
1897	511,132	510	1.00
1898	534,141	504	0.94
1899	548,039	521	0.95
1900	561,937	583	1.04

Source: See text.

* Per thousand employees.

Fatalities to Others, 1852–1900

Fatalities to others (trespassers and individuals killed at grade crossings) are available for Massachusetts and Great Britain from the 1850s on and are presented in Table A1.5. Once again, these data are likely to reflect considerable underreporting, but there is no way to correct them.

Disasters as Rare Events

As the text describes, the number of disasters for the 1840s and 1850s comes from Shaw.[6] Passenger train miles for 1839, 1849, and 1859 are in Table A1.1. These imply annual average growth rates of 17.3 percent and 14.2 percent a year, respectively. Using these I calculated that there were about 47.9 million passenger train miles from 1840 to 1849 and 190.6 million from 1850 to 1859. Both of these are probably overstated due to the existence of depressions in each decade. The incidence of disaster in the 1870s employs *Gazette* data on train accidents and passenger train mile calculations, described below.

Calculation of Train Miles, 1873–1881

Beginning in 1882, data on both freight and passenger train mile data are available from Poor's *Manual of Railroads* and the ICC.[7] To estimate train miles for 1873–81 that are contained in Table A1.7 and also employed to calculate the prevalence of disasters during 1877–80 reported in chapter 1, I did the following. First, I took Synder's[8] estimates of ton miles for 1873–81 and reduced them by 15 percent to make them consistent with Fishlow's work for earlier years. To estimate passenger miles I interpolated between Fishlow's 1870 estimate to 1882. Then I related the natural log of passenger or freight train miles (*PTRNMI* or *FTRNMI*; both in millions) for the

Table A1.4. **Passenger Fatality Rates, United Kingdom, 1847–1900**

Year	Passenger Journeys (thousands)	Length of Journey	Fatalities	Passenger Miles (thousands)	Fatality Rate*
1847	51,352	18.7	31	960,285	32.28
1848	57,965	16.3	20	944,831	21.17
1849	60,398	16.6	23	1,002,609	22.94
1850	63,842	17.2	32	1,096,178	29.19
1851	72,854	17.2	36	1,255,617	28.67
1852	83,392	16.4	32	1,370,792	23.34
1853	89,176	16.7	64	1,488,853	42.99
1854	102,287	15.9	31	1,622,050	19.11
1855	111,206	13.3	28	1,474,675	18.99
1856	118,596	15.4	26	1,822,048	14.27
1857	129,348	14.3	47	1,850,150	25.40
1858	139,699	13.3	51	1,853,624	27.51
1859	139,199	14.5	29	2,020,909	14.35
1860	163,000	12.9	45	2,107,590	21.35
1861	173,700	12.4	79	2,145,195	36.83
1862	180,400	12.4	35	2,236,960	15.65
1863	204,635	11.3	35	2,310,329	15.15
1864	229,272	10.8	36	2,476,138	14.54
1865	251,862	10.3	36	2,591,660	13.89
1866	274,294	10.2	31	2,786,827	11.12
1867	287,688	10.0	36	2,871,126	12.54
1868	304,136	9.7	62	3,115,280	21.02
1869	311,528	10.0	39	3,115,280	12.52
1870	336,545	9.4	100	3,170,254	31.54
1871	375,221	9.4	57	3,514,320	16.21
1872	422,875	9.3	127	3,937,812	32.25
1873	455,320	9.3	160	4,215,353	37.96
1874	477,500	9.2	211	4,394,910	48.01
1875	509,900	9.2	134	4,665,585	28.72
1876	534,400	9.1	144	4,860,902	29.62
1877	549,300	9.0	126	4,966,771	25.37
1878	564,900	9.0	125	5,077,321	24.62
1879	565,400	8.9	160	5,051,284	31.68
1880	614,300	8.9	147	5,454,984	26.95
1881	625,600	8.8	111	5,521,546	20.10
1882	654,700	8.8	131	5,743,028	22.81
1883	683,300	8.7	126	5,957,009	21.15
1884	694,600	8.7	141	6,049,966	23.31

Continued

Table A1.4. Continued

Year	Passenger Journeys (thousands)	Length of Journey	Fatalities	Passenger Miles (thousands)	Fatality Rate*
1885	697,100	8.7	107	6,071,741	17.62
1886	725,700	8.7	·98	6,320,847	15.50
1887	733,500	8.7	126	6,388,785	19.72
1888	742,900	8.7	113	6,470,659	17.76
1889	775,000	8.7	185	6,750,250	27.41
1890	817,400	8.7	121	7,119,554	17.00
1891	845,200	8.7	111	7,361,692	15.08
1892	864,600	8.7	151	7,530,666	20.05
1893	872,700	8.7	113	7,601,217	14.87
1894	911,500	8.7	121	7,939,165	15.24
1895	980,200	8.7	87	8,537,542	10.19
1896	980,600	8.7	98	8,541,026	11.47
1897	1,030,900	8.7	135	8,979,139	15.03
1898	1,062,600	8.7	162	9,255,246	17.50
1899	1,106,400	8.7	165	9,636,744	17.12
1900	1,142,700	8.7	142	9,952,917	14.27

Source: See text.

* Per billion passenger miles.

1882–1909 period to the log of passenger or ton miles (*PM* or *TM*; both in millions) and a time trend with the following results.

$$\text{Log}(PTRNMI) = 8.777 + 0.190 \text{Log}(PM) + 0.031\, Trend$$
$$(2.12) \qquad\qquad (4.63)$$
$$R^2 = 0.98, N = 28, D.W. = 1.72, Rho = 0.884$$
$$\text{Log}(FTRNMI) = 1.264 + 0.665 \text{Log}(TM) - 0.021\, Trend$$
$$(7.07) \qquad\qquad (3.09)$$
$$R^2 = 0.96, N = 28, D.W. = 1.80, Rho = 0.821$$

I then employed these equations and the passenger and ton mile data to estimate passenger and freight train miles for 1873–81.[9]

Fatality Rates and National Estimates, 1882–1889

These rely on estimates of fatalities to passengers and employees derived from the *Railroad Gazette* and the ICC. Both the passenger and ton mile data are from the ICC. I estimated employment as follows. I first related ICC employment figures (*E*, in thousands) from 1889 to 1909 to passenger and ton miles (*PM* and *TM*, in millions) and a trend (*Time*) with the following result.

$$\text{Log}(E) = -11.482 + 0.786 \text{Log}(TM) + 0.241 \text{Log}(PM) - 0.023\, Trend$$
$$(5.54) \qquad\qquad (1.76) \qquad\qquad (2.55)$$
$$R^2 - 0.98, N - 21, D.W. - 1.44, Rho - 0.576$$

Tables A1.5. **Other Fatalities, Great Britain and Massachusetts, 1852–1900**

Year	Great Britain			Massachusetts		
	Other Fatalities	Others per Train Mile	Others per Capita*	Other Fatalities	Others per Train Mile*	Others per Capita*
1852	63	1.04	2.30	39	7.16	39.24
1853	82	1.21	2.98	33	5.62	31.73
1854	80	1.10	2.89	31	4.88	28.70
1855	93	1.26	3.35	36	5.81	31.80
1856	109	1.40	3.89	39	6.41	33.88
1857	95	1.14	3.37	34	5.93	29.04
1858	95	1.09	3.35	32	5.95	26.87
1859	99	1.06	3.47	35	5.72	28.90
1860	89	0.87	3.09	29	4.24	23.56
1861	77	0.73	2.66	25	3.92	20.19
1862	82	0.76	2.80	49	7.44	39.36
1863	51	0.44	1.73	43	6.36	34.32
1864	83	0.64	2.79	45	5.78	35.71
1865	64	0.46	2.17	53	6.64	41.83
1866	84	0.59	2.79	55	6.09	42.21
1867	68	0.46	2.21	60	5.99	44.78
1868	67	0.44	2.18	58	5.67	42.09
1869	131	0.83	4.24	51	4.14	35.99
1870	81	0.48	2.60	82	6.14	56.28
1871	434	2.42	13.73	73	4.98	48.86
1872	386	2.06	12.10	95	5.59	62.01
1873	439	2.26	13.68	90	4.49	57.29
1874	425	2.14	13.08	87	4.31	54.00
1875	391	1.89	11.92	83	4.09	50.24
1876	715	3.36	21.47	75	3.64	44.72
1877	697	3.21	20.74	92	4.42	54.02
1878	619	2.81	18.26	106	4.94	61.31
1879	430	1.92	12.50	66	2.90	37.59
1880	454	1.90	13.12	88	3.53	49.36
1881	476	1.93	13.68	97	3.56	53.47
1882	442	1.72	12.56	98	3.37	53.12
1883	502	1.88	14.14	176	5.65	93.77
1884	460	1.70	12.99	120	3.72	62.86
1885	420	1.54	11.67	120	3.51	61.79
1886	432	1.57	11.90	121	3.32	60.56

Continued

Tables A1.5. Continued

	Great Britain			Massachusetts		
Year	Other Fatalities	Others per Train Mile	Others per Capita*	Other Fatalities	Others per Train Mile*	Others per Capita*
1887	395	1.40	10.79	149	3.78	72.47
1888	422	1.46	11.47	148	3.48	69.98
1889	478	1.57	12.85	166	3.45	76.29
1890	478	1.54	12.75	151	3.32	67.44
1891	304	0.95	8.04	163	3.33	71.21
1892	488	1.50	12.81	167	3.25	71.37
1893	462	1.45	12.00	218	3.80	91.14
1894	542	1.64	13.93	161	3.03	65.82
1895	514	1.53	13.08	176	3.39	70.40
1896	505	1.43	12.75	190	3.20	74.28
1897	554	1.51	13.85	177	3.19	67.61
1898	555	1.46	13.77	173	3.07	64.58
1899	591	1.50	14.49	139	2.46	50.71
1900	552	1.37	13.43	166	2.81	59.18

Source: See text.

* Rates are per million train miles and per million of population.

Table A1.6. **Estimates of Total U.S. Railroad Passenger and Worker Fatalities, 1882–1890**

Year	Passenger Miles (billions)	Ton Miles (billions)	Estimated Employment (thousands)	*Gazette* Fatalities Passenger	*Gazette* Fatalities Employee	Estimated Fatalities Passenger	Estimated Fatalities Employee
1882	7.7	39	499	104	276	281	1,797
1883	8.5	44	548	132	341	315	1,988
1884	8.8	45	545	89	300	261	1,870
1885	9.1	49	579	85	222	255	1,620
1886	9.7	53	606	115	292	294	1,846
1887	10.6	62	683	207	406	390	2,161
1888	11.2	65	709	138	434	418	2,155
1889	11.6	60	731	108	336	328	1,951
1890	11.9	79	773	172	569	291	2,269

Sources: Passenger and ton miles from ICC; Gazette fatalities from *Railroad Gazette.* For estimated employment and fatalities see text.

Table A1.7. **Train Accidents by Cause, 1873–1900 (annual averages)**

	1873–1877	1878–1882	1883–1887	1888–1892	1893–1897	1898–1900
Train miles	434	511	565	781	850	914
Total accidents	1,055	1,100	1,347	2,083	1,674	2,438
Total collisions	295	417	548	959	691	1,096
Rear collisions	150	275	342	464	324	460
Butting collisions	96	121	174	286	151	235
Other collisions	44	21	32	209	217	401
Total derailments*	709	646	728	1,031	912	1,268
Total road defects	149	116	191	175	121	110
Broken rail	71	48	68	49	39	24
Spread rail	31	34	66	47	29	28
Broken bridge	24	27	32	38	21	26
Broken switch	9	3	13	27	18	19
Broken frog	5	2	11	11	7	5
Other	9	2	1	3	6	9
Total equipment defects*	76	79	108	170	200	290
Broken wheel	22	28	33	40	40	65
Broken axle	32	36	49	56	76	106
Broken truck	10	11	15	28	25	37
Coupling/drawbar	4	1	4	13	23	25
Fall of brake beam	3	0	5	16	15	15
Air brake	0	0	0	0	0	8
Total operating negligence	97	91	84	124	8	140
Misplaced switch	76	77	69	71	47	48
Derailing switch	0	0	0	0	4	13
Trackmen's negligence	9	5	5	8	7	10
Runaway engine	2	2	2	10	7	18
Open draw	4	3	4	3	4	2
Other	6	4	5	32	29	48
Total track obstructions	158	128	157	178	158	130
Animals	48	39	33	53	31	27
Snow or ice	20	13	17	9	12	6
Washout	28	21	21	19	16	19
Landslide	6	7	16	29	26	30
Accidental	38	32	32	21	26	22
Malicious	15	13	30	39	40	21
Other	3	3	8	8	6	5

Sources: Train miles prior to 1882 author's estimates. "Train Accidents in 1900," *Railroad Gazette* 33 (Feb. 15, 1901): 112–13.
*Includes some uncategorized.

Using this equation I estimated employment figures presented in Table A1.6. I derived the estimates of fatalities via a similar procedure. The *Railroad Gazette* fatality data overlap ICC estimates from 1889 through 1901. Accordingly, I derived the following relationship, where *Iccfat* and *Rrgfat* are fatalities reported by each organization, and *W* is a dummy that is zero for passenger and one for worker fatalities.

$$\text{Log}(Iccfat) = 3.590 + 0.436\text{Log}(Rrgfat) + 1.466\,W$$
$$(4.43) \qquad\qquad (10.81)$$
$$R^2 = 0.97,\ N = 26,\ D.W. = 1.37,\ Rho = -0.10$$

Train Accidents, 1873–1900

These data in Table A1.7 are from the *Railroad Gazette*'s annual compilations. The *Gazette* always claimed that it reported a fairly complete list of *serious* train accidents—especially those resulting in fatalities to passengers. In 1890, after the ICC had began to require reporting of casualties from train accidents, the *Gazette* compared them to its data. Its estimate of passenger fatalities was only four fewer than that of the ICC while it reported more fatalities to workers, probably due to differences in definitions. In 1898 its figures equaled 96 percent of passenger fatalities reported to the ICC but only 81 percent of worker fatalities, suggesting that the paper missed a number of minor but lethal freight collisions.[10]

Appendix 2
Casualties and Accidents from Interstate Commerce Commission Statistics, 1888–1965

The basic data to measure railroad safety from 1888 on derive largely from the statistics collected by the ICC. This appendix presents the series from which most of the tables and graphs in the text derive. It also contains the rail failure data derived from the AREA and employed in chapters 7–10.

From 1888 to 1900 the ICC gathered no statistics on numbers of train accidents but it presented railroad casualties (fatalities and injuries) in its *Statistics of Railways*. From 1889 on these data are divided into injuries and fatalities to employees, passengers, and others. The data are presented for the United States as a whole and for ten geographic divisions. Employees are sometimes divided into broad groups of occupations (trainmen, switchmen, flagmen, watchmen) while casualties to "others" include trespassers and nontrespassers. Casualties are also classified by broad class and cause of accident (casualties from collisions, from train and nontrain accidents; casualties from car coupling).[1]

Beginning in 1901, under authority granted by the Accident Reports Act of that year, the commission also began to gather data on train accidents and a separate set of casualty statistics, both of which it presented in its quarterly (later yearly) *Accident Bulletin*. Thus numbers of collisions, derailments, and other train accidents date from 1901 (fiscal 1902), as does the definition of a reportable train accident, which was one doing damage of $150 and/or resulting in a casualty. Until 1910, the casualty statistics presented in the bulletin include only those to passengers and employees that result from movement of trains. Employee injuries were defined as those causing incapacitation for more than three days in the succeeding ten.[2]

There are many deficiencies in these early data. Those in *Statistics of Railways* do not separate casualties to employees on duty from those to employees not on duty. Employee injuries are grossly underreported and casualties by type of accident do not always agree with those in the *Bulletin*.

Beginning in 1910 the commission ended the dual system of reporting and thereafter all statistics on accidents and casualties derive from its *Accident Bulletin*. In that year it divided reportable casualties into three groups. There were those arising from train accidents (and the $150 damage figure was now revised to include the cost of clearing wrecks). Train service accidents were those arising from movement of trains but doing less than $150 of damage, while there were casualties from nontrain accidents, such as passengers who fell in stations or workers injured in shops. In the text I have grouped accidents differently. Because contemporaries placed so much emphasis on collisions and derailments, I treat them separately. All other accidents and injuries, including some train accidents such as boiler explosions, I call "little accidents."

The ICC revised the definition of what constituted a reportable accident or injury many times from 1910 on, and its enforcement of reporting accuracy also varied. In the period under consideration the most substantial revisions occurred in 1957 and data from that period on are not entirely comparable with those of earlier years.[3]

Passenger Casualties

Table A2.1 contains passenger casualty rates per billion passenger miles, from 1888 to 1965. Over the years the ICC presented passenger casualties using different definitions of who constituted a passenger and what constituted an accident. For the sake of continuity I have chosen a series from the appendix to the commission's *Accident Bulletin*, for these are the only data that span the entire period under consideration. Passengers exclude individuals carried under contract (e.g., postal employees) while the data include only those passengers killed in train and train service accidents. They exclude a small number of passengers killed or injured in nontrain accidents, for example, by falling down the stairs in a station.

For the early years I have been unable to discover the precise definition for either fatalities or injuries. By 1901 a passenger injury included anyone hurt sufficiently to miss a day of work while anyone who died within 24 hours of the accident counted as a fatality. Since the commission presented the data in Table A2.1 as one series, and since it usually noted any inconsistencies, I assume that these definitions obtained from the beginning.

A more serious difficulty is that the definition the commission employed counted as an injury anyone who died more than 24 hours after the accident. In 1922 the commission began to require separate reporting of such fatalities but never included them in official data. To determine subsequent fatalities for years prior to 1922 I estimated the relationship of subsequent to total (official) fatalities for 1922–65 and used the equation to calculate estimates for 1890–1921.[4]

$$Subfat_p = 22.30 + .062\,Total_p - 0.32\,Trend$$
$$(5.54) \qquad (5.24)$$
$$R^2 = 0.71,\ N = 44,\ D.W. = 1.93,\ Rho = 0.082$$

The statistical significance of the trend has two implications. First, underreporting was higher in the earlier years. Second, the decline in passenger fatality rates exceeded slightly the measured decline, probably as a result of improving medical care.

Employee Casualties

Table A2.2 presents basic data on fatalities and injuries per million manhours from 1889 through 1965. They, too, derive mostly from the appendices to the commission's *Accident Bulletin*. They also exclude subsequent fatalities, which the ICC began to publish in 1920. However, they include employees not on duty. This seems appropriate, as these included men cutting across freight yards or traveling on

Table A2.1. **Passenger Fatality and Injury Rates per Billion Passenger Miles, Selected Causes, 1889–1965**

Year	Fatalities				Injuries		
	All Causes	Including Subsequent Fatalities	Derailments	Collisions	All Causes	Derailments	Collisions
1889	26.83	30.38	2.42	9.26	185.74	33.67	38.51
1890	24.14	27.43	3.21	1.69	204.64	56.46	20.51
1891	22.81	25.93	3.82	4.59	309.25	65.17	48.51
1892	28.14	31.43	3.07	10.18	245.23	56.57	58.59
1893	21.01	23.75	1.55	4.78	226.93	54.40	54.26
1894	22.67	25.54	1.12	9.24	212.33	42.13	53.68
1895	13.95	16.49	1.39	0.66	194.86	29.37	32.66
1896	13.87	16.25	1.00	1.76	220.17	46.06	42.53
1897	18.11	20.80	3.26	4.32	228.03	31.57	50.91
1898	16.52	18.98	2.54	2.84	220.10	38.49	46.26
1899	16.38	18.71	1.44	4.18	235.90	50.85	55.86
1900	15.52	17.64	1.43	4.05	257.37	29.93	78.75
1901	16.25	18.32	3.17	3.17	287.43	43.22	84.02
1902	17.52	19.50	1.88	6.60	339.41	60.64	116.71
1903	16.97	18.89	2.10	5.64	393.53	69.71	138.22
1904	20.12	22.17	4.70	7.57	415.59	64.86	154.31
1905	22.56	24.66	6.34	8.32	439.37	121.47	146.76
1906	14.26	15.81	1.91	3.54	427.70	93.02	142.89
1907	22.01	23.95	5.74	6.96	470.47	134.13	155.02
1908	13.10	14.44	1.58	3.51	397.35	102.36	134.20
1909	8.69	9.76	1.03	2.47	354.22	84.17	93.30
1910	10.02	10.29	2.16	1.76	385.18	80.43	120.54
1911	9.01	10.00	1.23	2.08	362.69	76.02	99.30
1912	8.54	9.51	2.05	1.54	450.86	123.08	128.97
1913	10.12	11.16	0.75	3.53	460.72	107.65	116.79
1914	9.19	10.29	1.66	1.27	550.20	127.48	123.49
1915	6.16	6.97	1.30	1.21	338.06	69.54	54.92
1916	7.19	8.01	1.43	1.90	209.04	43.43	57.17
1917	7.62	8.41	0.30	2.56	192.06	43.62	59.93
1918	11.04	12.02	1.24	4.83	169.08	54.15	45.18
1919	5.89	6.51	0.26	1.83	160.84	37.99	52.96
1920	4.89	5.42	0.53	1.09	162.03	44.68	42.56
1921	2.81	3.38	0.59	1.90	149.65	30.50	32.13
1922	5.17	5.56	0.70	1.98	186.48	37.47	41.52
1923	3.53	4.36	0.84	0.24	168.09	35.33	33.19
1924	3.68	4.70	0.82	0.30	164.73	30.60	27.14

Continued

Table A2.1. Continued

Year	Fatalities				Injuries		
	All Causes	Including Subsequent Fatalities	Derailments	Collisions	All Causes	Derailments	Collisions
1925	4.53	4.87	1.69	0.61	155.42	30.77	25.35
1926	4.04	4.51	1.01	1.07	143.72	16.46	34.28
1927	2.16	2.87	0.12	0.18	134.59	17.43	26.54
1928	2.52	3.00	0.25	0.25	126.46	24.72	18.19
1929	2.95	3.30	1.12	0.03	139.74	6.48	16.46
1930	1.67	1.86	0.22	0.04	117.87	15.96	12.58
1931	1.64	2.19	0.14	0.05	121.92	10.58	11.12
1932	1.41	2.00	0.00	0.06	138.78	11.96	11.31
1933	3.05	3.73	0.55	0.86	154.08	14.42	25.35
1934	2.16	2.54	0.66	0.00	141.19	15.31	5.47
1935	1.40	1.84	0.00	0.00	135.99	10.97	8.59
1936	1.56	1.78	0.00	0.04	144.17	14.47	17.99
1937	1.26	1.66	0.08	0.04	133.35	11.70	9.60
1938	3.60	3.88	1.99	0.42	138.89	10.48	10.39
1939	1.76	2.07	0.40	0.13	146.17	18.36	14.26
1940	3.44	3.82	1.05	1.72	140.41	9.74	24.69
1941	1.43	1.70	0.54	0.03	128.58	21.49	18.23
1942	2.07	2.42	0.09	0.61	80.23	7.20	14.03
1943	3.01	3.26	1.87	0.47	70.00	14.93	11.82
1944	2.62	2.78	0.95	0.94	60.24	8.70	10.18
1945	1.56	1.71	0.24	0.46	60.84	8.51	12.33
1946	1.82	2.01	0.40	0.60	87.66	6.33	17.78
1947	1.63	1.85	0.59	0.00	111.94	16.49	9.57
1948	1.33	1.41	0.12	0.34	110.03	13.17	8.76
1949	1.05	1.37	0.17	0.00	93.70	6.03	6.69
1950	5.66	5.95	0.31	4.37	129.44	14.63	29.95
1951	4.33	4.39	2.48	1.15	111.84	28.20	10.59
1952	0.71	0.94	0.00	0.00	80.86	5.17	2.14
1953	1.55	1.61	0.63	0.03	104.86	16.38	4.83
1954	1.02	1.09	0.17	0.00	103.14	9.93	5.22
1955	0.84	1.02	0.14	0.00	107.22	9.81	4.34
1956	2.02	2.09	1.20	0.39	126.56	17.69	18.39
1957	0.58	0.62	0.00	0.12	60.43	3.86	0.00
1958	2.62	2.70	1.89	0.09	69.89	8.84	10.26
1959	0.45	0.54	0.05	0.00	61.25	7.02	2.08
1960	1.50	1.64	0.75	0.00	68.74	9.77	4.32

Continued

Table A2.1. Continued

	Fatalities				Injuries		
Year	All Causes	Including Subsequent Fatalities	Derailments	Collisions	All Causes	Derailments	Collisions
1961	0.84	0.89	0.30	0.00	92.92	18.96	5.02
1962	1.36	1.36	1.20	0.00	105.84	26.65	9.03
1963	0.70	0.76	0.00	0.16	115.18	8.69	35.59
1964	0.44	0.49	0.00	0.00	81.50	11.33	6.13
1965	0.63	0.63	0.00	0.00	68.12	6.88	3.67

Sources: Interstate Commerce Commission, *Accident Bulletin,* Appendix A, and text tables.

trains going and coming to work. Down to 1961, reported injuries include only workers who lost more than three days of work in the ten days following the accident. Thereafter all lost workday injuries are included. Finally, as noted, there were reporting changes associated with the Accident Report Act of 1910 and subsequently from time to time.

In the period to World War II I have dealt with many of these matters elsewhere,[5] and I have reproduced these data with one modification. Here I have employed the following equation to estimate the relation between reported and subsequent fatalities, 1920 to 1965, and used the results to estimate subsequent fatalities for earlier years. The equation below suggests that such fatalities would have averaged over 9 percent of the reported totals in 1920 and would have exceeded this figure in earlier years. The statistically significant negative trend probably reflects improvements in medical care.[6]

$$Subfat_W = 60.12 + 0.072\,Total_w - 0.690\,Trend$$
$$(11.17) \qquad (3.50)$$
$$R^2 = 0.94, N = 46, D.W. = 2.09, Rho = 0.201$$

In the 1930s the ICC became alarmed that the carriers' Safety First contests were causing them to underreport injuries. To check on such underreporting, beginning in 1936 the commission began to require reporting of 1 to 3 day accidents. It claimed that those carriers with low rates of "official" (reported) casualties had high rates of 1 to 3 day casualties, suggesting that they pressured men to go back to work to avoid having to report an injury. The ratio of 1 to 3 day injuries to official injuries also rose steadily, which the commission also alleged reflected underreporting. Yet I have shown that carriers with low rates of reported casualties also had low rates of 1 to 3 day casualties.[7] In addition, average lost workdays associated with reported accidents declined modestly during these years (see below). Thus if carriers did underreport official injuries they did not do so in a way that bulged less severe injuries.

Other critics have also claimed that the carriers' injury data were grossly misleading. In 1940 a committee of the railroad section of the National Safety Council compared the ratio of injuries to fatalities across carriers and found large differences. It assumed fatalities were correctly reported and charged that differences in the ratios reflected reporting differences. In 1962 the Railroad Retirement Board made a similar claim, presenting a graph showing major fluctuations in the ratio of injuries to fatalities for all carriers.[8]

There clearly was underreporting of work injuries, but other forces were also at work shaping the ratio of injuries to fatalities. For one thing, there is no reason to believe that the ratio should remain constant with fluctuations in fatalities and manhours. I estimated the following nonlinear relationship between injuries on the one hand (I), and fatalities (F) and manhours (MH) on the other, which accounts for much of the observed change in the ratio.

$$Ln(I) = 7.90 + 0.77 {}^{*}10^{-3}F - 0.11 {}^{*}10^{-6}F^2 + 0.96 {}^{*}10^{-3}MH - 0.91 {}^{*}10^{-7}MH^2$$
$$(3.23) \qquad (2.75) \qquad (4.94) \qquad (3.84)$$
$$R^2 = 0.99, \ D.W. = 1.22$$

This equation fits the data quite well, and while it is not entirely clear *why* such regularities should have existed, that they *did* suggests that fluctuations in the ratio of injuries to fatalities, both over time and across carriers, were not simply a matter of reporting. However, this equation still underpredicts injuries from about 1924 to 1929 and then overpredicts them in the mid-1930s. It seems plausible that much of this rise and fall reflects the spread of safety work. It probably was true that the rise in injuries reflected better reporting as carriers discovered that they had first to record accidents if they wished to reduce them. Similarly, much of the collapse in the 1930s probably resulted because early safety work that stressed worker training and behavior rather than workplace modification was more effective at reducing injuries than fatalities. Some suggestive evidence for this contention is that from 1924 to 1934 the ratio of injuries to fatalities fell from 7 to 4 in train accidents, and from 32 to 20 in train service accidents. But it *declined from 322 to 87* in nontrain accidents. Nontrain accidents involved mostly those in shops and yards, which was where most carriers began safety work and where it was most easily enforced.

Even if one adjusts for subsequent fatalities and less than four-day injuries, worker injury or fatality rates cannot be compared with those from other industries. In other areas of the economy reportable injuries, as defined by the American Standards Institute rule Z16.1, were simply those that "arise out of or in the course of employment." Reportable railroad injuries, however, were those "arising from the *operation of a railroad*" (emphasis added). As narrowly construed by the ICC these excluded certain activities (e.g., public relations), certain accidents (resulting from use of company trucks on a highway), and certain injuries (e.g., from repetitive use of tools). For such reasons, railroad work injury and fatality rates are probably biased downward compared to those of other industries.[9]

In the 1920s researchers who were dissatisfied with the use of injury frequency rates developed severity rates. These involved measuring lost workdays of accidents, or assigning them in some cases, and computing lost workdays per million manhours. Although the ICC never computed severity rates, it did provide severity data for train and train service (but not nontrain) injuries and fatalities. In doing so it employed the standard methods for assigning lost time to fatalities and permanent total injuries that were employed by the Bureau of Labor Statistics. I have used these to create a severity measure that is in Table A2.3. Because my series employs all manhours in the denominator, but only the lost workdays from train and train service injuries and fatalities in the numerator, it has a downward bias,

Table A2.2. **Fatality and Injury Rates, All Workers, and Trainmen, per Million Manhours, Selected Causes, 1889–1965**

Year	Fatality Rate	Including Subsequent Fatalities	Derailment Fatality Rate	Collision Fatality Rate	Injury Rate	Adjusted Injury Rate**	Fatality Rate*
	All Workers						Trainmen
1889	0.91	1.00	0.06	0.08	9.2	17.58	8.52
1890	1.04	1.14	0.05	0.08	9.48	18.01	9.52
1891	1.07	1.17	0.08	0.12	10.55	19.69	9.57
1892	0.97	1.06	0.06	0.09	10.76	20.02	8.88
1893	1.03	1.13	0.06	0.09	12.01	21.93	8.72
1894	0.73	0.81	0.04	0.06	9.44	17.95	8.26
1895	0.75	0.83	0.05	0.06	10.67	19.87	6.45
1896	0.75	0.83	0.05	0.07	12.12	22.11	6.59
1897	0.67	0.74	0.05	0.06	10.94	20.30	6.05
1898	0.74	0.81	0.05	0.07	11.96	21.87	6.68
1899	0.77	0.84	0.05	0.07	12.14	22.14	6.46
1900	0.72	0.89	0.06	0.08	12.65	22.91	7.30
1901	0.79	0.86	0.06	0.10	12.18	22.21	7.35
1902	0.80	0.87	0.06	0.11	13.58	24.34	7.42
1903	0.98	0.99	0.07	0.14	16.26	28.38	8.09
1904	0.91	0.96	0.07	0.11	16.75	29.11	8.96
1905	0.82	0.84	0.07	0.09	16.28	28.40	8.10
1906	0.82	0.89	0.07	0.10	16.04	28.05	9.20
1907	0.92	0.99	0.07	0.11	17.77	30.63	7.98
1908	0.75	0.81	0.06	0.07	18.09	31.10	6.78
1909	0.57	0.51	0.04	0.04	16.28	28.41	4.87
1910	0.68	0.73	0.05	0.07	19.13	32.65	5.41
1911	0.69	0.75	0.05	0.06	24.09	32.22	5.16
1912	0.67	0.73	0.05	0.05	26.35	34.83	4.81
1913	0.67	0.72	0.04	0.05	30.87	40.00	4.63
1914	0.62	0.68	0.04	0.04	31.64	40.86	4.63
1915	0.44	0.49	0.03	0.02	28.62	37.42	3.15
1916	0.51	0.55	0.02	0.03	30.54	39.62	4.14
1917	0.58	0.63	0.03	0.04	31.6	40.82	4.30
1918	0.59	0.64	0.04	0.05	27.02	35.59	4.39
1919	0.42	0.46	0.03	0.03	25.77	34.15	3.02
1920	0.47	0.51	0.03	0.03	27.06	35.63	3.60
1921	0.36	0.40	0.02	0.01	25.71	34.09	2.15
1922	0.39	0.42	0.03	0.02	27.63	36.29	2.27
1923	0.42	0.46	0.02	0.02	31.43	40.63	2.61
1924	0.34	0.38	0.02	0.02	28.02	36.73	1.99

Continued

Year	Fatality Rate	Including Subsequent Fatalities	Derailment Fatality Rate	Collision Fatality Rate	Injury Rate	Adjusted Injury Rate**	Trainmen Fatality Rate*
	All Workers						**Trainmen**
1925	0.36	0.39	0.03	0.02	26.8	35.34	2.14
1926	0.37	0.40	0.01	0.02	24.55	32.74	2.08
1927	0.34	0.37	0.02	0.01	19.15	26.45	1.98
1928	0.32	0.34	0.02	0.01	16.91	23.79	1.61
1929	0.34	0.37	0.01	0.02	14.38	20.75	1.88
1930	0.27	0.29	0.01	0.01	9.85	15.20	1.49
1931	0.23	0.25	0.01	0.01	7.97	12.81	1.20
1932	0.25	0.28	0.01	0.01	7.76	12.54	1.31
1933	0.25	0.28	0.02	0.01	7.41	12.09	1.26
1934	0.24	0.27	0.03	0.01	7.53	12.25	1.21
1935	0.26	0.30	0.02	0.01	7.28	11.92	1.29
1936	0.28	0.32	0.02	0.02	8.67	13.70	1.53
1937	0.26	0.29	0.02	0.02	8.88	13.96	1.44
1938	0.23	0.26	0.02	0.01	7.35	12.01	1.21
1939	0.22	0.25	0.03	0.01	7.21	11.83	1.13
1940	0.23	0.26	0.01	0.01	7.23	11.84	1.26
1941	0.28	0.30	0.02	0.02	8.90	14.98	1.70
1942	0.30	0.33	0.01	0.03	10.84	17.95	1.77
1943	0.29	0.32	0.01	0.02	12.73	20.86	1.59
1944	0.28	0.31	0.01	0.02	12.65	21.15	1.64
1945	0.26	0.28	0.01	0.02	12.80	21.20	1.49
1946	0.21	0.23	0.01	0.02	11.41	19.65	1.14
1947	0.23	0.25	0.01	0.02	10.72	18.93	1.16
1948	0.18	0.21	0.01	0.01	9.48	16.95	0.99
1949	0.16	0.17	0.01	0.01	8.06	14.50	0.78
1950	0.14	0.16	0.01	0.01	8.30	14.88	0.65
1951	0.15	0.17	0.01	0.02	8.60	15.56	0.77
1952	0.14	0.16	0.01	0.01	7.95	14.41	0.67
1953	0.13	0.14	0.01	0.01	7.71	14.31	0.58
1954	0.10	0.11	0.00	0.00	7.57	13.43	0.36
1955	0.12	0.14	0.00	0.01	8.27	14.69	0.52
1956	0.13	0.13	0.00	0.01	8.65	15.31	0.53
1957	0.09	0.10	0.00	0.01	5.77	—	0.43
1958	0.10	0.12	0.01	0.01	7.40	—	0.44
1959	0.10	0.12	0.00	0.00	8.10	—	0.36
1960	0.12	0.13	0.01	0.00	8.21	—	0.48

Table A2.2. Continued

| Year | All Workers | | | | | | Trainmen |
	Fatality Rate	Including Subsequent Fatalities	Derailment Fatality Rate	Collision Fatality Rate	Injury Rate	Adjusted Injury Rate**	Fatality Rate*
1961	0.09	0.11	0.00	0.00	13.20***	13.20	0.30
1962	0.13	0.15	0.01	0.01	13.81	13.81	0.51
1963	0.12	0.15	0.01	0.00	14.29	14.29	0.51
1964	0.14	0.16	0.00	0.02	14.80	14.80	0.43
1965	0.14	0.16	0.00	0.01	14.49	14.49	0.42

Sources: ICC, *Accident Bulletin,* various years; author's calculations.

Note: Reporting changes make data marked—not comparable with previous figures.

* Rate is per thousand trainmen.

** Adjusted for nonreporting prior to 1911 and including 1 to 3 day injuries estimated prior to 1936, see my *Safety First.*

*** All lost work day injuries reported from 1961 on.

but it does indicate trends. Reporting changes in 1961 make comparisons before and after that year impossible, however. The data clearly demonstrate that temporary injuries were becoming less severe in the 1920s and 1930s, which seems inconsistent with the allegations of underreporting, but that the improvement stopped by World War II. The broader measure of severity improved until the mid-1950s.[10]

Postwar Safety and Profitability

The postwar years reveal an inverse relationship between some aspects of safety and profitability. For 1959–65, for a large sample of companies, I relate three measures of safety—derailments, collisions, and worker casualty rates—to the return on assets. The equations for collisions control for train miles while for derailments I employ the more appropriate car miles. The casualty rate equations include man-hours as a measure of company size. In no case were collisions significantly affected by profitability, while profits were statistically significant in four of the seven years for derailments and five of the seven for casualty rates. These results probably understate the impact of profitability on safety because company derailments were affected by other companies' profits due to freight car interchange and because casualties were reported with error. Both problems bias the results toward zero. Typical results are in Table A2.4.[11]

Train Accidents

As noted, official data on train accidents begin in 1901 (fiscal 1902). For the first few years they include only collisions and derailments, but beginning in 1909 they include "miscellaneous train accidents including boiler explosions." For many years such an accident required damage of $150, which makes the numbers sensitive to changes in the price level. The commission first acknowledged this fact in 1919 in quarterly accident bulletins, but it made no change in the requirement until 1948, when the threshold was raised to $250, and then in a number of jumps to $750 in 1957, where it remained to the end of the ICC era.[12]

The constant reporting threshold of $150 resulted in a ballooning of the data on collisions and especially derailments during the inflations of World War I and

Table A2.3. **Work Injury Severity, Train and Train-Service Accidents Only, 1934–1965**

| Year | Temporary Injuries | | Permanent Injuries | | | |
	Number	Average Duration*	Number	Average Duration*	Fatalities**	Severity Rate***
1934	6,424	32	201	2,260	475	1.53
1935	6,351	32	221	2,089	517	1.64
1936	8,488	30	259	2,166	646	1.81
1937	8,812	29	246	2,290	603	1.63
1938	6,098	29	192	2,096	425	1.39
1939	6,605	31	200	2,378	442	1.38
1940	7,549	32	228	2,441	501	1.50
1941	10,634	30	311	2,419	691	1.80
1942	15,707	26	372	2,389	893	2.00
1943	21,158	24	409	2,541	913	1.90
1944	23,267	22	382	2,387	912	1.79
1945	23,788	23	412	2,585	831	1.74
1946	18,700	24	364	2,547	605	1.45
1947	18,127	24	328	2,539	627	1.46
1948	15,533	25	342	2,708	505	1.29
1949	11,343	27	245	2,573	361	1.09
1950	11,928	27	247	2,805	312	1.06
1951	13,217	27	243	2,524	361	1.11
1952	11,465	28	236	2,699	310	1.05
1953	10,439	28	227	2,554	265	0.94
1954	8,974	28	185	2,450	179	0.78
1955	10,186	28	174	2,756	235	0.95
1956	10,720	28	178	2,340	210	0.87
1957	8,179	29	162	2,171	206	0.86
1958	7,597	29	168	2,239	200	1.00
1959	8,150	28	145	2,166	188	0.95
1960	7,727	27	122	2,430	212	1.06
1961	17,351	24	250	1,394	160	1.13
1962	18,858	24	278	1,210	200	1.39
1963	19,095	25	302	1,291	156	1.29
1964	19,681	22	260	1,448	168	1.31
1965	18,400	23	221	1,474	154	1.27

Sources: ICC, *Accident Bulletin,* various years; author's calculations.

* Number of days.

** Fatalities include subsequent fatalities.

*** Severity is lost work days per million manhours.

II. On the other hand, the sharp increases in the reporting threshold after World War II artificially reduced reported accidents. To adjust for these changes I have estimated the impact of inflation and reporting requirements separately on collisions and derailments for two periods: 1902–21 and 1940–65. The variable "Correct" is a price index for railroad construction costs for 1914–56 extended forward and backward using McCusker's index, and divided by the reporting threshold. It is set to one for the first year of each period. The results are in Table A2.5.[13]

Table A2.4. **Company Profitability and Safety, 1959–1965***

	1959	1960	1961	1962	1963	1964	1965
				Derailments			
Return on assets	−284.25	−305.70	−345.43	−400.64	−86.56	−110.44	−232.74
	(2.59)	(2.36)	(2.87)	(2.37)	(0.58)	(1.11)	(−1.35)
N	76	75	74	73	74	73	62
R^2	0.67	0.62	0.60	0.65	0.62	0.64	0.53
				Log of Casualty Rate			
Return on assets	−8.35	−5.98	−10.03	−7.39	0.94	−6.57	−9.33
	(2.38)	(1.90)	(3.88)	(1.57)	(0.13)	(2.77)	(2.93)
N	76	73	73	72	74	73	60
R^2	0.16	0.14	0.24	0.12	0.07	0.19	0.31

* Casualty rates are worker injuries and fatalities per million manhours. All equations also contain a constant term; derailment equations control for car miles, and casualty-rate equations control for manhours. Figures in parentheses are *t*-ratios and are heteroskedastic-consistent.

Table A2.5. **Impact of Inflation and Reporting Requirements on Collisions and Derailments, 1902–1965**

	1902–1921		1940–1965	
Variable	Log (collisions)	Log (derailments)	Log (collisions)	Log (derailments)
Constant	−12.74	0.0096	−12.16	−5.17
Log (car miles)	—	0.912	—	1.34
		(2.63)		(7.52)
Log (train miles)	3.15	—	2.61	—
	(8.52)		(9.39)	
Log (correct)	1.71	1.178	0.673	0.863
	(10.81)	(4.34)	(6.46)	(9.34)
Time	−0.10	0.0016	0.047	0.012
	(10.10)	(0.11)	(5.98)	(2.32)
R^2	0.88	0.94	0.98	0.97
D.W.	1.88	1.18	1.71	1.75
N	20	20	26	26

Note: Figures in parentheses are *t*-ratios.

Collisions and derailments adjusted for inflation and reporting changes are simply the reported numbers with the impact of the correction factor (the variable "Correct" times its coefficient) subtracted. They are reported in Table A2.6.

Table A2.6. **Collisions, Derailments, and Other Train Accidents, 1902–1965**

Year	All Train Accidents	Collisions			Derailments		
		Reported	Adjusted	Adjusted/ Train Miles	Reported	Adjusted	Adjusted/ Car Miles
1902	8,675	5,042	5,042	5.57	3,633	3,633	0.39
1903	10,643	6,167	5,811	6.11	4,476	4,297	0.44
1904	11,291	6,436	6,065	6.22	4,855	4,661	0.47
1905	11,595	6,224	5,865	5.83	5,371	5,156	0.50
1906	13,455	7,194	6,649	6.20	6,261	5,931	0.52
1907	15,458	8,026	6,880	6.05	7,432	6,685	0.56
1908	13,034	6,363	5,556	5.08	6,671	6,078	0.55
1909	9,670	4,411	3,999	3.72	5,259	4,917	0.43
1910	11,779	5,861	4,933	4.17	5,918	5,257	0.41
1911	13,984	5,605	4,717	3.93	6,260	5,561	0.43
1912	15,743	5,483	4,373	3.65	8,215	7,032	0.54
1913	15,526	6,477	5,075	4.11	9,049	7,653	0.54
1914	15,006	5,241	4,036	3.33	8,565	7,158	0.52
1915	10,387	3,538	2,678	2.37	6,849	5,656	0.44
1916	13,990	5,737	3,418	2,.85	8,253	5,782	0.38
1917	19,435	7,497	3,111	2.51	9,991	5,461	0.34
1918	24,695	8,715	2,579	2.20	13,568	5,878	0.39
1919	25,596	6,904	1,694	1.52	15,897	6,054	0.42
1920	36,313	10,110	1,892	1.58	22,477	7,109	0.46
1921	21,251	5,102	1,452	1.33	13,615	5,741	0.46
1922	21,594	5,611			13,155		
1923	27,497	7,115			16,708		
1924	22,368	5,166			14,259		
1925	20,785	5,166			12,756		
1926	21,077	5,572			12,606		
1927	18,976	4,803			11,370		
1928	16,949	4,302			9,938		
1929	17,185	4,435			9,821		
1930	12,313	2,979			6,967		
1931	8,052	1,913			4,554		
1932	5,770	1,265			3,321		
1933	5,623	1,219			3,291		
1934	6,023	1,317			3,489		

Continued

Table A2.6. Continued

Year	All Train Accidents	Collisions			Derailments		
		Reported	Adjusted	Adjusted/ Train Miles	Reported	Adjusted	Adjusted/ Car Miles
1935	6,551	1,251			4,031		
1936	8,286	1,767			4,926		
1937	8,412	1,910			4,941		
1938	5,682	1,201			3,272		
1939	6,074	1,527			3,224		
1940	7,106	1,869	1,869	2.11	3,723	3,723	0.25
1941	9,401	2,899	2,755	2.81	4,866	4,559	0.25
1943	16,061	4,989	4,166	3.54	8,286	6,575	0.28
1944	16,258	4,867	4,051	3.41	8,673	6,853	0.28
1945	16,892	4,789	3,882	3.39	9,397	7,179	0.32
1946	15,556	4,334	3,275	3.12	8,497	5,932	0.29
1947	16,816	4,451	3,140	3.01	9,404	6,010	0.28
1948	11,893	3,112	2,915	2.91	6,791	6,245	0.30
1949	8,597	2,077	2,070	2.33	4,867	4,845	0.27
1950	10,211	2,431	2,388	2.71	5,980	5,845	0.30
1951	11,077	2,656	2,650	2.97	6,611	6,592	0.32
1952	10,085	2,514	2,602	3.04	5,783	6,044	0.30
1953	8,976	2,319	2,460	2.95	5,011	5,405	0.27
1954	7,497	1,796	1,901	2.47	4,109	4,421	0.24
1955	8,716	2,244	2,494	3.19	4,857	5,561	0.27
1956	8,447	2,393	2,632	3.41	5,369	6,067	0.30
1957	4,106	1,273	2,184	3.00	2,684	5,365	0.28
1958	3,662	976	1,639	2.51	2,579	5,016	0.29
1959	4,047	1,082	1,810	2.81	2,850	5,516	0.31
1960	4,016	989	1,639	2.64	2,918	5,580	0.32
1961	4,149	982	1,613	2.72	2,671	5,048	0.30
1962	4,378	999	1,629	2.75	2,830	5,298	0.31
1963	4,822	1,092	1,767	2.97	3,170	5,880	0.34
1964	5,317	1,229	1,972	3.26	3,399	6,233	0.35
1965	5,966	1,380	2,190	3.62	3,869	6,999	0.39

Source: Interstate Commerce Commission, Accident Bulletin. Adjusted collisions and derailments are corrected for inflation and reporting via procedures described in the text, and are also expressed per million train or car miles.

Fatalities to Trespassers and at Grade Crossing

Fatalities to trespassers in Table A2.7 also come from the appendix of ICC accident bulletins down to 1956 and from annual tables thereafter. They are probably considerably undercounted, although there is no way to know for certain. Unlike passengers and workers there is no obvious way to measure exposure for trespassers. Accordingly, I present the raw data as well as the figures relative to both train miles and population. Although the ICC also collected information on injuries to trespassers I have not presented them, as they are likely to be even more untrustworthy. Once again the official data exclude subsequent fatalities, which the ICC reported separately from 1922 on. Statistical analysis suggests that they averaged about 6 percent of reported deaths. Unlike such fatalities for passengers and employees these data exhibit no significant trend, perhaps because injured trespassers were unlikely to get the benefits of improving railroad medical care. To calculate them for earlier years I used the following equation fitted to the data for 1922–65:

$$Subfat_T = 7.63 + 0.063\,Total_T - 0.41\,Trend$$
$$(5.50) \qquad\qquad (.57)$$
$$R^2 = 0.96,\ N = 44,\ D.W. = 2.2,\ Rho = 0.604$$

Fatalities from grade crossing accidents in Table A2.7 are from the same source. They include a small number of individuals who were found to be trespassing and so there is some overlap with those data. Curiously, the ICC never collected information on subsequent fatalities from crossing accidents and so these are likely to be at least as serious an undercount as existed for trespassers. Here again I have presented the totals as well as rates relative to train miles and population. As can be seen trespassing fatalities per capita or per train mile rise slightly before collapsing about World War I while those at grade crossings peak in the late 1920s.

Rail Failures

The AREA began to collect statistics on rail failures in 1908. They are expressed as cumulative failures over a five-year period per hundred track miles for a given year's new rail purchases. These data for 1908 through 1964 (the last available) are in Table A2.8 and are also normalized for ton miles over the five-year period and expressed as an index. Table A2.8 also contains AREA figures on transverse fissures, first reported in 1919. These data are total failures during a year irrespective of the year the rail was rolled. In 1929, with the advent of Sperry's detector, the AREA began to divide failures into those that occurred in service and detected failures. Although these failures represent rails of many different ages, I have expressed them relative to ton miles of traffic during the year reported as a partial control for changes in wear.

The Social Costs of Railroad Output

The data on fatalities and injuries are in the form of risks to exposed groups, but an alternative perspective on safety is also useful. Railroad casualties are a byproduct of transportation; they do not figure into economists' productivity estimates and are usually undervalued in railroad cost accounts. Casualties per unit of output are a measure of some of the social costs of transportation, and they need not mimic measures of risk. For example, if better safety came as a result of transferring resources from output to safety it would be possible for passenger and worker risks to decline

Table A2.7. **Fatalities from Trespassing and at Grade Crossings, 1890–1965**

Year	Fatalities	Trespassing			Grade Crossings		
		Fatalities* per Train Mile	Including Subsequent Fatalities	Fatalities* per Capita	Fatalities	Fatalities* per Train Mile	Fatalities* per Capita
1890	3,062	4.25	4.52	48.6	402	0.56	6.38
1891	3,465	4.60	4.89	53.9	564	0.75	8.77
1892	3,603	4.49	4.78	54.8	568	0.71	8.65
1893	3.673	4.35	4.63	54.8	596	0.71	8.90
1894	3,720	4.81	5.12	54.5	571	0.74	8.36
1895	3,631	4.73	5.04	52.2	408	0.53	5.86
1896	3,811	4.69	4.99	53.8	615	0.76	8.69
1897	3,919	4.90	5.21	54.3	575	0.72	7.96
1898	4,063	4.80	5.11	55.3	657	0.78	8.94
1899	4,040	4.69	4.99	54.0	674	0.78	9.01
1900	4,346	5.08	5.40	57.1	730	0.85	9.59
1901	4,601	5.25	5.58	59.3	831	0.95	10.71
1902	4,403	4.85	5.16	55.6	827	0.91	10.44
1903	5,000	5.26	5.59	62.0	898	0.94	11.14
1904	5,105	5.24	5.57	62.1	808	0.83	9.83
1905	4,865	4.84	5.14	58.1	838	0.83	10.00
1906	5,381	5.01	5.33	62.9	929	0.87	10.87
1907	5,612	4.93	5.24	64.5	934	0.82	10.74
1908	5,489	5.00	5.32	61.9	837	0.76	9.44
1909	4,944	4.60	4.89	54.6	735	0.68	8.12
1910	4,864	4.11	4.37	52.6	839	0.71	9.08
1911	5,284	4.41	4.68	54.5	992	0.83	10.24
1912	5,434	4.54	4.82	57.0	1,032	0.86	10.83
1913	5,558	4.49	4.77	57.2	1,125	0.91	11.57
1914	5,471	4.51	4.79	55.2	1,147	0.95	11.57
1915	5,084	4.49	4.77	50.6	1,086	0.96	10.81
1916	4,847	4.05	4.30	47.5	1,652	1.38	16.20
1917	4,243	3.42	3.63	41.1	1,969	1.59	19.08
1918	3,255	2.78	2.95	31.5	1,852	1.58	17.95
1919	2,553	2.29	2.43	24.4	1,784	1.60	17.07
1920	2,166	1.81	1.92	20.3	1,791	1.50	16.82
1921	2,481	2.28	2.42	22.9	1,705	1.57	15.71
1922	2,430	2.21	2.35	22.1	1,811	1.65	16.46
1923	2,779	2.31	2.45	24.8	2,268	1.88	20.27
1924	2,556	2.18	2.31	22.4	2,149	1.84	18.83
1925	2,584	2.18	2.32	22.3	2,206	1.86	19.05
1926	2,561	2.12	2.25	21.8	2,491	2.06	21.22

Continued

		Trespassing				Grade Crossings	
Year	Fatalities	Fatalities* per Train Mile	Including Subsequent Fatalities	Fatalities* per Capita	Fatalities	Fatalities* per Train Mile	Fatalities* per Capita
1927	2,726	2.32	2.46	2.29	2,371	2.02	19.92
1928	2,487	2.14	2.28	20.6	2,568	2.21	21.31
1929	2,424	2.08	2.21	19.9	2,485	2.13	20.40
1930	2,409	2.19	2.32	19.6	2,020	1.83	16.41
1931	2,489	2.57	2.75	20.1	1,811	1.87	14.60
1932	2,577	3.11	3.31	20.6	1,525	1.84	12.22
1933	2,892	3.65	3.90	23.0	1,511	1.91	12.03
1934	2,697	3.28	3.51	21.3	1,554	1.89	12.29
1935	2,786	3.35	3.55	21.9	1,680	2.02	13.20
1936	2,801	3.09	3.25	21.9	1,786	1.97	13.94
1937	2,654	2.84	3.01	20.6	1,875	2.01	14.56
1938	2,360	2.85	3.00	18.2	1,517	1.83	11.69
1939	2,352	2.75	2.90	18.0	1,398	1.63	10.68
1940	2,095	2.36	2.48	15.9	1,808	2.04	13.71
1941	2,195	2.24	2.35	16.5	1,931	1.97	14.51
1942	2,013	1.82	1.87	15.0	1,970	1.78	14.71
1943	1,755	1.49	1.55	13.1	1,732	1.47	12.91
1944	1,550	1.31	1.35	11.7	1,840	1.55	13.86
1945	1,592	1.39	1.46	12.0	1,903	1.66	14.37
1946	1,635	1.56	1.63	11.7	1,851	1.76	13.21
1947	1,480	1.42	1.47	10.3	1,791	1.72	12.49
1948	1,445	1.44	1.51	9.9	1,612	1.61	11.03
1949	1,287	1.45	1.53	8.7	1,507	1.70	10.13
1950	1,215	1.38	1.44	8.0	1,576	1.79	10.42
1951	1,142	1.28	1.32	7.4	1,578	1.77	10.29
1952	1,043	1.22	1.28	6.7	1,407	1.65	9.04
1953	1,044	1.25	1.30	6.6	1,494	1.79	9.44
1954	870	1.13	1.18	5.4	1,303	1.69	8.08
1955	867	1.11	1.14	5.3	1,446	1.85	8.80
1956	818	1.06	1.09	4.9	1,338	1.73	8.00
1957	742	1.02	1.05	4.4	1,371	1.88	8.05
1958	711	1.09	1.12	4.1	1,271	1.94	7.33
1959	641	0.99	1.03	3.6	1,203	1.87	6.79
1960	617	0.99	1.03	3.4	1,364	2.20	7.58
1961	624	1.06	1.10	3.4	1,291	2.18	7.05
1962	617	1.04	1.10	3.3	1,241	2.09	6.68

Continued

Year	Trespassing				Grade Crossings		
	Fatalities	Fatalities* per Train Mile	Including Subsequent Fatalities	Fatalities* per Capita	Fatalities	Fatalities* per Train Mile	Fatalities* per Capita
1963	571	0.96	1.00	3.0	1,302	2.19	6.91
1964	619	1.02	1.07	3.2	1,543	2.55	8.07
1965	634	1.05	1.09	3.3	1,534	2.54	7.93

Sources: ICC, *Accident Bulletin,* Appendix A; author's calculations
* Rates are per million train miles or U.S. population.

Table A2.8. **Rail Failures, All Causes and Transverse Fissures, 1908–1964**

Year	All Causes		Transverse Fissures			
			Numbers		Per Billion Ton Miles	
	Five-Year Rate*	Index** per Ton Mile	Service	Detected	Service	Detected
1908	398.1	654				
1909	277.8	438				
1910	198.5	294				
1911	176.3	253				
1912	107.1	150				
1913	91.9	120				
1914	74	92				
1915	82.4	101				
1916	105.4	121				
1917	137	144				
1918	125.4	124				
1919	115.7	109	1,544		42.05	
1920	119.6	105	1,843		44.55	
1921	98.9	88	2,115		68.33	
1922	110	101	24,72		72.24	
1923	114.1	105	3,277		78.73	
1924	110.7	100	3,257		83.10	
1925	110.7	100	4,139		99.16	
1926	131.3	110	4,620		103.25	
1927	112.4	91	4,998		115.69	
1928	76.4	61	5,458		125.16	
1929	121.2	94	5,760	478	127.95	10.62
1930	60	47	5,257	1,454	136.26	37.69

Continued

Table A2.8. Continued

| Year | All Causes | | Transverse Fissures | | | |
| | | | Numbers | | Per Billion Ton Miles | |
	Five-Year Rate*	Index** per Ton Mile	Service	Detected	Service	Detected
1931	67.4	57	4,795	2,686	154.14	86.35
1932	64.1	60	4,624	2,430	196.51	103.27
1933	73.5	76	4,463	4,894	178.06	195.25
1934	35.8	42	4,994	5,514	184.76	204.00
1935	51	64	5,067	8,169	178.64	288.01
1936	50.7	62	5,924	9,126	173.63	267.48
1937	23.9	27	7,024	10,995	193.60	303.05
1938	32.6	36	5,722	13,027	196.05	446.33
1939	36.8	39	5,915	14,484	176.37	431.87
1940	37.2	37	4,882	15,372	130.06	409.52
1941	44.8	41	6,069	21,915	127.08	458.88
1942	38	30	7,407	29,848	115.56	465.65
1943	34.9	23	7,795	36,071	106.76	494.03
1944	33.3	19	6,976	31,978	94.20	431.79
1945	28.8	15	5,507	30,813	80.49	450.39
1946	25.1	13	4,238	25,831	71.23	434.18
1947	25.5	13	3,801	29,364	57.78	446.34
1948	11.5	6	3,166	38,445	49.38	599.67
1949	15.8	9	2,398	32,396	45.32	612.27
1950	19.6	11	1,767	30,303	29.87	512.26
1951	16.3	9	1,639	31,333	25.22	482.17
1952	13.2	7	1,320	30,178	21.36	488.36
1953	15.6	9	1,207	26,138	19.82	429.22
1954	10.9	6	913	29,376	16.53	531.98
1955	6.9	4	767	24,578	12.23	392.06
1956	10.1	6	837	25,718	12.85	394.94
1957	7.4	4	693	25,439	11.14	409.05
1958	9.3	5	600	23,848	10.82	430.05
1959	9.8	5	493	20,303	8.52	350.88
1960	9.6	5	506	19,460	8.79	338.22
1961	10.6	6	506	21,831	8.94	385.51
1962	11.1	7	506	21,894	8.49	367.49
1963	14.6	8	506	21,554	8.09	344.81
1964	18.4	10	506	23,256	7.64	351.25

Source: AREA Proceedings; author's calculations.

* Five-year failure rates are per 100 track miles.

** Five-year failure rate divided by ton miles over the period expressed as an index, 1925=100.

*** Annual transverse fissures discovered in service or detected and divided by ton miles.

even as fatalities per unit of output rose. Such a situation would imply rising social costs of transportation.

Table A2.9 constructs a partial measure of social costs of transportation in the form of fatalities per unit of transportation services; a more complete measure would include other social costs such as injuries and environmental damages. As the table reveals, by this measure the social costs of railroad transport fell from the 1890s on, suggesting, as the text argues, that safety cannot be separated from generalized technical change.

Table A2.9. **Social Costs of Railroad Output, 1888–1965**

Year	Fatalities per $Million of Output*	Index 1929 = 100	Year	Fatalities per $Million of Output*	Index 1929 = 100
1888	6.74	284.4	1916	4.01	169.1
1889	6.98	294.5	1917	4.15	175.1
1890	7.09	299.2	1918	3.66	154.6
1891	7.63	321.7	1919	2.99	126.2
1892	7.22	304.6	1920	2.91	122.8
1893	7.06	297.7	1921	2.49	104.9
1894	6.81	287.2	1922	2.45	103.4
1895	6.43	271.0	1923	2.57	108.4
1896	6.29	265.2	1924	2.45	103.2
1897	6.38	269.1	1925	2.49	105.2
1898	6.13	258.5	1926	2.51	105.7
1899	6.04	254.7	1927	2.54	107.0
1900	5.86	247.4	1928	2.42	102.1
1901	5.94	250.6	1929	2.37	100.0
1902	5.59	235.6	1930	2.33	98.3
1903	6.02	253.7	1931	2.49	105.1
1904	5.97	251.6	1932	2.79	117.5
1905	5.41	228.1	1933	2.81	118.3
1906	5.40	227.9	1934	2.66	112.1
1907	5.68	239.6	1935	2.70	113.8
1908	5.14	216.6	1936	2.43	102.5
1909	4.27	180.2	1937	2.44	102.8
1910	4.35	183.3	1938	2.25	95.1
1911	4.59	193.5	1939	1.91	80.6
1912	4.74	199.9	1940	1.98	83.5
1913	4.50	189.9	1941	1.83	77.0
1914	4.13	174.1	1942	1.47	62.2
1915	3.95	166.4	1943	1.23	51.8

Table A2.9. Continued

Year	Fatalities per $Million of Output*	Index 1929 = 100	Year	Fatalities per $Million of Output*	Index 1929 = 100
1944	1.16	48.8	1955	0.94	39.7
1945	1.24	52.2	1956	0.85	35.8
1946	1.47	61.9	1957	0.80	33.5
1947	1.41	59.3	1958	0.87	36.9
1948	1.24	52.2	1959	0.77	32.6
1949	1.22	51.5	1960	0.89	37.5
1950	1.15	48.4	1961	0.88	36.9
1951	1.11	46.8	1962	0.86	36.3
1952	0.97	41.0	1963	0.85	35.9
1953	0.98	41.4	1964	0.96	40.5
1954	0.95	40.2	1965	0.93	39.2

*Fatalities from all causes divided by passenger and ton miles weighted by current rates and deflated by McCusker's price index, 1860 = 100, from John J. McCusker, *How Much Is That in Real Money: A Historical Price Index for Use as a Deflator of Money Values in the Economy of the United States* (Worcester: American Antiquarian Society, 1992).

Abbreviations

AAR	Association of American Railways
AES	Association of Engineering Societies
AIME	American Institute of Mining Engineers
APHA	American Public Health Association
ARA	American Railway Association
ARABE	American Railway Association Bureau of Explosives
AREA	American Railway Engineering Association
ARMM	American Railway Master Mechanics Association
ARRJ	*American Railroad Journal*
ART	*American Railway Times*
ASCE	American Society of Civil Engineers
ASME	American Society of Mechanical Engineers
ASTM	American Society of Testing Materials
CTC	Centralized Traffic Control
DAB	*Dictionary of American Biography*
EN	*Engineering News*
HML	Hagley Museum and Library
ICC	Interstate Commerce Commission
ICE	Institution of Civil Engineers
JFI	*Journal of the Franklin Institute*
LE	*Locomotive Engineering*
MBRC	Massachusetts Board of Railroad Commissioners
MCB	Master Car Builders Association
MG	Manuscript Group
MR	*Modern Railroads*
NA	National Archives
NARC	National Association of Railway Commissioners
NCAB	*National Cyclopedia of American Biography*
NSC	National Safety Council
NYT	*New York Times*
PRRC	Pennsylvania Railroad Collection
PSA	Pennsylvania State Archives
RA	*Railway Age*
REM	*Railway Engineering and Maintenance*
RG	*Railroad Gazette*
RLC	*Railway Locomotives and Cars*

RME	*Railway Mechanical Engineer*
RMM	*Railway Master Mechanic*
RR	*Railway Review*
RRC	Reading Railroad Collection
RS	*Railway Surgeon*
RSI	*Railway Signaling*
RTS	*Railway Track and Structures*
USRA	United States Railroad Administration

Notes

Introduction

1. Thomas Johnson, ed., *The Complete Poems of Emily Dickinson* (Boston: Little, Brown, 1960), number 585. Nathanial Hawthorne, "The Celestial Railroad," in *Hawthorne's Short Stories*, ed. Newton Arvin (New York: Knopf, 1946), 234–50.
2. Quotation from "The Chatsworth Wreck," author unknown, in *Great Poems from Railroad Magazine*, ed. Robert Wayner (New York, 1968).
3. Wolfgang Schivelbusch, *The Railway Journey* (New York: Berg, 1977).
4. Fatality rates for airlines in the 1990s averaged 0.2 per billion passenger miles; in 1907 railroad passenger fatalities were about 22 per billion passenger miles.
5. Paul Slovic, "Perception of Risk," *Science* 236 (Apr. 17, 1987): 280–85.
6. A good review of risk amplification and stigmatization is James Flynn et al., eds., *Risk, Media and Stigma* (London: Earthscan, 2001).
7. For a brief discussion of the idea of a risk transition see James Mitchell, "Human Dimensions of Environmental Hazards: Complexity, Disparity, and the Search for Guidance," in *Nothing to Fear: Risks and Hazards in American Society*, ed. John Kirby (Tucson: University of Arizona Press, 1990), 131–78.
8. For a superb discussion of the origins of increasing life expectancy see Robert Fogel, *The Escape from Hunger and Premature Death, 1700–2100: Europe, America, and the Third World* (New York: Cambridge University Press, 2004).
9. For a more complete development of this approach see Walter Oi, "On the Economics of Industrial Safety," *Law and Contemporary Problems* 38 (Summer–Autumn 1974): 669–99. See also W. Kip Viscusi, *Fatal Tradeoffs* (New York: Oxford University Press, 1992).
10. For the importance of information in shaping business decisions see Naomi Lamoreaux, Daniel Raff, and Peter Temin, "Beyond Markets and Hierarchies: Toward a New Synthesis of American Business History," *American Historical Review* 108 (Apr. 2003): 404–33. For differences in organizational capabilities see Richard Langlois and Nicholas Foss, "Capabilities and Governance: The Rebirth of Production in the Theory of Economic Organization," *Kyklos* 52 (June 1999): 201–17. Joel Mokyr, *The Lever of Riches: Technological Creativity and Economic Progress* (New York: Oxford University Press, 1990), quotation on 3–4.
11. On the importance of technological change in reducing social costs see Arnulf Grübler, *Technology and Global Change* (New York: Cambridge University Press, 1998).
12. Langlois and Foss, "Capabilities and Governance," stress the importance of tacit knowledge. See also Jeremy Howells, "Tacit Knowledge, Innovation, and Tech-

nology Transfer," *Technology Analysis and Strategic Management* 8 (June 1996): 91–106. Richard Easterlin, "Why Isn't the Whole World Developed?" *Journal of Economic History* 41 (Mar. 1981): 1–19, discusses the importance of education in technology diffusion. Paul David, *Technical Choice, Innovation and Economic Growth* (London: Cambridge University Press, 1975) stresses the localized nature of technological change. The idea that engineers are inherently technology critics I get from Henry Petroski, *Invention by Design* (Cambridge, Mass.: Harvard University Press, 1996). See also William Baumol, *The Free-Market Innovation Machine* (Princeton: Princeton University Press, 2002), ch. 3. On railroad technological change generally see Steven Usselman, *Regulating Railroad Innovation* (New York: Cambridge University Press, 2002).

13. Thomas Haskell, "Capitalism and the Origins of the Humanitarian Sensibility, Part I," *American Historical Review* 90 (Apr. 1985): 339–61. Aaron Wildavsky, *Searching for Safety* (New Brunswick, 1988), 205, makes the point that "the secret of safety lies in danger."

14. For analyses of voluntarism as a response to modern environmental problems see John Maxwell et al., "Self-Regulation and Social Welfare: The Political Economy of Corporate Environmentalism," *Journal of Law and Economics* 42 (Oct. 2000): 583–617, and works cited therein.

15. On induced technological change see Vernon Ruttan and Yujiro Hayami, "Toward a Theory of Inducted Innovation," *Journal of Development Studies* 20 (July 1984): 203–23, and Vernon Ruttan, *Technology, Growth, and Development: An Induced Innovation Perspective* (New York: Oxford University Press, 2001), ch. 4.

16. JoAnne Yates, *Control by Communication: The Rise of System in American Management* (Baltimore, 1987). In ch. 4 Yates briefly discusses communication for safety. Henry Petroski, *To Engineer is Human: The Role of Failure in Successful Design* (Vintage, 1992). Theodore Porter, *Trust in Numbers: The Pursuit of Objectivity in Science and Public Life* (Princeton: Princeton University Press, 1995).

17. Crisis of control is from James Beniger, *The Control Revolution* (Cambridge, Mass.: Harvard University Press, 1986), ch. 6. Charles Perrow, *Normal Accidents* (New York: Basic, 1984). For a review that also criticizes Perrow's emphasis on the technology see Larry Hirschorn, "On Technological Catastrophe," *Science* 228 (May 17, 1985): 846–47. See also James Reason, *Managing the Risks of Organizational Accidents* (Burlington, Vt.: Ashgate, 1997).

18. "Rules for Travelers," *Scientific American* 5 (May 4, 1850): 1.

Chapter 1. In the Beginning

Epigraphs: Captain Douglas Galton, *Report to the Lords of the Committee of Privy Council for Trade and Foreign Plantations on the Railways of the United States* (HMSO, 1857), 26. Alfred Bunn, *Old England and New England in a Series of Views Taken on the Spot* 1 (London, 1853), 282.

1. For the beginnings of the B&O see James Dilts, *The Great Road: The Building of the Baltimore and Ohio, the Nation's First Railroad, 1828–1853* (Stanford: Stanford University Press, 1993), ch. 1. For the concept of a technological style see Thomas Hughes, "Evolution of Large Technological Systems," in *The Social Construction of Technological Systems,* ed. Wiebe Bijker et al. (Cambridge, Mass.: MIT Press, 1987), 51–83.

2. Franz Anton Ritter von Gerstner, *Early American Railroads (Die innern Com-*

municationen [1842–1843]), ed. Frederick Gamst (Stanford: Stanford University Press, 1997), 284.

3. Dionysius Lardner, *Railway Economy* (New York: Harpers, 1850), quotation on 334.

4. Charles Ellett, "Exposition of the Causes Which Have Conduced to the Failure of Many Railroads in the United States," *ARRJ* 14 (Feb. 1, 1842): 78–84. John C. Trautwine, "Remarks on the Injudicious Policy Pursued in the Construction and Machinery of Many Railroads in the United States," *JFI* 33 (May 1842): 307–16 and (June 1842): 370–80, quotation on 380, italics in original. "Extracts from a Report of H. Allen to the Board of Directors of the South Carolina Canal and Railroad Company, Feb. 6, 1832," *ARRJ* 1 (Mar. 17, 1832): 180. Not all contemporaries asserted that American methods were best in America while European methods were best in Europe. Zerah Colburn and Alexander Holley, *The Permanent Way and Coal Burning Locomotive Boilers of European Railways* (privately printed, 1858) argued that the saving in operating costs would justify American carriers in following the European lead and investing more in their permanent way. Some writers also claimed that American methods would be economical in Europe. For general discussions of these issues see Howard White, "European Railways—as they Appear to an American Engineer," ASCE *Transactions* 3 (1873): 61–66. See also Edward Bates Dorsey, "English and American Railroads Compared," ASCE *Transactions* 15 (1886): 1–77, 733–90, and Robert Gordon, "On the Economical Construction and Operation of Railways in Countries where Small Returns Are Expected, as Exemplified by American Practice," ICE *Proceedings* 85 (1886): 54–85.

5. Good discussions of early practice are also contained in J. L. Ringwalt, *Development of Transportation Systems in the United States* (Philadelphia, 1888), 82–87. See also J. Elfreth Watkins, "Development of American Rail and Track," ASME *Transactions* 22 (1890): 209–32. British signaling is described in William Preece, "On Railway Telegraphs and the Application of Electricity to the Signaling and Working of Trains," ICE *Proceedings* 22 (1862–63): 167–92; Richard Rapier, "On the Fixed Signals of Railways," ICE *Proceedings* 38 (1873–74): 194–247; Richard Blythe, *Danger Ahead* (London: Newman Neame, 1951). "Railroad Management," *ARRJ* 26 (Sept. 3, 1853): 564–65. "Railroad Accidents," *Scientific American* 3 (Mar. 25, 1848): 210.

6. Lardner, *Railway Economy,* 334 and 336. "Report of A. A. Dexter and C. E. Detmold, Civil Engineers to the Committee on the Preliminary Survey of the Upper Route of the Columbia Railroad," *ARRJ* 3 (Nov. 15, 1834): 706–7.

7. Lardner, *Railway Economy,* 336. On American locomotives see David Barnes, "Distinctive Features and Advantages of American Railway Practice," ASCE *Transactions* 29 (1893): 384–425. The best modern treatment is John White, *American Locomotives: An Engineering History 1830–1880* (Baltimore: Johns Hopkins University Press, 1997).

8. Lardner, *Railway Economy,* 337. For good surveys of the development of American rolling stock during the antebellum period see Ringwalt, *Development of Transportation Systems,* 90–103. The definitive modern treatments are John White, *The American Railroad Passenger Car* (Baltimore: Johns Hopkins University Press, 1978) and *The American Railroad Freight Car* (Baltimore: Johns Hopkins University Press, 1993). For British passenger cars see Hamilton Ellis, *Nineteenth Century Railway Carriages* (London: Modern Transport, 1949).

9. Galton, *Report to the Lords of the Committee of Privy Council for Trade and Foreign Plantations,* quotations on 3, 11, 12, 13, 14, 16, 26.

10. Charles Oliffe, *American Scenes: Eighteen Months in the New World,* trans. Ernest Falbo and Lawrence Wilson (Lake Erie College Studies, 1964), 108. Bunn, *Old England and New England* I, 282, 283, 291. Leon Beauvallet, *Rachel and the New World* (New York: Dix Edwards, 1856), quotations on 200, 295–96. Charles Dickens, *American Notes* (New York: Charles Scribners, 1911), 135. Aleksandr Lakier, *A Russian Looks at America,* trans. Arnold Schrier and Joyce Story (Chicago: University of Chicago Press, 1979), 74–75. Charles MacKay, *Life and Liberty in America* I (London: Smith, Elder, 1859), 216. Weld is quoted in August Mencken, *The Railroad Passenger Car* (Baltimore: Johns Hopkins Press, 1957), 121.

11. L.T.C. Rolt, *Red For Danger,* 4th ed. (London: David & Charles, 1982) provides a good history of British railroad accidents and safety. For a more complete analysis of the kind of induced innovation discussed in the text see Vernon Ruttan, *Technology, Growth and Development, An Induced Innovation Perspective* (New York: Oxford University Press, 2001), ch. 4.

12. The quotation is from "How Are Railroads to be Managed?" *ARRJ* 27 (Dec. 23, 1854): 808. The discussion of railroad work culture is based on the following. James Ducker, *Men of the Steel Rails* (Lincoln: University of Nebraska Press, 1983). Walter Licht, *Working for the Railroad: The Organization of Work in the Nineteenth Century* (Princeton: Princeton University Press, 1983). Frederick Gamst, *The Hoghead: An Industrial Ethnology of the Locomotive Engineer* (New York: Holt, Rinehart, and Winston, 1980), and his "The Railroad Apprentice and the 'Rules': Historic Roots and Contemporary Practices," in *Apprenticeship: From Theory to Method and Back Again,* ed. Michael Coy (New York: State University of New York Press, 1989), 65–86. I also employed the many reminiscences of railroad workers. See also Michale Taillon, "'What We Want is Good Sober Men': Masculinity, Respectability and Temperance in the Railroad Brotherhoods, c. 1870–1910," *Journal of Social History* 36 (Winter 2002): 319–38.

13. Railways in the United Kingdom killed about 1.5 individuals per million dollars of revenue (1£ = $4.90) compared with 7.75 on Massachusetts lines, using dollars of 1860 purchasing power.

14. The signaling practice associated with these accidents is from "Railroad Management," *ARRJ* 26 (Sept. 3, 1853): 564–65, and "Precaution Against Accidents," *ARRJ* 26 (Aug. 6, 1853): 502.

15. "Dreadful Collision of Railroad Cars," *Springfield* [Massachusetts] *Republican* (Oct. 9, 1841), 2. "Collision of Cars on the Boston & Worcester Railroad," *Springfield Republican* (Nov. 20, 1841), 2. For *Scientific American* see "Double vs. Single Track Railways," *ART* 4 (Jan. 8, 1852): 3.

16. Catherine Delaney is from New York Railroad Commissioners, *Annual Report, 1855* (Albany, 1856), 219.

17. Mrs. Walters is from New York Railroad Commissioners, *Annual Report, 1855* (Albany, 1856), 217. For the other accidents see New York Railroad Commissioners, *Annual Report, 1852* (Albany, 1853), 76–80.

18. "Breakage of Railway Axles," *ARRJ* 27 (Apr. 1, 1854): 202.

19. "Late Accident on the Erie Railroad," *ARRJ* 31 (July 24, 1858). 412; "The Trial by Frost of Iron on Railroads," *ARRJ* 29 (Feb. 23, 1856): 125. "Report of the Committee Appointed to Examine and Report the Causes of Railroad Accidents, the

Means of Preventing their Recurrence, &c," New York State Senate *Document* 13, 1853 (Albany 1853), 59. [Hereafter, New York Senate *Document* 13.]

20. Lakier, *A Russian Looks at America*, 74–75. "Such baseness" is from "Railroad Accidents," *ARRJ* 18 (May 8, 1845): 297. Henry Dawson, *Trips in the Life of a Locomotive Engineer* (New York: J. Bradburn, 1863), 124–27. "Queer Railroad Accident," *ARRJ* 19 (Dec. 5, 1846): 778. Casey is from Mencken, *The Railroad Passenger Car*, 111.

21. Bunn, *Old England and New England*, I, 291. Gratton is quoted in Mencken, *The Railroad Passenger Car*, 103–4.

22. New York Senate *Document* 13, 7. Dilts, *The Great Road*, 269.

23. "Railroad Homicides," *Harper's Weekly* 4 (June 23, 1860): 386. "Accidents on Railroads," *ARRJ* 28 (Oct. 27, 1855): 680. These accidents are all in "Returns of the Railroad Corporations of the State of Massachusetts for 1848," Senate *Document* 40 (Boston, 1849).

24. On the Board of Trade see Henry Parris, *Government and the Railways in Nineteenth Century Britain* (London: Routledge and Kegan Paul, 1965). New York Senate *Document* 13, 3.

25. Henry Petroski argues that failure is instructive in *To Engineer is Human: The Role of Failure in Successful Design* (New York: Vintage, 1985). The best source on the institutions shaping railroad safety is Steven Usselman, *Regulating Railroad Innovation* (New York: Cambridge University Press, 2002).

26. "Risks of Railway Travel," *ART* 4 (Apr. 29, 1852): 1. The *Standard* is quoted in "Responsibility for Accidents," *ART* 8 (July 31, 1856): 2.

27. Early laws are summarized in W. P. Gregg, *The Railroad Laws and Charters of the United States*, 2 vols. (Boston, 1851). For a discussion of early commissions see Edward C. Kirkland, *Men, Cities, and Transportation* (Cambridge, Mass.: Harvard University Press, 1948). "Report of the Committee on Railroads and Canals in their Examination in Relation to the Accidents Upon the Western Rail-road," Massachusetts State Senate *Document* 55 (Boston, 1842). New York Senate *Document* 13.

28. New York Senate *Document* 13, 41–42. Accident statistics are the sorts of "recipes" that Thomas Haskell notes underlay the rise of humanitarian movements. See his "Capitalism and the Origins of Humanitarian Sensibility, Part I," *American Historical Review* 90 (Apr. 1985): 339–61. Barbara Welke, *Recasting American Liberalism* (New York: Cambridge University Press, 2001), 10, also makes this point.

29. "Railroads in Connecticut," *ARRJ* 28 (July 7, 1855): 418–20. The *Republican* is quoted in *ART* 4 (Apr. 22, 1852): 1. New York Railroad Commissioners, *Annual Report, 1855* (Albany, 1856), i–viii.

30. This discussion of antebellum railroad accident law is based on the following. Isaac Redfield, *Practical Treatise Upon the Law of Railroads*, 2d ed. (Boston: Little, Brown, 1858). Christopher Patterson, *Railway Accident Law: The Liability of Railways for Injuries to the Person* (Philadelphia: T. & J. W. Johnson, 1886). James Ely, *Railroads and American Law* (Lawrence: University of Kansas Press, 2001), ch. 9. John Witt, *The Accidental Republic* (Cambridge, Mass.: Harvard University Press, 2004), ch. 4. "A Case Four Times Tried and Seven Times Appealed," *NYT* (Feb. 9, 1887), 1. For diversity jurisdiction see Edward Purcell, *Litigation and Inequality: Federal Diversity Jurisdiction in Industrial America* (New York: Oxford University Press, 1992). Earnings are from *Historical Statistics of the United States* series D-735.

31. Many early cases are reported in the *ARRJ* in a section entitled "Journal of Railroad Law." The quotation is from "Improvements—Adoption of them By Railroad Companies," *ARRJ* 33 (Jan. 14, 1860): 29. For cases that apply this rule see Hegeman vs. Western Railroad, 113 New York *Reports* 9, 1855, Smith vs. New York and Harlem Railroad, 19 New York *Reports* 127, 1859, and Alden vs. New York Central Railway, 26 New York *Reports* 102, 1862.

32. Norwalk is from "The South Norwalk Bridge Accident," file A-13-7, box 20, record group 1, New York, New Haven & Hartford Collection, University of Connecticut. Corning is from "Liability of Railroad Companies for the Carelessness of their Servants," *ARRJ* 22 (May 12, 1849): 295, and "Railway Decisions—Damages for Injury," *ART* 2 (May 1, 1850): 1. Earnings are from Robert Margo, *Wages and Labor Markets in the United States, 1820–1860* (Chicago: University of Chicago Press, 2000), Tables 3A.5 and 3A.12.

33. "Railroad Accidents," *ARRJ* 18 (Sept. 25, 1845): 619, "The Money Value of a Small Boy," *RG* 30 (Sept. 9, 1898): 643. Earnings are from *Historical Statistics of the United States* series D-735.

34. On the higher standard for children see Patterson, *Railway Accident Law,* 193–94, and Peter Karsten, *Heart vs. Head: Judge Made Law in Nineteenth Century America* (Chapel Hill: University of North Carolina Press, 1997), ch. 7. On the influence of sex see Welke, *Recasting American Liberalism.* Several boxes of accident and damage claim files of the Reading Railroad at the Hagley Museum and Library contain numerous examples of small payments to injured employees and others.

35. For the American Society of Civil Engineers see Daniel Calhoun, *The American Civil Engineer, Origins and Conflict* (Cambridge, Mass.: MIT Press, 1960). Haskell, "Capitalism and the Origins of Humanitarian Sensibility."

36. For the importance of the *Journal* and Poor's role see Alfred Chandler, *Henry Varnum Poor, Business Editor, Analyst, and Reformist* (Cambridge, Mass.: Harvard University Press, 1956).

37. For early coupler technology see White, *American Railroad Passenger Car,* ch. 7. Edwin Price, "Journal or Memoirs, 1929–1901," collection #70, National Museum of American History Archives, Smithsonian Institution.

38. "Safety Platforms—Bumpers and Couplings," *ARRJ* 36 (Nov. 28, 1863): 1124. "Security from Railroad Accidents," *ARRJ* 38 (Mar. 4, 1865): 201–2. The Erie is from "The Miller Platform," *ARRJ* 40 (Mar. 16, 1867): 245; New York's law from "Prevention of Accidents," *ARRJ* 40 (Aug. 10, 1867): 769. Ohio Commissioner of Railroads and Telegraphs, *Annual Report, 1872* (Columbus, 1873): 27. For the Allegheny Valley see "Miller's Platform, Buffers and Couplers—Telescoping of Cars Prevented," *ARRJ* 43 (June 4, 1870): 644–45. "The Miller Platform," *ARRJ* 43 (Aug. 13, 1870): 901. Diffusion from "Miller's Platforms, Buffers, and Couplers," *ARRJ* 47 (Dec. 26, 1874): 1633.

39. Quotation from "Miller's Platforms, Buffers, and Couplers," italics in original.

40. "J. K. Smith's Self-Acting Brakes for Railroad Cars," *ARRJ* 5 (Apr. 2, 1836): 193–95, which also contains the Reading's brake. "Griggs' Hand and Pulley Brake," *ARRJ* 21 (Oct. 28, 1848): 691–92; "Compressed Air Railway Brake," *ARRJ* 21 (Oct. 21, 1848): 680. The best surveys of the evolution of railway equipment are John White, *The American Railroad Passenger Car,* ch. 7, and *The American Railroad Freight Car,* ch. 7. For the dangers to brakemen see Henry Dawson, *Trips in the Life of a Locomotive Engineer* (New York: J. Bradburn, 1863), 79.

41. My description of the earliest form of brake is from W. V. Turner and S. V.

Dudley, *Development in Air Brakes for Railroads* (Pittsburgh: Westinghouse Air Brake Company, 1909). For more detail and subsequent modifications see Al Krug, "North American Freight Train Brakes Page," http//www.railway-technical.com/brake2.html [accessed June 14, 2005].

42. Henry G. Prout, *A Life of George Westinghouse* (New York: Arno, 1972), and Francis Leupp, *George Westinghouse: His Life and Achievements* (Boston: Little, Brown, 1918). The description of the vacuum brake is from ARMM, *Proceedings* 7 (1874): 251–54; the quotation is on 266–67.

43. Good discussions of the strengths and weaknesses of the various brakes are in ARMM, *Proceedings* 5 (1872), 7 (1874), and 8 (1875). The 75 percent figure is from White, *The American Railroad Passenger Car*, 551.

44. The Brimfield accident is from MBRC, *Fifth Annual Report, 1873* (Boston, 1874), 34–36. The train speed of 40 miles per hour at the time of impact is a guess, as there is inadequate information to estimate it. Doubling of freight train speed is from "Air Brakes and Freight Train Speed," *RG* 26 (Apr. 20, 1894): 286–87.

45. Early developments are from Watkins, "Development of American Rail and Track," and "Pennsylvania Railroad," *ARRJ* 37 (Feb. 27, 1864): 202. *Report of the President and Managers of the Philadelphia & Reading Railroad Co to the Stockholders* (Philadelphia, 1870). *Eighteenth Annual Report of the Board of Directors of the Pennsylvania Railroad* (Philadelphia, 1865), 10.

46. On rail joints see Benjamin H. Latrobe, "Remarks Upon the Defects of Railway Track and Their Remedy," *ARRJ* 24 (Mar. 1, 1851): 120–21. Origins of fish plates from "Discussion on Joints of Railways," *ARRJ* 30 (Aug. 8, 1857): 498–500, and Watkins, "Development of American Rail and Track." For their evolution see "The Growth of the Angle Bar," *RG* 45 (Nov. 6, 1908): 1284–85. On compound rails see "New Rail for Railroads," *ARRJ* 24 (Mar. 15, 1851): 161–63, and "Patent Compound Rail," *ARRJ* 24 (Apr. 12, 1851): 238–40.

47. "Railway Sketches," *ART* 6 (Sept. 21, 1854): 2; "Tyler's Patent Safety Switch," *ARRJ* 22 (Aug. 25, 1849): 538; "Best switch" is from "The Tyler Switch," *RG* 3 (May 6, 1871): 61. "Wharton Safety Switch," *ARRJ* 40 (June 8, 1867): 533–34. "The Wharton Patent Safety Railroad Switch," *RA* 4 (Dec. 25, 1879): 616–17.

48. The Franklin Institute study is reprinted and evaluated in Bruce Sinclair, *Early Research at the Franklin Institute: The Investigation into the Causes of Steam Boiler Explosions* (Philadelphia: Franklin Institute, 1966). Stame is from "Report on the Causes of Steam Boiler Explosions," *ARRJ* 22 (Dec. 15, 1849): 789. The electrical rod is from "Boiler Explosions," *ART* 11 (Sept. 3, 1859): 1. For subsequent investigations see "Explosion of the Locomotive 'Richmond,'" *ARRJ* 18 (Feb. 6, 1845): 91–92; "Explosion of the Locomotive 'Engineer,'" *ARRJ* 21 (Aug. 26, 1848): 554. The Schenectady Locomotive Works found low water the cause when it investigated the explosion of one of its locomotives on the Saratoga & Schenectady. See "The Explosion of the Boston," *ARRJ* 23 (Mar. 30, 1850): 210–12.

49. For the work of the English iron commission see Nathan Rosenberg and Walter Vincenti, *The Britannia Bridge: The Generation and Diffusion of Technological Knowledge* (Cambridge, Mass.: MIT Press, 1978). For it and the work of Wöhler see also Stephen Timoshenko, *History of the Strength of Materials* (New York: McGraw Hill, 1953). For contemporary American discussions see "Report of the Commissioners Appointed to Inquire into the Application of Iron to Railway Structures," *JFI* 49 (May 1850): 289–302, and (June 1850): 361–71. "An Account of

the Construction of the Britannia Bridge," *ARRJ* 23 (Jan. 19, 1850): 35–36. "Application of Iron to Railroad Structures," *ARRJ* 24 (Mar. 8, 1851): 147–48, 178–79, 213–14. Edward Philbrick, "Iron Girder Bridge for the Boston & Worcester Railroad," *JFI* 71 (Mar. 1861): 156–60.

50. Ignorance of quality is from "Accidents on Railroads," *ARRJ* 24 (Feb. 15, 1851): 97. Purchase on trust is from "The Quality of Railroad Car Wheels and Rails—How to Test Them," *ARRJ* 32 (Aug. 27, 1859): 550–51. For reporting of the English experiments see "Application of Iron to Railroad Structures," *ARRJ* 24 (Mar. 1, 1851): 147–48, 178–79, 213–14. The quotation is from New York Senate *Document* 13, 38–39.

51. "Kite's Patent Safety Beam," *ARRJ* 11 (Oct. 22, 1840): 242–43 is an editorial endorsement that also contains a testimonial from the Philadelphia, Wilmington & Baltimore. There is an advertisement in the same issue. For the New York & New Haven see "Necessary Precautions," *ARRJ* 24 (Oct. 11, 1851): 648. "Tapping Car Wheels," *RG* 25 (Feb. 24, 1893): 139 is a brief historical review. Boston & Lowell from "Railroad Management—Precautions for Safety," *ARRJ* 26 (Nov. 26, 1853): 764. New York roads are from New York Senate *Document* 13.

52. British tests are from "Testing Railway Axles," *ARRJ* 27 (Jan. 27, 1854): 61. For rolling of axles see "Railway Axles," *ARRJ* 26 (July 16, 1853): 400. Their use on the Pennsylvania and other lines is from "Rolled vs. Hammered Axles," *ARRJ* 33 (Nov. 24, 1860): 1034, which also claims that some companies dated their wheels and axles. For government tests see "Relative Strength of Different Specimens of American Iron," *ARRJ* 32 (Mar. 12, 1859): 169. Comparative tests are from "Test of Railroad Axles in Detroit," *ARRJ* 25 (Aug. 15, 1854): 522. *Report of the President and Managers of the Philadelphia & Reading Railroad Co., to the Stockholders, Jan. 11, 1875* (Philadelphia, 1875).

53. For the Utica & Schenectady and many other early lines see von Gerstner, *Early American Railroads*. The Western is from Stephen Salsbury, *The State, the Investor, and the Railroad* (Cambridge, Mass.: Harvard University Press, 1967), ch. 9. Charles O'Connell, "The Corps of Engineers and the Rise of Modern Management, 1827–1860," in *Military Enterprise and Technological Change,* ed. Merritt Roe Smith (Cambridge, Mass.: MIT Press, 1985).

54. "Report of Alexander Black, Commissioner, to the Stockholders of the South Carolina Canal and Railroad Company," *ARRJ* 2 (May 25, 1833): 322–23. For the Tonawanda and Boston & Providence see Von Gerstner, *Early American Railroads.*

55. Personal accountability (emphasis in original) is from Alfred D. Chandler, *The Railroads: The Nation's First Big Business* (New York: Harcourt Brace, 1965), 101–8. See also "Railroad Management," *ARRJ* 29 (Apr. 12, 1856): 225–26, which contains the second quotation. Alfred Chandler and Stephen Salsbury, "The Railroads: Innovators in Modern Business Administration," in *The Railroad and the Space Program,* ed. Bruce Mazlish (Cambridge, Mass.: MIT Press, 1965), 153–64.

56. The rules of the Baltimore & Susquehanna are from "Railway Police—Collisions," *ARRJ* 13 (Sept. 15, 1841): 161–67, quotation on 164. For the Ogdensburg see "Railroad Accidents," *ARRJ* 25 (Apr. 17, 1852): 243. Western & Atlantic from "Train Operation Sixty Years Ago," *RG* 55 (Aug. 22, 1913): 339–41. "Explosion of Locomotive Engine on the Camden & Amboy," *ARRJ* 22 (Mar. 9, 1850): 153. For the Central of Georgia see "How they Ran Trains Forty Years Ago," *RA* (May 13, 1898): 326. For the wreck on the Camden & Amboy see Robert Reed, *Train Wrecks* (New York: Bonanza, 1982), 155, and *Report of Executive Committee to*

the Directors of the Delaware and Raritan Canal and the Camden and Amboy Transportation Companies on the Accident of the Twenty-Ninth of Aug., 1855 (Trenton, 1855.) On early timetables and train rules see Warren Jacobs, "Early Rules and the Standard Code," Railway and Locomotive Historical Society *Bulletin* 50 (Oct. 1939): 29–55, and Ian Bartky, "Running on Time," *Railroad History* 159 (Autumn 1988): 19–38.

57. Gamst, "The Railroad Apprentice and the 'Rules.'" "Report of the Committee on Rail-roads and Canals in their Examination in Relation to Accidents Upon the Western Rail-road." Salsbury, *The State, the Investor, and the Railroad.* Bartky, "Running on Time," discusses the spread of watch inspection.

58. For early use of the telegraph in Great Britain see Preece, "On Railway Telegraphs and the Application of Electricity to the Signaling and Working of Trains." Joseph Bromley, *Clear the Tracks!* (New York: Whittlesey House, 1943), quotation on 4.

59. Erie's introduction of the telegraph is summarized in "Railroad Management," *ARRJ* 29 (Apr. 12, 1856). "The Late Disaster on the North Pennsylvania Railroad," *ARRJ* 29 (Apr. 26, 1856): 473. Use of the telegraph on other lines is from "Telegraphs on Railroads," *ARRJ* 31 (Oct. 23, 1858): 677.

60. "The Late Accident on the New York Central Railroad," *ARRJ* 29 (Sept. 13, 1856): 583, contains the accident and the claim that the Erie fired enginemen for such wrecks. For the B&O see Dilts, *The Great Road,* 270. David Lightner, *Labor on the Illinois Central* (New York: Arno, 1977), quotation on 110. For the Columbus & Xenia see "Prizes to Engineers," *ARRJ* 27 (Jan. 28, 1854): 52. The Nashville & Chattanooga and Western & Atlantic are from Edwin Price, "Journal or Memoirs," 25 and 48.

61. Price, "Journal or Memoirs," 43. Oliver Williamson, *The Economic Institutions of Capitalism* (New York: The Free Press, 1985), ch. 10. For the Alton accident see Herbert Pease, *Singing Rails* (New York: Thomas Y. Crowell, 1900), 39.

62. Lightner, *Labor on the Illinois Central,* 113. Licht, *Working for the Railroad,* ch. 4.

63. For the Providence & Worcester see Reed, *Train Wrecks,* 20.

64. "Railway Accidents," *ART* 5 (May 12, 1853): 2. "What Causes Railway Accidents," *ARRJ* 31 (Aug. 7, 1858): 505 also contains quotations from the *Evening Journal.* Seymour Dunbar, *A History of Travel in America* (Indianapolis: Bobbs-Merrill, 1915), vol. 3, ch. 46. Reed, *Train Wrecks,* chs. 1 and 2.

65. For a comparison of the ICC and *Gazette* data see "Our Accident Statistics," *RG* (June 13, 1890): 419–20, and untitled, *RG* (June 28, 1899): 541.

66. For the Boston & Worcester and Eastern see Robert Shaw, *Down Brakes* (New York: P. R. Macmillan, 1961), appendix. Charles Francis Adams Jr. also stressed luck in *Notes on Railroad Accidents* (New York: Putnam's, 1879), 11. A Poisson distribution is usually employed to compute the likelihood of rare events. The probability of there being sixteen accidents by chance alone if the expected frequency is 4 is given by $P(16) = (e^{-4} \times 4^{16})/16(!)$. This is about 3.75 in a million.

67. The reported antebellum disasters all involve passenger trains and so it is appropriate to express them relative to passenger train miles. However, a number of the disasters reported by the *Gazette* involved work trains, thus biasing upward the number of disasters relative to passenger train miles for the years following the Civil War.

68. Train size for 1839 calculated from data in Von Gerstner. Massachusetts data from "Returns of the Railroad Corporations of the State of Massachusetts for 1848," Senate *Document* 40 (Boston, 1849), 29–30. An example may clarify the

point in the text about the effect of increases in average train size. If train size were normally distributed with a mean of 55 passengers and a standard deviation of 24, only 2.5 percent of all trains would contain as many as 102 passengers (2 standard deviations above the mean). If the mean now rises to 64 (a 16% increase) with the same standard deviation, the number of trains holding at least 102 people will double.

Chapter 2. Off the Tracks

Epigraphs: E. E. Russell Tratman, "The Relations of Track to Traffic on American and Foreign Railways," New York Railroad Club *Proceedings* 7 (Dec. 19, 1896): 7–80, quotation on 7. "Railway Accidents to Persons—II," *EN* 27 (Feb. 20, 1892): 179–80.

1. For the Bradford wreck see MBRC, *Twentieth Annual Report, 1888* (Boston, 1889), appendix C. "The Haverhill Accident on the Boston and Maine Railroad," *EN* 19 (Jan. 14, 1888): 24. "The Boston & Maine Wreck," *RR* 28 (Jan. 21, 1888): 34–35.

2. For the weight and nearly everything else about freight cars see John White, *American Railroad Freight Car* (Baltimore: Johns Hopkins University Press, 1993). See also J. L. Ringwalt, *Development of Transportation Systems in the United States* (Philadelphia, 1888).

3. For comparisons of ICC and *Gazette* statistics see "Our Accident Statistics," *RG* (June 13, 1890): 419–20, and untitled, *RG* 31 (June 28, 1899): 541.

4. Train accidents and casualties are author's calculations derived from data in the *Railroad Gazette.* Passengers per train are for 1882 and were calculated from data in ICC Bureau of Statistics, "Railway Statistics Before 1890," *Statement No. 13251* (Washington, 1932).

5. "Train Accidents in 1895," *RG* 28 (Feb. 14, 1896): 112–13. The number of passengers per train declined from forty-seven to forty-one over this period. Had this decline not occurred, fatalities per train would have been 0.13 for derailments and 0.16 for collisions in the latter period. It is also unlikely that the decline in accident severity simply reflected more complete reporting of minor accidents, for the proportion of accidents with no casualties showed no change from the late 1870s to the late 1890s.

6. For a recent review of thinking about social capital see Partha Dasgupta, *Social Capital: A Multifaceted Approach* (Washington: World Bank, 2000). The quotation is from "The Railway Master Mechanics Association" *ART* 21 (Oct. 2, 1869): 318.

7. Gavin Wright, "Can a Nation Learn? American Technology as a Network Phenomenon," in *Learning By Doing in Markets, Firms and Countries,* ed. Naomi Lamoreaux et al. (Chicago: University of Chicago Press, 1999), 295–326.

8. On standards see Charles Kindleberger, "Standards as Public, Collective, and Private Goods," *Kyklos* 36 (June 1983): 377–96. A listing of associations is in "The Development of Local Engineering Societies," *EN* 37 (Jan. 21, 1897): 41–42. "Focal Points" is from "Railway Clubs and Their Work," *RMM* 10 (Sept. 1887): 139–40. See also "The Rise of the Railroad Club," *EN* 31 (May 17, 1894): 410–11. "Among the Most Useful" is from "The Evlolution of the Railroad Club," *RG* 26 (Sept. 28, 1894): 670. "Much Good" is from "An English View of American Special Technical Associations," *RG* 15 (Mar. 30, 1883): 195–96.

9. For an analysis of technology sharing that stresses complementarities see William Baumol, *The Free Market Innovation Machine* (Princeton: Princeton University Press, 2002). Robert Allen, "Collective Invention," *Journal of Economic Behavior and Organization* 4 (Mar. 1983): 1–24. See also Peter Meyer, "Episodes of Collective Invention," U.S. Bureau of Labor Statistics *Working Paper 368* (Washington, 2003). The origins of the laboratories are from "Railway Chemical Laboratories," *RMM* 10 (Feb. 1887): 20. For the Pennsylvania see "Departments of Physical and Chemical Tests," box 661, PRRC. For biographical material on Dudley see ASTM, *Memorial Volume Commemorative of The Life and Life Work of Charles B. Dudley* (Philadelphia, 1911). Dudley's work on steel rails is discussed in Steven Usselman, *Regulating Railroad Innovation* (Cambridge: Cambridge University Press, 2002), ch. 6.

10. For the career of Lanza see *Who Was Who*, 1, 705; for Swain see *DAB*, S 1, 680, and *NCAB*, 12, 276. Swain's work for the Massachusetts railroad commission is noted below in ch. 5. For Goss see *NCAB*, 20, 16, and Merriman, *NCAB*, 23, 70–71. For Cooper see *NCAB*, 19, 261–62, and *DAB*, 4, 413; Chanute is in *NCAB*, 10, 212, and *DAB*, 4, 10.

11. William Acworth, *American and English Railways Compared* (Chicago: Railway News Bureau, 1898), 14.

12. For Forney see *DAB*, 6, 527–28, and *NCAB*, 22, 78. "Matthias Nace Forney," *RG* 44 (Jan. 17, 1908): 76–84, and "M.N. Forney," *EN* 59 (Jan. 23, 1908): 93–95, both discuss the significance of his work at the *Gazette*. See also "Autobiographical Notes by Matthias Nace Forney," *RG* 44 (Jan. 17, 1908): 82–84. Angus Sinclair is in *Who's Who*, 1, 1129; for Staufer see *DAB*, 17, 538, and *NCAB*, 9, 45. "Arthur Mellen Wellington," *EN* 33 (May 23, 1895): 337–38. See also *NCAB*, 11, 168, and *DAB*, 19, 634. The quotation is from Arthur Mellen Wellington, *Economic Theory of the Location of Railways* (New York: John Wiley, 1887), 1. For Camp see *NCAB*, 16, 392–94; for Fowler, *Who Was Who*, 1, 419; for Tratman, *Who Was Who*, 5, 730. Usselman, *Regulating Railroad Innovation*, chs. 5 and 9.

13. "Train Accidents in 1878," *RG* 11 (Jan. 24, 1879): 46. "Train Accidents in 1882," *RG* 15 (Feb. 16, 1883): 107. MBRC, *Fourth Annual Report, 1872* (Boston, 1873): 11–13.

14. MBRC, *Fifth Annual Report, 1873* (Boston, 1874), 145.

15. On the role of phosphorous in making rails brittle at low temperatures see Robert Thurston, *A Treatise on Iron and Steel* (New York: John Wiley, 1884), 558–60. For Huntington's remarks see "Chairs and Sub-sills—Something for Consideration," *RG* 3 (Nov. 4, 1871): 316. Alexander Holley, *American and European Railway Practice* (New York: D. Von Nostrand, 1860), 154.

16. Pennsylvania Railroad, *Thirty-Second Annual Report, 1879* (Philadelphia, 1879), 105. For the Grand Trunk see MBRC, *Fourth Annual Report, 1872* (Boston, 1873), 150–52. "American Practice as to Rails and Ballast," *RG* 17 (Jan. 16, 1885): 43–44. "Standard Track of American Railways," *EN* 35 (June 25, 1896): 410–12. "Standards of Track Construction on American Railways," *EN* 44 (Aug. 30, 1900): 142–49. "Gravel, Washed Gravel, and Broken Stone for Ballast," *RR* 49 (Apr. 10, 1909): 332–33.

17. "Really ought not" from untitled, *RG* 22 (Feb. 7, 1890): 94. Passengers' efforts from "Expansion of Rails by Heat," *RR* 38 (Oct. 8, 1898): 557. MBRC, *Third Annual Report, 1871* (Boston, 1872), xcii–xciv.

18. On tie preservation see "Preserving Cross-ties," *RG* 12 (May 21, 1880): 272–73. "The Treatment of Railroad Tie Timber," *RA* 34 (Sept. 12, 1902): 261–62; "The Crosstie Problem," *RA* 36 (July 17, 1903): 65.

19. Twenty-seven percent from "Why Do Rail Joints Break?" *RG* 17 (Apr. 3, 1885): 212. For the decline in broken joints see "Rail Joints," *RG* 24 (Oct. 14, 1892): 768–69. For tie plates see "Value of Tie Plates and Lock Spikes," *RR* 31 (Apr. 25, 1891): 271. "A Review—Tie Plates Up to Date," *RA* 20 (Oct. 11, 1895): 11–12, and "Tie Plates—Benefits and Results to be Obtained by Their Use," *RR* 38 (Sept. 24, 1898): 530–32. A good discussion of switches is W. B. Parsons, *Track, A Complete Manual of Maintenance of Way* (New York: Engineering News, 1886). For use of tie plates, guard rails, and safety switches see the surveys cited in n. 16.

20. For two of many such examples see "Bull Causes Wreck," *RR* 44 (July 16, 1904): 517, and "Bull Attacks a Train," 44 *RR* (Oct. 8, 1904): 715. Southern Minnesota from "Employees and Operations," *RR* 20 (Mar. 13, 1880): 130. "Report of the Connecticut Railroad Commissioners," *RG* 1 (July 9, 1870): 349. "Accidents to Trespassers," *RG* 19 (Dec. 23, 1887): 831. For a good discussion of the law of fencing see James W. Ely, *Railroads and American Law* (Lawrence: University of Kansas Press, 2001), ch. 5. See also Earl Hayter, "The Fencing of Western Railways," *Agricultural History* 19 (July 1945): 163–67, which is the source of the Texas & Pacific and Missouri, Kansas & Texas examples.

21. The poem is excerpted from "Things and Other Things," *RA* 34 (July 25, 1902): 92. For a similar but longer duet see "The 'Cow Coroner' Yearns," *RA* 17 (Apr. 15, 1892): 304.

22. "The Cattle Guard Problem," *EN* 30 (July 27, 1893): 78; untitled, *EN* 27 (June 9, 1892): 573.

23. Pennsylvania Railroad, *Eighteenth Annual Report, 1864* (Philadelphia, 1865), 10. For the Grand Trunk see MBRC, *Fourth Annual Report, 1872* (Boston, 1873), 150–52.

24. Quality problems are noted in MBRC, *Eighth Annual Report, 1876* (Boston, 1877), 133–34. Quotations from "Durability and Tests of Rails," *ART* 21 (May 1, 1869): 142, and "Accidents on American Railways," *RG* 1 (Aug. 27, 1870): 507–8. "Steel vs. Iron Rails," *ARRJ* 48 (Feb. 6, 1875): 165 notes use of combination rails on the Reading. "Cast Iron Rails," *ARRJ* 46 (Nov. 8, 1873): 1394. Philadelphia, Wilmington & Baltimore Railroad, *Thirty-Ninth Annual Report, 1876* (Philadelphia, 1877), 13. "Steel Rails," *ARRJ* 42 (May 22, 1869): 590.

25. Pennsylvania Railroad, *Twentieth Annual Report, 1866* (Philadelphia, 1867), 26. Chicago & Alton Railroad, *Eleventh Annual Report, 1873* (Chicago, 1874), 8. Ashbel Welch, "Comparative Economy of Iron and Steel Rails," *JFI* 60 (1870): 22–32, and his "On the Form, Weight, Manufacture, and Life of Rails," ASCE *Transactions* 3 (1874–75): 87–105, and "Comparative Economy of Steel Rails with Light and Heavy Heads," ASCE *Transactions* 10 (1881): 251–63. For modern discussions see Jeremy Atack and Jan Brueckner, "Steel Rails and American Railroads, 1867–1880," *Explorations in Economic History* 19 (1982): 339–59, and subsequent responses and replies.

26. The best discussion of companies' motives for introducing steel rails is Usselman, *Regulating Railroad Innovation,* ch. 6. He shows that companies replaced steel on main lines several times before first putting it on secondary track.

27. For a review of European literature see "The Strength of Iron in Construction," *RG* 9 (May 5, 1877): 202–3. "Effects of Punching Rails," *ART* 21 (Feb. 20, 1869): 23. "Poor Steel Rails," *RG* 17 (July 24, 1885): 467.

28. The economics of standards is discussed in Hemenway, *Industrywide Voluntary Product Standards.* Testing on the Pennsylvania is from James Dredge, *Pennsylvania Railroad* (London, 1879).

29. Robert W. Hunt, "Proposed Rail Sections," *American Institute of Mining Engineers* 17 (1889): 778–85. Hunt was an early and persistent advocate of the position that physical treatment of rails was more important than their chemistry. For his career see *NCAB*, 19, 17. "Blue Shortness," *RG* 23 (Aug. 7, 1981): 459, notes specifications forbid working at less than red heat. "Effect of Temperature on the Tensile Strength and Ductility of Metals," *RG* 24 (May 6, 1892): 323–24 et seq. Albert Sauveur, *The Metallurgy and Heat Treatment of Iron and Steel*, 2d ed. (Cambridge: Cambridge University Press, 1918). For Sauveur see *DAB*, 22, 594–95, and *NCAB*, 29, 135–36.

30. For the Philadelphia, Wilmington & Baltimore see "Steel Rails," *ARRJ* 42 (May 22, 1869): 590. "Rail Testing Machines," and "Tests of Bessemer Rails and Axles," *ART* 21 (Feb. 27, 1869): 67 and (July 10, 1869): 222. "Testing Machines Used on Railroads," *ARRJ* 49 (July 15, 1876): 896. Ward's remarks from "Rail Tests and Weight of Rails," *RG* 20 (Sept. 7, 1888): 583–84.

31. For an analysis that discusses the role of analogy see Trevor Pinch, "'Testing— One, Two, Three . . . Testing!': Toward a Sociology of Testing," *Science, Technology, & Human Values* 18 (Winter 1993): 25–41. Inappropriate tests are from "Tensile Tests for Rails," *RG* 18 (June 25, 1886): 445–46. Michigan Central from "Poor Steel Rails," *RG* 17 (July 24, 1885): 471–72, quotations on 471; "No one has yet been able" from "How to Get Good Rails," *RG* 21 (Jan. 25, 1889): 62–63, which also stresses the value of record keeping.

32. For the Philadelphia, Wilmington & Baltimore see its *Fortieth Annual Report, 1877* (Philadelphia, 1878), 40, and "Steel Rails," *ARRJ* 42 (May 22, 1869): 590. "Manufacturer has no control" from "Hunt's Rail Specifications," *Iron Age* 43 (Jan. 24, 1889): 120–21.

33. For the spread of testing see AREA *Proceedings* 2 (1901): 192. W. H. Hippel to E. C. Felton, Dec. 29, 1891, William H. Brown to Luther S. Bent, Feb. 27, 1893, box 4, PRRC, HML. James J. Hill to Walter Scranton, Aug. 23, 1892, file 133, box 12, 132.e.7.10.f, President's Subject Files, Great Northern Railroad Collection, Minnesota Historical Society.

34. "Broken Rails," *RG* 4 (Mar. 16, 1872): 119. "One prolific cause" is from "The Weight of Rails," *RG* 4 (Mar. 30, 1872): 138–39. For rail weight during the antebellum period see Holley, *American and European Railway Practice*, 161. "American Practice as to Rails and Ballast," *RG* 17 (Jan. 16, 1885): 43–44. "Standard Track of American Railways," *EN* 35 (June 25, 1896): 410–12. "Standards of Track Construction on American Railways," *EN* 44 (Aug. 30, 1900): 142–49.

35. "Track Inspection with Dudley's Dynagraph," *RG* 13 (May 20, 1881): 274–75. Its use on other lines is from "Annual Inspection of the Boston & Albany—the Dynagraph," *RG* 19 (Oct. 29, 1887): 695. "Plimmon H. Dudley, C. E., Ph.D.," *RR* 74 (Mar. 8, 1924): 465–66. "Standards of Track Construction on American Railways," *EN* 59 (June 4, 1908): 600–607.

36. "Light Rails," *RG* 17 (Mar. 13, 1885): 168–69, quotation on 169. "Cross Ties," *RG* 17 (Apr. 10, 1885): 230–31.

37. "Cross Ties," *RG* 17 (Apr. 10, 1885): 230–31, quotation on 231. In the text and the figure I have rounded some of the numbers. Wellington also noted that there were physical limits to the number of ties per mile and that rail weight bought durability as well as stiffness.

38. For premiums see "Track Inspection and Premiums," *RG* 17 (Nov. 13, 1885): 730. Inspection on the New York Central is from "Premium System of Track Inspection," *RG* 34 (Dec. 26, 1902): 979. The Pennsylvania is from William Sipes, *The*

Pennsylvania Railroad (Philadelphia, 1875), appendix. For the B&O see "Track Standards and Rules for the Government of Trackmen on the Baltimore & Ohio R.R.," *EN* 26 (Oct. 31, 1891): 416–17. "The Annual Inspection of the Boston & Albany," *RG* 30 (Nov. 4, 1898): 789. Premiums are also noted in "The Physical Characteristics of the Lake Erie & Western," *EN* 33 (Apr. 25, 1895): 279–80, and "Track Inspection on the L&N," *RR* 38 (Jan. 8, 1898): 4–5.

39. MBRC, *Twenty-Second Annual Report, 1890* (Boston, 1891), 113–56. Not all companies learned from the Old Colony's mistake. A section man was killed removing a jack from in front of an onrushing train on the Southern in 1908. "Accident with a Track Jack," *RR* 48 (Feb. 8, 1908): 107.

40. "Train Robbing and Train Wrecking," *RG* (July 7, 1893): 506. The robberies on the Southern Pacific and Union Pacific are from "Bon Arizona Derailment," and "Derailment at Saugus, California," both in Train Robbery Related Records, Mss C-A-372, Bancroft Library, University of California at Berkeley.

41. For Tratman see *Who Was Who,* 5, 730. E. E. Russell Tratman, "The Relations of Track to Traffic on American and Foreign Railways," New York Railway Club *Proceedings* 7 (Dec. 19, 1896): 7–80, quotation on 7. "Needed Improvements in Railway Track and Maintenance of Way," *EN* 52 (Sept. 8, 1904): 219–20, quotation on 220. "When track" is from "Railway Accidents to Persons—II," *EN* 27 (Feb. 20, 1892): 179–80. For an argument that European standards remained uneconomic see "The Construction and Maintenance of Railway Embankments," *EN* 48 (Oct. 9, 1902): 292–93. "The Physical Characteristics of the Lake Erie & Western R.R.," *EN* 33 (Apr. 25, 1895): 279–80. "Very large mileage" is from W. M. Camp, *Notes on Track* (Chicago, 1903), 153.

42. The inspections are in Missouri Railroad and Warehouse Commissioners, *Twenty Fourth Annual Report, 1899* (Jefferson City, 1900), and *Twenty-Fifth Annual Report, 1900* (Jefferson City, 1901). Victor Morse, *36 Miles of Trouble: The Story of the West River R.R.* (Brattleboro, Vt., 1959), 19.

43. Details on freight car size may be found in White, *The American Railroad Freight Car.*

44. MCB *Proceedings* 16 (1882): 47–55.

45. ARMM, *Annual Report* 13 (1880): 37–38. "Average practice" is from "The Standard Axle and Reasons for Adopting It," *RG* 11 (May 23, 1879): 284–85. "The Car Axle," *LE* 32 (Nov. 1919): 330–32. Barnes is from "The Strength of Car Axles," *RR* 35 (May 18, 1895): 270–71.

46. Empiricism is stressed in "The Breakage of Car Wheels," *RG* 13 (May 13, 1881): 264–65. For British accident experience see C. F. Allen, "Wheels and Axles, and Their Relation to Track," *RG* 20 (Mar. 23, 1888): 183–84. For one of many articles urging use of safety chains see "The Use of Safety Chains on Car Trucks," *RG* 5 (Nov. 8, 1873): 452–53. For car proportions see "American and English Freight Cars," *RG* 16 (Feb. 8, 1884): 101–2. George Swain, "The Proper Method of Comparing the Cost of Chilled-Iron and Steel-Tired Car Wheels," *Technology Quarterly* 1 (May 1888): 409–19.

47. For early foundry practice see George Fowler, "The Car Wheel," in Block Signal and Train Control Board, *Third Annual Report, 1910* (Washington, 1911), 135–75. The 30-ton car is from an untitled article, *RMM* 10 (Nov. 1887): 163. For other examples of gross overloading see "Cast Iron Wheels Under 50 Ton Cars," *RG* 33 (May 24, 1901): 352, and White, *American Railroad Freight Car,* 179. Three to one is from "Are Higher Grade Car Wheels Necessary?" *RR* 41 (Nov. 16, 1901): 737.

48. For different size wheels see "Inaccuracies of Chilled Car Wheels," *EN* 15 (Jan. 2, 1886): 22. "Irregularities in the Gage of Car and Engine Wheels," *RMM* 16 (May 1893): 78. The MCB report is quoted in "The Strength of Car Truck Bolsters," *RMM* 18 (Feb. 1895): 21. On use of hand brakes see T. A. Griffin to ed., *RG* 33 (June 14, 1901): 391. The position of side bearings is from "One Cause of Derailments," *RG* 42 (June 14, 1907): 823–24.

49. "Car Wheels," *RG* 15 (Apr. 20, 1883): 248, and "The Breakage of Car Wheels," *RG* 15 (June 1, 1883): 348–49. Ely is quoted in "The First Step Toward Safe Wheels," *RG* 20 (May 25, 1888): 337–38. On information and contracting problems generally see Oliver Williamson, *The Economic Institutions of Capitalism* (New York: The Free Press, 1985).

50. Yehuda Grunfeld, "The Effect of the Per Diem Rate on the Efficiency and Size of the American Railroad Freight Car Fleet," *Journal of Business* 32 (Jan. 1959): 52–73. For interchange rules see MCB *Proceedings* 6 (1872): 124–26. Poor maintenance is noted in its *Proceedings* 9 (1875): 74–76; complaints about repair prices are in *Proceedings* 13 (1879): 52–53. For the role of prices for cleaning brakes in causing wheel failures see "To Raise the Efficiency of the Air Brake," *RG* 33 (July 5, 1901): 486. Problems of improperly gauged wheels, racked trucks, and mismatched wheels are in "The Breakage of Car Wheels," *RG* 13 (May 13, 1881): 254–55, and "Wheel Flanges," *RR* 33 (Mar. 25, 1893): 181–82. The later study is "How Pooling Equipment Increased Car Maintenance Costs," *RR* 61 (July 30, 1921): 147. "No amount of care" is from "Cast Iron Wheel Failures and Inspection," *RR* 48 (June 27, 1908): 534.

51. For the Erie see "Cast Iron Wheels," *RR* 16 (Apr. 25, 1884): 194–95. Best and worst makers are from "Bad Wheels and Bad Wheel Makers," *RG* 17 (June 5, 1885): 360–61. Fifty percent from "Cheap Car Wheels," *RG* 17 (Apr. 24, 1885): 265–66. See also "Effect of Quality and Make of Wheels on Sharp Flanges," *EN* 20 (Nov. 10, 1888): 361. The *Review*'s remarks are in "The Boston & Maine Wreck," *RR* 28 (Jan. 21, 1888): 34–35. The accident is from an untitled news item, *RG* 19 (July 27, 1887): 484. Pennsylvania experience from MCB *Proceedings* 31 (1897): 130–31.

52. Manufacturing techniques from "Steel Axles," *EN* 51 (May 19, 1904): 471–72. For testing on the Pennsylvania see "Bessemer Steel Axles," *ARRJ* 42 (June 26, 1869): 706, and "Car Axle Tests on the Pennsylvania," *ARRJ* 56 (Feb. 1886): 336. "The Chemical Composition of Steel Axles," *RR* 42 (July 13, 1901): 483–84. "Steel Axles," *RR* 44 (May 21, 1904): 372–73; "The Car Axle," *LE* 32 (Nov. 1919): 330–32. "Breakage of Axles, *LE* 8 (Apr. 1895): 216–17, quotation on 216.

53. See "William Forsyth," *DAB*, 20, 237, and "George Gibbs," *NCAB*, E, 120.

54. MCB *Proceedings* 30 (1896): 149–65. See also "The Car Axle," *LE* 32 (Nov. 1919): 330–32.

55. For the arrangement with Purdue see "Report of Standing Committee on Tests of MCB Couplers," MCB *Proceedings* 34 (1900): 185–88. For Goss see *NCAB*, 20, 16. For faculty membership on the eve of World War I, with the dates they joined, see "Associate Members," MCB *Proceedings* 49 (1915): xxxi. For typical financial arrangements see "Minutes of the Executive Committee," MCB *Proceedings* 43 (1909): 532–33.

56. For Ely's letter see MCB *Proceedings* 22 (1888): 133. The standard is on 80–87. The test was hardly demanding. Consider a lot of a hundred wheels, of which half are defective. The probably of getting one is 0.5 and the probably of drawing three bad wheels is (about) $0.5^3 = 0.125$, or about a one in eight chance of

rejecting the lot. For European tests see "The Quality of Chilled Cast-Iron Wheels—Special Quality for Heavy Service," *RR* 41 (Sept. 7, 1901): 590–91.

57. For early testing on the Pennsylvania see Dredge, *Pennsylvania Railroad*, 88–89. Wheel tests are from "Drop Testing Machine for Car Wheels," *RA* 16 (July 25, 1891): 578. For other lines see "Testing Laboratories for Railways," ARMM *Annual Report* 24 (1891): 64–83 (which contains the results of testing on the B&O), and G. W. Beebe, "Cast Iron Wheels," *EN* 44 (Oct. 18, 1900): 266. Specifying wheel chemistry is from untitled, *RMM* 10 (Feb. 1887): 17. The Great Northern is from "A Railway Laboratory," *RA* 17 (Sept. 9, 1892): 709–10. For Barr see "The First Step." "Generally purchasing" is from "Cast Iron Wheels," *RG* 21 (July 5, 1889): 439.

58. An early complaint about testing is in P. H. Griffin, "The Manufacture of Chilled Wheels," *RG* 38 (June 16, 1885): 68–70. A plea for minutely drawn specifications that cites the Pennsylvania as a good example is "Wheel Tests," *RME* 20 (Apr. 1897): 47. Complaints about rigged tests are from "The MCB Coupler," *RG* 30 (Apr. 8, 1898): 253–54. Origins of the thermal test are from "The Chilled Cast Iron Wheel," *RG* 30 (Apr. 29, 1898): 303–4. For the Southern Pacific see "Report on Wheel Tests Made at Sacramento Shops," MCB *Proceedings* 26 (1892): 115–27. See also "Specification and Guarantee for Cast Iron Wheels," *EN* 37 (June 10, 1897): 367–68, which discusses both the thermal test and a safety versus durability trade-off as does G. R. Henderson, "The Manufacture of Car Wheels," ASME *Transactions* 20 (1898–99): 615–27. "Report of the [MCB] Committee on Standards of the Association," *RR* 32 (June 18, 1892): 396.

59. For the New York Car Wheel Works see "The Quality of Chilled Cast-Iron Wheels—Special Quality for Heavy Service," *RR* 41 (Sept. 7, 1901): 590–91. The Burlington is from "Cast Iron Car Wheels," *EN* 44 (Oct. 18, 1900): 266. "Without aid of chemistry" is from "Chilled Cast Iron Wheels," *RG* 34 (Mar. 28, 1902): 227. "Car Wheel Foundry Practice; Canadian Pacific," *RA* 49 (Oct. 7, 1910): 645–55.

Chapter 3. Collisions and the Rise of Regulation

Epigraphs: "Rear Collisions," *RG* 16 (Nov. 7, 1884): 804–5. Mark Twain, "The Danger of Lying in Bed," *Galaxy* 2 (Feb. 1871): 317–19.

1. "The Republic Disaster," *RR* 27 (Feb. 26, 1887): 118–19; "United States Railway Catastrophes," *RME* 63 (Mar. 18, 1887): 216–17. "Safety Appliances for Fast Trains," *RG* 19 (Mar. 4, 1887): 144–45.

2. For Adams's approach to regulation see Thomas McCraw, *Prophets of Regulation* (Cambridge, Mass.: Harvard University Press, 1984), ch. 1.

3. For the Revere disaster and the rule changes see MBRC, *Third Annual Report, 1871* (Boston, 1872), xcv–cv, and Appendix F. *Twenty-First Annual Report, 1889* (Boston, 1890), 33 and 35.

4. McCraw, *Prophets of Regulation.* quotation on 35. McCraw ignores the parallels between Adams's approach and the Accidents Reports Act. Garfield is from "Government Inspection of Railroad Accidents," *RG* 9 (Feb. 16, 1877): 74. See also U.S. 44th Cong., 2d sess, *Congressional Record* 5 (Jan. 31, 1877): 1139–40.

5. "Railroad Accidents and Railroad Commissioners," *RG* 15 (Sept. 21, 1883): 622, is a scathing indictment of railroad investigations of their own accidents. "Government Investigation and Inspection," *RR* 27 (Oct. 22, 1887): 61–62. "Government Investigation of Railway Accidents," *EN* 22 (Oct. 26, 1889): 398–99.

6. The number of passengers per train declined from forty-seven to forty-one over this period. Hence the proportion killed in a collision fell from about 0.5 percent to 0.3 percent.

7. Robert Wayner, *Great Poems from Railroad Magazine* (New York, 1968), 127.

8. "The Westinghouse Continuous Brake," *JFI* 113 (Jan. 1882): 64–66. For the Long Island see "The Westinghouse Brake Tests," *EN* 18 (Nov. 26, 1887): 385–86.

9. A simple discussion of kinetic energy is in Lewis Sillcox, *Mastering Momentum* (New York: Simmons Boardman, 1936), ch. 1, which contains the first equation. The second equation is from "Equated Emergency Stops—Burlington Brake Tests," *RG* 24 (Feb. 19, 1892): 139–41.

10. Ohio Commissioner of Railroads and Telegraphs, *Annual Report* (Columbus, various years). For car weight see John White, *American Railroad Passenger Car* (Baltimore: Johns Hopkins University Press, 1978). For driver brakes see MBRC, *Twenty-Second Annual Report, 1890* (Boston, 1891), 133. The reader should not confuse these figures with ICC data that report locomotives equipped with *train* brakes, not driver brakes. "The Mud Run Disaster," *RG* 20 (Oct. 19, 1888): 681–82 and 689; "The Mud Run Disaster," *LE* 1 (Nov. 1888): 11. "Some First Principles," *RG* 20 (Oct. 26, 1888): 705–6.

11. "A gentler way" is from "The Silver Creek Catastrophe," *RG* 18 (Sept. 17, 1886): 643; "decidedly mild" from same title (Sept. 24, 1886): 655.

12. The Erie is from New York Railroad Commissioners, *Annual Report*, various years. "Fast and Unusual Runs from 1880–1906," *RG* 41 (Nov. 2, 1906): 395. The express is from "The Highest Train Speed Ever Maintained Regularly," *LE* 10 (Nov. 1897): 822. Captain Douglas Galton, "On the Effect of Brakes Upon Railway Trains," Institution of Mechanical Engineers *Proceedings* (June 1878): 467–89; (Oct. 1878): 590–616; and (Apr. 1879): 170–218.

13. "The Westinghouse Brake Tests," *EN* 18 (Nov. 26, 1887): 385–86, quotation on 385. "The Brake Trials," *RG* 19 (June 3, 1887): 366–67. The Southern Pacific is from E. J. Urquhart to J. A. Fillmore, Aug. 3, 1888, box 2, Letcher Collection, Ms 268, Stanford University. The sleeping cars are from "Air Brakes—When and How They Fail," *RMM* 13 (Dec. 1890): 205.

14. "The Massacre at May's Landing—an Engineer's Confession," *RG* 12 (Aug. 20, 1880): 441. "Last 25 or 50 feet . . ." from "The Emergency Brake and the Garrison Accident," *RG* 29 (Nov. 19, 1897): 817. "A School for Brakes," *RG* 14 (Nov. 17, 1882): 713. "Air Brake Instruction," *RR* 35 (Sept. 7, 1895): 498.

15. The Boston & Albany is from MBRC, *Twenty-First Annual Report, 1889* (Boston, 1890), 40–41.

16. For the Old Colony see ch. 2. Swain's report is in MBRC, *Twenty-Second Annual Report, 1890* (Boston, 1891), 136–56. Gaetano Lanza, "The Application of Brakes to the Truck Wheels of a Locomotive," ASME *Transactions* 16 (1894): 69–74. "Brakes for High Speed Trains," *RG* 32 (Dec. 24, 1892): 809.

17. For a description of the trials see Charles Clark, "The Railroad Safety Movement in the United States: Origins and Development, 1869–1913" (Ph.D. diss., University of Illinois at Champaign, 1966).

18. "Report of Committee on Metal for Brake Shoes," MCB *Proceedings* 25 (1891): 50–68. "Report of Committee on Air Brake Tests," MCB *Proceedings* 27 (1893): 223–43. "Report of Committee on Air Brake Tests," and "Report of Committee on Laboratory Tests of Brake Shoes," MCB *Proceedings* 28 (1894): 32–62, 149.

19. Good discussions of the various tests are in Westinghouse Air Brake Company,

Air Brake Tests (Pittsburgh, 1905). See also Walter Turner and S. W. Dudley, *Development in Air Brakes for Railroads* (Westinghouse Air Brake Co., 1909), and "The High Speed Brake," *RG* 31 (Apr. 26, 1899): 299. "The Galton Westinghouse Experiments to Determine the Effect of Brakes," *RG* 19 (June 10, 1887): 381–83. The Fourth Avenue Tunnel is from "The High Speed Brake in Relation to an Accident," *RG* 34 (Jan. 17, 1902): 42. Diffusion of the high-speed brake is from an untitled article *RG* 36 (Jan. 29, 1904): 80. "Heavy passenger trains" from "The High Speed Brake," *LE* 21 (July 1908): 302.

20. Walter V. Turner, "The Air Brake as Related to Progress in Locomotion," *JFI* 170 (Dec. 1910): 461–94, quotation on 479.

21. "The Rio Disaster," *RG* 18 (Nov. 5, 1886): 758–59.

22. "The Accident at Kout's Station," *RG* 19 (Oct. 14, 1887): 673. Quotation from "The Crusade Against Car Lamps," *RMM* 9 (Apr. 1886): 50. George Gibbs, mechanical engineer of the Milwaukee, was also skeptical of the dangers from oil light. See his "Methods of Railway Car Lighting," *EN* 25 (Mar. 14, 1891): 246–47. A survey found that thirteen of twenty-four fires during passenger train wrecks came from stoves while one came from a lamp. See H. S. Haines, "Railroad Accidents, Their Causes and Prevention," *RG* 25 (June 30, 1893): 483–86.

23. "Heating Railroad Cars," *RG* 3 (Nov. 11, 1871): 331, contains a quote from the *Post*. "Eight Charred Victims," *NYT* (Jan. 15, 1882), 1. "The Terrible Car Stove," *NYT* (Oct. 30, 1886), 4, and "The Car Stove Once More," *NYT* (Feb. 6, 1887), 6, which contains the quotation. "Heating and Lighting Railroad Cars," *ARRJ* 55 (June 3, 1882): 370.

24. For some early history of car heating see "Western Railway Club Discussions on Steam Car Heating," *RR* 27 (Dec. 24, 1887): 733–34. "Heating Railroad Cars," *RG* 3 (Nov. 11, 1871): 331, contains the temperature differential. Drop testing the heater is reported in "The Drop Testing Machine," *RA* 47 (July 16, 1909): 88–89. The best modern discussion is John White, *The American Railroad Passenger Car* (Baltimore: Johns Hopkins University Press, 1978), ch. 5.

25. Massachusetts developments are recounted in MBRC, *Nineteenth Annual Report, 1887* (Boston, 1888), 55–65; *Twentieth Annual Report, 1888* (Boston, 1889), 1–18 and 267–300. *Twenty-Second Annual Report, 1890* (Boston, 1891), 6–10. Connecticut Railroad Commissioners, *Thirty-Fifth Annual Report, 1887* (Hartford, 1888), 41–45; *Thirty-Sixth Annual Report, 1888* (Hartford, 1889), 43–48. Maine Railroad Commissioners, *Thirty-Second Annual Report, 1890* (Augusta, 1891), 12–14. For company experiences see "Steam Heating in a Blizzard—The Necessities Imposed by Extreme Conditions," *RR* 28 (Jan. 28, 1888): 46–47, and "Steam Heating in Extreme Weather," *RR* 28 (Jan. 11, 1888): 71–72.

26. Lanza's report is in MBRC, *Twentieth Annual Report, 1888* (Boston, 1889), 267–300. "Steam Heating in Michigan," *RG* 21 (Jan. 18, 1889): 46. The comments on President Clark are from an untitled story, *EN* 25 (Mar. 7, 1891): 228–29.

27. White, *American Railroad Passenger Car*, 394, discounts the role of state pressure, claiming that the "railroads appear to have converted at their own pace." "Discreditable" is from Edward Kirkland, *Men, Cities and Transportation* (Cambridge, Mass.: Harvard University Press, 1948), 368. "The Present Status of Car Heating in the United States," and "Status of Car Heating in the United States," *RR* 31 (Apr. 11, 1891): 228–29 and 237–38, quotation on 237.

28. "The Reading Accident," and "The Shoemaker Disaster," *RG* 22 (Sept. 26, 1890): 666 and 670.

29. The Rock Island is from "Handling Freight Trains Partially Equipped with Air," *RMM* 21 (Jan. 1898): 7. "Report of the Committee on Trains Parting," MCB, *Proceedings* 31 (1897): 107–11.

30. Untitled ed., *RG* 33 (Sept. 10, 1901): 542.

31. The defective brake that caused a wreck is from "A Real Brake Failure and a Moral," *RG* 31 (Mar. 31, 1899): 229. A good discussion is Al Krug, "North American Freight Train Brakes Page." http://www.railway-technical.com/brake2.html [accessed June 14, 2005].

32. George Hackney to T. J. Walsh, June 5, 1884, and "H.R. Nickerson to All Employees—Newton," Dec. 23, 1884, Benjamin Johnson Papers, Cornell University. "The Master Car Builders and Brake Rigging," *RG* 21 (June 26, 1889), 431, claimed there was so much spring in parts "an important part of the piston travel is lost in taking up all this slack." For a similar view see "Brake Rigging—Its Care and Operation," *RG* 23 (Dec. 26, 1890): 890–91. For comments on foreign cars see ICC, *Fifteenth Annual Report, 1901* (Washington, 1902), 269.

33. Mud wasps are from "The Air Brake Encounters a New Enemy," Western Railway Club *Proceedings* 9 (1897): 218–29. "Miserable Condition" from "Defective Freight Car Brakes," *LE* 9 (Jan. 1896): 36. "A Yard Installation for Testing Air Brakes," *RG* 31 (Mar. 24, 1899): 203.

34. The percent of braked weight is from "The Air Brake and the Big Car," *RG* 30 (Nov. 25, 1898): 846. "The Making up of Freight Trains Partially Equipped with Air Brakes," *EN* 38 (Dec. 23, 1897): 411–12.

35. "The Westinghouse Air Brake Instruction Car," *RG* 31 (Jan. 13, 1899): 27. The instruction car on the Pennsylvania is from "Air Brake Instruction Train," *RG* 23 (Mar. 27, 1891): 215. "Southern Railway Air Brake Instruction Car," *LE* 9 (June 1896): 483. The International Correspondence School car is discussed in "Air Brake Department," *LE* 14 (Jan. 1901): 25. G. W. Rhodes, "Air Brakes and their Maintenance," Western Railway Club *Proceedings* 6 (1894): 144–57. A. M. Waitt, "Air Brake Equipment on Freight Trains," Western Railway Club *Proceedings* 8 (1895): 164–74. For instructions on handling trains see "Air Brake Men's Association," *RA* 27 (Apr. 21, 1899): 300–301. ICC, *Fifteenth Annual Report, 1901* (Washington, 1902), 273; *Seventeenth Annual Report, 1903* (Washington, 1904), 333. "New Amendment to the Safety Appliance Act," *RA* 35 (June 24, 1903): 1056. For Westinghouse's repair business see "The Repair of Air Brakes," *RA* 36 (July 31, 1903): 123.

36. "One of the . . ." from "The Westinghouse Frictional Draft Gear," *RA* 24 (Aug. 13, 1897): 666–68. "Service Performance of Friction Draft Gear," *RG* 33 (Nov. 29, 1901): 826. Buffing shocks are from "Underframing," *RA* 36 (May 23, 1902): 802–3. Number of cars from "Friction Draft Gear," *RG* 35 (Jan. 9, 1903): 30.

37. For the early history of British signaling see William Preece, "On Railway Telegraphs and the Application of Electricity to the Signaling and Working of Trains," ICE *Proceedings* 22 (1862–63): 167–92. George Hodgins, "Signals and Signaling," *LE* 16 (Nov. 1903): 517–18, 527–28. Richard Rapier, "On the Fixed Signals of Railways," ICE *Proceedings* 38 (1873–74): 142–94. Richard Blythe, *Danger Ahead* (London: Newman Neame, 1951). Stanley Hall, *History and Development of Railway Signaling in the British Isles,* 2 vols. (York, England: Friends of the National Railway Museum, 2000). Early U.S. practice is from Warren Jacobs, "Early Rules and the Standard Code," Railway and Locomotive Historical Society *Bulletin* 50 (1939): 29–55.

38. On single-track lines most companies used "absolute permissive" signaling.

This meant that an engineman facing a stop signal set by a preceding train might stop and then proceed, but he must take a siding and stay for any stop signal set by an oncoming train.

39. Welch is from J. Dutton Steele, "Railway Signals," ASCE *Transactions* 4 (1875): 147–61. For Welch's work see also "Safety Signals," *RG* 2 (Feb. 11, 1871): 460–61, and "An Early Report on Signals," Railway Signal Association *Proceedings* 6 (1909): 82–84.

40. ICC, *Report on Block Signal Systems* (Washington, 1907), quotation on 14. This source also contains good descriptions of the various controlled manual systems.

41. George Hodgins, "Signals and Signaling," *LE* 16 (Nov. 1903): 517–18, 527–28. J. A. Anderson, "The First Block Signal System in America," Railway Signal Association *Proceedings* 6 (1909): 13–19. Braman B. Adams, *The Block System of Signaling on American Railroads* (New York: Railroad Gazette, 1901). Ralph Scott, *Automatic Block Signals and Signal Circuits* (New York: McGraw Hill, 1908).

42. ARA, *The Invention of the Track Circuit* (New York, 1922). Block signals rapidly became more complex; by the early twentieth century companies were shifting from semaphores to lights for daytime use and developing signals with multiple aspects that governed speed control. For good, available modern discussions see "Signaling in the USA" (http//:www.trainweb.org/railwaytechnical/US-sig.html) and Carsten Lundsten, "North American Signaling Basics" (http//:www.lundsten. dk/us_signaling/movement.html) [accessed June 14, 2005].

43. "Train Accidents in 1878," *RG* 11 (Jan. 24, 1879): 46–47, quotation on 47. "Train Accidents in 1880," *RG* 13 (Feb. 11, 1881): 85–86, quotation on 85.

44. J. Patrick Desmond, *Ace Corson, Railroader, 1878–1960* (New York: Frederick Fell, 1964). Herbert Pease, *Singing Rails* (New York: Thomas Y. Crowell, 1900), ch. 1. New York Railroad Commissioners, *Annual Report, 1884* (Albany, 1885), 215–18. Charles B. George, *Forty Years on the Rail* (Chicago: R. R. Donnelley & Sons, 1887), 75. J. Harvey Reed, *Forty Years a Locomotive Engineer* (Prescott, Wash., 1913). Chauncey Del French, *Railroadman* (New York: Macmillan, 1938). U.S. Commissioner of Labor, *Fifth Annual Report, 1889, Railroad Labor* (Washington, 1890), ch. 3.

45. "On one road" and "on one a single note" from J. Dutton Steele, "Railway Signals," ASCE *Proceedings* 4 (1875): 147–55, quotations on 153. U.S. Commissioner of Railroads, *Annual Report, 1880* (Washington, 1881), 427, 447. Garfield is from "Diversity in Signals," *RG* 13 (Sept. 30, 1881): 539. The survey is from New York Railroad Commissioners, *Annual Report, 1883* (Albany, 1884), 38–41, quotation on 38. "Diversity . . ." is from "Signals: What Ought to be Done About Them?" *RG* 24 (Aug. 8, 1882): 508–9. "Lack of system . . . Philadelphia lawyer" is from "Train Rules," *RG* 15 (May 11, 1883): 292–93.

46. Harry Forman, *Rights of Trains,* rev. ed., ed. Peter Josserand (New York: Simmons Boardman, 1945), quotation on iii. The accident is from New York Railroad Commissioners, *Annual Report, 1883* (Albany, 1884), 289–95. "Complete confusion" from "Uniformity in Codes of Rules," *RG* 16 (Jan. 11, 1884): 28. Adoption of the code is from "The Time Convention," *RG* 21 (Apr. 5, 1889): 241–42.

47. "The Value of the Standard Code Illustrated," *RG* 24 (Nov. 4, 1892): 824.

48. "Some sort of an interlocking" from "Uniform Train Rules," *RG* 17 (Sept. 4, 1885): 568. "The Lessons of Three Collisions," *RA* 53 (July 26, 1912): 146–47.

49. Del French, *Railroadman,* 62, 203. "Almost like a revolution" from "The Costli-

ness of Accidents," *RG* 5 (Nov. 29, 1873): 481–82. "Slaughtered like sheep" from "How to Guard Against Stock Claims, Fire Claims, and Injuries to Track Employees," *RA* 48 (Nov. 14, 1908): 930–31.

50. "Our railroads ..." is from "Rear Collisions," *RG* 16 (Nov. 7, 1884): 804–5. "Spuyten Duyvil Catastrophe," *EN* 9 (Feb. 4, 1882): 40.

51. The Battle Creek accident is from untitled, *EN* 30 (Oct. 26, 1893): 325.

52. James Ducker, *Men of the Steel Rails* (Lincoln: University of Nebraska Press, 1983), ch. 2. "Wonderful restraining influence" is from "Railway Companies and Their Employees," *RA* 19 (May 16, 1894): 562–63, quotation on 562. For the carriers' benefit funds see Max Riebenack, *Railway Provident Institutions in English Speaking Countries* (Philadelphia, 1905), and U.S. Commissioner of Labor, *Twenty-Third Annual Report, 1908 Workmen's Insurance and Benefit Funds in the United States* (Washington, 1909), ch. 3.

53. For company physicals and state regulations governing color blindness see my "Train Wrecks to Typhoid Fever: The Development of Railroad Medicine Organizations, 1850 to World War I," *Bulletin of the History of Medicine* 75 (Summer 2001): 254–89. A survey of physicals is in ARA *Proceedings* 3 (1899–1902): 678–799.

54. Joseph Bromley, *Clear the Tracks!* (New York: Whittlesey House, 1943), quotation on 282–83.

55. For details on the Brown system see "Brown's Discipline," *RG* 29 (Oct. 1, 1897): 690–91. For its use see "Brown's Discipline on American Railroads," *RG* 63 (July 7, 1917): 19–22.

56. For an early call for block signaling see "Spuyten Duyvil Accident," *RG* 14 (Jan. 20, 1882): 42–44. "Accidents on British Railways in 1888," *EN* 22 (Sept. 21, 1889): 281.

57. "Three Thousand Miles of Block Signals," *RG* 24 (Feb. 5, 1892): 102. "Accidents on British Railways in 1888," *EN* 22 (Sept. 21, 1889): 281. Neil Weinstein, "Optimistic Biases About Personal Risks," *Science* 246 (Dec. 8, 1989): 1232–33.

58. "The Rowell Automatic Safety Stop," *EN* 21 (May 11, 1889): 428. For descriptions of the various early systems see Scott, *Automatic Block Signals and Signal Circuits*, "The Hall Block Signal," *RG* 22 (Sept. 13, 1890): 629, and "The Union Electric Signal System," *RG* 12 (Mar. 19, 1880): 153–54. "Three Thousand Miles of Block Signals," *RG* 24 (Feb. 5, 1892): 102.

59. The payoff to interlocking is from J. A. Peabody, "The Cost of Stopping a Train Compared with the Cost of Maintenance, Operation and Inspection of Interlocking Plants," *EN* 54 (Oct. 12, 1905): 372–73.

60. W. H. Elliott, "Derailments and the Conclusions to be Drawn Therefrom," Railway Signal Association *Proceedings* 2 (1901): 20–25. For similar findings see Charles Hansel, "A Moral and Physical Agent in Safe Railway Travel," *RR* 37 (Sept. 18, 1897): 538–39, 541–42.

61. The formula is from "The Relations Between Traffic and Disaster," *RA* 40 (Dec. 15, 1905): 746–47. The B&O is from "Another Serious Railway Accident," *EN* 20 (Oct. 13, 1888): 279.

62. The Old Colony is from "Train Dispatching on a Boston Road," *RG* 14 (Apr. 21, 1882): 235–36. Detentions on the Southern Pacific are from boxes 1 and 2, Letcher Railroad Collection, Ms 268, Stanford University. The quotation is from J. W. Younger to H. J. Small, Dec. 12, 1888, box 2, Letcher Railroad Collection, Ms 268, Stanford University. Causes of detention are from "Detentions on Railroads from Defects in Locomotives," *ARRJ* 67 (Nov. 1893): 548. Forney is from "Reports of Detentions and Defects of Locomotives," *ARRJ* 67 (Dec. 1893): 594.

See also George Fowler, *Locomotive Breakdowns, Emergencies and Their Remedies* (New York: Henley, 1903).

63. "Seven Persons Killed," *NYT* (Dec. 25, 1891), 1; "The Killing on the Central; A Terrible Accident on the New-York Central," *NYT* (Dec. 27, 1891), 4. "The Latest New York Central Collision," *EN* 27 (Jan. 2, 1892): 14–15. "America's Greatest Railroad—for Rear Collisions," *LE* 5 (Feb. 1892): 44.

64. "The Railroad Collision," *NYT* (Jan. 16, 1894), 4. "The Slaughter on the Delaware, Lackawanna, & Western," *LE* 7 (Feb. 1894): 56, which is the source of "right away." "Disastrous Collision on the Delaware, Lackawanna & Western," *RG* 26 (Jan. 19, 1894): 43. Gaetano Lanza, "The Application of Brakes to the Truck Wheels of a Locomotive," ASME *Transactions* 16 (1894): 67–75, reported experiments in which a four-car train running at 45 miles per hour stopped in about 860 feet.

65. "Distant Signals and the Fitchburg Collision," *EN* 28 (Oct. 20, 1892): 373.

Chapter 4. The Major Risks from Minor Accidents

Epigraphs: MBRC, *Second Annual Report, 1870* (Boston, 1871), 13. New York Railroad Commissioners, *Annual Report, 1883* (Albany, 1884), 37.

1. For risk attenuation see Roger Kasperson and Jeanne Kasperson, "The Social Amplification and Attenuation of Risk," *Annals of the American Academy of Political and Social Science* 545 (May 1996): 95–105.

2. An annual failure rate of 0.8 per thousand over twenty years is $[(1 - 0.0008)^{20} - 1] = 0.02$.

3. "Explosion of a Locomotive Boiler," *ART* 20 (Apr. 17, 1868): 127. Carrying excess pressure is from "Locomotive Explosions," *RG* 1 (June 4, 1870): 220. Gilbert Lathrop, *Little Engines and Big Men* (Caldwell, Idaho, 1954), 67. J. Harvey Reed, *Forty Years a Locomotive Engineer* (Prescott, Wash., 1913), 51. "A Broken Locomotive Parallel Rod," *EN* 18 (Nov. 5, 1887): 334–35.

4. "Wretchedly defective" is from "Boiler Construction," *RG* 4 (July 13, 1872): 298. ARMM, *Proceedings* 13 (1880): 72–74, quotations on 73 and 74. "Boiler Explosions," *RG* 12 (Jan. 23, 1880): 48; (Feb. 6, 1880): 76–78; (Feb. 13, 1880): 90–91. "Convenient scapegoat" is from untitled, *RG* 20 (Sept. 7, 1888): 590. "How long it will be" from untitled, *RMM* 2 (June 1887): 79.

5. ARMM, *Proceedings* 17 (1884): 54–65, 18 (1885): 33–40, 19 (1886), quotation on 9. Use of steel in driving axles is discussed in 24 (1891): 48–52. The use of steel in drive gear is from "A Broken Locomotive Parallel Rod." "Excessive Weights of Locomotive Machinery," *RA* 50 (Mar. 3, 1911): 389–90. On the complexities of steel see Thomas Misa, "Controversy and Closure in Technological Change: Constructing Steel," in *Shaping Technology/Building Societies,* ed. Weibe Bijker and John Law (Cambridge, Mass.: MIT Press, 1997), 109–39.

6. "Reduction in the Weight of Reciprocating Parts of Locomotives" *EN* 36 (Aug. 13, 1896): 99–100. "Cast Steel in Locomotive Construction," *RA* 23 (Jan. 29, 1897): 80–81. "Cast Steel in Locomotive Construction," *RR* 38 (Aug. 20, 1898): 455. "Cast Steel," *RG* 40 (Jan. 26, 1906): 75–76.

7. "The Vibration Test for Staybolts," *RA* 40 (Sept. 24, 1905): 343–34. For locomotive testing see Lawrence Fry, "The Locomotive Testing Plant and Its Influence on Steam Locomotive Design," ASME *Transactions* 47 (1925): 1267–93. For testing on the Pennsylvania see also Frederick Westing, *Apex of the Atlantics* (New York: Bonanza, 1974).

8. ARMM *Proceedings* 26 (1893): 68–87, quotation on 73.

9. ARMM *Proceedings* 26 (1893), quotation on 74. The specifications are in *Proceedings* 27 (1894): 68–78. For company specifications see "Steel for Boilers and Fireboxes," *EN* 37 (Jan. 21, 1897): 44–45.

10. Chauncy Del French, *Railroadman* (New York: Macmillan, 1938), 85. Angus Sinclair, *Locomotive Engine Running and Management,* 23d ed. (New York: John Wiley, 1915), ch. 10. Better lubrication and boiler washing are discussed in "Efficiency of Modern Locomotives Compared with Lighter Locomotives of Twenty Years Ago," *RR* 46 (Apr. 7, 1906): 265–67. "Locomotive Tire Fastenings," *RR* 30 (Nov. 15, 1890): 689. "Boiler Construction," *RG* 4 (July 13, 1872): 298 also discusses water problems. "Railway Master Mechanics' Report on Softening or Purification of Feedwater," *RA* 16 (June 20, 1891): 481, 490–91. Boiler washing is discussed in ARMM *Proceedings* 18 (1885): 138–43.

11. The *Jupiter* is from "The Mass RR. Commissioners on the Fall River Boiler Explosion," *RG* 13 (Jan. 14, 1881): 19. "Staybolt Inspection and Specifications," *RG* 25 (Jan. 28, 1893): 564–65. Testing is from "Partially Broken Staybolts," *RA* 24 (July 23, 1897): 597, and ARMM *Proceedings* 30 (1897): 200. For the Reading see untitled, *RG* 25 (May 19, 1893): 380.

12. Plugging staybolts is from "Tempting Boiler Explosions," *LE* 5 (Mar. 1892): 74, which also contains Sinclair's comments. "Inspection of Locomotive Staybolts," *RR* 77 (Dec. 5, 1925): 844–47.

13. Forney's paper is in ARMM *Proceedings* 14 (1881): 120–29. See also his remarks in *Proceedings* 19 (1886): 33. "Safety Boiler Checks," *RR* 31 (Aug. 22, 1891): 548–49. Del French *Railroadman,* 112–13. Front end steps and water glasses are from ARMM *Proceedings* 21 (1888): 162 and 26 (1893): 155–64.

14. MBRC, *Second Annual Report, 1871* (Boston, 1871), 13. New York Railroad Commissioners, *Annual Report, 1884* (Albany, 1884), 37.
   In 1890 there were about 153,000 trainmen; of these 68,000, or 44 percent, were firemen and enginemen and 85,000, or 56 percent, were other trainmen. Thus, $0.44(3.38) + 0.56x = 9.62$, so $x = 14.52$. Del French, *Railroadman,* 15.

15. The cow is from "Curious Accident to a Train Crew," *RR* 40 (Sept. 22, 1900): 531. The other injuries are from New York Railroad Commissioners, *Annual Report, 1868* (Albany, 1869), 300–305.

16. The verse is from Robert Wayner, *Great Poems from Railroad Magazine* (New York, 1968). Clark is quoted in U.S. Industrial Commission, *Report on Transportation* 4 (Washington, 1900), 111.

17. Thomas Aldridge is from New York Railroad Commisioners, *Annual Report, 1868* (Albany, 1869), 300–305. The wreath is from Frederick Gamst, *The Hoghead: An Industrial Ethnology of the Locomotive Engineer* (New York: Holt, Rinehart and Winston, 1980). The calculation is the resultant of the effect of gravity ($v^2 = 2gs$; $v = 30$ ft/sec) and forward velocity of 10 miles per hour ($v = 14.6$ ft/sec).

18. "Laws in Force in the Several States in Relation to Railway Safety Appliances," ARA *Proceedings* 1 (Apr. 8, 1891): 337–50. P. M. Andrews to Railroad Commission, Oct. 13, 1873, Accident vol. 10, Record Group 41 Connecticut Railroad Commission, Connecticut State Archives.

19. "In the Matter of the Fatal Accident to O.M. Wilmont," Mar. 8, 1894, box 00594443, Minutes of Meetings, 1892–1994, Vermont Board of Railroad Commissioners, Vermont State Archives.

20. "Bridge Warnings for Low Overhead Structures," American Association of Railway Superintendents of Bridges and Buildings *Proceedings* 7 (1897): 154–56.

William S. Huntington, "Accidents at Overhead Bridges Highway Crossings and Permanent Way," *ARRJ* 79 (Nov. 1885): 230–31.

21. "Ladder Steps—a Despicable Trick," *RR* 27 (Jan. 15, 1887): 30. The problem with fruit cars is from Benjamin Welch to H. J. Small, Aug. 11, 1888, box 2, Ms 268, Letcher Collection, Stanford University.

22. "Unnecessarily dangerous" is from "Dangerous Freight Cars," *RG* 2 (Mar. 4, 1871): 533. "The Perils of Coupling Cars," *RG* (Oct. 4, 1873): 402–3. New York Railroad Commissioner, *Annual Report, 1883* (Albany, 1884); all quotations on 37. Atlantic and Great Western from "Looking Out for the Brakeman's Hand," *RG* (Dec. 31, 1878): 605. West is from "Western Railway Club," *RR* 24 (Dec. 27, 1884): 662–63, where some companies also claimed that strict enforcement of rules would cause their better men to leave.

23. American Association of Railway Superintendents of Bridges and Buildings *Proceedings* 7 (1897): 176.

24. MCB *Proceedings* 5 (1871): 95, 13 (1879):104–9, 22 (1888): 18–29. "Utterly useless," from "Coupling Cars," *RG* (Mar. 21, 1874): 102. Untitled, *EN* 27 (June 16, 1892): 608.

25. On the link and pin and its alternatives see John White, *American Railroad Freight Car* (Baltimore: Johns Hopkins University Press, 1993), ch. 7. Also see my *Safety First* (Baltimore: Johns Hopkins University Press, 1997), ch. 1. "Accidents to Railroad Employees," *RG* (Sept. 30, 1881): 525–26. Using ICC data, Seung-Wook Kim and Price Fishback, "Institutional Change, Compensating Differentials and Accident Risk in American Railroading, 1892–1945," *Journal of Economic History* 53 (Dec. 1993): 796–823, find a risk premium for fatalities for this period of 2.8 percent and for injuries of 0.21 percent per million man-hours. Using data from the *Report of the Employers' Liability and Workmen's Compensation Commission,* 2 vols. (Washington, 1913) I obtained the following results:

$$\text{Log}(w) = 0.0013 DRate + 0.0005 Irate, \ R^2 = 0.0008, \ N = 720$$
$$\phantom{\text{Log}(w) = }(1.84) (2.34)$$

where *DRate* and *Irate* are death and injury rates per thousand workers. Figures in parentheses are t-ratios. The equations also include a constant and controls for five regions and thirty-three occupations. A death risk premium of 0.13 percent per thousand workers is about 0.046 percent per million manhours, or 1/61 of the Kim-Fishback finding. A skeptical modern assessment of wage premiums is in Peter Dorman and Paul Hagstrom, "Wage Compensation for Dangerous Work Revisited," *Industrial and Labor Relations Review* 52 (Oct. 1998): 116–35. Quotation from "Accidents to Railroad Employees," *RG* 13 (Sept. 28, 1881): 525–26.

26. Figures calculated from "Report of the Committee on Car Service," ARA *Proceedings* 1 (1886–92). Arthur Mellen Wellington, *Economic Theory of the Location of Railways,* rev. ed. (New York: John Wiley, 1887), 488–89.

27. "Spooner's Car Coupler," *RG* 7 (Nov. 20, 1875): 475. "The Potter Safety Draw Bar," *RG* 10 (June 28, 1878): 315. See also "Coupling Cars," *RG* 9 (June 20, 1877): 328, and White, *American Railroad Freight Car.* The role of the Pennsylvania is from Charles Clark, "The Railroad Safety Movement in the United States: Origins and Development, 1869–1893" (Ph.D. diss., University of Illinois at Champaign, 1966), 131.

28. For a discussion of the problem of locking in an inferior technology see Paul David, "Cleo and the Economics of QWERTY," *American Economic Review* 75 (May 1985): 332–37. For a critique of the qwerty example and a discussion of some of the ways firms avoid lock-in see Stan Liebowitz and Stephen Margolis, *The Economics of QWERTY: History, Theory, and Policy* (New York: New York University Press, 2002). The *Courant* is quoted in "Legislation for the Prevention of Accidents," *RG* 14 (Apr. 7, 1882): 211–12.

29. For details of brake diffusion, see Clark, "The Railroad Safety Movement," ch. 3.

30. Time of application is from "The Westinghouse Brake," *RG* 19 (June 10, 1887): 380–81. This discussion is based on Clark, "Railroad Safety Movement," quotation on 149.

31. For Harrison see Fred Israel, *The State of the Union Messages of the Presidents, 1790–1966*, 2 (New York: Chelsea House, 1967), 1647. Henry Cabot Lodge, "A Perilous Business and the Remedy," *North American Review* 154 (Feb. 1892): 189–95, quotations on 189 and 191.

32. Lorenzo Coffin, "Safety Appliances on the Railroads," *Annals of Iowa* 5 (Jan. 1903): 561–82. For more detailed contemporary discussions see Clark, "Railroad Safety Movement," ch. 10, who emphasizes railroad opposition, and my *Safety First*, ch. 1.

33. Theodore Porter, *Trust in Numbers* (Princeton: Princeton University Press, 1995). Thomas Haskell, "Capitalism and the Origins of Humanitarian Sensibility, part I," *American Historical Review* 90 (Apr. 1985): 339–61.

34. For a discussion of the effect of the law on the diffusion of couplers and brakes see my *Safety First*, ch. 1. On the problems of technological fixes see also Steven Usselman, "The Lure of Technology and the Appeal of Order: Railroad Safety Regulation in Nineteenth Century America," *Business and Economic History* 21 (1992): 290–99.

35. "Report of the Executive Committee on Automatic Freight Car Couplers," MCB *Proceedings* 21 (1887): 193. Waiving patent rights is noted in "The Master Car Builders Association," *RMM* 11 (July 1888): 108. Switching low- for high-priced couplers and "law of evolution" are from "The MCB Coupler Again," *RG* 31 (May 26, 1899): 372–73. Cast iron knuckles from "Home Made MCB Couplers," *EN* 25 (Apr. 4, 1891): 322–23.

36. Unchanged design is from "A Diagnosis of MCB Coupler Defects Based on Results in Service," *RG* 34 (May 23, 1902): 377–78.

37. "The Record of Janney Couplers in Freight Service," *RR* 30 (Feb. 15, 1890): 90. "Freight Car Couplers on the C., B., & Q.," *RG* 22 (June 6, 1890): 394–95. Forsyth's paper is reprinted in "The Strength of the MCB Coupler," *EN* 27 (Mar. 5, 1892): 232–33. The call for standardized tests is in "Standard Brake and Coupler Equipment," *EN* 27 (May 5, 1892): 456–57. "Tests of Master Car Builders' Couplers," Western Railway Club *Proceedings* 5 (Apr. 1893): 183–95, quotation on 195. MCB *Proceedings* 12 (1893): 123–77. "Tests of MCB Couplers," *RG* 25 (July 21, 1893): 545–47, reports drop tests at Altoona and tension tests at Watertown. An untitled editorial, *RG* 25 (July 21, 1893): 548, claimed the effect of the tests "has already been felt."

38. For the strategy see MCB *Proceedings* 39 (1905): 123–25. "The Coupler at the Convention," *RG* 31 (May 19, 1899): 354–55. "The Master Car Builders' Coupler," *EN* 41 (June 22, 1899): 396–97. The problem of keeping rejected couplers out of circulation is from "The M.C.B. Coupler," *RMM* 21 (July 1898): 90–91. The shift to steel is from "The MCB Coupler in Interchange," *RA* 40 (July 14, 1900):

34. The importance of the shift to steel is from "Freight Car Couplers," *RG* 33 (June 14, 1901): 415. The decline in defects is documented in the ICC's annual reports for 1905–10. "Development and Construction of Standard Couplers," *LE* 32 (Jan. 1919): 5–6.

39. "In Re Fatal Accident to Maynard Ryan . . . July 12, 1903," Complaints and Petitions, box 00599445, Vermont Board of Railroad Commissioners, Vermont State Archives.

40. ICC, *Operation of Trains on Heavy Grades* (Washington, 1907), quotations on 3, 5, 7. "Present condition" from "The Interstate Commerce Commission's Car Inspector," *RG* 34 (May 2, 1902): 322. Pennsylvania's policy from W. W. Atterbury to Edgar M. Clark, Feb. 19, 1921, box 553, and F. W. Hankins to Elisha Lee, Mar. 9, 1926, box 555, PRRC, HML.

41. Roger Grant, ed., *Brownie the Boomer* (DeKalb: Northern Illinois University Press, 1991), 64–65. ICC data that begin in 1916 reveal that the proportion of yard trainmen to other trainmen rose steadily from then on. Clark, "Railroad Safety Movement," epilogue, also stresses the role of yard work in raising injuries.

42. Numbers in the text were generated by taking the difference between actual casualty rates in each year and the average rate from 1890 to 1893 times actual employment.

43. New York Railroad Commissioners, *Annual Report, 1885* (Albany, 1886), xviii, emphasis added.

44. MBRC, *Second Annual Report, 1870* (Boston, 1871), 13.

45. "Passing Stations at Full Speed," *RG* 13 (Sept. 30, 1881): 533. The quotation is from an untitled editorial, *RG* 22 (Jan. 24, 1890): 96.

46. On the changing conceptions of freedom and liberty see Barbara Welke, *Recasting American Liberty* (New York: Cambridge University Press, 2001), ch. 1.

47. The survey is from "Provisions to Insure Safe Walking in and About Passenger Stations," AREA *Proceedings* 19 (1918): 248–70, quotation on 250. "Temptation to brakemen" is from "What is the Use of Car Steps?" *RA* 19 (Aug. 10, 1894): 445.

48. "Radical Treatment of the Cash Fare Evil," *RG* 25 (Nov. 10, 1893): 820. "Perceptible decrease" from "Gates on Passenger Cars," *RA* 19 (Mar. 30, 1894): 179. Extension to Illinois is from untitled, *RMM* 17 (May 1894): 72.

49. "On the New Haven System," *LE* 8 (Mar. 1895): 170. De Haven is quoted in "Gates on Cars," *RR* 38 (Aug. 13, 1898): 443–44. "The Car Gate System," *RR* 38 (Aug. 13, 1898): 443–44, 446–47. "Train Gates," *RR* 25 (Aug. 20, 1898): 461–62.

50. "Trespassers on the Track or Train," *RG* 15 (Oct. 26, 1883): 708. The Alabama court is from "Accidents to Trespassers," *RG* 19 (Dec. 23, 1887): 831.

51. Daily wages for 1871–80 from *Historical Statistics* D735–738. For 1885–1900 daily wages computed as annual earnings divided by 333. Fares are for all Massachusetts railroads from MBRC, *Thirty-First Annual Report, 1900* (Boston, 1901), 22. For statistical evidence that rising wages and the spread of automobiles reduced trespassing casualties see ch. 9.

52. An early study that differentiates hobos from tramps is Nels Anderson, *The Hobo: The Sociology of the Homeless Man* (Chicago: University of Chicago Press, 1923). Early popular treatments include Josiah Flynt, "Tramping with Tramps," *The Century* 47 (Nov. 1893): 99–108, and "The Tramp at Home" (Feb. 1894): 517–26. See also Glen Mullin, *Adventures of a Scholar Tramp* (New York: Century, 1925). Gregory Woirol, ed., *In the Floating Army* (Urbana: University of Illinois Press, 1992). Roger Bruns, *Knights of the Road* (New York: Methuen, 1980). Gypsy Moon, *Done and Been* (Bloomington: Indiana University Press, 1996).

53. Jack London, *The Road* (New York: Macmillan, 1907). "Most Serious Railway Accident," *EN* 40 (Oct. 27, 1898): 257.

54. The Rio Grande is from "Automatic Train Robber Killer," *LE* 14 (Apr. 1901): 158–59.

55. New York Railroad Commissioners, *Annual Report, 1888* (Albany, 1889), xvi. MBRC, *Twentieth Annual Report, 1888* (Boston, 1889), 50–52, quotation on 50.

56. Vermont Railroad Commission, *Tenth Biennial Report, 1904–1906* (Montpelier, 1907), 99.

57. "It seems not improbable" from MBRC, *Twenty-Second Annual Report, 1890* (Boston, 1891), 7. "Severely punished" is from MBRC, *Twentieth Annual Report, 1888* (Boston, 1889), 51. "So rare" is on 50; retaliation against the carriers for efforts to prevent trespassing is noted on 53.

58. Wilhelm Hoff and Felix Schwabach, *North American Railroads* (New York, 1906), 210. New York Railroad Commissioners, *Annual Report, 1883* (Albany, 1884), 27. MBRC, *Twentieth Annual Report, 1888* (Boston, 1889), 64.

59. New York Railroad Commissioners, *Annual Report, 1883* (Albany, 1884), 31–35.

60. The Maine case is from "Railroad Law," *RG* 20 (Dec. 14, 1888): 824.

61. Rochester is from Augustus Locke et al., *Report of an Investigation into the Subject of the Gradual Abolition of the Crossing of Highways by Railroads at Grade* (Boston, 1889), 70–71. For the Pennsylvania see the untitled article, *EN* 20 (Nov. 24, 1888): 408–9.

62. Connecticut Railroad Commissioners, *Thirty-Fourth Annual Report, 1886* (Hartford, 1887), 28–34; *Thirty-Fifth Annual Report, 1887* (Hartford, 1888), 29–33; *Thirty-Sixth Annual Report, 1888* (Hartford, 1889), 30–41. The 11 percent figure is computed from data on 41.

63. The hearings are reported in "A Dangerous Grade Crossing," *RG* 19 (Feb. 4, 1887): 79.

64. Locke, *Report of an Investigation,* quotations on 16 and 21. See also Augustus Locke, "The Abolition of Grade Crossings Between Railroads and Highways," AES *Journal* 14 (May 1895): 422–35.

65. "The Massachusetts Crossing Report," *EN* 21 (Feb. 23, 1889): 169–70.

66. Robert Adam, "History of the Abolition of Railroad Grade Crossings in the City of Buffalo," Buffalo Historical Society *Proceedings* 8 (1905): 149–254. The reduction in fatalities is on 254. See also "The Buffalo Crossing Question," *EN* 18 (Oct. 29, 1887): 313–14, and "Apportionment of Expense, Grade Crossing Removal, at Buffalo, N.Y.," *EN* 33 (Jan. 31, 1895): 79, and "The Removal of Grade Crossings in Buffalo, N.Y.," *EN* 34 (Oct. 24, 1895): 266–68. The number removed is from "Grade Crossing Elimination," *EN* 61 (Apr. 1, 1909): 348–49. "No general plan" from untitled, *RG* 41 (Aug. 17, 1906): 129.

67. Crossings removed are from "Grade Crossing Elimination," *EN* 61 (Apr. 1, 1909): 348–49. "Report on the Abolition of Grade Crossings at Cleveland," *EN* 54 (Feb. 7, 1901): 98–99. The decline in fatalities in Chicago is from Chicago Track Elevation Department, *Track Elevation within the Corporate Limits of the City of Chicago to Dec. 31st 1908* (Chicago, 1909).

Chapter 5. Engineering Success and Disaster

Epigraphs: Theodore Cooper, "American Railroad Bridges," ASCE *Transactions* 21 (1889): 50–51, and discussion, 589.

1. Based on Joint Committee of the Ohio Legislature, *Report Concerning the Ashtabula Bridge Disaster* (Columbus, 1877); Stephen Peet, *The Ashtabula Di-*

saster (Chicago, 1877); Ohio Commissioner of Railroads and Telegraphs, *Annual Report, 1877*, 22, lists the death toll as eighty-nine. Charles MacDonald, "The Failure of the Ashtabula Bridge," ASCE *Transactions* 6 (1877): 74–87; "A Terrible Disaster," *NYT* (Dec. 30, 1876), 1; "The Lake Shore Disaster," *NYT* (Dec. 31, 1876), 1; "The Ashtabula Calamity," *NYT* (Jan. 1, 1877), 1. "The Ashtabula Bridge," *RG* 9 (Feb. 23, 1877): 86–87.

2. "Committee Report of the Ohio Assembly on the Ashtabula Bridge," *EN* 4 (May 26, 1877): 133. Cooper's remarks are in "On the Failure of the Ashtabula Bridge," *EN* 4 (Sept. 9, 1877): 240–41. A modern assessment concludes that the bridge collapsed due to fatigue failure in an angle block. D. A. Gasparini and Melissa Fields, "Collapse of the Ashtabula Bridge on Dec. 29, 1876," *Journal of Performance of Constructed Faculties* 7 (May 1993): 109–25. The factor of safety as then defined was the ratio of the breaking strength of the member to its calculated stress.

3. For biographical information on Cooper see *NCAB*, 19, 261–62, and *DAB*, 3, 413–14.

4. "The Ashtabula Bridge," *EN* 4 (Jan. 20, 1877): 19. "Wooden Trestles," *EN* 18 (Aug. 13, 1887): 113. George S. Morison, "Advance in the Design of Bridge Superstructure," *EN* 30 (July 27, 1893): 80–81. On the evolution of American railroad bridges see the following: Carl Condit, *American Building Art, the Nineteenth Century* (New York: Oxford University Press, 1960), chs. 3–4; George Danko, "The Evolution of the Simple Truss Bridge 1790–1860: From Empiricism to Scientific Construction" (Ph.D. diss., University of Pennsylvania at Philadelphia, 1979); Llewellyn Edwards, *A Record of History and Evolution of Early American Bridges* (Orono: University of Maine Press, 1959); Henry Grattan Tyrrell, *History of Bridge Engineering*, 4th ed (Chicago, 1911); J.A.L. Waddell, *Bridge Engineering* (New York: John Wiley, 1916), vol. 1; Charles Schneider, "The Evolution and Practice of American Bridge Building," ASCE *Transactions* 54 (1905): 213–34; Theodore Cooper, "American Railroad Bridges," ASCE *Transactions* 21 (1889): 1–59; Larry Kline and Anthony Thompson, "The Evolution of American Railroad Bridges, 1830–1994," in *Symposium on Railroad History*, ed. Anthony Thompson, 3 (1994): 71–93.

5. Morison, "Advance in the Design of Bridge Superstructure." For further comparisons of British and American bridge construction techniques see also "Bridge Building in England and the United States," *EN* 29 (Dec. 15, 1892): 572, and "Some Fundamentals of American Bridge Building," *RG* 31 (July 14, 1899): 508.

6. Cooper, "American Railroad Bridges." Cooper's contacts are evidenced in box 1, file 1, Theodore Cooper Papers, Cornell University. A later survey by the ICC revealed that Cooper's estimates were too low. See "The Second Interstate Commerce Report," *EN* 24 (Aug. 20, 1890): 168–69. Unfortunately there is insufficient detail in the ICC data to modify Cooper's estimates.

7. J. Parker Snow, "Wooden Bridge Construction on the Boston & Maine Railroad," AES *Journal* 15 (July 1995): 31–39. MBRC, *Twenty-Ninth Annual Report, 1897* (Boston, 1898), 16–18.

8. I have modified Appleton's data by adding his category "hurricane" to "freshet," and "floor broken by train" to "knocked down." Most of the bridges that Stowell classified as made of unknown material" were wood according to the editor of *Engineering News.*

9. In "The Bridge Failures of Eleven Years," *Engineering News* calculated that there were 24,450 iron and 15,250 wooden bridges over 20 feet long in 1889. It left out

shorter spans as likely to be culverts, which were excluded from the accident statistics. Over the two years 1888–89, there were eleven failures of iron bridges and forty-two failures of wood and unknown materials, which the *News* assumed to be wood. The implied annual failure rates are [11/24,450]/2 and [42/15,250]/2, or about one in 4,445 for iron and one in 726 for wood. In "Bridge Accidents in the United States and Canada in 1896," *EN* 37 (Feb. 11, 1897): 93, Stowell estimated that for 1896, inclusion of trestles increased the number of failing wood structures by 36 percent. If these proportions held true in 1888–89 the implied failure rate for wooden structures would be one in 534. Cooper's efforts are from Courney Boyle [Board of Trade] to Theodore Cooper, Jan. 12, 1889, box 1, file 1, Theodore Cooper Papers, Cornell University.

10. Robert C. Reed, *Train Wrecks, A Pictorial History of Accidents on the Main Line* (New York: Bonanza, 1982), ch. 7, stresses collapses, although he also notes the range of ways in which bridges failed. Robert Shaw, *Down Brakes: A History of Railroad Accidents, Safety Precautions, and Operating Practices* (London: P. R. Macmillan, 1961), stresses square falls. Henry Petroski, *Engineers of Dreams* (New York: Vintage, 1995) emphasizes design errors in bridge failures.

11. "Wooden Trestles," *EN* 18 (Aug. 13, 1887): 113. Squire Whipple, *Bridge-Building; Being the Author's Original Work, Published in 1847, with an Appendix . . .* (Albany, 1869). Herman Haupt, *General Theory of Bridge Construction* (New York, 1853[?]). What were called strain diagrams in fact calculated the stress on bridge members from some assumed load. For the metallurgical investigations of Fairbairn, Wöller, and others, see Stephen Timoshenko, *History of the Strength of Materials* (New York: McGraw Hill, 1953), and Nathan Rosenberg and Walter Vincenti, *The Britannia Bridge: The Generation and Diffusion of Technological Knowledge* (Cambridge, Mass.: MIT Press, 1978).

12. Appleton, "Railroad Bridge Accidents," *EN* 5 (Feb. 21, 1878): 59–61; W. H. Barlow, commenting on "Thomas Clarke, "The Design Generally of Iron Bridges of Very Large Span for Railway Traffic," Institution of Civil Engineers *Proceedings* 54 (1877–78): 179–247, quotation on 218. "A Criminal Structure," *RR* 33 (Mar. 4, 1893): 130.

13. "Rider's Iron Bridge," *ARRJ* 21 (Dec. 2, 1848): 769; 775–76 contains the testimonial from Allen and Jervis. "Accident on the Erie Railroad," *ARRJ* 23 (Aug. 24, 1850): 534; Squire Whipple, "Iron Bridges," *ARRJ* 20 (Nov. 27, 1847): 754–55; Squire Whipple, "The Breaking of the Iron Bridge on the New York an Erie Railroad," *ARRJ* 23 (Sept. 21, 1850): 594–95; John A. Roebling, "The Breaking of Rider's Iron Bridge on the New York and Erie Railroad," *ARRJ* 23 (Sept. 28, 1850): 609–10, emphasis in original. For a modern discussion see Victor Darnell, "The Pioneering Iron Trusses of Nathaniel Rider," *Construction History* 7 (1991): 69–81.

14. James Ward, ed., *Southern Railroad Man: N.J. Bell's Recollections of the Civil War Era* (DeKalb: Southern Illinois University Press, 1994), 33, 39. Engine weight from J. E. Greiner, "What Is the Life of an Iron Bridge?" ASCE *Transactions* 34 (1895): 294–307. "Cheap and nasty" is from "Report of Committee Appointed by Western Society of Engineers [on Bridge Legislation] Mar. 5, 1890," AES *Journal* 10 (Nov. 1891): 517–25.

15. "Heavy Bridges and Economy," *RG* 18 (Dec. 1 1886): 674–75. See also C. M. Barber, "Old Bridges Under New Loads," AES *Journal* 5 (Mar. 1886): 159–63.

16. Albert Robinson, "Relative Cost of Heavy vs. Reinforced Bridges," *EN* 30 (Sept. 21, 1893): 237–38.

17. Central Railroad bridge from "Dangerous Railroad Bridges," *EN* 14 (Dec. 12, 1885): 276–77. "Old Fink and Bollman . . ." is from "Knocked Down," *EN* 24 (Aug. 22, 1890): 173. The speaker to the ASCE is Greiner, "What Is the Life of an Iron Bridge?" 296. The wreck is from "Bridge Failures in 1886," *EN* 17 (Apr. 30, 1887): 287–88.

18. Colburn's remarks are in the discussion of James Mosse, "American Timber Bridges," ICE *Proceedings* 22 (1862–63): 305–26, quotation on 319. For a bridge that collapsed as a result of poorly adjusted tension rods see New York Railroad Commissioners, *Annual Report, 1884* (Albany, 1885), 207–10. "Railway Manslaughter," *ART* 11 (Aug. 13, 1859): 2. Bell is from Ward, *Southern Railroad Man*, 94.

19. For the Erie see Wolcott C. Foster, *A Treatise on Wooden Trestle Bridges According to the Present Practice on American Railroads* (New York: John Wiley, 1891), 80–82. "Form for Reports of Bridge Inspectors on the Buffalo, Rochester & Pittsburgh Railway," *EN* 19 (May 12, 1888): 380–81.

20. Foster, *A Treatise on Wooden Trestle Bridges*, 4. Mansfield Merriman, "The Farmington Bridge Disaster," *EN* 5 (Jan. 31, 1878): 39. For biographical details see *NCAB*, 23, 70–71. For a similar assessment see also the reports of the civil engineer Albert Hill, "The Tariffville Disaster," *RG* 10 (Jan. 25, 1878): 41, and (Feb. 1, 1878): 52. The quotation is from "Verdict of the Coroner's Jury on the Tariffville Bridge Accident," *RG* 10 (Mar. 1, 1878): 109. See also Connecticut Railroad Commissioners, *Twenty-Sixth Annual Report, 1879* (Hartford, 1879), 1–17.

21. "The Chatsworth Disaster," *EN* 18 (Aug. 20, 1887): 126–28, and "The Lesson of the Chatsworth Disaster," 131. For similar views about double-headers see the untitled editorial, *RR* 27 (Oct. 29, 1887): 624–25. Air brakes are discussed in "The Chatsworth Disaster," *RG* 19 (Aug. 10, 1887): 544–45.

22. "The Size of Railway Culverts," and "The Accident on the Michigan Southern Railway," *ART* 11 (July 9 and 30, 1859): 2 and 2. Charles Folsom, "Railroad Washouts," AES *Journal* 5 (June 1886): 304–9. Donald Jackson, "Nineteenth Century American Bridge Failures: A Professional Perspective," *Proceedings of the Second Historic Bridges Conference, Mar. 1988* (Columbus, 1988), 113–25, states (117), "Failures resulting from washouts could certainly be categorized as bridge collapses but they did not relate to the design or construction of the truss proper."

23. This recounting is based on Dow Helmers, *Tragedy at Eden* (Pueblo, 1971).

24. "The Chester Bridge Disaster," *EN* 30 (Sept. 7, 1893): 192; quotations from "The Last Great Bridge Disaster," *EN* 30 (Sept. 7, 1893): 195–96. Testimony at hearings to the Massachusetts Board of Railroad Commissions is in "The Chester Bridge Disaster," *EN* 30 (Sept. 14, 1894): 219–21. See also "The Chester Bridge Disaster Finding," *EN* 30 (Sept. 28, 1983): 255–56, and MBRC, *Twenty-Fifth Annual Report, 1893* (Boston, 1894), Appendix B.

25. Ewing Matheson, "English vs. American Bridges," *RG* 6 (Apr. 4, 1874): 119–20. Of course British railroad bridges did fall down occasionally. See Marion Pinsdorf, "Engineering Dreams into Disaster: History of the Tay Bridge," *Railroad History* 179 (Aug. 1998): 89–116.

26. Matheson, "English vs. American Bridges." The discussion of design is by W. Shelford and A. H. Shield, "On Some Points for the Consideration of English Engineers with Reference to the Design of Girder Bridges," British Association for the Advancement of Science *Report* 56 (1886): 472–82. The time of erection is from Edward Howland, "Iron Bridges and Their Construction," *Lippencott's*

Magazine of Popular Literature and Science 11 (Jan. 1873): 9–26. For a case study of the role of contracting and other organizational matters affecting bridge safety see Eda Kranakis, "Fixing the Blame: Organizational Culture and the Quebec Bridge Collapse," *Technology and Culture* 45 (July 2004): 487–518.

27. Matheson, "English vs. American Bridges." The quotation from *The Engineer* is from "A Lesson For American Bridge Engineers," *EN* 31 (Feb. 15, 1894): 134.

28. Charles Bender, "Comparison of the Merits of the Mode of Building Iron Truss Bridges in America with the System Used in Europe," *RG* 6 (Apr. 11, 1874): 129–30, (Apr. 18, 1874): 139–40, and (Apr. 25, 1874): 149–51. Forney's assessment is in "English Versus American Iron Bridges," *RG* 6 (Apr. 25, 1874): 152–53.

29. Appleton, "Railroad Bridge Accidents," 60, emphasis in original.

30. Donald Jackson, "Nineteenth Century American Bridge Failures," blames knockdowns on the lack of guardrails and ignores the matter of pin versus rivet construction, claiming that "the lack of these guardrails can certainly be considered a major engineering deficiency, but it does not relate to structural problems with the design" (quotation on 116).

31. For the Norwalk disaster see Charles Francis Adams, *Notes on Railroad Accidents* (New York: Putnams, 1879). Vermont Railroad Commission, *Fourth Annual Report, 1859* (Rutland, 1859), 4; *Sixth Annual Report, 1861* (Rutland, 1861), 6–9.

32. MBRC, *Fourth Annual Report, 1872* (Boston, 1873), 124–33. Ohio Commissioner of Railroads and Telegraphs, *Sixth Annual Report, 1872* (Columbus, 1873), 34–34; *Seventh Annual Report, 1873* (Columbus, 1874), Appendix A. The Connecticut commissioners are reported in "The Framingham [*sic*] Bridge Disaster," *EN* 5 (Jan. 24, 1878): 26. The engineer's report is in "Field Notebook, 1877–1881," Records of the Railroad Commission, Record Group 41, Connecticut State Library. Vose's remarks from "Testing Railroad Bridges," *ARRJ* 51 (Mar. 2, 1878): 302. For biographical information on Vose see *DAB*, 19, 294. The Cahaba Bridge accident is from William Thomas, *Lawyering for the Railroad* (Baton Rouge: Louisiana State University Press, 1999), 120–24.

33. The exchange of correspondence took place in 1888 at the time the bridge was built. It was later published in "A Remarkable Blunder in Bridge Construction," *EN* 58 (Sept. 19, 1907): 317–18, the occasion being the collapse of the Quebec bridge that was also thought to be a result of poor design.

34. "On the Means of Averting Bridge Accidents," ASCE *Transactions* 4 (1875): 123–35, and "Discussion," 208–22. The committee originally had ten members but only seven signed the final reports. The committee's demise is in "Minutes of the Twenty-Fourth Annual Meeting, Nov. 1, 1876," ASCE *Proceedings* 2 (1877): 146–47.

35. The cost to the Lake Shore is from an untitled editorial, *EN* 17 (Mar. 19, 1887): 181. Boller is quoted in "On the Failure of the Ashtabula Bridge," *EN* 4 (Aug. 25, 1877): 225–26. For the Ohio investigation see n. 1 above. Wisconsin Railroad Commissioner, *Fourth Annual Report, 1877* (Madison, 1878), 14–23. "Field Notebook, 1877–1881," Records of the Railroad Commission, Record Group 41, Connecticut State Archives.

36. "The Ashtabula Bridge," *EN* 4 (Jan. 20, 1877): 19; "Minutes of the Ninth Annual Convention, Apr. 24–30, 1877," ASCE *Proceedings* 3 (1877): 45–46, and "Minutes of the Meeting of Oct. 3, 1877," ASCE *Proceedings* 3 (1877): 86–88.

37. The interaction between engineering societies and the technical press is revealed in Thomas C. Clarke, "Lowthorp on the Role of Cast and Wrought Iron

in Bridge Construction," *ARRJ* 43 (Dec. 17, 1870): 1405–6, which is a critique of a paper presented to the ASCE. John Griffin and Thomas Clarke, "Loads and Strains on Bridges," ASCE *Transactions* 1 (1872): 93–105, note that many engineers still focused on the breaking strength of metal rather than the elastic limit. See also Octave Chanute, "Factors of Safety," *EN* 7 (Jan. 31, 1880): 41–42. For a discussion of ways to treat live loads see Henry Seaman, "The Launhardt Formula, and Railroad Bridge Specifications," ASCE *Transactions* 41 (1899): 140–65, and "Working Stresses for Railroad Bridges," *RG* 30 (Nov. 4, 1898): 797–98, which discusses Seaman's paper. See also E. Herbert Stone, "The Determination of the Safe Working Stress for Railway Bridges of Wrought Iron and Steel," ASCE *Transactions* 41 (1899): 466–502. For tests of impact see "Report of Committee 15—Iron and Steel Structures," AREA *Proceedings* 12 (1911): 12–300. On the use of standard train loads to proportion bridges see Theodore Cooper, "Train Loadings for Railroad Bridges," ASCE *Transactions* 31 (1894): 174–219. Testing is discussed in B. L. Marsteller, "Inspection of Iron Bridges and Viaducts," AES *Journal* 8 (Jan. 1889): 7–12.

38. Isaac Hinckley to president, May 3, 1849, box 19, superintendent's letters, Providence & Worcester Railroad Collection, Dodd Research Center, University of Connecticut.

39. "For the early history of bridge contracting see n. 26 above. For some typical specifications see "Western Union Railroad," *EN* 5 (Jan. 31, 1878): 40 and 47; "General Specifications for a Wrought Iron Railway Draw-Bridge . . . for the Chicago, Milwaukee & St. Paul Railway . . .," *EN* 6 (Jan. 18, 1879): 21–22; "General Specifications for Iron Bridges [on the Erie]," *EN* 6 (May 31, 1879): 74–175. Clarke, "The Design Generally of Iron Bridges," contains specifications for the Cincinnati Southern and reports from consulting engineers on tests. Theodore Cooper, *General Specifications for Iron Railroad Bridges and Viaducts* (New York: Engineering News, 1884). For the influence of Cooper's specifications see "Working Stresses for Railroad Bridges," *RG* 30 (Nov. 4, 1898): 797–98. "A Series of Failure Tests of Full-Sized Compression Members Made for the Pennsylvania Lines West of Pittsburgh," *EN* 58 (Dec. 26, 1907): 685–95.

40. George Vose, "Bridge Disasters in America—the Causes and the Remedy," *RG* 12 (June 25, 1880): 339–41, and (July 2 1880): 355–57. These later formed the basis of his *Bridge Disasters in America* (Boston: Lee & Shepherd, 1887). See also his "Dangerous Highway Bridges," *EN* 6 (Jan. 18, 1879): 20–21, (Feb. 1, 1879): 37–38, and (Feb. 8, 1879): 45–46, and "Factors of Safety, Danger, and Ignorance," *EN* 7 (Jan. 17, 1880): 23–24. See also David Simmons, "Bridge Building on a National Scale: The King Iron Bridge and Manufacturing Company," *Industrial Archeology* 15 (Jan. 1989): 23–39.

41. New York Railroad Commissioners, *Annual Report, 1883* (Albany, 1884), 296-301; *Annual Report, 1884* (Albany, 1885), xix–xxi. "State Bridge Inspection in New York," *EN* 17 (Apr. 30, 1887): 284–85. Both Ohio and Massachusetts reported similar management lapses. Ohio Commissioner of Railroads and Telegraphs, *Twelfth Annual Report, 1878* (Columbus, 1879), 21; MBRC, *Nineteenth Annual Report, 1887* (Boston, 1888), 38.

42. For biographical information on Stowell see *Who Was Who,* 1. The quotation is from "State Bridge Inspection in New York," *EN* 17 (Apr. 30, 1887): 284–85.

43. *Report of the Railroad Commissioners of the State of New York on Strains on Railroad Bridges of the State* (Albany, 1891).

44. For New York Central bridges see George Gray, "Notes on Early Practice in Bridge Building," ASCE *Transactions* 37 (1897): 1–15. The *Gazette* began collecting such statistics in 1873. Howard Miller, "Truss Failures Reconsidered," *Technology and Culture* 22 (Oct. 1981): 849–50, argues that it missed many bridge failures, basing his claim on the criticisms contained in several articles in *Engineering News*. He apparently failed to note "The Railroad Gazette Accident Record," *EN* 17 (May 7, 1887): 303, in which Stowell claimed that "I have found the *Gazette*'s tables to be generally reliable," and the journal's editors apologized to the *Gazette* for their erroneous criticism. In fact, users of the *Gazette*'s statistics often missed bridge failures because they were listed by cause of train accident.

45. "Light Bridges and Bridge Accidents," *RG* 18 (Nov. 5, 1886): 755.

46. Arthur Wellington, *The Economic Theory of the Location of Railways* (New York: John Wiley, 1887), 1–3, quotation on 1. "Bridge Failures in 1886," *EN* 17 (Apr. 30, 1887): 287–88. For Wellington's career see *NCAB*, 11, 168–70, and "Arthur Mellen Wellington," *EN* 33 (May 23, 1895): 337–38.

47. A deck truss carries the tracks on the top chord.

48. "The Facts in Regard to the Woodstock Disaster," *EN* 17 (Feb. 12, 1887): 105–6. Vermont Railroad Commission, *First Biennial Report, Dec. 1, 1886–June 30, 1888* (Boston, 1888), 91–100.

49. Vermont Railroad Commission, *Second Biennial Report, June 30, 1888 to June 30, 1890* (Burlington, 1890), 105–21, quotation on 106.

50. Vermont Railroad Commission, *Third Biennial Report, June 30, 1890 to June 30, 1892* (Burlington, 1892). The Central Vermont is noted in *Fourth Biennial Report, June 30, 1892 to June 30, 1894* (Rutland, 1894), 40–41. The quotation is from *Fifth Biennial Report, June 30, 1894 to June 30, 1896* (Rutland, 1896), 13–15.

51. Untitled editorial, *EN* 17 (Feb. 12, 1887): 108. In fact, Latimer had patented the device. "The Latimer Safety-Guard Patent," *EN* 17 (May 7, 1887): 199. But as the editors pointed out, the same result could be easily achieved without fear of infringement.

52. For early evidence of concern with inadequate flooring see "Railroad Bridges," *ARRJ* 26 (Sept. 10, 1853): 582, which complains about the "want of side protection." MBRC, *Thirteenth Annual Report, 1881* (Boston, 1882), Appendix F; Ohio Commissioner of Railroads and Telegraphs, *Annual Report, 1884* (Columbus, 1885), 182–83. "The Best Safeguard Against Woodstock Disasters," *EN* 17 (Feb. 12, 1887): 112–13.

53. "Some Strong and Weak Points in Railroad Bridges," *EN* 17 (Mar. 12, 1887): 170–71.

54. For a partial listing see "All at Once and Nothing First," *EN* 18 (Nov. 3, 1887): 171. "Inside and Outside Guard Rails," *EN* 19 (Jan. 28, 1888): 60–61 and 64–65; "Fall of the Apple River Bridge," *EN* 19 (Mar. 31, 1888): 248, which contains the quotation. "A Riveted Bridge in a Collision," *EN* 19 (Apr. 14, 1888): 298. "The Flat Creek Trestle Disaster," *EN* 22 (Aug. 31, 1889): 198–99. "Knocked Down," *EN* 24 (Aug. 23, 1890): 173.

55. George H. Thomson, "American Bridge Failures: Mechanical Pathology Considered in Relation to Bridge Design," *Engineering* 48 (Sept. 14, 1888): 252–53 and 294.

56. Cooper, "American Railroad Bridges," 50–51.

57. Discussion of Cooper, "American Railroad Bridges," 589, 592, 598.

58. "The Consequences of a Cow," *EN* 17 (June 11, 1887): 377. The second quotation is from "Fall of the Apple River Bridge."

59. "Masterpiece of engineering" is from untitled, *EN* 17 (Mar. 19, 1887): 181. "Of far from good iron" is from an untitled editorial, *EN* 17 (Mar. 19, 1887): 188. See also A. G. Robbins, "The Bussey Bridge," *Technology Quarterly* 1 (Sept. 1888): 68–72. "The Second Contributing Cause to the Bussey Bridge Disaster," *EN* 17 (Apr. 30, 1887): 285–87. "The Testimony as to the Brakes at the Fall of the Bussey Bridge," *EN* 17 (Apr. 30, 1887): 289. MRBC, *Nineteenth Annual Report, 1887* (Boston, 1889), 26–28, 38–52 and Appendix C.

60. The report of the commissioners to the legislature, with testimony, is from "The Bussey Bridge Disaster," *RR* 17 (May 7, 1887): 261–62. "More About Bridge Testing," *RG* 19 (Apr. 8, 1887): 233; "The Massachusetts Law Concerning Railroad Bridges," *RG* 19 (July 15, 1887): 465. MBRC, *Twentieth Annual Report, 1888* (Boston, 1889), 26–38. *Twenty-First Annual Report, 1889* (Boston, 1890), 11–21.

61. "Precaution Against Accident," *RR* 27 (Oct. 8, 1887): 583. "Trenton Falls Bridge, Adirondack & St. Lawrence Railroad," *EN* 29 (Apr. 13, 1893): 344. Untitled editorial, *EN* 42 (Aug. 10, 1899): 88. "Guard Rails and Deck Construction for Railway Bridges," *EN* 62 (Sept. 9, 1909): 270–76. The surveys are in "Report of Committee 7—On Wooden Bridges and Trestles," AREA *Proceedings* 14 (1913): 652–76, 1136–43, quotation on 1138, and 15 (1914): 402–5.

62. "Progress in Bridge Building," *RR* 44 (June 30, 1904): 556–57. A. J. Himes, "On Classification of Existing Bridges," AREA *Proceedings* (1907): 361–75, quotation on 364. "Two Large Plate Girder Railway Bridges," *EN* 51 (Feb. 18, 1904): 166–67. For bridge building techniques at the turn of the century see Charles Fowler, "Some American Bridge Shop Methods," and "Machinery in Bridge Erection," *Cassier's Magazine* 17 (Jan. and Feb. 1900): 200 215 and 327 44. Markley's remarks are in "Report of Committee 7—Bridges and Trestles," AREA *Proceedings* 2 (1901): 172–73.

63. Quotation from "Precaution Against Accident," 583. "Methods of Making Annual Inspections of Bridges and Culverts," *EN* 50 (Oct. 29, 1903): 394–95. See also B. W. Guppy, "On Maintenance of Existing Metal Bridges," AREA *Proceedings* 8 (1907): 369–75, and E. H. McHenry, *Engineering Rules and Instructions, Northern Pacific Railway* (New York, 1899), ch. 5.

64. For a well-documented instance where a bridge guard averted disaster see "Report of Committee 7—On Wooden Bridges and Trestles," AREA *Proceedings* 14 (1913): Appendix H. "Train Accidents in 1900," *RG* 33 (Feb. 15, 1901): 112–13. In 1896, Stowell found twenty-nine bridge accidents, or about as many as occurred in the mid-1880s, but with more bridges in 1896, the rate of failure must have been lower. ICC *Accident Bulletins* do not separate bridge failures from other forms of accident until 1917.

65. That failure is instructive is argued by Petroski, *To Engineer is Human: The Role of Failure in Successful Design* (New York: St. Martins, 1985).

Chapter 6. Coping with the Casualties

Epigraphs: Motto from Robert G. Paterson, *Robert Harvey Reed: A Sanitary Pioneer in Ohio* (Columbus: Ohio Public Health Association, 1935), 46; Perkins is quoted in Paul Black, "Reluctant Paternalism: Employee Relief Activities of the Chicago, Burlington & Quincy Railroad in the 19th Century," *Business and Economic History* 6 (1977): 120–34, quotation on 130.

1. A. G. Ellingwood, "Reminiscences of a Collision," *RS* 1 (Nov. 6, 1894): 287.
2. No comprehensive secondary literature on railroad medical programs exists. Early discussions of some programs are in Emory Johnson, "Railway Relief Departments," U.S. Department of Labor *Bulletin* 8 (Washington, 1897): 39–57, and W. H. Baldwin, "Railroad Relief and Beneficiary Associations," American Economic Association *Papers and Proceedings of the Twelfth Annual Meeting* (1899): 213–30. See also W. H. Allport, "American Railway Relief Funds," *Journal of Political Economy* 20 (Jan. 1912): 49–78, and (Feb. 1912): 101–34.

   A survey of the extent of such programs as of 1930 is in Pierce Williams, *The Purchase of Medical Care Through Fixed Periodic Payment,* National Bureau of Economic Research 20 (New York, 1932). Modern writers such as Henry Selleck, *Occupational Health in America* (Detroit: Wayne State University Press, 1962), treat railroad medical programs briefly. See too John Duffy, *The Healers: A History of American Medicine* (Champaign: University of Illinois Press, 1979); Ronald Numbers, "The Third Party: Health Insurance in America," in *The Therapeutic Revolution,* ed. Morris Vogel and Charles Rosenberg (Philadelphia: University of Pennsylvania Press, 1979); William Rothstein, *American Physicians in the 19th Century: From Sects to Science* (Baltimore: Johns Hopkins University Press, 1972, 1992); Ira Rutkow, "Railway Surgery," *Archives of Surgery* 128 (Apr. 1993): 458–63; and Paul Starr, *The Social Transformation of American Medicine* (New York: Basic, 1982). Stuart Brandes, *American Welfare Capitalism* (Chicago: University of Chicago Press, 1970) largely ignores railroad medical programs, as do more recent histories of railroad labor such as Walter Licht, *Working for the Railroad* (Princeton: Princeton University Press, 1983) and James Ducker, *Men of the Steel Rails* (Lincoln: University of Nebraska Press, 1983). For details of individual programs see Black, "Reluctant Paternalism"; Henry J. Short, *Railroad Doctors, Hospitals, and Associations: Pioneers in Comprehensive, Low Cost Medical Care* (Upper Lake, Calif., 1986); J. Roy Jones, *The Old Central Pacific Hospital* (Western Association of Railway Surgeons, 1960); and Logan Eib, "Pacific System's Big Human Repair Plant," Southern Pacific *Bulletin,* 10 (Dec. 1921): 3–5.
3. Early details of railroad doctors are from "The Railway Surgeon as a Practical Pioneer of Industrial and Traumatic Surgery," *RS* 34 (Nov. 12, 1928): 190–92; G. J. Northrop, "Railway Surgery with the Duties of the Chief and Local Surgeons," *RS* 1 (June 19, 1894): 25–29; and E. R. Lewis, "The Evolution of Railroad Surgery," *RS* 1 (Sept. 25, 1894): 227–33. Lowman's appointment is from I. P. Klingensmith, "The Care of the Injured by the Pennsylvania Railroad," National Association of Railway Surgeons *Proceedings* 5 (1892): 46–49. Early mining operations also found it expeditious to provide medical care for employees. See Larry Lankton, *Cradle to Grave: Life, Work, and Death at the Lake Superior Copper Mines* (New York: Oxford University Press, 1991), ch. 11.
4. The exchange with Carr is from J. E. Wooten to W. C. Wheeler, Jan. 14, 1873, and J. B. Ramsey to W. C. Wheeler, Jan. 15, 1873, box 174, Accident and Damage Claims, 1873, RRC, HML. William McKenzie to J. E. Wooten, Sept. 29, 1873, emphasis on original; McKenzie to Wooten, Apr. 8, 1874, box 177, Accident and Damage Claims, 1874, RRC, HML.
5. The Central of Georgia had a structure like that of the Reading as late as 1885. See S. S. Herrick, "Railway Medical Service," *ARRJ* 59 (Feb. 1886): 328–29. By 1892 the company had changed to a fee-bill system. The origins of the Central Pacific organization are from E. R. Lewis, "The Evolution of Railway Surgery,"

and Herrick, "Railway Medical Service," *ARRJ* 59 (Oct. 1885): 195–96, which contains the quotation, and claims that the Marine Hospital Service was the model for the CP organization. Also see the following. Short, *Railroad Doctors.* Jones, *The Old Central Pacific Hospital.* Eib, "Pacific System's Big Human Repair Plant."

6. S. S. Herrick, "Railway Medical Service," *ARRJ* 59 (Oct. 1885): 195–96; (Nov. 1885): 232–33; (Dec. 1885): 264–65; (Jan. 1886): 297–98; (Feb. 1886): 328–29. Lewis, "Evolution of Railway Surgery." Herrick describes the Northern Pacific as a beneficial association, but the only benefits it provided except for medical care were funeral expenses. See also "Northern Pacific Beneficial Association," *RS* 8 (Dec. 1901): 210–12. For other hospital associations see "The Atchison Railroad Employees' Association," *RR* 24 (Mar. 15, 1884): 135, and "Chronology of Hospital Affairs," file U-7-F1, Union Pacific Archives, Omaha, Nebr. The survey is George Chaffee, "The Railway Employees' Hospital Association," *RS* 2 (Sept. 22 1896): 193–97. R. W. Corwin, "History of the Colorado Fuel and Iron Company's Hospital," *RS* 8 (Feb. 1902): 270–71. "A Hospital System on the Chicago, Milwaukee, & St. Paul," *RA* 19 (Feb. 2, 1894): 61. Short, *Railroad Doctors.* For hospital systems established by mining companies see Lankton, *Cradle to Grave.* Richard Langlois, "The Vanishing Hand: The Changing Dynamics of Industrial Capitalism," *Industrial and Corporate Change* 12 (Apr. 2003): 351–85.

7. The Lackawanna is from Albert Sellenings, "First Aid to the Injured," New York Railroad Club *Proceedings* 16 (Dec. 1905): 7–19. For the Pennsylvania's early support of hospitals see Charles Pugh to S. M. Prevost, Mar. 20, 1885, and [illegible] to Charles Pugh, Mar. 4, 1889, who also notes that employees could choose their hospital. Both in microfilm roll 84, board file 86, PRRC, HML. For later activities see F. W. Hankins to W. W. Atterbury, Oct. 27, 1920, box 485, PRRC, HML. For the Delaware & Hudson see Lewis Carr to F. P. Gutlius, Oct. 7, 1919, and Gutlius to Carr, Dec. 8, 1919, box 17, series 2, Delaware & Hudson Collection, New York State Archives.

8. For early history of the Pennsylvania see Klingensmith, "The Care of the Injured." For the early programs see Herrick, "Railway Medical Service."

9. That is, $\{\exp[(0.0013 \times 8) + (0.0005 \times 101)] - 1\} \times \$630 = \$40$.

10. Michigan Bureau of Labor, *Eleventh Annual Report, 1893* (Lansing, 1894). Kansas Bureau of Labor and Industry, *Sixteenth Annual Report, 1900* (Topeka, 1901). Kansas over-sampled relatively well paid transportation workers. I weighted the savings and home ownership of these workers by their national shares of employment and then weighted the average of all other workers combined by their residual share of employment.

11. Peter Karsten, *Heart vs. Head: Judge-Made Law in Nineteenth Century America* (Chapel Hill: University of North Carolina Press, 1997), has argued persuasively that the employees did not always lose compensation suits. The point remains, however, that few cases went to court; of those, comparatively few workers won large sums, and legal fees took a large proportion of such awards.

12. The B&M and New York, Providence & Boston are from MBRC, *Twenty-Third Annual Report, 1891* (Boston, 1892), Appendix H. The Pennsylvania is from chief claim agent to E. H. Seneff, June 7, 1920, box 1186, PRRC, HML.

13. Elizabeth Russell to J. E. Wooten, Nov. 19, 1874, box 177, and Frederick Marty to D. C. Reinhart, Feb. 29, 1876, box 188, both in RRC, HML. Financing the

hospital is from "Celebrated Steam Engineers—John E. Wooten," *LE* 22 (Apr. 1909): 161–63. Average earnings are from *Historical Statistics,* series D-735.

14. "Common practice" is from testimony of Emory Johnson, U.S. Industrial Commission, *Report on Transportation* 4 (Washington, 1900), 152. E. P. Dela-hay to A. J. Cassatt, Jan. 26, 1898, file 43/32, box 29, PRRC, PSA. "When a train" is from MBRC, *Twenty-Third Annual Report, 1891* (Boston, 1892), 214. For the Southern Pacific see George Fabers to A. J. Stevens, Dec. 7, 1887, box 1, Letcher Railroad Collections, Ms 268, Stanford University. The Santa Fe examples are from Ducker, *Men of the Steel Rails,* ch. 2. See also Licht, *Working for the Railroad,* ch. 5.

15. For the union plans see Emory Johnson, "Brotherhood Relief and Insurance of Railway Employees," Bureau of Labor Statistics *Bulletin* 17 (Washington, 1898); Samuel Lindsay, "Report on Railway Labor in the United States," U.S. Industrial Commission *Report* 17 (Washington, 1901), 709–1135; J. B. Kennedy, "The Beneficiary Features of the Railway Unions," in *Studies in American Trade Unionism,* ed. Jacob Hollander and George Barnett (New York: Henry Holt, 1907), 321–50; and U.S. Commissioner of Labor, *Twenty-Third Annual Report, 1908: Workmen's Insurance and Benefit Funds in the United States* (Washington, 1909). For brief discussion of industrial life insurance see George McNeill, *A Study of Accidents and Accident Insurance* (Boston, 1900), and John Witt, *The Accidental Republic* (Cambridge, Mass.: Harvard University Press, 2004), ch. 3.

16. "To the Employees of the Baltimore and Ohio Railroad Company and Its Branches and Divisions" (n.p., n.d.; copy in New York Public Library). Baltimore and Ohio Employees Relief Association, *Act of Incorporation, Prospectus, Constitutions and By-laws of its Relief, Annuity, Savings Fund and Building Features* (Baltimore, n.d.). Baltimore and Ohio Employees Relief Association, *First Annual Report, 1881* (Baltimore, 1882).

17. That the Pennsylvania was studying some form of relief system is revealed in "Employees Fund," Oct. 13, 1875, microfilm roll 84, board file 91, PRRC, HML. Quotations on the value of medical examiners from J. A. Anderson to Advisory Committee of Relief Fund, Feb. 2, 1892, microfilm roll 85, board file 91, PRRC, HML. Labor unrest also played a role in motivating the Pennsylvania's relief department. See General Manager Frank Thompson to President George B. Roberts, Aug. 23, 1881, microfilm roll 84, board file 91, PRRC, HML. For a survey of relief departments see Max Riebenack, *Railway Provident Institutions in English Speaking Countries* (Philadelphia, 1905). Even where these programs were ostensibly voluntary, employees often felt pressured by officers to join, probably in order to keep costs low. See U.S. Industrial Commission, *Report on Transportation* 4 (Washington, 1900): 43–71.

18. Black, "Reluctant Paternalism." The Pinkerton involvement is revealed in Edward Gaylor to James Landis, July 14, 1894, box 998, RRC, HML. The quotation on discharging nonjoiners is from C. M. Lawler to A. A. McLeod, Mar. 12, 1889, box 998, RRC, HML.

19. On opposition to corporate medicine see James Burrow, *AMA, Voice of American Medicine* (Baltimore: Johns Hopkins University Press, 1963), ch. 9, and Starr, *The Social Transformation of American Medicine,* 202–5. Neither *Railway Surgeon* nor the *Journal of the American Medical Association* reveals any hostility of the AMA toward railroad programs, which suggests that its opposition

was more selective than these writers suggest. The only complaints about low fees I have discovered are in S. R. Miller, "Resolution," Association of Surgeons of the Southern Railway *Proceedings of the Twelfth Annual Meeting* (1907): 43, which resulted because the company's fees had not kept up with inflation. G. G. Dowdall [chief surgeon of the Illinois Central], "So-Called 'Contract Practice,'" *RS* 21 (June 1915): 345–52, and discussion, 352–55, claimed that the Illinois Central hospital organization was not contract medicine. It is the only reference to contract medicine I found in that journal.

20. Samuel Lindsay, "Railway Employees in the United States," U.S. Bureau of Labor Statistics *Bulletin* 37 (Washington, 1901). See also Riebenack, *Railway Provident Institutions in English-Speaking Countries*.

21. Montana Bureau of Agriculture, Labor, and Industry, *First Annual Report* (Helena, 1893), Tables 14–16. I aggregated the value of individuals' death benefits from all sources and regressed them on age, marital status (*Married*), daily earnings (*Earn*), and membership in a union or beneficial society (*Benorg*). *Married, Union,* and *Benorg* are 0,1, dummies with membership in each category being 1. Figures in parentheses are t-ratios. The results were as follows.

$$Benefits = -1019.06 - 2.70Age + 431.36Married + 676.46Earn$$
$$(0.16) \qquad (1.77) \qquad\qquad (5.44)$$
$$+ 704.04Union) + 669.27Benorg \; R^2 = 0.15, \; N = 386$$
$$(2.96) \qquad\qquad (2.99)$$

The determinants of life insurance were as follows, where *UnDeath* and *BnDeath* are union and benefit program death benefits in dollars. Figures in parentheses are *t*-ratios.

$$Life \; Insurance = -1058.74 + 9.80Age - 77.84Married +$$
$$(0.69) \qquad (0.37)$$
$$630.05Earn - 0.204UnDeath + 0.011BnDeath \; R^2 = 0.09, \; N = 386$$
$$(6.05) \qquad\; (2.07) \qquad\qquad (0.01)$$

22. C. B. Stemen, "The History of Railway Surgical Associations," *RA* 17 (June 17, 1892): 478.

23. Membership is from "Railway Surgery," *RA* 16 (Jan. 3, 1891): 13. The origins are from "Department of Railway Surgery," *Medico Legal Journal* 12 (1894): 180. For Reed's activities see Paterson, *Robert Harvey Reed*. Decline in the organizations is discussed in Ira Rutkow, "Railway Surgery: Traumatology and Managed Health Care in 19th Century United States," *Archives of Surgery* 128 (Apr. 1993): 458–63.

24. W. B. Outten, "Railway Relief Organizations in the United States," *RS* 2 (Oct. 8, 1895): 223–24. Pritchard is quoted in Clark Bell, "Railways and Railway Surgeons," *RS* 1 (July 17, 1894): 73–77, quotation on 74.

25. Pritchard is quoted in Bell, "Railways and Railway Surgeons," on 74.

26. "Age of mollusks" from R. Harvey Reed, "Surgical Department," *RA* 18 (Feb. 10, 1893): 127. Other quotations from R. Harvey Reed, "Surgical Department," *RA* 18 (June 30, 1893): 521–22. The 86 percent is based on data in U.S. Railroad Administration, *Survey and Recommendations of the Committee on Health and Medical Relief* (Washington, 1920).

27. Reed, "Surgical Department," *RA* 18 (June 30, 1893): 521–22. "I am viewed" is in the discussion of I. P. Klingensmith, "The Care of the Injured," 49–50.

28. Clark Bell, "The True Field of Duty of the Railway Surgeon," *Medico Legal Journal* 12 (1894): 374–86. Cole is quoted in Bell, "The True Field," 380. Clark Bell, "The Expert Witness," *RS* 2 (July 30, 1895): 103–5. For brief discussions of expert witnesses and the Medico-Legal Society see James Mohr, *Doctors and the Law: Medical Jurisprudence in Nineteenth-Century America* (New York: Oxford University Press, 1993), chs. 14 and 15.

29. Alexander Cochran, "The Relationship Existing Between the Legal and Hospital Departments of Railways," *RS* 9 (Aug. 1902): 61–63. The proportion reporting to claims department from U.S. Railroad Administration, *Survey and Recommendations.*

30. Patients on the Missouri Pacific from "News and Notes," *RS* 3 (July 1, 1896): 91. For Massachusetts General see George Gay, "Abuse of Medical Charity,'" reprinted in *Caring for the Working Man: The Rise and Fall of the Dispensary,* ed. Charles E. Rosenberg (New York: Garland, 1989): 179–248. Baltimore and Ohio Railway Employees Relief Association, *First Annual Report* (Baltimore, 1880). *Statement of the Workings of the Railroad Hospital at Sacramento, California for the Year 1883* (Sacramento, 1884).

31. "An empty sleeve . . ." is from H. T. Bahnson, "The Status of the Southern Railway Surgeon," Association of Surgeons of the Southern Railway Company, *Twentieth Annual Meeting* (1916): 5–15, quotation on 14.

32. Huntington's work is described in J. Roy Jones, *Memories, Men and Medicine* (Sacramento, 1950), 140–43. J. Roy Jones, *The Old Central Pacific Hospital,* and Henry Short, *Railroad Doctors.* For the Gulf, Colorado & Santa Fe see Patricia Benoit, *Men of Steel, Women of Spirit: History of the Santa Fe Hospital 1891–1991* (Temple, Tex., 1991), ch. 4. "Professor Roentgen's Discovery," *RS* 2 (Feb. 25, 1896): 470–72. Joel Howell, *Technology in the Hospital* (Baltimore: Johns Hopkins University Press, 1996), ch. 4. For a modern discussion of the reception of Listerism in the United States see Thomas Gariepy, "The Introduction and Acceptance of Listerism: Antisepsis in the United States," *Journal of the History of Medicine* 49 (Apr. 1994): 167–206.

33. Benoit, *Men of Steel,* ch. 4. Logan Eib, "The Southern Pacific's Big Human Repair Plant." A. Tyroler, "The Santa Fe Hospital in Los Angeles Provides a Practical Cause for Thanksgiving," *The Santa Fe Magazine* 20 (Nov. 1926): 21–26, quotation on 24.

34. On welfare capitalism see Brandes, *American Welfare Capitalism.* David Brody, *Workers in Industrial America,* 2d ed. (New York: Oxford University Press, 1993), ch. 2, argues that welfare capitalism had more worker support than most historians would like to admit. For the rail unions' views of the benefit plans see, for example, "Testimony of Mr. Frank P. Sargent" [Brotherhood of Locomotive Firemen], U.S. Industrial Commission, *Report on Transportation* 4 (Washington, 1900), 64–95.

35. The Burlington executive is N. T. Gurnsey, "The Relief Department of the Chicago, Burlington and Quincy Ry," *RR* 45 (July 22, 1905): 544–46. See also testimony of Hugh Bond of the B&O in Senate Committee on Interstate Commerce, *Hearings on Liability of Employers, May 3–8, 1906* (Washington, 1906), 128–29. Texas railroads from O. W. Karn to C. H. Markham, June 28, 1915, file U-7-F1, Union Pacific Archives, Omaha, Nebr.

36. W. H. Elliott, "The Organization of a Railway Surgical Department," National Association of Railway Surgeons *Proceedings* 5 (1892): 31–36. For similar views see J. F. Pritchard, "Surgical Department," *RA* 18 (Nov. 3, 1893): 814. The Freeman case is from "Imposture as a Profession," *RS* 1 (Mar. 26, 1895): 528–33, quotation on 528.

37. This discussion is based on Michael Trimble, *PostTraumatic Neurosis: From Railway Spine to Whiplash* (New York: John Wiley, 1981). For Erichsen's views see John Eric Erichsen, *The Science and Art of Surgery* 7th ed., 1 (Philadelphia, 1878), 560–79. Broader discussions of the significance of railway spine are in Wolfgang Schivelbusch, *The Railway Journey* (New York: Berg, 1986), ch. 9, and Ken Dornstein, *Accidentally on Purpose* (New York: St. Martins, 1996), ch. 5. R. Harvey Reed, "Surgical Department," *RA* 18 (June 30, 1893): 521–22. For Cowan's remarks see "Surgical Department," *RA* 18 (Oct. 12, 1893): 768–69. Arthur Dean Bevan, "Real and Alleged Injuries of the Spine," *RS* 2 (Aug. 13, 1895): 121–28.

38. Bevan, "Real and Alleged Injuries of the Spine," quotations on 124 and 125. For a similar interpretation see Eric Caplan, *Mind Games* (Berkeley: University of California Press, 1998), ch. 1. Bevan was articulating a widely shared view. See J. H. Green, "Hypnotic Suggestion in its Relation to the Traumatic Neuroses," National Association of Railway Surgeons *Proceedings* 5 (1892): 106–13, and "A Case That Ought to Have Been One of Railway Spine," *RA* 19 (Dec. 13, 1894): 711. For Bevan's career see *NCAB*, 31, 282–83.

39. Baltimore and Ohio Employees Relief Association, *Fourth Annual Report, 1884* (Baltimore, 1884), 8. On the activities of manufacturing company medical departments see National Industrial Conference Board, *Health Services in Industry,* Report 34 (New York, 1921), National Industrial Conference Board, *Medical Care of Industrial Workers* (New York, 1926), and "Medical and Hospital Service for Industrial Employees," *Monthly Labor Review* 24 (Jan. 1927): 7–19. Angela Nugent, "Fit for Work: The Introduction of Physical Examinations in Industry," *Bulletin of the History of Medicine* 57 (Winter 1983): 578–95, claims these later examinations were rather perfunctory. Both she and Alan Derickson, "'On the Dump Heap': Employee Medical Screening in the Tri-State Zinc Lead Industry, 1924–1932," *Business History Review* 62 (Winter 1988): 656–77, argue that physicals benefited companies rather than workers.

40. For Jeffries's role in Massachusetts see "Tests for Color Blindness," *RR* 23 (Mar. 17, 1883): 153. Jeffries's views are contained in "Defective Vision Among Railway Employees," *RR* 23 (Sept. 8, 1883): 518 and 526–28. See also his *Color Blindness: Its Dangers and its Detection* (Boston, 1883). The broad public support is revealed by hundreds of petitions sent to the Board of Health signed by men (but no women) from a broad range of occupations. See boxes 3 and 4, Health Department, Record Group 16, Connecticut State Library; box 3 also contains many letters of support for the men's position from superintendents and general managers. The quotations are from William Carmalt to G. W. Chamberlin, Oct. 26, 1880, box 3, Health Department, Record Group 16, Connecticut State Library (emphasis in original).

41. Jeffries's views of the Massachusetts and Connecticut laws, and his acceptance of Thomson's compromise, are from his "Defective Vision." Thomson's test as administered on the Reading is outlined in William Thomson, "Color Blindness Among Railroad Employees," *RR* 27 (Sept. 3, 1887): 511, and his "The Sight

and Hearing of Railway Employees," *Popular Science Monthly* 26 (Feb. 1885): 433–41. By 1895 the Burlington was also using Thomson's methods, see "Surgical Department," *RA* 20 (Aug. 23, 1895): 417. The use of tests is from ARA *Proceedings* 3 (1899–1902): 689–798. "Well nigh universal" from untitled editorial, *RA* 49 (Sept. 2, 1910): 383.

42. For the ARA-approved examination see its *Proceedings* 4 (1903–6): 361–64. Baltimore and Ohio Employees Relief Association, *Fourth Annual Report, 1884* (Baltimore, 1884), 8. On the origins of the North Western see Ralph Richards, "The Benefits of a Surgical Department," *RS* 9 (Feb. 1903): 241–45. See also Ralph Richards, "Physical Fitness of Railway Employees," *Medico Legal Journal* 17 (1899–1900): 816–26. That examinations were widespread is from ARA *Proceedings* 3 (1899–1902): 681–799. The Burlington is from J. C. Bartlett to W. Hoffman, May 21, 1897, box 998, RRC, HML. The Delaware & Hudson only instituted physicals for laborers in 1926, when it began to get large numbers of workmen's compensation claims. "Annual Report of Assistant to the General Manager for Personnel, 1926," box 4, Delaware & Hudson Collection, New York State Library.

43. Baltimore and Ohio Employees Relief Association, *Second Annual Report, 1883* (Baltimore, 1883), 10–11; *Third Annual Report, 1884* (Baltimore, 1884), 8.

44. For the Pennsylvania see J. A. Anderson to Advisory Committee of Relief Fund, Feb. 2, 1892, microfilm roll 85, board file 91, PRRC, HML. "In a general way" is from S. S. Herrick, "Railway Medical Service," *ARRJ* 59 (Nov. 1885): 232.

45. For the Pennsylvania's circular see S. W. Latta, "Water Supply for Drinking Purposes and Water Closets on Railroads," APHA *Papers and Reports of the Seventeenth Annual Meeting* (1889): 145–50. Charles B. Dudley, "The Ventilation of Passenger Cars on Railroads," *JFI* 144 (July 1897): 1–16. R. Harvey Reed, "Railway Surgery," *RA* 16 (Jan. 3, 1891): 13.

46. Nancy Tomes, *The Gospel of Germs* (Cambridge, Mass.: Harvard University Press, 1998), chs. 1 and 2. "Who knows . . ." is from discussion by John Hurty of "Report of the Committee on Car Sanitation," APHA *Papers and Reports of the Twenty-Fifth Annual Meeting* (1897): 245–61, quotation on 261. For Hurty's career see *NCAB*, 22, 370–71, and Ralph Williams, *The United States Public Health Service, 1798–1950* (Washington: U.S. Public Health Service, 1951), 137–38.

47. The issue is *RS* 2 (Mar. 24, 1896). The survey is in "Report of the Committee on Car Sanitation," APHA *Papers and Reports of the Twenty-Seventh Annual Meeting* (1899): 395–409. "Unless the railroads" is from "Sanitation of Passenger Cars," *ARRJ* 71 (Dec. 1897): 428–30, quotation on 429.

48. For both the Big Four and Illinois Central see "Sanitation of Passenger Cars." For the Pennsylvania see "Department of Physical and Chemical Tests (Historical)," box 661, and "Report of the Association of Transportation Officers," Oct. 31, 1913, box 408; both in PRRC, HML.

49. Charles B. Dudley, "The Dissemination of Tuberculosis as Affected by Railway Travel," APHA *Papers and Reports of the Thirty-Third Annual Meeting* (1905): 187–97. Thomas Crowder, "The Problem of Car Sanitation," APHA *Papers and Reports of the Thirty-Fourth Annual Meeting* (1906): 278–93. Tomes, *Gospel of Germs,* ch. 4.

50. "Report of the Committee on Hygiene and Sanitation," *RS* 18 (Apr. 1912): 329–32, quotation on 329. "Report of the Committee on Hygiene and Sanitation," *RS* 19 (Jan. 1913): 201–8, quotation on 201.

51. "Report of the Committee on Sanitation of Public Conveyances," Transactions of the Eleventh Annual Conference of State and Territorial Health Officers with the USPHS, *Public Health Bulletin* 63 (1913): 51–54. U.S. Railroad Administration, *Survey and Recommendations*. The four railroad surgeons were A. Z. Dunott, of the Western Maryland; George Cale, of the St. Louis & San Francisco; Thomas R. Crowder, from Pullman; and Henry M. Bracken of the Rock Island. For the code see "Standard Railway Sanitary Code," USPHS, *Public Health Reports* 35 (July 23, 1920): 1749–61.

52. For the survey see "Report of the Committee on Car Sanitation." See also A. J. McCannel, "Drinking Water for the General Public on Railway Trains," *RS* 18 (Jan. 1912): 213–19. "Railroad Water Supplies in Minnesota," USPHS, *Public Health Reports* 29 (May 15, 1914): 1222–45.

53. Norfolk & Western from "Problems in Interstate Health Work," Transactions of the Seventeenth Annual Conference of State and Territorial Health Officers with the USPHS, *Public Health Bulletin* 105 (1920): 68–85, which also contains the quotation on 68. The 60 percent figure is from Arthur Gorman, "Drinking Water Supplied to Trains," *RS* 30 (Jan. 1924): 89–93.

54. Free typhoid vaccination on the Illinois Central and Northern Pacific from Arthur Collins, "Antityphoid Prophylaxis Among Railway Employees," *RS* 22 (Mar. 1916): 265–69. The SP is from Philip Brown, "Tuberculosis in a Railway Health Insurance Program," *American Journal of Public Health* 25 (June 1935): 741–48. The Rock Island's efforts to combat typhoid and hookworm are from I. F. Crosby, "Preventive Medicine for Our Employees," *RS* 20 (Apr. 1914): 283–85, and T. B. Bradford, "Hookworm Disease," *RS* 20 (May 1914): 333–37. For the Canadian Pacific see "Anti-Typhoid Inoculation," *EN* 75 (Mar. 16, 1916): 530. "Typhoid as an Accident Under Compensation Law," *EN* 75 (Mar. 2, 1916): 426–27. For Tennessee Coal and Iron see Lloyd Noland, "Problems of Administration in Industrial Surgery," *Journal of the American Medical Association* 99 (Oct. 18, 1932): 1215–18.

55. For antimalarial work by the Illinois Central see Transactions of the Fifteenth Annual Conference of State and Territorial Health Offices with the U.S. Public Health and Marine-Hospital Service, *Public Health Bulletin* 93 (1918): 74. H. N. Old, "Malaria Control on the Central of Georgia Railway," Transactions of the Third Annual Conference of Malaria Field Workers, *Public Health Bulletin* 125 (1922): 89–93. H. W. Van Hovenberg, "How Cotton Belt Cut Malaria Rate 97 Per Cent in Nine Years," *REM* 22 (Oct. 1926): 382–90. Don Hofsommer, "St. Louis Southwestern Railway's Campaign Against Malaria in Arkansas and Texas," *Arkansas Historical Quarterly* 62 (Summer 2003): 182–93.

56. W. E. Estes, "Two Hundred and Sixty-four Amputations with Deductions," National Association of Railway Surgeons *Proceedings* 5 (1892): 149–51. W. E. Estes to A. A. McLeod, Sept. 3, 1892, box 998, RRC, HML.

57. This history is based on Charles R. Dickson, "The Progress of the First Aid Movement," *RS* 9 (Nov. 1902): 156–57. The manual is Charles R. Dickson, *First Aid in Accidents* (Chicago, 1901).

58. "A Hospital Car for the Southern Pacific Ry," *EN* 55 (Jan. 11, 1906): 32–33, and "Lehigh Valley Hospital Car," *RG* 40 (Jan. 12, 1906): 42–43. A. F. Jonas, "Union Pacific Emergency Association," *RS* 8 (Sept. 1901): 97–99; "The surgeon gives freely" is from A. F. Jonas, "Method of Instruction of First Aid to the Injured," *RS* 8 (Jan. 1902): 221–26, quotation on 222; Charles Dickson, "The Progress of the First Aid Movement," *RS* 9 (Nov. 1902): 156–62; Jonas's quotation on 158.

Significantly, Dickson subtitled his manual "What to Do and What Not to Do in Case of Injury."

59. Charles R. Dickson, "The Progress of the First Aid Movement," quotations on 157, 159–60. For the review of the Pennsylvania program see J. A. Bower et al. to E. B. Hunt, June 6, 1916, box 408, PRRC, HML.

60. For the Northern Pacific see "First Aid and Emergency," *RS* 9 (Apr. 1903): 319–20, 349–50. "Pennsylvania's Plan for First Aid to the Injured," *RA* 37 (Apr. 1, 1904): 725. A survey of first aid on two hundred carriers is reported in "Emergency Surgical Boxes," *RG* 33 (Apr. 26, 1901): 280. The Buffalo, Rochester & Pittsburgh is from "First Aid to the Injured in Railroad Service," *RS* 8 (Sept. 1901): 114–18.

Chapter 7. Safety Crisis and Safety First

Epigraphs: Braman B. Adams, "The Superintendent, the Conductor, and the Engineman," *RG* 34 (June 18, 1902): 564. ICC, *Report on Block Signal Systems* (Washington, 1907), 20.

1. The increase in derailments from 1897 to 1900 is based on *Railroad Gazette* figures. The post-1902 rise is from ICC reports and the two cannot be compared. "A Train Falls 75 Feet," *NYT* (Sept. 28, 1903), 1. Don Helmers, *Tragedy at Eden* (Pueblo, 1971).

2. The Hodges accident is from "A Rule That Would Prevent Forgetfulness of Train Orders," *RA* 37 (Oct. 7, 1904): 498–99.

3. "The Westfield Collision," *RG* 35 (Feb. 6, 1903): 96. Training on the Central is from "Preventable Collisions of a Year," *RG* 35 (July 3, 1903): 556.

4. "Who is the Murderer," *Washington Post* (Dec. 27, 1904), 6. "Being butchered" is quoted in "The Causes of Railway Collisions," *RA* 43 (Jan. 25, 1907): 109–10. "Harvest of Death," *World's Work* 13 (Apr. 1907): 8711–12. "American Railway Slaughters and British Railway Safety," *Scientific American* 88 (Mar. 28, 1903): 226–27. On comparisons with the Spanish American War see "Fallacies of the Deadly Parallel," *RA* 30 (July 20, 1900): 47–48. For the Gettysburg comparison and a critique see "Railway Accidents—Causes and Remedy," *RA* 44 (Aug. 9, 1907): 172–73.

5. The calculation is 54,032 × 5 = 270,160, which is 0.36 percent of the U.S. population in 1900 and 1/.0036 = 278.

6. The changing conception of accidents is based on Thomas Haskell, "Capitalism and the Origins of Humanitarian Sensibility, Parts I and II," *American Historical Review* 90 (1985): 339–61, 547–66. See also his *Emergence of Professional Social Science* (Champaign: University of Illinois Press, 1977), chs. 1 and 11, and Lawrence Friedman, "Civil Wrongs: Personal Injury Law in the Late 19th Century," *American Bar Foundation Research Journal* 12 (1987): 351–76, and Randolph Bergstrom, *Courting Danger: Injury and Law in New York City 1870–1910* (Ithaca: Cornell University Press, 1992), ch. 7. William Thomas, *Lawyering for the Railroad: Business, Law, and Power in the New South* (Baton Rouge: Louisiana State University Press, 1999), ch. 5. An analysis that sees Progressive-era regulation as a response to business efforts to thwart the liability system is Edward Glaeser and Andrei Schleifer, "The Rise of the Regulatory State," *Journal of Economic Literature* 41 (June 2003): 401–25. Needless to say this does not fit the facts of railroad safety very well.

7. For the *Tribune* see "Railway Accidents—Causes and Remedy." Frank Haigh

Dixon, "Railroad Accidents," *Atlantic Monthly* 99 (May 1907): 577–90. Edward Moseley, "Railroad Accidents in the United States," *American Monthly Review of Reviews* 30 (Nov. 1904): 593–96. "Out of control" is from "The Commission on Block Signaling," *RA* 37 (Jan. 1, 1904): 4–5.

8. For Roosevelt's proposals see Fred Israel, *The State of the Union Messages of the Presidents, 1790–1966* (New York: Chelsea House, 1967), vols. 2 and 3. John J. Esch, "Should the Safety of Employees and Travelers on Railroads be Promoted by Legislation?" *North American Review* 179 (Nov. 1904): 671–84.

9. "June Accidents," *RG* 34 (Aug. 8, 1902): 627. The Seaboard is from untitled, *RG* 40 (Mar. 2, 1906): 193.

10. U.S. 59th Cong., 1st sess., House Committee on Interstate and Foreign Commerce, *Hearings on H.R. 4438, H.R. 16676, and H.R. 18671 to Limit the Hours of Service of Railroad Employees* (Washington, 1906). U.S. 59th Cong., 2d sess., House Committee on Interstate and Foreign Commerce, *Hearings on S. 5133 and H.R. 24373 to Limit the Hours of Service of Railroad Employees* (Washington, 1907). U.S. 59th Cong., 2d sess., House Committee on Interstate and Foreign Commerce, *Hearings to Limit the Hours of Service of Railroad Employees,* 2 vols. (Washington, 1907, 1908). See also my *Safety First: Technology, Labor, and Business in the Building of American Work Safety, 1870–1939* (Baltimore: Johns Hopkins University Press, 1997), ch. 5.

11. "Exceedingly large number" from "News Item," *RG* 44 (Mar. 8, 1908): 324. "Noticeably deficient" is from Block Signal and Train Control Board, *Second Annual Report* (Washington, 1909), 10; the Illinois Central is on 50. For LaFollette's figures see U.S. 59th Cong., 2d sess., "Hours of Labor of Railroad Employees," *Congressional Record* 41 Part 1 (Jan. 9, 1907): 811–19.

12. "Electric Headlights," *RA* 48 (Apr. 29, 1910): 1074; "Government Regulation of Headlights," *RA* 56 (June 17, 1914): 1454; "The High-Power Headlight Rule," *RR* 60 (Jan. 6, 1917): 5–8. Bureau of Railway Economics, *The Arguments for and Against Train-Crew Legislation,* Miscellaneous Series Bulletin 18 (Washington, 1915); *The Arguments for and Against Limitation of Length of Freight Trains,* Miscellaneous Series Bulletin 23 (Washington, 1916).

13. "Government Control of Safety Appliances," *RA* 48 (Jan. 28, 1910): 171–72; "Harmonious Railway Action Regarding Safety Appliances," *RA* 49 (Oct. 21, 1910): 728; "Safety Appliance Standards," *RA* 49 (Nov. 11, 1910): 910. "The Proposed U.S. Safety Appliance Standards," *RR* 50 (Aug. 20, 1910): 806–8. "The Present Status of Clearance Legislation," *RA* 57 (Aug. 28, 1914): 377–80.

14. For details on the legislation and its impact see my "Safe and Suitable Boilers: The Railroads, the Interstate Commerce Commission, and Locomotive Safety, 1900–1945," *Railroad History* 171 (Autumn 1994): 23–44.

15. This shift in emphasis to managing men is discussed in Monte Calvert, *The Mechanical Engineer in America, 1830–1910* (Baltimore: Johns Hopkins Press, 1967), ch. 12. H. Raynar Wilson, "British Railway Methods and Management with Special Reference to Safety in Operation," *EN* 56 (Aug. 30, 1906): 218–20.

16. "Good Discipline as a Factor of Safety in Train Operation in England," *EN* 38 (Aug. 12, 1897): 105–6. "Some Lessons of Recent Collisions," *EN* 49 (Feb. 15, 1903): 126–27. The absence of distant signals is from ICC, *Report on Block Signals* (Washington, 1907), 46.

17. Braman B. Adams, "The Superintendent, the Conductor, and the Engineman," *RG* 34 (June 18, 1902) and (Sept. 14, 1902): 563–65 and 865.

18. Endorsement of a law requiring blocking is in "Prevention of Collisions," *RG* 36 (Dec. 16, 1904): 633.

19. Advice from the ARA is from U.S. 61st Cong., Senate Committee on Interstate Commerce, *Hearing on H.R. 3649, Accident Reports, Feb. 5, 1910* (Washington, 1911), 11. The board is identified in "Block Signals in 1907," *RG* 44 (Dec. 27, 1907): 906–7. Absence of central authority and critique of piecemeal legislation are from Block Signal and Train Control Board, *Fourth Annual Report* (Washington, 1912), 12–14. "The primary cause . . . ," "the American is not reared . . . ," and "so serious . . ." are from *Third Annual Report* (Washington, 1911), 25–27. "Devoted the same . . . ," "some central authority . . . ," and "publicity often cures . . ." from *Fourth Annual Report* (Washington, 1912), 11, 12, and 20. "Conditions of society . . ." from *Final Report* (Washington, 1912), 22.

20. W. H. Boardman, "What Are We Going to Do about Railroad Accidents," *RG* 44 (Jan. 31, 1908): 144–45.

21. Ralph Richards, "The Safety First Movement on American Railways," in Second Pan-American Scientific Conference *Proceedings* 11 (Washington, 1917): 326–35, quotation on 328. "The Campaign Against Accidents on the Chicago & North Western," *RA* 49 (Sept. 2, 1910): 391–93. For an extended discussion see my "Safety First Comes to the Railroads, 1910–1939," *Railroad History* 162 (Spring 1992): 6–33.

22. Douglass North, *Institutions, Institutional Change, and Economic Performance* (New York: Cambridge University Press, 1990), ch. 8. George Bradshaw, "The Safety First Movement," Western Railway Club *Proceedings* 25 (Feb. 18, 1913): 240–89, quotations on 242.

23. Meeting with midlevel managers first is from Ralph Richards, *What the Safety Committees of the Chicago & North Western Railway Have Done for the Conservation of Men* (Chicago, 1912), quotations on 11 and 13. R. H. Newbern to S. C. Long, Mar. 26, 1914, box 485, PRRC, HML.

24. "Report of the Committee on Transportation," ARA *Proceedings* 6 (May 1912): 818–21, quotations on 820 and (May 1913): 16–17. Numbers of carriers from Chicago & North Western Railway Co., *Report of the Central Safety Committee* (Chicago, 1913).

25. Bryce Stewart and Murray Latimer, *Railroad Pensions in the United States and Canada* (Industrial Relations Counselors, 1929). Ripley is from "Discipline and Signal Systems," *RA* 43 (Jan. 11, 1907): 32–33. The Alton and New York, Chicago & St. Louis are from Adams, "The Superintendent." "Instruction of Queen & Crescent Train Men," *LE* 15 (Apr. 1902): 162–63. "Erie Employment Bureau," *RG* 37 (Dec. 30, 1904): 688–89. "Record Discipline on the B&O," *RG* 56 (Jan. 16, 1914): 113–18.

26. For emphasis on discipline and supervision see Ralph Richards, *Railroad Accidents, their Cause & Prevention* (Chicago, 1906), and A. Hunter Boyd, "Accidents Attributable to the Carelessness of Employees," Steam Railroad Section NSC *Proceedings* 1 (1912): 135–46. "Accident Statistics on the Pennsylvania," *RA* 57 (Nov. 13, 1914): 913. Boardman, "What are We Going to do About Railroad Accidents," suggests Kruttschnitt may have originated surprise checking. Its use on the Illinois Central is from an untitled item, *RA* 50 (Mar. 24, 1911): 673–74. "The Pennsylvania Railroad's Safety Campaign," *RA* 56 (Feb. 20, 1914): 361–65. For the Cincinnati Southern see "Operation of a Single Track Railway Under the Block System," *EN* 54 (Sept. 28, 1905): 322–23. The Big Four is from

"Concerning the Railway Signal Situation," *RR* 47 (Feb. 7, 1907): 93. The slow spread of surprise checking is from "The Baker Bridge," *RG* 39 (Dec. 8, 1905): 526. The Lehigh Valley is from "Surprise Tests of Signals," *RR* 47 (Aug. 24, 1907): 750–51.

27. Boardman, "What are We Going to Do About Railroad Accidents." "Derailment of Pennsylvania Special—Report of Committee on the Cause," *RG* 42 (Mar. 1, 1907): 272–73. J. Kruttschnitt, "General Letter No 25," file U-8-C1, Union Pacific Archives, Omaha, Nebr. (emphasis in original). W. L. Park, "Publicity for Accidents," Western Railway Club *Proceedings* 21 (1909): 189–223. For a modern discussion of trust and stigma see James Flynn et al., eds., *Risk Media and Stigma* (London: Earthscan, 2001).

28. Surveys of use of record discipline are in "Brown's Discipline on American Railroads," *RA* 63 (July 6, 1917): 19–21. "Discipline on the Chicago, Burlington & Quincy," *RA* 56 (June 12, 1914): 1321–23. "Steady Employment of Section Men on the Long Island," *RR* 56 (Apr. 17, 1914): 539–40.

29. Robert Lovett to A. L. Mohler, Nov. 22, 1915, file U-8-C1, Union Pacific Archives.

30. "The Progress of Block Signaling on American Railways," *EN* (Aug. 21, 1902): 129, estimated that "from one-fourth to one-third of American railway traffic is now moved under protection of block signals."

31. Better glass is from "New York Railroad Club; Signal Glass," *RA* 56 (Feb. 27, 1914): 432. See also "A Century of Signal Glass Developments," *Railway Signaling* 42 (May 1949): 310–13. "Automatic Block Signaling on Single Track," *RG* 38 (Jan. 5, 1905): 1–2. "Degrees in Block Signal Protection," *RG* 38 (Feb. 24, 1905): 154.

32. "Steel Passenger Cars," *RA* 32 (Sept. 13, 1901): 232–33. "Steel Cars in Wrecks," *RA* 32 (Sept. 20, 1901): 295. "Stronger Passenger Cars," *RA* 36 (Nov. 27, 1903): 717. The need for stronger cars is from untitled, *EN* 51 (Apr. 14, 1904): 352, untitled, *EN* 52 (July 2, 1904): 71, and "Yielding Resistance in Trains of Steel Passenger Cars," *RA* 39 (Apr. 28, 1905): 666.

33. On the development of steel cars see John White, *The American Railroad Passenger Car* (Baltimore: Johns Hopkins University Press, 1978), ch. 2. For specifications of the Pennsylvania cars see "Steel Passenger Cars for Pennsylvania Trains," *RR* 47 (June 3, 1907): 481–83. See also "Steel Passenger Cars," *RG* 42 (June 14, 1907): 834–24. The evolution of car design is discussed in "Steel Postal Car Design," *RA* 51 (Nov. 17, 1911): 983–84, and "Proposed Postal Car Design," *RA* 51 (Dec. 1, 1911): 1086. "The Steel Passenger Cars in the Odessa Wreck on the St. Paul," *RA* 52 (Jan. 26, 1912): 152–53. "New End Construction for Pullman Cars," *RA* 55 (Aug. 8, 1913): 231–32. "A Collapsible Platform and Vestibule," *RA* 54 (Jan. 24, 1913): 142–44. For a review of the development of steel cars see "The Steel Passenger Car and Existing Passenger Equipment, I–III," *RR* 53 (Oct. 25, 1913): 999, (Nov. 8, 1913): 1040–41, (Nov. 22, 1913): 1084–85, (Dec. 6, 1913): 1128–29, and 54 (Jan. 3, 1914): 20–21. "All Steel Passenger Car Equipment for the Erie Railroad," *RR* 60 (June 8, 1917): 785–86.

34. I fitted the equations below to data for 1902–40 with a correction for serial correlation (observations are unavailable between 1917 and 1921) employing (the log of) fatalities per derailment or collision, passengers per train, percent of cars with steel or steel frame, a trend, and a postwar dummy, *pw*, that is zero before 1921 and 1 for 1921 on, which also interacts with all other variables. Only the interaction with steel cars and the trend is shown. Figures in parentheses are t-ratios.

$$\text{Log}(Fat/derail) = 2.05 - 0.042Pass/Trn - 0.018PctSteel + 0.072Trend - 2.25PW$$
$$(4.43) \qquad\qquad (3.76) \qquad\qquad (3.87) \qquad\qquad (5.07)$$
$$+ 0.014PWSteel - 0.058PWTrend \; R^2 = -0.03, N = 34, D.W. = 1.29$$
$$(1.71) \qquad\qquad (2.03)$$
$$\text{Log}(Fat/collision) = 0.740 - 0.013Pass/Trn - 0.004PctSteel + 0.016Trend - 0.947PW$$
$$(3.56) \qquad\qquad (3.54) \qquad\qquad (1.75) \qquad\qquad (4.99)$$
$$- 0.019PWSteel + 0.055PWTrend \; R^2 = 0.44, N = 34, D.W. = 1.78$$
$$(3.54) \qquad\qquad (3.31)$$

By 1916, 31 percent of cars were steel and the predicted effects are $\exp(-0.018 \times 31) - 1 = -0.43$ and $\exp(-0.004 \times 31) - 1 = -0.12$. However, the positive trends suggest that much or all of the gain was offset, probably by increases in train speed.

35. For improved freight brakes see "The Improved Air Brake," *RA* 45 (Dec. 4, 1908): 1466–67, quotation on 1466, and "The 'K' Triple," *LE* 19 (Dec. 1906): 562–63. "The Control of Modern Passenger Trains as Achieved by the 'PC' Equipment," *RR* 51 (July 15, 1911): 635–36. Clasp brakes on the Reading and Central of New Jersey are from "Clasp Brakes for Passenger Cars," *RA* 53 (Aug. 16, 1912): 279–81. S. W. Dudley, "Brake Tests on the Pennsylvania," *RA* 56 (Feb. 13, 1914): 311–13. A good review of developments is "Modern Air Brake Equipment as Applied to Steam Roads," *RR* 53 (June 14, 1913): 572.

36. "Accident Record, Union Pacific Railroad," *RR* (Jan. 4, 1913): 15–16. Untitled, *RA* 50 (Mar. 24, 1911): 673–74, discusses the Illinois Central. "Safety on the Illinois Central," *RA* 50 (June 16, 1911): 1411. "Discipline on the Chicago, Burlington & Quincy," *RA* 56 (June 12, 1914): 1321–23. ICC, *Twenty-Sixth Annual Report, 1912* (Washington, 1913), 66.

37. Ivy Lee to Samuel Rea, May 18, 1909, box 120, PRRC, HML. "Our Railways are Safest," *New York American* (May 20, 1912), 1. Phoebe Snow is from B. A. Botkin and Alvin Harlow, *A Treasury of Railroad Folklore* (New York: Crown, 1953), 392.

38. Legal developments are discussed in "The Steel Passenger Car and Existing Passenger Equipment."

39. The reports of the Special Committee are reprinted in "Progress in Steel Passenger Car Equipment," *RR* 51 (Aug. 5, 1911): 689, "Steel and Steel Underframe Passenger Equipment," *RR* 52 (Oct. 5, 1912): 912, "Steel Passenger Train Equipment," *RA* 55 (Aug. 15, 1913): 269–70. The article was "Steel Passenger Train Car Situation," *RA* 55 (Nov. 21, 1913): 947–52. See U.S. 63d Cong., 2d sess., House Committee on Interstate and Foreign Commerce, *Hearings on Bills Relative to Safety of Railroads: The Use of Steel Cars II, Dec. 17, 1913 and V, Jan. 19, 1914* (Washington, 1914), 58.

40. The cost figure is from U.S. 63d Cong., 2d sess., House Committee on Interstate and Foreign Commerce, *Hearings on Bills Relative to Safety of Railroads: Block Signals, Automatic Stops, Uniform Train Rules III Jan. 15, 1914* (Washington, 1914), exhibit 1. "Where Government Control of Railroads is Needed," *RG* 39 (Nov. 3, 1905): 406–7, calls for federal accident investigation, while "A Railroad Accident Investigation Bureau," *RG* 42 (Feb. 22, 1907): 231–32, claimed it would provide the carriers with public relations benefits. "The Lessons of Three Collisions," *RA* 53 (July 26, 1912): 146–47. "Report of the State Commission on the Western Springs Wreck, C, B & Q R.R.," *RR* 52 (Aug. 17, 1912): 767–68. The New Haven general manager is quoted in "Report on the Stanford Wreck," *RR* 53 (July 19, 1913): 689–90. ICC, *Twenty-Sixth Annual Report, 1912*

(Washington, 1913), 62–66, quotation on 65. *Twenty-Seventh Annual Report, 1913* (Washington, 1914), 59–71.

41. Beginning in 1919 the ICC briefly took note of the effect of inflation on accident reporting but did nothing about it. See *Accident Bulletin 71* (Washington, 1919), 7, as well as *Accident Bulletin 72* (Washington, 1920), 7, and *Accident Bulletin 74* (Washington, 1922), 15.

42. "Some apprehension" is from "Quality of Chilled Cast-Iron Wheels," *RR* 41 (Sept. 7, 1901): 590–91. "Rumors Afloat" is from "The Cast Iron Wheel," *RA* 32 (July 12, 1901): 20. The quotation is from A. Feldpauche to E. D. Nelson, Sept. 2, 1903. The findings are from an attachment to a memo by superintendent of motive power [R. N. Durborow] to A. P. Gest, Oct. 29, 1904, and [?] to R. N. Durborow, Aug. 16, 1904. All in box 770, PRRC, HML.

43. For flange size and wheel taper see "Effect of Coning on Sharp Flanges," *RA* 53 (July 19, 1912): 92–93, and "Improvements in Chilled Car Wheels," *RA* 53 (Sept. 20, 1912): 500–501. "Flange Wear of Wheels as Affected by Anti-Friction Side Bearings," *RR* 48 (May 23, 1908): 414–15. MCB *Proceedings* 36 (1902): 246–47. "Flange Lubrication," *RR* 51 (July 8, 1911): 609–19.

44. For new specifications see MCB *Proceedings* 43 (1909): 182–215. Use of braking power as a guide to wheel size is noted in "Car Wheel Failures," *RA* 54 (May 9, 1913): 1021–22. Problems with refrigerator cars are from "Convention of the Master Car Builders' Association," *RR* 55 (June 19, 1915): 832–35. A good analysis of the economics of standards is David Hemenway, *Industrywide Voluntary Product Standards* (Cambridge, Mass.: Ballinger, 1975).

45. Quotation from "The Wheel Situation," *RG* 44 (Feb. 14, 1908): 204–5. Tests at Purdue are described in "Stresses in the Plates in Cast Iron Wheels," *RR* (Apr. 4, 1914): 539–40. Manufacturers' testing is described in "The Chilled Iron Car Wheel," *RR* 57 (Feb. 2, 1917): 270–71. That the Bureau of Standards was studying wheels at the behest of the association is from remarks of its president G. W. Lyndon. See "Supplemental Report of Standing Committee on Car Wheels," MCB *Proceedings* 50 (1916): 195–205. See also J. M. Snodgrass and F. H. Guilder, "An Investigation of the Properties of Chilled Iron Car Wheels," University of Illinois Engineering Experiment Station *Bulletin* 129 (May 1922). The Norfolk & Western wreck is from untitled, *RA* 44 (Nov. 22, 1907): 709. For regulatory worries see W. J. Morrison, "Has the Cast Iron Wheel Reached its Limit of Usefulness?" Southern and Southwestern Railroad Club *Proceedings* 11 (Mar. 1912): 6–13. ICC, *Twenty-Fifth Annual Report, 1911* (Washington, 1912), 91, *Twenty-Eighth Annual Report, 1914* (Washington, 1915), 56, and *Twenty-Ninth Annual Report, 1915* (Washington, 1916), 57.

46. For Hyde Park see an untitled editorial, *RR* (Aug. 3, 1912): 722. Elmer Howson, "Recent Developments in Track Construction," AREA *Proceedings* 16 (1915): 1–29. "Need for Data Regarding Track Construction," *RA* 54 (Mar. 21, 1913): 700.

47. Howson, "Recent Developments," notes improvements in ballast. "The Treatment of Railroad Tie Timbers," *RA* 34 (Sept. 12, 1902): 261–62; "The Crosstie Problem," *RA* 36 (July 17, 1903): 65. "Treated Tie Data from the Santa Fe," *RG* 39 (July 21, 1905): 63–64. "The Value of Tie Preservation," *RG* 44 (Jan. 10, 1908): 44. "Report of the Committee on Ties," AREA *Proceedings* 9 (1908): 695–797. The survey is reported in "Bridges and Trestles," *RR* 41 (Mar. 16, 1901): 181. That tie plates were still experimental is from "Tie Plates on American Railways," *RA* 49 (Dec. 9, 1910): 1104. "Development in the Use of Screw Spikes," *RA* 54 (Mar. 14, 1913): 499–504. "Care of Ties Pays Large Returns on the Lackawanna," *REM*

24 (Feb. 1928): 59–61. The Lakeview wreck is from J. D. Farrell to R. S. Lovett, May 13, 1913, file U-8-C1, Union Pacific Archives.

48. "Report of Examination of the Colorado and Southern Ry. and Main Line of the Fort Worth and Denver City Ry., July 5 1900," folder 2157, box 46, RG 513, Denver & Rio Grande Collection, Colorado Historical Society. ICC, *Report of the Chief Inspector of Safety Appliances Covering His Investigation of an Accident Which Occurred on the New Orleans & Northeastern Railroad Near Eastabuchie, Miss., May 6, 1912* (Washington, 1912). ICC, *Report of the Chief Inspector of Safety Appliances Covering His Investigation of an Accident Which Occurred on the Central of Georgia Railway Near Clayton. Ala., Nov. 13, 1913* (Washington, 1914), quotation on 13.

49. Cushing is quoted in "Report of the Committee on Rail," AREA *Proceedings* 6 (1905): 191. The B&O experience is from remarks of its chief engineer J. R. Onderdonk, in "General Discussion of Steel Rails," ASTM *Proceedings* 8 (1908): 121–22. A. C. Shand to Powell Stackhouse, Dec. 29, 1906, Stackhouse to Shand, Jan. 23, 1907, and Shand to Stackhouse, Feb. 13, 1907, which contains the quotation; all in box 4, PRRC, HML.

50. "Rails Broken by Flat Wheels," *RG* 52 (May 10, 1912): 1031–32. For manufacturers' views see "General Discussion on Steel Rails," ASTM *Proceedings* 7 (1907): 87–130, and 8 (1908): 109–27, and the "Report of the Committee on Standard Rail and Wheel Sections," ARA *Proceedings* 3 (Oct. 1907 and Apr. 1908): 175–78 and 312–17.

51. "Proposed Standard Specifications for Steel Rails," ASTM *Proceedings* 1 (1899–1900): 100–103. Hunt is from "General Discussion of Steel Rails," ASTM *Proceedings* 7 (1907): 96–99, quotation on 96. "Heavy Rail Sections in America," *Engineering* (London) 84 (Nov. 15, 1907): 688–89. "Rail Specifications," *RA* 44 (Oct. 11, 1907): 489–90. "Report of the Committee on Rail," AREA *Proceedings* 3 (1902): 202–3, notes the Pennsylvania had adopted the shrinkage clause in 1901; *Proceedings* 5 (1904): 465–67, 6 (1905): 182–92, and 7 (1906): 549–55.

52. "Amazing Increase in Broken Rails," *NYT* (Apr. 26, 1907), 5. The quotation is from "Defective Rails," *NYT* (May 13, 1907), 8. Dexter Marshal, "The Problem of the Broken Rail," *McClures* 29 (Aug. 1908): 428–33. "Broken Rails and Railroad Accidents," *Scientific American* 96 (Apr. 20, 1907): 326; "The Peril of the Broken Rail," *Scientific American* 96 (May 18, 1907): 409–11; "Steel Rails and the Public," *Outlook* 86 (July 6, 1907): 486–87.

53. "For Better Rails," *RG* 42 (May 17, 1907): 667. "The Railroads and the Rail Mills," *Iron Age* 79 (May 2, 1907): 1354. See also H. V. Wille, "Greater Loads on Rails," *Iron Age* 80 (Oct. 3, 1907): 922–24.

54. "Philadelphia & Reading Specifications for Rails," *RA* 45 (Feb. 28, 1908): 269. The Great Northern's experience is from H. H. Parkhouse to Lewis W. Hill, Nov. 30, 1908, President's Subject Files, 132.E.12.6.F, box 58, Great Northern Collection, Minnesota Historical Society. For the New York Central see "Ferrotitanium in Steel Rails," *Iron Age* 83 (Mar. 25, 1909): 988–89. Harriman lines from "Steel Rail Peril to be Inquired Into," *NYT* (May 11, 1907), 1. The Pennsylvania's rail committee is from E. C. Felton to W. W. Atterbury, May 21, 1907, and Theodore Ely to Rail Committee, June 8, 1907, box 4, PRRC, HML. Work of the committee may be followed in "Minute Book of the Rail Committee," R.G. 286, PRRC, PSA.

55. "New Rail Specifications and Sections," *RA* 44 (Oct. 11, 1907): 489–90. "Report of the Committee on Standard Rail and Wheel Sections." Failures on the Har-

riman lines are from "Report of the Committee on Rail, Appendix A," AREA *Bulletin* 106 (Dec. 1908): 107–25.

56. For Wickhorst's career see Harold Lane, ed., *Biographical Dictionary of Railway Officials of America* (New York: McGraw Hill, 1913), 588–89. For a good overview of the committee's work see "Review of Rail Investigations, 1910–1914," AREA *Proceedings* 16 (1915): 411–31, and M. H. Wickhorst, "American Research Work on Rails," *Iron Age* 90 (Sept. 12, 1912): 614–16.

57. Robert Hunt, "Recent Developments in the Inspection of Steel Rails," AIME *Transactions* 44 (1912): 269–77, and his "Manufacture of Steel Rails," AIME *Transactions* 62 (1920): 174–81. The 60 percent figure is from C. W. Genett, "Inspection of Rails at the Mill," *RA* 63 (Sept. 21, 1917): 51–52. That the nick and break test was rarely used is from Charles Gennet, "How the Quality of Rails Can Be Improved," *RA* 77 (July 12, 1924): 63–65.

58. "Report of the Committee on Rails and Equipment," NARC *Proceedings* 24 (1912): 261–337, quotation on 261. Indiana Railroad Commission, *Seventh Annual Report, 1912* (Indianapolis, 1913), 104–86. "A New Method of Testing Rails," *Scientific American Supplement* 72 (Dec. 23, 1911): 403–4.

59. ICC, *Report of Accident on the Line of the Lehigh Valley Railroad Near Manchester, N.Y., Aug. 25, 1911* (Washington, 1912), 13. "The Lehigh Valley Rail," *RA* 52 (Feb. 16, 1912): 267.

60. Farrel's testimony is in U.S. 62d Cong., 2d sess., House Committee on Investigation of United States Steel Corporation, *Hearings* 4 (Washington, 1913), 2680–83. A. W. Gibbs to Rail Committee, Jan. 27, 1912, box 5, PRRC, HML. For evidence on carbon content of rails see "Comparison of Chemical Constituents of Steel Rails from 1870 to Date," *RA* 53 (Oct. 11, 1912): 684–86. Robert Lovett to A. L. Mohler, Mar. 31, 1916, file U-13-3b, Union Pacific Archives.

61. For Gary's remarks see "For the Improvement of Rail Quality," *Iron Age* 89 (Mar. 7, 1912): 632–35, quotation on 632. ICC, *Report of the Chief Inspector of Safety Appliances Covering His Investigation of an Accident Which Occurred on the Great Northern Railway Near Sharon, N. Dak., Dec. 30, 1911 . . .* (Washington, 1912) and *Report of the Chief Inspector of Safety Appliances Covering His Investigation of an Accident which Occurred on the Wabash Railroad Near West Lebanon, Ind., Mar. 7, 1912 . . .* (Washington, 1912).

62. ICC, *Twenty-Sixth Annual Report, 1912* (Washington, 1913), 62. "No Government Inspection of Steel Rails," *Iron Age* 90 (Dec. 26, 1912): 1512–13. Indiana Railroad Commission, *Seventh Annual Report, 1912* (Indianapolis, 1913), 104–86, quotation on 106. "The Rail Situation on the Harriman Lines," *RA* 51 (Dec. 29, 1911): 1310–11.

63. The P&LE is from "Catching up with the Locomotives," *REM* 24 (July 1928): 328–35. For locomotive testing see Lawrence Fry, "The Locomotive Testing Plant and its Influence on Steam Locomotive Design," ASME *Transactions* 47 (1925): 1267–93. "Atlantic Type Locomotives on the Pennsylvania," *RG* 56 (Feb. 20, 1914): 356–59. "Vanadium Steel Driving Axles and Frames," *RA* (June 11, 1913): 1254. For five-year failure rates per hundred miles of track from 1908 through 1921 see "Report of the Rail Committee," AREA *Proceedings* 29 (1928): 558.

64. C. W. Baldridge, "A Review of the Performance of Steel Rails in American Railways," AREA *Proceedings* 34 (1933): 705–29, also minimized the significance of the shift from Bessemer to open hearth steel.

65. "Derailment at Warrior Ridge," *RA* 52 (Feb. 23, 1912): 347. "Telephones on American Railways," *RA* 52 (May 24, 1912): 1152–53.

66. "Frank McManamy," *NCAB* 33, 462. "Frank McManamy—Interstate Commerce Commissioner," *LE* (June 1923): 194–95. "Government Boiler Inspection," *RMM* 37 (Apr. 1913): 173–74. "Boiler Inspection Report," *RMM* 38 (Dec. 1914): 534. The Grand Rapids and Indiana is from J. R. Fitzpatrick to B. Fitzpatrick, Oct. 24, 1913, and James Weir to A. G. Pack, July 27, 1920, Locomotive Inspection Policy Correspondence Files, box 1, Record Group 134, NA.

67. "Inspection of Locomotive Staybolts," *RR* 77 (Dec. 5, 1925): 844–47. "A Non-Breaking Water Glass Gauge Glass," *RA* 34 (Oct. 24, 1902): 414. "Low Water Boiler Tests," *RR* 52 (June 29, 1912): 617–19; "Comparative Tests of Locomotive Boilers Equipped with the Jacobs-Shupert and the Radial Stay Fireboxes," and "The Jacobs-Shupert Firebox Tests," *RR* 53 (Feb. 8, 1913): 118–19 and 124–25. "An Unusual Locomotive Crown Sheet Accident," *RR* 61 (Apr. 9, 1921): 574–75.

68. C. Herschel Koyl, "The Work of Railroad Men on the Problem of Water for Steam Boilers," *RG* 34 (Mar. 30, 1900): 200–201. "Water Purification for Locomotive Boilers," *RA* 35 (Mar. 13, 1903): 357–58. "Water Softening on the Union Pacific," *RG* 36 (June 24, 1904): 62–63. The Santa Fe is from "The Care of Locomotive Boilers at Terminals and While in Service," *EN* 56 (Oct. 4, 1906): 354–55. "Walschaert Valve Gear," *EN* 56 (Sept. 20, 1906): 305. "Vanadium Steel and its Application to Locomotive Construction," and "Service Record of Vanadium Steel Main and Side Rods," *RR* 53 (Sept. 20, 1913): 894–95, and (Nov. 1, 1913): 1023–24.

69. "The Safety Movement on the Norfolk & Western Ry," and "Results of the Safety Movement on the Norfolk & Western Railway," *RR* 56 (Oct. 21, 1916): 552–55, and (Nov. 11, 1916): 672. The North Western is from [?] to R. C. Richards, Mar. 29, 1918, box A-1, series 1, Record Group 14, USRA, NA, and "Safety First Results on the C&N.W. Ry," *RR* 62 (June 29, 1918): 958–59. "Safety First on the Baltimore & Ohio R.R.," *RR* 56 (Oct. 14, 1916): 507–9. "The Pennsylvania Railroad's Safety Campaign," *RA* 56 (Feb. 20, 1914): 361–65. "Safety on the Pennsylvania Railroad," *RR* 54 (Feb. 7, 1914): 218–19. "Practical Results of the Safety Campaign on the New York Central Lines," *RR* 55 (Oct. 17, 1914): 475.

70. "Simply another way" is from "An Object Lesson in Safety First," *Railroad Trainman* 32 (Oct. 1915): 985–86, quotation on 985. "Federal Plan for Conducting Safety Work in Effect on the Baltimore & Ohio," *B&O Employees' Magazine* 6 (Aug. 1918): 5–7. For the Northwestern Pacific see "Operates Two Years Without a Fatality," *RR* 78 (Jan. 9, 1926): 122–25.

71. The "No Accident Weeks" are discussed in R. H. Aishton to Northwestern Railroads, Aug. 12, 1919, box 1, series 73, Record Group 14, USRA, NA. The figures on safety meetings, etc., are in "Supplement No. 1 to Safety Section Bulletin No. 14," box 1, series 73, Record Group 14, USRA, NA. The decline in casualties is from "Results of No Accident Drive," Nov. 20, 1919, box A-1, series 1, Record Group 14, USRA, NA. The unions' views are in a series of resolutions in box 13, series 81, Record Group 14, USRA, NA.

72. Chicago & North Western Railway, *Report of the Central Safety Committee* (Chicago, 1913).

73. "Trespassing on Pennsylvania Railroad Property," *EN* 65 (Jan. 11, 1911): 59. Whiting is from "Trespassers Killed and Injured on Railways—Who Are They?" *RA* 52 (Mar. 8, 1912): 432–83. The larger survey results are reprinted in NARC *Proceedings* 29 (1917): 195–96.

74. "Darius Miller Urges Co-operation of Governors to Reduce Trespassing," *RA* 52 (Mar. 8, 1912): 440–41; "Frisco Asks Governors to Co-operate in Reducing

Trespassing," *RA* 52 (Apr. 26, 1912): 968–69. "Why 5,000 Trespassers are Killed Yearly," *RA* 53 (Dec. 20, 1912): 1173–78, quotation on 1174. "Trespassers Killed and Injured on the New York Central," *RA* 55 (Aug. 15, 1913): 267–69.

75. John Droege, *Passenger Terminals and Trains* (New York: McGraw-Hill, 1916). "Provisions to Insure Safe Walking In and About Passenger Stations," AREA *Proceedings* 19 (1918): 248–70. Doubling passenger miles per journey does not necessarily double them relative to exits and entrances to trains because of the need to change trains on a given journey.

76. The over half assertion is in "Report of the Committee on Grade Crossings," NARC *Proceedings* 27 (1915): 42–49. Some of the novel dangers resulting from automobiles are also discussed in its *Proceedings* 28 (1916): 34–40, and in "Report of the Special Committee on the Prevention of Accidents at Grade Crossings," ARA *Proceedings* 4 (May 1916): 19–23.

77. NARC *Proceedings* 27 (1915): 42–49, quotation on 42. A. H. Rudd to W. G. Coughlin, May 23, 1914, and A. B. Clark to G. W. Snyder, June 12, 1914, box 422, PRRC, HML. The meeting is discussed in its *Proceedings* 28 (1916): 34–40. For the railroads' view see "Report of the Special Committee on the Prevention of Accidents at Grade Crossings," ARA *Proceedings* 4 (May 1916): 19–23, and (Nov. 1916): 122–24. Lewis Carr to J. T. Lorree, July 26, 1922, box 17, series 2, Delaware & Hudson Collection, New York State Archives. For comments that the automobiles increased derailment risks see "The Automobile Peril," *RA* 61 (July 28, 1916): 135. The Pennsylvania's tests are reported in J. K. Johnson to G. W. Snyder, Sept. 15, 1914, box 422, PRRC, HML.

78. Both the surveys are reported in "To Promote Safety at Highway Crossings," *RA* 59 (Dec. 17, 1915): 1119–20. For a larger survey with similar findings see "Carelessness as a Cause of Grade Crossing Accidents," *RA* 56 (May 1, 1914): 974–75.

79. The Long Island's efforts are in "To Promote Safety at Highway Crossings." Gate crashing is from an untitled article, *RA* 60 (Jan. 14, 1916): 71. Frank V. Whiting [claim agent, New York Central Railroad], "Stop, Look, Listen," *Outlook* 104 (Aug. 23, 1913): 927–33.

Chapter 8. Lobbying for Regulation

Epigraphs: "Informal Discussion of Moving Explosives," May 19, 1905, file 41, box 408, PRRC, HML, 10, and James McCrea to Taft, June 3, 1908, Presidential Correspondence of A. J. Cassatt and James McCrea, 1899–1913, file 62/10, box 49, MG 286, PRRC, PSA.

1. The Crestline disaster is from an untitled article in *EN* 50 (Nov. 12, 1903): 442. For discussions of institutional innovations see Lance Davis and Douglass North, *Institutional Change and American Economic Growth* (Cambridge: Cambridge University Press, 1971) and North, *Institutions, Institutional Change, and Economic Performance* (Cambridge: Cambridge University Press, 1990). For an up-to-date review of the literature see Vernon Ruttan, *Technology, Growth, and Development* (New York: Oxford University Press, 2001), ch. 4.

2. Ellis Hawley, "Herbert Hoover, the Commerce Secretariat and the Vision of an 'Associative State,' 1921–1928," *Journal of American History* 61 (June 1974): 116–40, quotation on 118, and his "Three Facets of Hooverian Associationalism: Lumber, Aviation and the Movies," in *Regulation in Perspective,* ed. Thomas McCraw (Cambridge, Mass.: Harvard University Press, 1981), 95–123.

3. Statistics on explosives production are from U.S. Bureau of the Census, *Manufacturing, 1905,* Part I (Washington, 1907), and Part IV (Washington, 1908). The first available figures on use by sector are for 1912 from Albert Fay, "Production of Explosives in the United States," U.S. Bureau of Mines *Technical Paper* 69 (Washington, 1912), Table 2.

4. The Wells Fargo explosion is from Arthur Van Gelder, *History of the Explosives Industry in America* (New York: Arno, 1972), 326–27. The 1866 law is in *U.S. Statutes at Large* 14 (1868): 81. Brief debate on the bill focused on recent explosions. See *Congressional Globe,* 39th Cong., 1st sess., 2635. The Worcester explosion is from MBRC, *Second Annual Report, 1870* (Boston, 1871), 27–28. "A Nitro Glycerin Accident," *RG* 10 (Jan. 11, 1878): 21. "A Terrible Accident," *RG* 15 (Aug. 24, 1883): 564. "Havoc Caused by the Explosion of a Car of Giant Powder," *RA* 13 (May 18, 1888): 317. The 3,300 figure is from the remarks of James McCrea, "Report of the Committee on Transportation of Explosives," ARA *Proceedings* 4 (Oct. 25, 1915): 406.

5. On the importance of the railroads and specifically the Pennsylvania in the development of modern management see Alfred Chandler, *The Visible Hand: The Managerial Revolution in American Business* (Cambridge, Mass.: Harvard University Press, 1977).

6. On the origins of the Pennsylvania's regulations see Charles B. Dudley, "Remarks," *EN* 61 (Apr. 1, 1909): 341.

7. Placing the sign is from U.S. 60th Cong., 1st sess., House Committee on Interstate and Foreign Commerce, *Hearings on HR 7557, To Promote the Safety of Transportation of Explosives* (Washington, 1908), 47. The disasters are from "Report of the Committee on Conducting Transportation on the Transportation of High Explosives," Apr. 29, 1895, box 407, PRRC, HML. See also ARABE, *Annual Report of the Chief Inspector, 1909* (New York, 1910), appendix 2 (hereafter, *Annual Report*). For other companies' regulations as of 1905, most of which are described as similar to those of the Pennsylvania, see "Compilation of Replies to Circular 603," ARA *Proceedings* 4 (1903–6): 470–78.

8. "Report of the Committee on Conducting Transportation on the Transportation of High Explosives," quotation on 3.

9. The naphtha explosion is reported in James McCrea to A. J. Cassatt, May 13, 1902, and McCrea, "Memorandum," May 13, 1902 (which contains the estimates of fatalities and injuries), Presidential Correspondence of A. J. Cassatt and James McCrea, 1899–1913, file 31/31, box 20, MG 286, PRRC, PSA. See also "Many Deaths from Naphtha Explosion," *NYT* (May 13, 1902), 1. Charles Perrow, *Normal Accidents* (New York: Basic, 1984), ch. 3.

10. Manufacturers of explosives are from an ARA survey in "Report of the Committee on Transportation of Explosives," ARA *Proceedings* 4 (Oct. 25, 1905): 446–51. The carrier with the next largest number of manufactures on its lines was the Central Railroad of New Jersey, with sixteen. Usselman, *Regulating Railroad Innovation,* chs. 7 and 9.

11. There is an immense literature on the conditions under which private firms will supply public goods. The classic is Ronald Coase, "The Lighthouse in Economics," *Journal of Law and Economics* 17 (Oct. 1974): 357–76.

12. "Report of the Committee on Conducting Transportation on the Transportation of High Explosives," quotations on 6.

13. For the Senate version of the bill, introduced by Senator Stephen B. Elkins of West Virginia, see U.S. 58th Cong., 2d sess., Senate Committee on Inter-

state Commerce, *Hearings on S 4319, Transportation of Explosives* (Washington, 1904). For the House bill, introduced by James Sherman of New York, see U.S. 58th Cong., 2d sess., House Committee on Interstate and Foreign Commerce, *Hearings on HR 11964, Prohibiting Common Carriers Engaged in Interstate Commerce From Transporting Gunpowder and Other High Explosive Compounds Over Their Lines Except Under Certain Conditions* (Washington, 1904). In the hearings on HR 11964 (4), McCrea said, "in the preparation of this bill I endeavored to go into the matter as fully as I could." One manufacturer, the Masurite Corporation, appeared to oppose the regulation, claiming that it would jeopardize the company's trade secrets. The ICC took no public stance on the bill. For British regulations see "The Explosives Act of 1875," 38 Vict. C 17, and *Annual Report of Her Majesty's Inspector of Explosives, 1875* (London: HMSO, 1876).

14. House Committee on Interstate and Foreign Commerce, *Hearings on HR 11964,* quotations on 4, 6, and 8.

15. House Committee on Interstate and Foreign Commerce, *Hearings on HR 11964,* quotation on 20.

16. On DuPont as a powder maker see Alfred Chandler and Stephen Salsbury, *Pierre DuPont and the Making of the Modern Corporation* (New York: Harper & Row, 1971), and Norman Wilkinson, *Lammot DuPont and the American Explosives Industry, 1850–1884* (Charlottesville: University of Virginia Press, 1984). For the company's early interest in safety see Donald Stabile, "The DuPont Experiments in Scientific Management: Efficiency and Safety, 1911–1919," *Business History Review* 61 (Aug., 1987): 365–86. Remarks of the DuPont representative are in House Committee on Interstate and Foreign Commerce, *Hearings on HR 7557,* 48.

17. "Report of the Committee on Transportation of Explosives," ARA *Proceedings* 4 (Apr. 5, 1905): 356.

18. For the Harrisburg disaster see ICC, *Accident Bulletin 16* (Washington, 1905), 10–11. Its cost is from ARABE, *Annual Report, 1909* (New York, 1910), appendix 2.

19. Untitled editorial in *EN* (May 25, 1905): 553. "Train Hit Dynamite; 163 Reported Dead," *NYT* (May 11, 1905), 1; "20 Dead, 100 Hurt in Dynamite Wreck," *NYT* (May 12, 1905), 1. The editorial is "Transporting Explosives," *NYT* (May 13, 1905), 8. A search of the *Reader's Guide to Periodical Literature* reveals no interest in the regulation of explosives on the part of any of the popular press. Neither the *New York Tribune* nor the *World* editorialized on the Harrisburg disaster, although both reported it fully. In Boston both the *Globe* and *Herald* followed similar procedures. In Cleveland, the *Plain Dealer* editorialized in "The Harrisburg Disaster" (May 12, 1905), 8, that the accident was unavoidable. Lack of interest by both the popular press and labor groups probably reflected the comparatively few accidents that resulted from explosives. ICC data reveal over 120,000 injuries and fatalities from railroad operation in 1907, only 132 of which occurred from transporting explosives.

20. The Pennsylvania's review, which lasted all afternoon and yielded a forty-eight-page transcript, is "Informal Discussion of Moving Explosives," May 19, 1905, file 41, box 408, PRRC, HML. "Was an accident" is from "Official Explains Wreck," *NYT* (May 11, 1905), 2. The Atterbury quotation is from "Informal Discussion of Moving Explosives," on 7.

21. "Informal Discussion of Moving Explosives," quotations on 8 and 10.

22. "Informal Discussion of Moving Explosives." The revised regulations are

"General Notice 174A," dated June 19, 1904, which superseded "General Notice 174," dated Sept. 25, 1899, both in box 411, PRRC, HML.

23. These developments are reported in "Report of the Committee on Transportation of Explosives," ARA *Proceedings* 4 (Oct. 25, 1905): 404–7. In the same volume see also "Laws—High Explosives," 463–68, and for railroad regulations "Compilation of Replies to Circular 603," 470–78. For biographical data on Drinker see *NCAB*, 15, 114; for McKenna see *Who Was Who*, I, 315; for Munroe, see *NCAB*, 29, 334. "To the Committee on Transportation of Explosives of the American Railway Association," and "Report of the Committee of Experts," ARA *Proceedings* 4 (Oct. 25, 1905): 446–51, and 452–55.

24. That Roosevelt was involved in the decision to delegate Dunn to the ARA is from Edgar Marburg, "Charles B. Dudley, Biographical Sketch," in ASTM, *Memorial Volume* Commemorative of the Life and Lifework of Charles Benjamin Dudley (Philadelphia, 1911), 11–42. For a biographical sketch of Dunn see ARABE, *Annual Report, 1936* (New York, 1937), 14–15. I have been unable to discover whether Dunn's work as an ordnance officer had made him known to DuPont or other manufacturers.

25. For state and city regulations see "Laws—High Explosives," 463–68, and for railroad regulations "Compilation of Replies to Circular 603," ARA *Proceedings* 4 (Oct. 25, 1905): 463–68 and 470–78. "Report of the Committee on Transportation of Explosives," ARA *Proceedings* 4 (Oct. 25, 1905): 404–7. That these rules generated no opposition at all probably resulted because they were voluntary and placed most of the burdens on shippers.

26. "Report of the Committee on Transportation of Explosives," ARA *Proceedings* 4 (Mar. 19, 1906): 595–98.

27. The original bills submitted by the ARA were HR 16011 and S4844. See "Report of the Committee on Transportation of Explosives," ARA *Proceedings* 4 (Mar. 19, 1906): 596. The 1866 law is reprinted in House Committee on Interstate and Foreign Commerce, *Hearings on HR 11964*. House Committee on Interstate and Foreign Commerce, *Hearings on HR 7557*.

28. Probably the decision to limit the scope of the bill reflected its sponsors' realization that inclusion of "other hazardous substances" might well have stirred up the entire chemical industry, as indeed occurred when the bureau later extended its rules to cover such substances (see below).

29. For the revised bill with amendments see U.S. 59th Cong., 2d. sess., House Committee on Interstate and Foreign Commerce, *Shipment of Gunpowder, etc.* Report 6746 (Washington, 1907). This bill became the basis for S 2611 and HR 7557 submitted by Representative James Sherman and Stephen Elkins in Dec. 1907 to the 60th Congress, 1st sess. McCrea's assessment is from McCrea to Dudley, Mar. 12, 1908, Presidential Correspondence of A. J. Cassatt and James McCrea, 1899–1913, file 65/33, box 50, MG 286, PRRC, PSA. For lobbying efforts of the Pennsylvania see also John Cassels and S. C. Neale to McCrea, Dec. 5, 1907, S. C. Neale to McCrea, Jan. 14, 1908, and identical letters from McCrea to Elkins, Senator John Kean, and Senator Foraker, all on Mar. 12, 1908. All in Presidential Correspondence of A. J. Cassatt and James McCrea, 1899–1913, file 65/33, box 50, MG 286, PRRC, PSA. For the hearings see U.S. 60th Cong., 1st sess., House Committee on Interstate and Foreign Commerce, *Hearings on HR 7557*. The ICC took no public stance on the bill.

30. For the sequence of ICC regulations and bureau rules see "Report of the Committee on Transportation of Explosives," ARA *Proceedings* 5 (Oct. 19, 1908):

477–78. The rules and regulations are codified in ARA, *The American Railway Association Rules and the Interstate Commerce Commission Regulations for the Transportation of Explosives and the American Railway Association Regulations for the Transportation of Inflammable Articles and Acids* (1909). For the extension of authority to regulate other hazardous substances see ICC, *Thirty-Fourth Annual Report, 1920* (Washington, 1920), 74–77, and *Thirty-Fifth Annual Report, 1921* (Washington, 1921), 47–48.

31. For the lobbying effort see Charles B. Dudley to Alfred Gibbs, Feb. 27, 1907, James McCrea to William Howard Taft, Mar. 7, 1907, and McCrea to Taft, June 3, 1908, which is the source of the quotation. All in Presidential Correspondence of A. J. Cassatt and James McCrea, 1899–1913, file 62/10, box 49, MG 286, PRRC, PSA.

32. Membership is from ARABE, *Annual Report, 1907* (New York, 1908). Nonmembers are from an untitled article in the *RG* 44 (Apr. 10, 1908): 494–95.

33. The quotation is from ARABE, *Annual Report, 1907* (New York, 1908), 9. In 1913, in part to present a coordinated front to deal with the bureau and other regulatory bodies, the manufacturers formed the Institute of Makers of Explosives. Van Gelder, *History of the Explosives Industry,* Appendix 1.

34. B. W. Dunn, "Origin and Work of the Bureau of Explosives, American Railway Association," *EN* 61 (Apr. 1, 1909): 340–41.

35. This story is taken from Dunn's remarks in "Report to the Committee on Transporting Explosives," ARA *Proceedings* 5 (Sept. 26, 1907): 86–88. For damages see "Explosions," *RG* 43 (Aug. 16, 1907), General News Section: 185, and untitled, *RG* 43 (Nov. 15, 1907): 577. The quotation is from ARABE, *Annual Report, 1907* (New York, 1908), 7.

36. The statistics remained unofficial. The ICC never published data on explosions or the casualties therefrom. Of course the condemned shipments represented a tiny fraction of all explosives shipped—1,468 kegs of powder amounted to about 0.02 percent of the total.

37. The quotation is from ARABE, *Annual Report, 1907* (New York, 1908), 5. ARABE, *Annual Report, 1909* (New York, 1910), 9.

38. The annual report of the chemical laboratory was routinely included in the bureau's annual reports. Impact tests at Altoona and tests for exudation by the bureau's chemical laboratory appear in ARABE, *Annual Report, 1908* (New York, 1909), 11–12 and 20–29. Improvements in waterproof wrappers for dynamite and tests of acetylene tanks are in ARABE, *Annual Report, 1909* (New York, 1910), appendix 3.

39. ARABE, *Annual Report, 1910* (New York, 1911), appendix 10.

40. Membership is reported in bureau annual reports. "Share in the direction" is from ARABE, *Annual Report, 1908* (New York, 1909), 10. McCrea to E. B. Thomas [president of the Lehigh Valley], Nov. 9, 1907, and general manager to W. H. White [president of the Richmond, Fredericksburg & Potomac], June 8, 1908, Presidential Correspondence of A. J. Cassatt and James McCrea, 1899–1913, file 68/27, box 51, MG 286, PRRC, PSA. Dunn's circular is in ARABE, *Annual Report, 1912* (New York, 1913), 5.

41. ARABE, *Annual Report, 1909* (New York, 1910), 7–8. "Report of the Committee on Transportation of Explosives," ARA *Proceedings* 5 (Apr. 19, 1909): 619 and (May 18, 1909): 623.

42. McCrea to W. W. Atterbury, Apr. 11, 1911, Presidential Correspondence of A. J. Cassatt and James McCrea, 1899–1913, file 62/10, box 49, MG 286, PRRC, PSA.

With little fanfare, this informal arrangement whereby the ICC essentially delegated rule-making power to a private association was given formal congressional approval in 1921. For a discussion of the 1921 law see U.S. 96th Cong., 1st sess., Senate Committee on Commerce, Science, and Transportation, *Hazardous Materials Transportation: A Review and Analysis of the Department of Transportation's Regulatory Program* (Washington, 1979).

43. ARABE, *Annual Report, 1908* (New York, 1909), 10–11. For the meeting with cardboard producers see ARABE, *Annual Report, 1914* (New York, 1915), appendix 6.

44. See "Report from Chemical Laboratory," ARABE, *Annual Report, 1913* (New York, 1914), appendix 2.

45. U.S. Bureau of the Census, *Annual Survey of Manufactures* (Washington, various years).

46. The quotation is from ARABE, *Annual Report, 1908* (New York, 1909), 4. The number of placarded cars is from "Report of the Committee on Transportation of Explosives," ARA *Proceedings* 5 (Oct. 19, 1908): 480–82, and (May 18, 1909): 621.

47. Revisions of the flashpoints and quantities are from ARABE, *Annual Report, 1909* (New York, 1910), 5. Revision of the proposed rules for the location of gasoline storage tanks is from ARABE, *Annual Report, 1919* (New York, 1920), 91–111.

48. The drug manufacturer is from ARABE, *Annual Report, 1908* (New York, 1909), 4–5. The embargo is from ARABE, *Annual Report, 1914* (New York, 1915), 14.

49. The circular and organizational chart are from ARABE, *Annual Report, 1919* (New York, 1920), appendix 6a. The quotation is from ARABE, *Annual Report, 1920* (New York, 1921), 4.

50. The tests of gas cylinders are in ARABE, *Annual Report, 1914* (New York, 1915), appendix 5. For acid carboys see ARABE, *Annual Report, 1910* (New York, 1911), 8. The beginnings of the test department are from *Annual Report, 1919* (New York, 1920), 7; its expanded duties are from *Annual Report, 1921* (New York, 1922), 3.

51. The tests are described in ARABE, *Annual Report, 1920* (New York, 1921), 5–6; for the improvements see *Annual Report, 1921* (New York, 1922), appendix 3, and *Annual Report, 1922* (New York, 1923), appendix 3.

52. The data for 1914 are from ARABE, *Annual Report, 1914* (New York, 1915), table 1. For the Ardmore disaster see ARABE, *Annual Report, 1915* (New York, 1916), table 5 and appendix 4.

53. ARABE, *Annual Report, 1916* (New York, 1917), 18, and *Annual Report, 1918* (New York, 1919), 4–5.

54. ARABE, *Annual Report, 1919* (New York, 1920), 6. ARA Mechanical Division *Proceedings* (1930): 197–233.

55. The contract with the Bureau of Standards is from ARA Mechanical Division *Proceedings* (1927): 423. ARABE, *Annual Report, 1927* (New York, 1928), 7–8; *Annual Report 1931* (New York, 1932), 6.

56. "Col. B. W. Dunn Dies," *NYT* (May 11, 1936), 19. ARABE, *Annual Report, 1936* (New York, 1937), 14–15. U.S. Congress, Office of Technology Assessment, *Transportation of Hazardous Materials* (Washington, 1986), ch. 4. See also U.S. 96th Cong., 1st sess., Senate Committee on Commerce, Science, and Transportation, *Hazardous Materials Transportation.*

57. David Vogel, "The 'New' Social Regulation," in McCraw, *Regulation in Perspec-*

tive, 155–86, notes that before the 1960s such regulations arose mostly at the state and local levels. Hawley, "Three Facets of Hooverian Associationalism" stresses the role of ideology.

58. For the changes in public policy toward railroad safety and hazardous substances in the 1960s see Office of Technology Assessment, *Transportation of Hazardous Materials,* Senate Committee on Commerce, Science, and Transportation, *Hazardous Materials Transportation,* and Ian Savage, *The Economics of Railroad Safety* (Boston: Kluwer, 1998), ch. 2.

Chapter 9. Private Enterprise and Public Regulation

Epigraphs: Both quotations from "Great Northern Train Control Petition Denied," *RA* 79 (Oct. 31, 1925): 805–8.

1. On the role of unions in safety on the Santa Fe see H. R. Lake, "Some Pro's and Con's on Railroad Safety Methods," ARA Safety Section *Proceedings* 10 (1930): 101–16. Representatives of the brotherhoods attended safety meetings on the New Haven from at least the late 1920s on. See Minutes of Midland Division, Apr. 15, 1930, box 25, Secretary Records, New York, Haven, and Hartford Railroad Collection, Record Group 1, Thomas Dodd Library, University of Connecticut at Storrs. For Aishton's career see "Association of American Railroads Organized," *RA* 97 (Sept. 29, 1934): 365–69. For more detail see my "Safety First Comes to the Railroads," *Railroad History* 166 (Spring 1992): 6–33.

2. "Safe and Efficient Hand Brakes," *The Railroad Trainman* 47 (Jan. 1930): 6–10.

3. Adjusting the numbers for inflation would reduce the increase in 1916 a bit and magnify the growth in costs in the 1930s. I have not made the adjustment because the evidence suggests the carriers based decisions on nominal, not inflation-adjusted values. J. C. Rose to E. H. Seneff, June 7 1920, box 1186, PRRC, HML. Alfred Chandler stresses organizational learning in many of his writings. See, for example, Alfred Chandler et al., *Big Business and the Wealth of Nations* (New York: Cambridge University Press, 1997), ch. 2. See also William Lazonick and Thomas Brush, "The Horndal Effect in Early U.S. Manufacturing," *Explorations in Economic History* 22 (Jan. 1985): 53–96.

4. "Taking the Ballyhoo out of Safety Methods," *RA* 86 (May 4, 1929): 1027. Norfolk & Western Railway Company, *Safety First* 15 (Roanoke, 1915). Quotations from "Injuries to Employees First Nine Months, 1925–1926," "Safety on the Pennsylvania Railroad in 1928," and "Meeting of the Regional Safety Committee . . . Sept. 2, 1927," all in box 635, PRRC, HML. "Pennsylvania Puts Force Behind its Safety Rules," *REM* 27 (July 1931): 646–47. For the North Western, Wabash, and Union Pacific see E. B. Hall, "Prevention of Accidents in the Motive Power Department," H. S. Caswell "Safety in Shop Operation," and George Warfel, "Discussion," all in NSC Railroad Section *Transactions* 18 (1929): 859–63, 868–69, and 882–83.

5. "Practice Growing of Paying Old Age Pensions," *RA* 75 (July 14, 1923): 65–68. Bryce Stewart and Murray Latimer, *Railroad Pensions in the United States and Canada* (Industrial Relations Counselors, 1929). "Group Insurance as Applied to Railroads," *RA* 72 (May 2, 1922): 1061–62, quotation on 1061; "Free Insurance on D&H," *RA* 72 (Jan. 14, 1922): 182. "Southern Pacific Provides Insurance for Employees," *RA* 75 (Nov. 3, 1923): 827–28. "How to Start Personnel Work on a Railroad," *RA* 76 (Mar. 15, 1924): 745–46. "Why Track Laborers Quit," *RA* 64 (Apr. 26, 1918): 1079–80.

6. Turnover is resignations and dismissals as a proportion of employment. Annual Reports of Safety Department, series 4, boxes 1–33, Delaware & Hudson Collection, New York State Archives. The survey is from Federal Coordinator of Transportation, *Unemployment Compensation for Transportation Employees* (Washington, 1936). H. G. Hassler, "Accident Prevention in a Steel Car Shop," ARA Safety Section *Proceedings* 6 (1926): 224–31, quotation on 225. For the Lackawanna see E. A. Koschinske, "Accident Prevention in a Large Locomotive Shop," ARA Safety Section *Proceedings* 10 (1930): 134–39, quotation on 138. The Reading is from "Car Riders on the Reading Eschew Carelessness," *RA* 87 (Oct. 19, 1929): 923. For the Pennsylvania see "A Truly Safe Yard," *RA* 83 (Nov. 5, 1927): 901–2.

7. I regressed the number of permanent disabilities on fatalities for the 1934–65 period with a correction for serial correlation. The data are for train and train service accidents only. Figures in parentheses are t-ratios.

$$Permanent = 55.80 + 0.366 Fatalities + 3.82(Trend) \; R^2 = 0.80, N = 32, D.W. = 2.03,$$
$$(4.88) \qquad\qquad (2.28)$$

8. On power reverse gears see "Better Engine Equipment," *Locomotive Engineers Journal* 65 (Feb. 1931): 85–86, "The Hearings on the Power Reverse Gear," *Locomotive Engineers Journal* 65 (Dec. 1931): 895–96, "United States et al., v. Baltimore & Ohio Railroad Co. et al.," *293 U.S. 454* (1937), "Rail Executives Meet in Chicago to Take up Reported Settlement with Unions on Power-Reverse Gears in Engines," *NYT* (Sept. 24, 1936), 35, and "Power Reverse Gears Ordered for Engines," *NYT* (June 15, 1937), 33. For lighter-weight reciprocating parts see "Modern Locomotive Equipment-I," *RME* 111 (Aug. 1937): 348–51. For thermic siphons see my "Energy Conservation on Steam Railroads: Institutions, Markets, Technology, 1889–1940," *Railroad History* 177 (Autumn 1997): 7–42.

9. "A Crown Sheet Failure Without an Explosion," *RA* 70 (Apr. 8, 1921): 885–86. "Recent Types of Freight and Passenger Locomotives for Canadian National Railways," *RR* 75 (Nov. 8, 1924): 744–45. "4-6-4 Type Locomotives on the Canadian Pacific," *RME* 105 (Apr. 1931): 167–71. "Can Boiler Explosions Due to Low Water Be Prevented?" *RA* (Mar. 21, 1936): 499–501. C. Herschel Koyl, "Treating Water Reduces Boiler Troubles," *RA* 66 (Apr. 25, 1919): 1053–56. "Railroad Water Problems Yield to Technical Advances," *RA* 95 (July 1, 1933): 44–46. "Embrittlement in Boilers," *RME* 114 (May 1940): 182–85.

10. On companies' failure to install appliances such as siphons see Robert A. Le Massena, "Design It Yourself Locomotive," *Railroad History* 182 (Spring 2000): 23–57. On water gauges see my "Safe and Suitable Boilers: The Railroads, the Interstate Commerce Commission and Locomotive Safety, 1900–1945," *Railroad History* 171 (Autumn 1994): 23–44. The Bureau of Locomotive Inspection occasionally noted the value of such devices but never pressed for them.

11. "Yard Operations Made Safe by Floodlighting," *RA* 84 (Feb. 18, 1928): 393–95. "Hump Yard Operates Without Car Riders," *RA* 77 (Nov. 15, 1924): 895–97. "Car Retarders in Hump Yards Effective in Winter," *RA* 78 (May 9, 1925): 1143–45. "Economics of Car Retarders," *RA* 85 (July 7, 1928): 22–24.

12. "Report on Birdsell Collision," *RA* 67 (Dec. 6, 1918): 1018–19.

13. An additional motive for the introduction of large cranes was the desire to lay longer and heavier rail sections that were cheaper to maintain and safer, because they were heavier and had fewer joints.

14. AAR Safety Section *Proceedings* 20 (1940): 198–200. A good summary of the new equipment is "25 Years Development in Work Equipment," *REM* 37 (June 1941): 412–21. For the Burlington see "Safe-Guarding Work Equipment," *REM* 31 (Mar. 1935): 159–67.

15. "Committee Report on Motor Cars," and "Control of Motor Car Operations," American Railway Bridge & Building Association *Proceedings* 36 (1926): 99–117, and 38 (1928): 101–21. "Taking Chances with Track Cars," *RA* 83 (July 9, 1927): 43–44; "Report of Derailment at Westfield N.Y., Apr. 29," *RA* 83 (July 9, 1927): 68; and "Lack of Precautions Results in Motor Car Accident," *RA* 83 (Oct. 8, 1927): 687.

16. The equation is fitted to all class one railroads for 1936 and 1939 and is:

$$Passenger\ rate = 23.66 + 9.94(Worker\ rate) + 38.53\ Year\ R^2 = 0.02,\ N = 170$$
$$(2.31) \qquad\qquad (0.62)$$

where *Passenger rate* is the passenger casualty rate per billion passenger miles and *Worker rate* is the worker casualty rate per million manhours. Figures in parentheses are t-ratios. The equation yields an elasticity at the mean of about 0.73, meaning that a 10 percent cut in the worker casualty rate was associated with a 7.3 percent cut in the passenger casualty rate.

17. The lack of a negative accident/safety relationship is discussed by Casey Ichniaowski, "Fuzzy Frontiers of Production: Evidence of Persistent Inefficiency in Safety Expenditures," National Bureau of Economic Research *Working Paper* 1366 (Cambridge, 1984). For graphs similar to those in the text see Frederic Scherer, *New Perspectives on Economic Growth and Technological Innovation* (Washington: Brookings, 1999), ch. 2, and Vernon Ruttan, *Technology, Growth, and Development* (New York: Oxford University Press, 2001). Statistical analysis confirms that there was no association between casualties, output, and resources. I estimated the following equations for 1923 and 1939 using heteroskedasticity-robust techniques for the seventy-seven carriers that were extant on those dates and also in the 1950s. Capital is investment in plant and equipment while output is a priced-weighted sum of ton and passenger miles. Figures in parentheses are t-ratios.

$$Log(Output_{23}) = -1.28 + 0.019Log(Casualties_{23}) + 0.619Log(Capital_{23})$$
$$(0.20) \qquad\qquad (3.83)$$
$$+ 0.409Log(Manhours_{23})\ R^2 = 0.94,\ N = 77$$
$$(1.85)$$
$$Log(Output_{39}) = -2.26 - 0.267Log(Casualties_{39}) + 0.657Log(Capital_{39})$$
$$(2.18) \qquad\qquad (4.72)$$
$$+ 0.571Log(Manhours_{39})\ R^2 = 0.92,\ N = 77$$
$$(3.23)$$

Clearly in 1923 interfirm differences in output are unrelated to casualties while in 1939 the relationship has the "wrong" sign, suggesting that both safety and productivity reflected generalized organizational capabilities. The decline in the constant term reflects the impact of depression on productivity.

18. For the ICC's views see its *Twenty-Sixth Annual Report, 1912* (Washington, 1913), 64–66; *Twenty-Seventh Annual Report, 1913* (Washington, 1914), 67–71;

Twenty-Eighth Annual Report, 1914 (Washington, 1915), 53–56; *Thirty-First Annual Report, 1917* (Washington, 1918), 47.

19. Many of these accidents are discussed in Robert Shaw, *Down Brakes* (London: P. R. Macmillan, 1961). "The Interstate Commerce Commission Report on the Amherst Wreck," *RR* (May 20, 1916): 716–17. "Mount Union Collision," *RA* 62 (Apr. 6, 1917): 730–33. "The Shepherdsville Collision," *RA* 64 (Jan. 4, 1918): 88–89. "The Circus Train Disaster," *RA* 65 (July 5, 1918): 5. "The Accident at Nashville," *RA* 65 (July 12, 1918): 78. "The Accident at Porter Indiana," *RA* 70 (Mar. 11, 1921): 539.

20. "Permissive Blocking on Eight-Mile Sections," *RA* 77 (Aug. 2, 1924): 184–85. "Little better than a farce," from "The Lesson of the Last Collision," *RA* 72 (Feb. 11, 1922): 357. "The Accident at Porter Indiana," *RA* 70 (Mar. 11, 1921): 539.

21. 69 ICC *Report* (1922), 253–79. For more detail see my "Combating the Collision Horror, The Interstate Commerce Commission and Automatic Train Control," *Technology and Culture* 34 (Jan. 1993): 49–77.

22. For a brief history of train control see ARA Committee on Automatic Train Control, *Bulletin 1 Automatic Train Control, Development and Progress* (Nov. 1930), and "History of Automatic Train Control," *RSI* 17 (Mar. 1924): 105–7. "Automatic Train Control," *RA* 53 (Aug. 30, 1912): 373–74. "The Automatic Train Stop—I," *RA* 57 (July 10, 1914): 42–43. "First step" is from "Report of Automatic Train Control Committee," *RR* 66 (Jan. 17, 1920): 100–101, quotation on 101.

23. This categorization is based on USRA, *Annual Report of Walker D. Hines Director General of Railroads, 1919, Automatic Train Control Committee* (Washington, 1920). The best description of the various systems is in F. L. Dodson, "Some Fundamentals of Train Control," *RR* 72 (Mar. 23, 1923): 544–50.

24. W. P. Borland, "Automatic Train Control—Government Viewpoint," *Journal of the Western Society of Engineers* 26 (Feb. 1921): 33–41, quotation on 35.

25. That most safety investments also had a productivity payoff is stressed in "The Automatic Train Control Order," *RA* 72 (Feb. 4, 1922): 308. A good economic and technical critique is A. H. Rudd, "Automatic Train Control," *Engineers and Engineering* 43 (Jan. 1926): 20–29.

26. 69 ICC *Report* (1922), 253–79, quotations on 272. "The ICC Train Control Hearing," *RSI* 17 (May 1924): 195–97, and (June 1924): 234–42. The Wheeling & Lake Erie is from ARA Committee on Automatic Train Control, *Bulletin 1* (n. p., 1930), 135.

27. "Results of Train Control Test on Pennsylvania," *RSI* 17 (Oct. 1924): 391–93, and "Continuous Train Control, Pennsylvania RR," *RR* 75 (Nov. 1, 1924): 654–57. For hearings on the second order see "42 Roads Object to Train Control Order," *RA* 76 (May 17, 1924): 1209–14, and "Hearing on Automatic Train Control," *RR* 74 (May 10, 1924): 853–54; (May 17, 1924): 889, 893–96; (May 24, 1924): 927–29, 933–34. 91 ICC *Report* (1924), 426–51.

28. For McManamy's career see *NCAB*, 33, 462.

29. "Great Northern Train Control Petition Denied," *RA* 79 (Oct. 31, 1925): 805–6, quotation on 806.

30. The carriers' change in strategy is from "Minutes of Meeting of Roads Responding to ICC Docket 13413, in the Matter of Automatic Train Control, . . . Feb. 28, 1928, . . . ," box 1131, PRRC, HML. 148 ICC *Report* (1928), 188–210, quotation on 201.

31. The figures are from ARA *Bulletin 1*. Because train control was usually in-

stalled on high-density lines it probably protected more than 15 percent of passenger miles. "ICC Approves Pennsylvania Cab Signaling," *RSI* 24 (Mar. 1931): 87–88; "ICC Permits Great Northern to Discontinue Train Control," *RSI* 25 (Feb. 1932): 42–43. "ICC Permits Cab Signals in Lieu of Automatic Train Control on the Union Pacific," *RSI* 25 (Sept. 1932): 231–32. "B&M Train Control Petition Granted," *RSI* 26 (Dec. 1933): 333–34.

32.   The equation based on company data in 1922 and 1929 and weighted by 1929 locomotive miles is

$$ARATE_{29} = 1.07 + 0.257 ARATE_{22} - 2.77 AUTOSIG_{29} + 0.591 MANSIG_{29}$$
$$(3.65) \qquad (7.54) \qquad (0.83)$$
$$+ 0.202 DENSITY_{29} + 3.35 ATC_{29} \ R^2 = 0.37, N = 152$$
$$(3.71) \qquad (1.34)$$

where the *ARATE* variables are collisions per million locomotive miles; *AUTOSIG* and *MANSIG* are automatic and manual signals per mile of track; *DENSITY* is locomotive miles per mile of track; and *ATC* train control per mile of track. Figures in parentheses are t-ratios. For Rudd's remarks see "American Railway Engineering Association Report of Committee X—Signaling Practice," *RA* 80 (Mar. 9, 1926): 633–36, quotation on 635. The Southern, the New York Central, and a few other carriers also began voluntary installations and by 1930 these covered about 5,000 miles of track. All in all the ARA estimated that both mandated and voluntary train control covered about 32 percent of all passenger train miles by that date.

33.   ARA Mechanical Section *Proceedings* (1919): 235–43; (1928): 505–9. The ICC catalogued accidents from faulty brakes under those resulting from equipment failure and those due to employee negligence. In a typical year such as 1928 there were 4,302 collisions and 9,938 derailments. Of these brake failures or negligence in their operation accounted for about 3.7 percent of the total. There were 30 collisions and 242 derailments as well as 250 "miscellaneous" accidents that were brake-related. Together these caused one fatality and thirty injuries.

34.   "A 100 Car Test of the Automatic Straight Air Brake," *RA* 65 (July 26, 1918): 173–76. "Report on the Automatic Straight Air Brake," *RA* 66 (Mar. 28, 1919): 840–42.

35.   "Investigation of Power Brakes and Appliances for Operating Power Brake Systems," 91 ICC *Report* (1924), 481–534. "The Commission's Report on Power Brakes," *RR* 75 (Aug. 16, 1924): 251–57. ICC, *Report of the Director of the Bureau of Safety in re Investigation of an Accident Which Occurred on the Pennsylvania Railroad at Altoona, PA on Nov. 29, 1925* (Washington, 1926), 11.

36.   ICC Bureau of Safety, *Annual Report, 1926* (Washington, 1926), 3. ICC Bureau of Safety, *Annual Report, 1933* (Washington, 1933), 3. J. T. Wallis to W. W. Atterbury, Dec. 10, 1920, box 553, PRRC, HML. Johnson's biography is from "Memorandum, Power Brake Investigation," box 555, PRRC, HML. That Westinghouse recommended him is from A. L. Humphrey [president of Westinghouse] to J. T. Wallis, Sept. 26, 1924, box 555, PRRC, HML. ARA Mechanical Section *Proceedings* (1929): 774–77. For a good description see "New AB Brake Will Reduce Costs of Operation and Facilitate Freight Movement," *RA* 94 (Jan. 28, 1933): 98–103. The empty and load brake is from "Some Recent Air Brake Developments," *RA* 98 (June 29, 1935): 1043.

37. "Economic Study of 'AB' Freight Brake," *RA* 96 (Apr. 21, 1934): 579–80. The diffusion of the AB brake may be followed in ICC Bureau of Safety, *Annual Report*, various years.

38. The Erie and New Haven are from "Report on Collision at Atlantic Mass.," *RA* 96 (Apr. 14, 1934): 550–51. "Speeding Up Trains," *RA* 86 (June 8, 1929): 1313–14, reports increasing speed on many passenger trains. The equations from ch. 7 (n. 34) fitted to data for 1902–40 with *PW* a dummy equal to 1 for 1921 on yielded Log (*Fat/derail*) = − 0.018*PctSteel* +0.014*PWSteel*, and Log (*Fat/collision*) = − 0.004*PctSteel* − 0.019*PWSteel*. Since the percent of steel and steel frame cars rose 55 points from 42 to 97 percent over this period, the predicted impact for derailments is (exp(− 0.018 + 0.014) × 55) − 1 = − 0.20. For collisions the results are exp(− 0.004 − 0.019)*55 − 1 = 0.72. As in the prewar years, however (chapter 7), the positive trends offset these gains, probably reflecting increases in speed.

39. "Revolutionizing Transportation," *RA* 80 (June 19, 1926): 1898–99.

40. The organizations are from W. H. Winterrowd, "Research and Steam Locomotive Development," ARA Mechanical Division *Proceedings* (1937): 100–106.

41. R. L. Kleine, "Freight Car Inspection Pit," ARA Mechanical Division *Proceedings* (1930): 136–43.

42. A good discussion of the arch bar truck is in John White, *The American Railroad Freight Car* (Baltimore: Johns Hopkins University Press, 1993). "Derailment at Warrior Ridge," *RA* 52 (Feb. 23, 1912): 347. For accidents on the Pennsylvania see "Effect of Arch Bar Failures on Freight Cars . . . Jan. 1, 1926 to Mar. 31, 1927," box 556, PRRC, HML. For the Rock Island see ARA Mechanical Division *Proceedings* (1930): 197–223. ICC Bureau of Safety, *Annual Report, 1935* (Washington, 1935), 6–7.

43. T. H. Symington, "Freight Car Derailments," *RA* 83 (Oct. 22, 1927): 763–67. J. M. Snodgrass and F. H. Guldner, "An Investigation of the Properties of Chilled Iron Car Wheels," University of Illinois Engineering Experiment Station *Bulletin* 129 (Urbana, 1922). "Chilled Wheel Design Improved," *RR* 78 (June 5, 1926): 1007–8. ARA Mechanical Division *Proceedings* (1928): 922–25. G. E. Doke, "Who Invented the Wheel?" Southern and Southwestern Railway Club *Proceedings* 21 (May 1931): 11–16. "Report of the [ARA Mechanical Section] Committee on Wheels," *RA* 87 (July 6, 1929): 50–53.

44. "Arthur Newell Talbot," *NCAB*, 33, 94–95. A general review of Talbot's work is in AREA, *Stresses in Railroad Track—the Talbot Reports* (Chicago, 1980). On rail canting see "A Valuable Work," and "Canting of Rail Inward and Taper of Tread of Wheel," *RA* 80 (Mar. 11, 1926): 713 and 732. For the impact on locomotive design see "Another Practical Application of Scientific Studies," *RA* 78 (June 6, 1925): 1375.

45. "Record breaking" is from "Better Track," *RA* 83 (Aug. 6, 1927): 241. "Cotton Belt Reconstructs Line in Texas," *RA* (Nov. 9, 1930): 1071–76. The Rio Grande is from "Cashing in on Improvements, Part II," *RA* (Nov. 9, 1929): 1085–88, quotation on 1085. For the Texas & Pacific see "Improved Facilities Produce Savings, Part I," *RA* 88 (Oct. 18, 1930): 786–88.

46. A. N. Reece, "Economical Selection of Rail," AREA *Proceedings* 31 (1930): 1495–1553. For rail weight, length, and section see "Rail—the Foundation of Progress," *REM* 36 (Sept. 1940): 558–73. The Kansas City Southern is from "What Is the Economic Weight of Rail?" *RA* 88 (Mar. 24, 1930): 1231–37. R. S. Cochrane,

"The Maintenance of Track Joints," *REM* 19 (Oct. 1923): 412–13. C. B. Bronson, "A Review of Current Improvements in Rails and Fastenings," *REM* 21 (Feb. 1925): 55–56. "Kinked Track on Norfolk Southern Causes Derailment," *REM* 38 (Feb. 1942): 103–4.

47. "Impressions of a Group of European Railway Maintenance of Way Engineers on Their Tour of U.S. Railroads," AREA *Proceedings* 52 (1951): 784–87. "Another Big Year for the Roadway," *RA* 84 (Jan. 7, 1928): 3.

48. "Northern Pacific Main Lines 76 Percent Signaled," *RA* 78 (Mar. 7, 1925): 545–48. "Comparative Frequency and Cost of Accidents Before and After the Installation of Automatic Block Signals on the Denver & Rio Grande Western," Railway Signal Association *Proceedings* 33 (1936): 13–15. That the Northern Pacific was the first to use slide detectors is from "Slide Detector Fences as Track Protection," *RA* 93 (Aug. 6, 1932): 174. "Taking the Danger out of Slides, *REM* 32 (Jan. 1936): 22–26. "Protective Devices Increase Safety of Operation on the SP," *REM* 36 (June 1940): 374–75. For Thomas Kendreck see Youngstown State University Oral History Collection, interview 950, available at http://www.maag.ysu.edu/oralhistory/oral_hist.html [accessed June 15, 2005].

49. ICC, *Report of the Chief of the Division of Safety Covering the Investigation of an Accident Which Occurred on the Long Island Railroad Near Central Islip, N. Y., on Apr. 15, 1918* (Washington, 1918); the call for "concerted action" is on 45. See also *Report of the Chief of the Division of Safety Covering the Investigation of an Accident Which Occurred on the Galveston, Harrisburg & San Antonio Railway Near Iser, Tex., Jan. 31, 1916* (Washington, 1917). ICC Bureau of Safety, *Annual Report, 1917* (Washington, 1918), 16.

50. R. Trimble to J. T. Wallis, Apr. 28, 1917, box 665, PRRC, HML, contains the background to Waring's study and the claim that the steel company wanted any findings to remain confidential. F. M. Waring, "Investigation of Transverse Fissures in Failed Rails," AREA *Proceedings* 20 (1919): 614–17. F. M. Waring and E. K. Hofammann, "Deep Etching of Rails and Forgings," ASTM *Proceedings* 19 (1919): 183–97.

51. Howard's views are contained in his accident reports to the ICC, in his comments, and in "The Shattered Zones in Certain Steel Rails with Notes on the Interior Origin of Transverse Fissures," ASTM *Proceedings* 20 (1920): 44–69. See also ICC, *Report on the Formation of Transverse Fissures and Their Prevalence on Certain Railroads* (Washington, 1923). That the ICC was initially interested in Waring's work is revealed in its Bureau of Safety, *Annual Report, 1919* (Washington, 1919), 19–21.

52. "Joint Investigation of Transverse Fissures," AREA *Bulletin* 263 (Jan. 1924): 432–37. The decision to collect data on transverse fissures is from "Report of the Committee on Rails," AREA *Proceedings* 23 (1922): 621–24. For Hoover's vision of the business/government partnership see Ellis W. Hawley, "Herbert Hoover, the Commerce Secretariat, and the Vision of an 'Associative State,' 1921–1928," *Journal of American History* 61 (June 1974): 116–40.

53. Funding problems are discussed in "Study of the Effect of Various Intensities and Repetitions of Wheel Loads upon Rails," AREA *Proceedings* 27 (1926): 580–85. Minutes of the meeting are "On the Examination of Transverse Fissures with Special Reference to Conditions at Their Nuclei," Jan. 22, 1925, and May 20, 1925 (copy of typescript from ICC library in author's possession).

54. Technological complementarities and focusing devices are discussed in Nathan Rosenberg, *Inside the Black Box* (Cambridge: Cambridge University

Press, 1982), and his *Perspectives on Technology* (Cambridge: Cambridge University Press, 1976). Sandberg later described his findings in C. P. Sandberg et al., "Effect of Controlled Cooling and Temperature Equalization on Internal Fissures in Rails," *Metals and Alloys* 3 (Apr. 1932): 89–92, quotation on 91. Sandberg's work is also discussed in Cecil Allen, "The Modern Application of the Sandberg Sorbitic Rail Treatment," International Railway Congress *Bulletin* 13 (Oct. 1931): 846–59. John Freeman and Willard Quick reported findings similar to Sandberg's in "Tensile Properties of Rail and Some Other Steels at Elevated Temperatures," Research Paper 164, U.S. Bureau of Standards *Research Journal* 4 (Apr. 1930): 549–91.

55. For the origins of Sperry's interest in transverse fissures see Thomas P. Hughes, *Elmer Sperry, Inventor and Engineer* (Baltimore: Johns Hopkins University Press, 1971), 287–89, who seems, however, to confuse transverse fissures with piped rails. Elmer Sperry, "Non-Destructive Detection of Flaws," *Iron Age* 122 (Nov. 15, 1928): 1214–17, quotation on 1215. For a more complete description of Sperry's development process see David Barnes, "Transverse Fissure Detecting Devices" AREA *Proceedings* 28 (1927): 967–74, and Sperry Rail Service, *Rail Defect Manual* (n.p., 1949). Tapping the rail is from W. C. Barnes, "Results of First Year's Operation of the ARA Rail Fissure Detector Car," AREA *Proceedings* 31 (1930): 1472–74.

56. W. C. Barnes, "Detection of Transverse Fissures in Track," AREA *Bulletin* 315 (Mar. 1929): 1236–41. "Rail Flaw Detector Greatly Improved in Five Years," *RA* 95 (Nov. 4, 1933): 655–58. "Sperry Detector Car Introduces Pre-Energization," *RA* 97 (Oct. 13, 1934): 435–36. "Install Flaw Detector at Rail Sawing Plant," *RME* (Dec. 1930): 555–57. The patent squabble may be followed in Sperry Rail Detector Car Collection, # 497, National Museum of American History Archives, Smithsonian Institution.

57. "Report of the Committee on Rail," AREA *Proceedings* 29 (1928): 558–59, and 30 (1929): 1232–33. John B. Emerson, "A Further Discussion of Transverse Fissures," *RA* 80 (Feb. 20, 1926): 465–66. For similar views, see W. C. Cushing, "The Genesis of the Transverse Fissure," AREA *Bulletin* 315 (1929): 227–94.

58. ICC, *Report of the Director of the Bureau of Safety in Re Investigation of an Accident Which Occurred on the St. Louis-San Francisco Railway Near Victoria, Miss., on Oct. 27, 1925* (Washington, 1927). ICC, *Report of the Director of the Bureau of Safety In Re Investigation of an Accident Which Occurred on the Missouri Pacific Railroad at West Junction, Kans., on June 25, 1928* (Washington, 1929). *Report of the Director of the Bureau of Safety In Re Investigation of an Accident Which Occurred on the Pennsylvania Railroad at Onley, Va., on Dec. 1, 1929* (Washington, 1930).

59. "Report of the Committee on Rail," AREA *Proceedings* 31 (1930): 1450. "Report of the Committee on Rail," AREA *Proceedings* 33 (1932): 557–58, and "Rail Investigations," ARA *Proceedings* (1930): 60–62. Moore briefly described his earlier work in "Rails Investigation," *Metal Progress* 53 (June 1948): 828–31, which contains the figure of one in five hundred wheel loads. Herbert F. Moore, "Progress Report of the Joint Investigation of Fissures in Railroad Rails," AREA *Bulletin* 376 (June 1935), quotations on 7 and 13. An interpretation stressing the importance of the Illinois work is Michael Duffy, "The Standard Rail Section, Transverse Fissures and Reformed Mill Practice, 1911–1955," *Journal of Mechanical Working Technology* 4 (Sept. 1980): 285–305.

60. The discovery of the role of hydrogen is discussed in S. Epstein, "Progress in

Steel Making Reported from Germany: A Correlated Abstract," *Metals and Alloys* 6 (Aug. 1935): 219–22, and C. A. Zapffe and C. E. Sims, "Hydrogen, Flakes, and Shatter Cracks, A Correlated Abstract, Parts I–IV," *Metals and Alloys* 11 (May 1940): 145–51; (June 1940): 177–83; 12 (July 1940): 44–52; (Aug. 1940): 145–48. That hydrogen is the main, although not exclusive, cause of transverse fissures remains the modern view. See "Rail Research and Development," AREA *Bulletin* 644 (1972): 1–38.

61. "Recent Advances in Rail Manufacture," *RA* 100 (June 13, 1936): 940–44.

62. The Norfolk & Western and B&O are from "Effect of Automobile Travel," *RA* 77 (Oct. 11, 1924): 643–44. See also "Railway Passenger Travel," *RA* 79 (July 4, 1925): 2–3. Gregory Thompson, *The Passenger Train in the Motor Age: California's Rail and Bus Industries, 1910–1941* (Columbus: Ohio State University Press, 1993), argues that most bus traffic did not come at the expense of railroads.

63. The ICC accident categories are those from getting off and on trains and struck or run over by a train not at a public crossing. Rates fell from 1.9 to 0.6 per billion passenger miles from 1922–23 to 1938–39. Data on automobile passenger miles are from James Nelson, *Railroad Transportation and Public Policy* (Washington: Brookings Institution, 1959), Table 5.

64. "Rubber tramps" is from an interview of Fred Thompson in Studs Terkel, *Hard Times* (New York: Pantheon, 1970), 307.

65. Quotation from Joe Morrison in Terkel, *Hard Times,* 123. The Santa Fe is from U.S. Children's Bureau, *Twentieth Annual Report, 1932* (Washington, 1932), 6.

66. Separate regressions for trespassing fatalities per locomotive mile on track and train for 1926–50 yield the following. *TrnFt* and *TrkFt* are train and track fatalities, *U/LF* is the unemployment rate, *Autos* automobile registrations in thousands, and *RW* the real transportation wage (Money wages divided by average passenger fares). Figures in parentheses are t-ratios.

$$\mathrm{Log}(\mathrm{TrnFt}) = -3.97 + 0.274\mathrm{Log}(U/LF) + 0.533\mathrm{Log}(Autos) - 1.52\mathrm{Log}(RW)$$
$$(3.41)(1.94)(3.91)$$
$$-0.037\,Time$$
$$(4.06)$$
$$R^2 = 0.98, N = 25, D.W. = 1.73$$
$$\mathrm{Log}(TrkFt) = 5.14 + 0.081\mathrm{Log}(U/LF) - 0.403\mathrm{Log}(Autos) - 0.612\mathrm{Log}(RW) +$$
$$(1.64)(1.68)(2.27)$$
$$0.005\,Time$$
$$(0.50)$$
$$R^2 = 0.96, N = 25, D.W. = 2.06$$

As can be seen, train fatalities are far more sensitive to the fall in real wages and the rise in unemployment than are track fatalities.

67. "Sunday Crossing Accidents in Nine States," *RA* 75 (Aug. 4, 1923): 219. "Black Diamond Express Derailed by an Automobile," *RA* 72 (May 20, 1922): 1176–77. "Highway Crossing Protection," *RA* 73 (Oct. 7, 1922): 636. Ralph Budd, "A Million Miles of New Railway Travel," *RA* 100 (Apr. 18, 1936): 659–61, quotation on 660. "On the other hand" is from "Pilot Height and Crossing Accidents," *RME* 105 (Sept. 1932): 362.

68. "The Careful Crossing Campaign" *RA* 73 (Nov. 11, 1922): 907. "Educating Drivers as to Hazards," *RA* 82 (May 28, 1927): 1593–94.

69. "Highway Crossing Protection," *RA* 73 (Oct. 7, 1922): 636. Similarly in the

1950s automobile producers defined accidents as the results not of unsafe cars but the "nut behind the wheel." See Joseph Gusfield, *The Culture of Public Problems* (Chicago: University of Chicago Press, 1981).

70. H. A. Rowe, "Remarks," ARA Safety Section *Proceedings* 8 (1928): 75. "Grade Crossing Desperadoes," *RA* 77 (Dec. 27, 1924): 1149. "Statement of R. V. Fletcher, Vice President Association of American Railroads," in U.S. 78th Cong., 2d sess, House Committee on Roads, *Hearings on Federal Aid for Post-War Highway Construction, H.R. 2426* (Washington, 1944), 528–39, quotation on 532.

71. "Engineer's View of the Crossing Fool," *Literary Digest* 86 (Sept. 26, 1925): 25. "Gray Hairs for Casey Jones," *Saturday Evening Post* 209 (Apr. 10, 1937): 8–9. "The Deadly Grade Crossing," *American City* 30 (Jan. 1924): 24–25.

72. For a critique of the carelessness explanation see "Modernizing Crossing Protection with New Safety Features," *RSI* 30 (June 1937): 343–44. Fred Lavis, "Grade Crossings: The Money Value of a Car Minute," *Annals* 133 (Sept. 1927): 172–77. George Barton, "Modernizing Crossing Protection with New Safety Features," *RSI* 30 (June 1937): 343–44. Warren Henry, "Efficiency of Crossing Protection," *RA* 103 (Dec. 11, 1937): 835–37.

73. California Railroad Commission, *Bulletin of the Progress of the Commission's Investigation of Railroad Grade Crossings in the State* (Sacramento, 1917). New York Public Service Commission, *Annual Report, 1939,* I, 179–95. "Nashville, Chattanooga & St. Louis Railway v. Walters, Commissioner of Highways et al.," *294 U.S. 405–435* (1935), quotation on 422.

74. Funding allocations are from NARC, *Proceedings* (1933): 489–94. "Dividing Cost of Highway Crossing Signals," *RA* 91 (Dec. 5, 1931): 851, and "Provisions of State Laws," box 328, PRRC, HML.

75. New York Public Service Commission, *Annual Report, 1940,* I (Albany, 1940), 68.

76. Economists have traced the origins of cost benefit analysis to French engineers and more recently to the work of the U.S. Bureau of Reclamation but have ignored its application in highway research. See, for example, Peter Sassone and William Schaffer, *Cost Benefit Analysis: A Handbook* (New York: Academic, 1978), and Robert Ekelund, *Secret Origins of Modern Microeconomics: Dupuit and the Engineers* (Chicago: University of Chicago Press, 1999). For an early example of cost benefit analysis applied to grade crossings see Philip Rice, "Railroad Crossing Protection," Institute of Traffic Engineers, *Fifth Annual Meeting* (1934): 41–55. California Railroad Commission, *Report on the Grade Crossing Situation of Public Streets, Roads and Highways with Steam and Electric Interurban Railroads in the State of California, Dec. 1, 1932* (Sacramento, 1933).

77. Henry's formula is in "Illinois to Install Crossing Protection at State Expense," *RA* 97 (Sept. 22, 1934): 345–46. The equation in the text is from L. E. Peabody and T. B. Dimmick, "Accident Hazard at Grade Crossings," *Public Roads* 22 (June 1941): 123–28, 144–45. Both formulas are entirely benefit based.

78. For reviews of the various formulas see "Method of Classifying Highway-Railway Crossings," AREA *Proceedings* 50 (1949): 244–51, and A. Kenneth Beggs, "The Railway-Highway Grade Crossing Problem" (Stanford, Stanford Research Institute, 1952). A comprehensive modern assessment is California Public Utilities Commission, "The Effectiveness of Automatic Protection in Reducing Accident Frequency and Severity at Public Grade Crossings" (San Francisco, 1974).

79. U.S. Public Roads Administration, *Highway Statistics Summary to 1945* (Washington, 1947), 57.

80. A considerable, although inconclusive, literature exists on the political economy of New Deal spending. See Gavin Wright, "The Political Economy of New Deal Spending: An Econometric Analysis," *Review of Economics and Statistics* 56 (Feb. 1974): 30–38, and John Wallis, "The Political Economy of New Deal Spending, Yet Again: A Reply," *Explorations in Economic History* 38 (Apr. 2001): 305–14.

81. The equation fitted to data for all states plus Washington, D.C., for 1935 is

$$ERA = 1072 + 49.21A - 0.073A^2 \quad R^2 = 0.66, N = 49,\text{ where } ERA \text{ is in thousands}$$
$$(6.41)\quad(3.30)$$

of dollars and A is accidents. Figures in parentheses are t-ratios.

82. The Pennsylvania's calculations are from William D. Wiggins to J. F. Deasy, Dec. 3, 1940, box 328, PRRC, HML. I regressed grade separations for each year from 1926 (the first year they are available) through 1942 (the last year of grade crossing appropriations in the federal aid to highway program) on a trend and dummy using a maximum likelihood correction for serial correlation with the following results. Figures in parentheses are t-ratios.

$$Grade\ separations = 454.15 - 27.77\,Trend + 350.00\,Dummy_{36-42}\quad R^2 = 0.47, N = 16,$$
$$D.W. = 1.59 (2.21)(3.83)$$

83. I regressed annual crossing fatalities on automobile miles traveled, train miles (both in millions), crossings, and a trend for the years 1925 (the first year for which crossing data are available) through 1965. The following results were typical. Figures in parentheses are t ratios.

$$\text{Log}(fatalities) = 6.21 + 0.0011(Train\ miles) + 0.14 \times 10^{-5}(Vehicle\ miles) +$$
$$(9.33)(5.89)$$
$$0.3 \times 10^{-5}(Crossings) - 0.023\,Trend\quad R^2 = 0.84, N = 41, D.W. = 1.97$$
$$(1.82)(6.72)$$

Between 1925 and 1941 train miles fell by 214 (thousand) while vehicle miles rose by 211,266 (million) and crossings increased 24,690. Figures in the text are, for example, $\exp(-214 \times 0.0011) - 1 = -21$ percent decline in fatalities from the reduction in train miles.

84. "How the Erie Has Reduced Highway Crossing Accidents," *RA* 97 (Aug. 18, 1934): 215–17. Driver learning is from "Modern Highway Protection Reduces Operating Costs," *RA* 91 (Nov. 14, 1931): 737–39.

Chapter 10. Safety in War and Decline

Epigraphs: Frank Cizek, "Accident Prevention," *RME* 117 (Sept. 1943): 407–12, quotation on 410. "The I.C.C. Signal Order," *RA* 122 (June 28, 1947): 1295–96.

1. New entrants from U.S. Railroad Retirement Board, *Annual Report, 1943* (Washington, 1944), ch. 5.

2. For numbers of collisions, derailments, and other train accidents from the 1930s through 1945 see ICC, *Accident Bulletin* 115 (Washington, 1946), Tables 37 and 38. For the Buie, Stockton, and Philadelphia wrecks see Robert Shaw, *Down Brakes* (London: P. R. Macmillan, 1961). For Buie see also "Transverse Fissures

Found Cause of Wreck," *RME* 40 (Feb. 1944): 144. For other wrecks on the ACL see "Rail Fissure in Joint Causes Serious Accident," and "Fissured Rail Derails Fast Passenger Train," *RME* 37 (Aug. 1941): 537 and (Sept. 1941): 611–12.

3. The *Silver Meteor* is from "Report on Accident at Kittrell NC," *RA* 113 (Sept. 9, 1942): 383. "Rear End Accident Kills 32," *RA* 119 (Aug. 18, 1945): 313. Bagley is from Shaw, *Down Brakes.*

4. Of course negligence can often be prevented by better equipment. Concern with superintendence is from "What is Happening to the Railroad Safety Record?" *RA* 112 (Feb. 28, 1942): 437–38. Accidents from poor supervision are from "Increase of Train Accidents," *RA* 113 (Sept. 26, 1942): 474. The superintendent is from "How to Have Adequate Manpower," *RA* 113 (Aug. 15, 1942): 270–72. "Accident on D & S. L.," *RSI* 35 (Oct. 1942): 584.

5. "An Empty Water Glass Requires Decisive Action," *RME* 117 (June 1943): 263–64. "Reducing Boiler Explosions," *RME* 117 (Aug. 1943): 353–55. "How to Guard Against Low Water," *RME* 117 (Oct. 1943): 470–73. "The Curse of Carelessness," *RA* 113 (Sept. 12, 1942): 400–401. The *Age* indicted poor discipline in "Are Boiler Explosions Really Due to Carelessness?" *RA* 114 (June 12, 1943): 1166–67. While the immediate cause of boiler explosions from low water reflected human error, they also resulted from poorly designed locomotives, as noted earlier.

6. "Train Kills Ten Trackmen," *RME* 38 (Dec. 1942): 902. "Passenger Train Kills Nine Men in Section Gang," *RME* 39 (Oct. 1943): 754.

7. ICC, *Accident Bulletin* 110 (Washington, 1942) published employees and fatalities by age, occupation, and class of work (yard, freight, or passenger) to trainmen from train service accidents for 1941. Statistical analysis employing controls for occupation and class (not shown) yields the following relationship:

$$Fr = 2.93 - 0.051 Age; R^2 = 0.05, N = 72, D.W. = 1.86.$$
$$(2.65)$$

Figures in parentheses are t-ratios. If this relationship held true for all workers it implies a $(-0.051 \times -5) = 0.255$ point rise in the fatality rate, or about a 19 percent increase from the sample mean of 1.37 per thousand workers.

8. "Erie Teaches Supervisors to Teach," *RA* 114 (Jan. 16, 1943): 198–201. The Rio Grande is from "Lessons from Boiler Explosions," *RA* 115 (July 10, 1943): 47–49. For the Milwaukee see "If There's No Water Kill the Fire," *RA* 115 (Dec. 18, 1943): 978–80. The Lehigh Valley is from "Boiler Explosions," *RA* 116 (Feb. 12, 1944): 339. See also "Non-Train Accidents," NSC Railroad Section *Transactions* 32 (1944): 537–42.

9. Note that ton miles = tons per train × train miles, while tons per train = tons per car × cars per train. For examples of how these affected individual carriers see "Modern Power Saves Train Miles," *RA* 119 (Dec. 18, 1945): 936–40, which discusses the St. Louis Southwestern, and "Illinois Central: A War-Time Weapon," *RA* 119 (Oct. 20, 1945): 636–38.

10. For interwar productivity increases see John W. Kendrick, *Productivity Trends in the United States* (Princeton: Princeton University Press, 1961), Table G-III. For the postwar years see Kendrick, *Postwar Productivity Trends in the United States, 1948–1969* (New York: Columbia University Press, 1973), Tables A-35 and A-61. A more restrained assessment is National Commission on Productivity, *Improving Railroad Productivity* (Washington, 1973).

11. "Definitely not" is from "The AAR Research Laboratory," *RA* 125 (Oct. 9,

1948): 48–49. "Pitifully small" from "What's Ahead in Railroad Equipment?" *RME* 136 (Jan. 1962): 19."What's New in AAR Research?" *RA* 142 (June 17, 1957): 52–53. For the failure to do operations research see "Research on the Railroads," *MR* 18 (June 1963): 74–78.

12. The power brake rules could only be changed for the purpose of improving safety. The argument in the text is not that federal control over brake maintenance was or is inappropriate; only that it was not a very important safety matter. "Full Crew Laws Gain Ground," *RA* 147 (Aug. 3, 1959): 9–10.

13. See, for example, ICC, *Sixty-Ninth Annual Report, 1955* (Washington, 1956), 142, and *Seventieth Annual Report, 1956* (Washington, 1957), 174–75.

14. Section 26 of the Transportation Act of 1920 allowed the commission to require automatic train control or "other safety devices" but it refused to construe this wording as allowing it to require block signals. For a similar critique of the ICC's safety work see U.S. Railroad Retirement Board, *Safety in the Railroad Industry* ch. 4, and U.S. 89th Congress, 2d sess., House Committee on Government Operations, *Interstate Commerce Commission Operations (Railroad Safety), H.R. 1452* (Washington, 1966). That regulatory commissions have a life cycle from youth to senility is from Marver Bernstein, *Regulating Business by Independent Commission* (Princeton: Princeton University Press, 1955).

15. "Report of the Special Committee on Trespassing," AAR Safety Section *Proceedings* 31 (1951): 100–103. I regressed annual trespassing fatalities on train miles, car registration (*Autos*), the real transportation wage (money wages divided by average railroad fares), and a trend for 1941–65 employing a maximum likelihood correction for serial correlation. Figures in parentheses are t-ratios. The following results were typical:

$$\text{Log}(\textit{Fatalities}) = 9.34 + 0.888\text{Log}(\textit{Trainmiles}) - 0.077\text{Log}(\textit{Autos})$$
$$(3.64) \qquad\qquad\qquad (1.02)$$
$$- 1.04\text{Log}(\textit{RW}) - 0.0004\,\textit{Trend}\ R^2 = 0.96,\ N = 25,\ D.W. = 1.71$$
$$(5.29) \qquad\qquad (0.03)$$

Car registration is rarely significant but it is also strongly correlated with wages. These findings suggest that the main forces reducing trespassing deaths in the postwar years were rising incomes, the decline of railroads, and the growth of large trains, which also reduced train miles.

16. I fitted the following equation to worker on-duty fatality rates, 1922 to 1965, with passenger and freight train miles in millions (*Ptm* and *Ftm*), a war dummy (for 1940–45), a postwar dummy (0 through 1957 and 1 thereafter), and an interaction between the trend and the postwar dummy. Figures in parentheses are t-ratios. The results indicate that fatality rates declined at 4.9 percent a year before 1957 but rose thereafter. Although reporting requirements changed in that year this cannot explain the rise after 1957. The negative effect of passenger train miles does not imply that such work was risk free but rather that it reduced average risks.

$$\text{Log}(\textit{FR}) = 0.701 - 4.60\textit{Postwar} - 0.049\,\textit{Trend} + 0.138\textit{War} + 0.066\textit{Postwar} \times \textit{Trend}$$
$$(4.03) \qquad\quad (9.69) \qquad\quad (2.62) \qquad\quad (4.26)$$
$$- 0.0022\textit{Ptm} + 0.0023\textit{Ftm}\ R^2 = 0.96,\ N = 44,\ D.W. = 2.11$$
$$(3.11) \qquad\quad (5.18)$$

17. See Tiller v. Atlantic Coast Line Railroad Co., *318 U. S. 54* (1943). Walter Kintz, "Statement," Aug. 5, 1952, box 17, Lehigh & Hudson River Railroad Collection, MG 199, Railroad Museum of Pennsylvania. "Cost of Employee Death Claims, 1931–1951," box 342, PRRC, HML. Railroad wages and material prices rose 80 percent over the same period. See Bureau of Railway Economics, *Index of Average Unit Prices of Railway Material and Supplies [and] Indexes of Charge Out Prices and Wage Rates*, Series Q-MPW-14 (Washington, 11957).

18. "Santa Fe Accident Prevention Plan," NSC Railroad Section *Transactions* 47 (1959): 9–12. For the ACL see "Safety Men Aim for Perfection," *RA* 142 (Apr. 15, 1957): 9–10. "On the Frisco Safety Is a Family Affair," *RA* 137 (Dec. 20, 1954): 30–32. The Milwaukee is from L. J. Benson, "Remarks," AAR Safety Section *Proceedings* 28 (1948): 54–56.

19. "Katy Cuts Reportable Injuries 43%," *MR* 6 (Aug. 1951): 95–96, 98. For the Rock Island see "Spark Human Interest in Safety," *MR* 7 (Oct. 1952): 135–36. "Frisco's Safety Program Gets Results," *MR* 6 (Sept. 1950): 33, 35–36, quotation on 35. AAR Safety Section *Proceedings* 28 (1948): 76. John Banks, "Remarks," AAR Safety Section *Proceedings* 28 (1948): 76. Howard E. Banta, Chair Local 1000 Brotherhood of Railroad Trainmen, to R. C. Diamond, superintendent B&O, Feb. 12, 1955. Diamond to Banta, Mar. 11, 1955, and Banta to Diamond June 29, 1959. All in box 16, Brotherhood of Railroad Trainmen Collection, 5149, Keehl Center, Cornell University.

20. A good discussion of maintenance requirements is in Robert Aldag, "Culture Clash: Diesel vs. Tradition," *Railroad History* (Millennium Special 2000): 89–99. John McCall, "Dieselization of American Railroads: A Case Study," *Journal of Transport History* 6 (Sept. 1985): 1–17, discusses the impact of the diesel on the Santa Fe. For the problem of higher wheel loads see "Diesels and Rail Lubricators," *RTS* 50 (July 1954): 46–49.

21. ICC Bureau of Locomotive Inspection, *Forty-Fourth Annual Report of the Chief Inspector, 1955* (Washington, 1955), quotation on 4.

22. Accident data are from ICC Bureau of Locomotive Inspection, *Thirty-Ninth Annual Report of the Chief Inspector, 1950* (Washington, 1951). "Locomotive Uncoupler Makes Switching Safer," *MR* 7 (Feb. 1952): 132. Problems with jewelry are from C. M. House, "Safety in Diesel Maintenance," NSC Railroad Section *Transactions* 40 (1952): 22–25. "Fire Protection for Diesel Locomotives," *RA* 119 (Oct. 27, 1945): 684–85, and "Fires in Diesel Locomotives," *RME* 120 (June 1946): 299–300. "How Spectrographic Analysis Controls Diesel Engine Maintenance," *RME* 126 (Apr. 1952): 65–68. Crankcase protection from Paxton-Mitchell advertisement, *MR* 10 (Oct. 1955): 132. "Eye Injuries—What One Transportation Department Does About Them," *RA* 131 (Sept. 24, 1951): 45–47.

23. The ICC divided total injuries and fatalities from locomotives into four subgroups: members of train crews, roundhouse and shop employees, other employees, and nonemployees. Because most of the accidents occurred to train crews I concentrate on this group. A statistical analysis that estimates the impact of diesels on employment is Gilbert Yochum and Steven Rhiel, "Employment and Changing Technology in the Postwar Railroad Industry," *Industrial Relations* 29 (Fall 1990): 479–503.

24. The data in the text are from ICC Bureau of Locomotive Inspection, *Annual Report of the Chief Inspector*, various years. A regression analysis of trainmen's

fatality rates (per thousand workers) from locomotive accidents against total locomotives, the percent defective, miles per locomotive, and percent non-steam for 1927–54 with a correction for serial correlation confirms the contention that diesels yielded modest risk reductions. The equation is:

$$FR = 0.031 - 0.805 \times 10^{-6}(Locos) - 0.900 \times 10^{-3}\%NS$$
$$\quad (0.72) \qquad\qquad\qquad\qquad (2.44)$$
$$+ 0.0014PctDef + 0.198 \times 10^{-5}Mileage$$
$$\quad (1.25) \qquad\quad (2.41)$$
$$R^2 = 0.36, N = 27, D.W. = 1.90$$

Figures in parentheses are t-ratios. Using this equation, an increase in the percent nonsteam from 1.6 percent to 56.3 percent yields $(54.7 \times -0.900 \times 10^{-3}) =$ a 0.049 point decline in the fatality rate.

25. "Lackawanna Develops Photoelectric Crane Protection," *RME* 126 (Apr. 1952): 87–90. For painting and lighting see "Baltimore & Ohio Streamlines its Shops," *RME* 120 (Mar. 1946): 136–37, and "Color and Light for the Shop," *RME* 122 (Jan. 1948): 29–31.

26. The calculations are from E. B. DeVilbiss to Charles D. Young, Apr. 19, 1947, box 342, PRRC, HML. The quotation is from Charles D. Young to Martin W. Clement, Apr. 21, 1947, box 342, PRRC, HML. "Economics of Retarder Equipped Yards for Classification Switching," AREA *Proceedings* 56 (1955): 328–31.

27. "The Standardization of Hand Brakes," AAR Safety Section *Proceedings* 30 (1950): 101–7. The dangers of old-style brakes are from L. E. Hoffman, "Remarks," AAR Safety Section *Proceedings* 29 (1949): 88 91.

28. For the Norfolk & Western see "Accidents Under Control," *MR* 6 (Mar. 1951): 114–16. "New SP Cabooses Stress Safety," *RA* 151 (Oct. 2, 1961): 14–15. L. A. Villella, "Track Maintenance," NSC Railroad Section *Transactions* 45 (1957): 11–13. For the Lackawanna see J. P. Hiltz, "Remarks," AAR Safety Section, *Proceedings* 32 (1952): 169–72.

29. Had total train miles in 1965 been unchanged but had they been 62 percent freight as they were in 1957 instead of 71 percent freight, there would have been 56 (million) more passenger and fewer freight train miles. Using the coefficients for freight and passenger train miles in the equation in n.16 above yields $\exp(-0.0023 \times 56) - 1 = -0.12$ and $\exp(0.0022 \times -56) - 1 = -0.12$ and so the shift would have reduced worker fatality rates by 24 percent.

30. Kunsman is from Works Manager to H. T. Cover, May 20, 1946, Injury Cases 1925–1947, box 635, PRR, HML. For a survey and critique of railroad safety programs see U.S. Railroad Retirement Board, *Safety in the Railroad Industry* (Washington, 1962), ch. 2. "Report of Interdepartmental Committee on Personal Injuries," June 15, 1951, box 342, PRRC, HML. The injury from a fall and settlement are from J. M. Symes to Ethelbert W. Smith, May 5, 1949, box 342, PRRC, HML.

31. The survey is from Railroad Retirement Board, *Safety in the Railroad Industry*, ch. 2. C. E. Hightower, "Remarks," AAR Safety Section *Proceedings* 30 (1950): 107–8.

32. J. P. Hiltz, "Remarks," AAR Safety Section *Proceedings* 32 (1952): 169–72, quotation on 170.

33. Hill is from "Non-Train Accidents," NSC Railroad Section *Transactions* 31

(1943): 618–41, quotation on 620. "What Radio on M/W Equipment is Doing to the DT&I," *RTS* 52 (June 1955): 38–41.

34. For the lack of set-offs see W. E. Rader to R. H. Burkett, Oct. 11, 1939, file 3, box 2, William Doble Papers, Collection 5182, Keehl Center, Cornell University Library. An accident that resulted because the track car would not trip a crossing signal is reported in L. R. Bloss to F. M. Brown, May 23, 1950, file 13, box 2. Refusal to provide a lineup is from R. M. Smith to W. E. Green, Jan. 17, 1945, file 7, box 2. "New Automatic Block on the Katy," *RSI* 42 (May 1949): 306–9. "AAR Studying Track Car Safety," *RA* 134 (Aug. 30, 1954): 9. U.S. 85th Cong., 2d sess., Senate Committee on Commerce, *Safety Regulation of Railroad Track Motorcars,* Hearings on S. 1729, May 26, 1958 (Washington, 1958).

35. A good survey of economic deregulation is Ronald Braeutigan, "Consequences of Regulatory Reform in the American Railroad Industry," *Southern Economic Journal* 59 (Jan. 1993): 468–80. See also Richard Saunders, *Main Lines: Rebirth of the North American Railroads, 1970–2002* (DeKalb: Northern Illinois University Press, 2003).

36. The survey is from "Where Do We Stand on M of W?" *MR* 17 (Mar. 1962): 71. Complaints about maintenance are voiced in U.S. 91st Cong., 1st sess., Senate Committee on Commerce, *Hearings on Federal Railroad Safety Act of 1969* (Washington, 1969). "Too much junk" is from J. M. Symes to Ethelbert W. Smith, Jan. 8, 1948; "any recommendation" from Damon O'Connell to Symes, Aug. 19, 1947, both in No Accident Month file, box 342, PRRC, HML. Edwin Mansfield, "Innovation and Technical Change in the Railroad Industry," in *Transportation Economics* (New York: Columbia University Press, 1965), 169–96.

37. For 1964 I found $\text{Log}(C) = -6.57 ROA$ (for details see Table A2.4)

$$(2.77)$$

where *C* is the casualty rate and *ROA* the return on assets. The standard deviation of *ROA* is 0.0257; a 2-standard deviation increase implies a decrease in casualties of $\exp(-6.57 \times 0.0514) - 1$, or -29 percent.

38. For the payoff to automatic signals on the North Western see "Why Grade Crossings are Safer," *RA* 147 (Aug. 10, 1959): 12–13. Warren Hedley, "The Achievement of Grade Crossing Protection," NSC Railroad Section *Transactions* 37 (1949): 39–46.

39. "ACL Works to Eliminate Grade Crossing Accidents," *RA* 131 (Aug. 27, 1951): 41–43, and "Educating Drivers to Reduce Crossing Accidents," *RA* 133 (Sept. 8, 1952): 54. "ACL Crossing Hazards Reduced by Using Reflective Letters," *RA* 138 (Jan. 17, 1955): 22. "How Reflection Helps Visibility," *RA* 138 (Mar. 14, 1955): 59–63.

40. Warren Hedley, "Second Report on the Achievement of Grade Crossing Protection," AREA *Proceedings* 53 (1952): 985–1003. Table 17 presents relative accident quotients by type of protection, controlling for exposure, with crossbucks set to equal 100. The ranking of fourteen for automatic gates means that at a given crossing gates would yield 14 percent as many accidents. They are, therefore $(100/14) = 7.14$ times as effective.

41. "Grade Crossing Deaths Cut 97.4%," *RA* 136 (Mar. 22, 1954): 88–89. "Facts About Greater Safety at Grade Crossings," *RA* 141 (July 2, 1956): 26–28, 36.

42. Quotation in "Crossing Safety Is Everybody's Job," *RA* 121 (Nov. 30, 1946): 912–13, emphasis in original. Although the Federal Aid Highway Act of 1944 provided funds for guarding of railroad crossings and fixed the railroads' share of costs at 10 percent or less, federal law covered only about one-fifth of all roads.

"ACL Works to Eliminate Grade Crossing Accidents," *RA* 131 (Aug. 27, 1951): 41–43. The B&O is from "Educating Drivers to Reduce Crossing Accidents," *RA* 133 (Sept. 8, 1952): 54. Figures on funding for guarding are available in "Why Grade Crossings Are Safer," *RA* 147 (Aug. 10, 1959): 12–14.

43. As noted in ch. 9 (n. 83) I found the following relationship between annual crossing fatalities, automobile miles traveled, train miles (both in millions), crossings, and a trend for the years 1925–65.

$$\text{Log}(fatalities) = 6.21 + 0.0011(Train\ miles) + 0.14 \times 10^{-5}(Vehicle\ miles) + 0.3 \times 10^{-5}(Crossings) - 0.023\,Trend$$

In the period from 1945 to 1965, train miles fell about 25.7 (million) per year, which implies $(\exp 0.0011 \times -25.7) - 1 = -2.8$ percent, while crossings fell by 438 per year and $\exp(0.3 \times 10^{-5} \times -438) - 1 = -0.13$ percent. Vehicle miles rose by 30,356 (million) per year, which implies $\exp(0.14 \times 10^{-5} \times 30,356) - 1 = +4.3$ percent. Collectively, then, the decline in train miles and crossings and the rise in vehicle travel would have raised fatalities 1.37 percent a year but this was more than offset by the trend, probably due to better guarding, yielding a net decline of about 0.93 percent a year.

44. A. E. Schulman and C. E. Taylor, *Analysis of Nine Years of Railroad Accident Data, 1966–1974* (Washington: AAR, 1976), report that train accidents rose from 1966 to 1974 after adjusting for changes in ton miles, reporting, and inflation.

45. For the *Red Arrow* and *Spirit of St. Louis* see Shaw, *Down Brakes,* and http://dotlibrary.specialcollection.net/ [accessed June 15, 2005].

46. "Much Ahead in Modern Signaling," *RA* 122 (Jan. 4, 1947): 39–42. "Signaling Order Issued by ICC," *RSI* 40 (July 1947): 420–22. "Territories Involved in the ICC Signal Order," *RSI* 40 (Aug. 1947): 469–70. "Railroads Request Relief from the ICC Signal Order," *RSI* 40 (Sept. 1947): 549–52. "ICC Holds Signal Hearing," *RSI* 40 (Oct. 1947): 631–34. That the order was similar to the brotherhoods' views is from "ICC Orders More Signaling on High Speed Lines," *RA* 122 (June 21, 1947): 1264. Quotation from "The ICC Signal Order," *RA* 122 (June 28, 1947): 1295–96.

47. For companies that reduced speed see "Millions for Signals," *Trains* 11 (Nov. 1950): 45–49. Statistics on signals and train control are from ICC, *Tabulation of Statistics Pertaining to Signals, Interlocking, Automatic Train Control . . . [in the] United States* (Washington, various years).

48. "ICC Orders Power Brakes on All Freight Cars," *RA* 118 (June 9, 1945): 1027–28.

49. "Advancements in the Braking of High Speed Trains," *RA* 108 (May 18, 1940): 851–54. "AAR Squeeze Tests Two ACF Cars," *RA* 138 (Mar. 21, 1955): 50–51. "Now the UP Mainliners Use Disc Brakes," *RA* 143 (Dec. 9, 1957): 30–31, 36. For truck guides see "Safety Features in Streamline Operation," *RME* 115 (Sept. 1941): 352. "They Protect the Passengers," *RA* 129 (Oct. 14, 1950): 6, notes tightlock couplers.

50. The clippings are all in "Long Island Collisions," box 342, PRRC, HML. Sending the article to congressmen is from James Lyne [Editor, *RA*] to J. M. Symes, Dec. 8, 1950, box 342, PRRC, HML.

51. ICC, "Ex Parte No. 179, Accident at Woodbridge, N.J" (http://dotlibrary.specialcollection.net/ [accessed June 15, 2005]). The cost of passenger fatalities

is from "Status of Wreck Claims as of May 21, 1954" and "Measures to be Taken to Prevent Recurrence of Passenger Train Accidents," both in box 342, PRRC, HML. For the New York Central see "Speed Tapes Determine Schedules," *RA* 139 (July 18, 1955): 22–25. Speed control is from "Monthly News Condensation," *MR* 6 (Apr. 1951): 11–12.

52. Sperry advertisement, *RTS* 54 (Nov. 1958): 7. The text figures do not include tests done by AAR or other railroad-owned cars.

53. A brief history of train radio is in "Train Communication Hot in '44," *RA* 118 (Jan. 6, 1945): 42–44. "Channels for Train Radio," *RA* 118 (Jan. 20, 1945): 191–92. "Inductive Train Communication," *RSI* 37 (Nov. 1944): 625–28. "Train Communication in Road Service on Duluth, Missabe & Ironton," *RA* 123 (Aug. 9, 1947): 58–59. "Train Communication on the Pennsylvania," *RA* 123 (Sept. 27, 1947): 516–19. "Radio Warning Signal," *RA* 120 (Feb. 9, 1946): 327. "Train Communication Installed on the Bessemer and Lake Erie," *RA* 109 (July 20, 1940): 114–17.

54. "Fewer Wayside Observers Make Automatic Hazard Detection Necessary," *RSI* 52 (Oct. 1959): 28–31, "Electric Eye Catches High Loads on NYC," *RSI* 53 (Aug. 1960): 25–26. "Radio Flashes Flood Warnings," *RSI* 46 (Dec. 1956): 887. "Economics of Dragging Equipment Detectors," AAR Signal Section *Proceedings* 46 (1949): 55–57.

55. "Sound Testing Searches out Rail Defects Within Limits of Joint Bars," *RA* 128 (Mar. 4, 1950): 55–59, and "Supersonic Testing of Rail Ends," *REM* 48 (June 1952): 570–71. "Portable Instrument Detects Flaws in Rail at Joint Bars," *MR* 5 (Aug. 1950): 88. "C&O Installing Fleet of Axle Flaw Detectors," *RME* 129 (Sept. 1955): 74–76. For the Southern see "Ultrasonic Testing Cuts Journal Failures," *MR* 14 (1959): 151–54.

56. Photo-elasticity is from "The Polariscope—A New Aid in Rail Join Design," *RA* 112 (Jan. 10, 1942): 157–59, and "Research and Machines—Key to Track Maintenance in War," *RA* 112 (Apr. 11, 1942): 745–48. "Suggests Change in Rail Design as Result of Investigations," *RA* 113 (Dec. 12, 1942): 963–67. "AREA Adopts Three New Rail Designs," *RA* 121 (Nov. 9, 1946): 752–54. "Why Western Roads Like Three New Rails," *MR* 15 (Sept. 1959): 117–20.

57. For the Pennsylvania see "Web Failures and Their Prevention," *RTS* 50 (Nov. 1954): 49–51. "Maintenance Labor vs. Stronger Track," *RA* (Apr. 9, 1946): 724–25, suggests that rising labor costs encourage the use of heavier rail. "Studies Show Economies of Heavier Rail Sections," *RA* 128 (Mar. 18, 1950): 533. "What's Ahead for Welded Rail?" *RA* 142 (Feb. 18, 1947): 8–9. The B&O and C&O are from "Two Roads Put 'Science' into the Track," *MR* 19 (Feb. 1964): 53–54.

58. "Inspection Methods Extend the Life of Parts," *RA* 121 (May 1947): 239. For the Pennsylvania see "Railroad X-ray Laboratory," *RME* 123 (July 1949): 396–400. "Cobalt 60 for Testing Castings," *RA* 133 (Sept. 15, 1952): 76–77. "D & RGW's Atomic Detectives," *MR* 10 (Apr. 1955): 63–64.

59. "Stress Analysis with Brittle Coating," *RME* 120 (Mar. 1946): 126–27. The UP is from "Chemical Inspection of Metal Parts," *MR* 7 (Apr. 1952): 129–30. The survey of testing is from "Car Lubrication," *RME* 125 (Nov. 1951): 93–94. For an even larger proportion of defective axles see "C&NW Intensifies Inspection of Equipment Parts," *RME* 123 (Oct. 1949): 556–59. For the range of materials tested on the Santa Fe and Pennsylvania see "Magnaflux Test of Diesel Locomotive Parts," *RME* 120 (June 1946): 304–7. "How the Pennsylvania Uses Non Destructive Testing," *RA* 135 (Dec. 28, 1953): 48–51. For efficiency studies on

the Pennsylvania see "On Two Jobs Methods Study Saves $670,000 Annually in Equipment Maintenance," *RA* 140 (Feb. 6, 1956): 54–55.

60. For Timkin see "Research to Improve Axle Design," *RA* 125 (July 10, 1948): 103–4. The quote is from "Are We Really Serious About Hot Boxes?" *RA* 132 (Feb. 11, 1952): 68–70.

61. "Pennsy Proves its Plypaks," *RA* 139 (Sept. 12, 1955): 46–48. "On the way out" from "Hot Boxes—Causes and Cures," *RA* 139 (Nov. 7 1955): 26. "Hot Box Progress," *MR* 11 (Feb. 1956): 51–52.

62. The Rock Island is from "Electronic Hot Box Detector Ordered," *RA* 141 (May 14, 1956): 8. "How Hot Boxes Can Stop Train Automatically," *RA* 134 (Feb. 9, 1953): 84–85. The story of the C&O is from "New Angle on Hot Box Detector," *RA* 143 (July 22, 1957): 31–32. "GE Offers New Hotbox Detector," *RA* 147 (Oct. 5, 1959): 14. "Seaboard Airline Has Talking Hotbox Detector," *RSI* 54 (July 1961): 22–23. "New York Central Develops Hot Box Alarm," *RA* 120 (Mar. 9, 1946): 504–5.

63. "The AB Load Compensating Brake," *RME* 123 (Mar. 1949): 124–29.

64. The ban on cast iron is from "Mandatory Date," *RA* 142 (Mar. 4, 1957): 6–7. "Here's the AAR X-2 Wheel," *RME* 129 (May 1955): 50–54. "Wheel Checker Detects Broken Flanges," *RME* 127 (Apr. 1953): 8. For companies installing wheel flange detectors see "Railroad Signaling Will Pay its Way," *RA* 144 (Jan. 20, 1958): 26–28.

65. Carrier returns on investment are from "1958 Review of Railway Operations," *RA* 146 (Jan. 19, 1959): 63–73, and subsequent annual numbers. Equipment installations are from "Signaling Construction in 1959," *RSI* 53 (Jan. 1960): 20, and "Signaling and Communications," *RSI* 54 (Jan. 1961): 18.

66. J. M. Symes to E.W.S., Jan. 8, 1948, No Accident Month file, box 342, PRRC, HML. Deferred maintenance on the Pennsylvania from U.S. 85th Cong., 2d sess., Senate Committee on Transportation, *Problems on Railroads,* Part 1 (Washington, 1958), 74–144.

67. "Long Freight Trains," *RA* 135 (Dec. 14, 1953): 90–99. "Heavy Cars and the Steel Highway," *MR* 17 (Sept. 1962): 102–6. "Wheel Loads and Rail Failures," *RA* 158 (Feb. 15, 1965): 23, 27, 36. National Transportation Safety Board, *Broken Rails: A Major Cause of Train Accidents,* Report No. NTSB-RSS-71-1 (Washington, 1974).

68. Devra Golbe, "Product Safety In a Regulated Industry, Evidence from the Railroads," *Economic Inquiry* 21 (Jan. 1983): 39–52. For 1962 I estimated *Derailments* $= -400.64ROA$ (for details see Table A2.4), where *ROA* is the decimal return on assets. An increase in *ROA* by 0.04 implies $0.04 \times -400.64 = -16$, or about a 37 percent decline from the mean. As a check on the claim that the carriers did not trade safety for output, I estimated equations (not shown) between the log of derailments and the log of output (passenger and ton miles) with controls for capital and manhours for each year from 1960 through 1965. In no case is there a statistically significant relationship between derailments and output. I found no statistically significant impact of profitability on collisions. The Pennsylvania wreck is from ICC, *Railroad Accident Investigation Report No. 4002, the Pennsylvania Railroad Company, Harrison N.J. July 24, 1963* (Washington, 1963). Available at http://dotlibrary.specialcollection.net/ [accessed June 15, 2005].

69. "Rail Tank Car Explodes," *NYT* (Sept. 5, 1956), 16. "Railroad Toll Rises to Five," *NYT* (Jan. 21, 1959), 35. Norman Turner, *The Train Wreck Disaster at Meldrim, Georgia on June 28 1959* (privately printed, 2003). "Derailed Train Afire," *NYT* (July 8, 1959), 59.

70. U.S. 90th Cong., 2d sess., House Committee on Interstate and Foreign Commerce, *Hearings on H.R. 16980, Federal Standards for Railroad Safety* (Washington, 1968), quotation on 1. U.S. 91st Cong., 1st sess., Senate Committee on Commerce, *Hearings on S. 1933, S. 2915, and S. 3061, Federal Railroad Safety Act of 1969* (Washington, 1970), quotations on 4.

Conclusion

1. Both quotations from "Some Current Lessons from the Railway Safety Record," *RA* 99 (July 20, 1935): 65–67.

2. Aaron Wildavsky, *Searching for Safety* (New Brunswick: Transaction, 1988), ch. 3.

3. For a brief discussion of business-industry relationships before World War II see David Mowery and Nathan Rosenberg, *Paths of Innovation* (Cambridge: Cambridge University Press, 1998), ch. 2.

4. The regulation literature is vast. An excellent survey is Thomas McCraw, "Regulation in America," *Business History Review* 49 (Summer 1975): 159–83. See Thomas McCraw, ed., *Regulation in Perspective* (Cambridge, Mass.: Harvard University Press, 1981). James Harvey Young, *Pure Food* (Princeton: Princeton University Press, 1989) also surveys the business role in public policy. That regulation reflected increasing ineffectiveness of earlier approaches is in Edward Glaeser and Andrei Schleifer, "The Rise of the Regulatory State," *Journal of Economic Literature* 41 (June 2003): 401–25.

5. Douglass North, *Institutions, Institutional Change, and Economic Performance* (New York: Cambridge University Press, 1990), ch. 11, argues that institutional change is strongly path-dependent. The initial focus of safety organizations on rules, discipline, and behavior, combined with their apparent success, made change extremely difficult.

Appendix 1. Nineteenth-Century Railroad Accident and Casualty Statistics

1. The available years are 1874, 1884, 1889, 1895, 1898, and 1901. See Great Britain, Board of Trade, *General Report Upon the Accidents that Have Occurred on the Railways of the United Kingdom During the Year 1901* (HMSO, 1902), 22.

2. I used the "FORCST" routine in TSP to generate the estimates. As a check on these figures I employed an equation from G. R. Hawke, *Railways and Economic Growth in England and Wales* (Oxford: Clarendon, 1970), 267, that uses receipts to generate employment. It yields substantially larger numbers for the 1861–73 period and greatly over-predicts for later periods.

3. Frances Neison, "Analytical View of Railway Accidents," *Quarterly Journal of the Statistical Society* 16 (Dec. 1853): 289–337.

4. Hawke, *Railways and Economic Growth in England and Wales,* Table II.02.

5. Edward Bates Dorsey, "English and American Railroads Compared," ASCE *Transactions* 15 (1886): 1–78, Table 2.

6. Robert B. Shaw et al., "Shaw's All-Time List of Notable Railroad Accidents, 1831–2000," *Railroad History* 184 (Spring 2001): 37–45.

7. ICC Bureau of Statistics, "Railway Statistics Before 1890," *Statement No. 31251* (Washington, 1932).

8. Carl Snyder, *Business Cycles and Business Measurements* (New York: Macmillan, 1927), Appendix Table 2.

9. The estimating procedure employs the "FORCST" command in TSP.

10. "Our Accident Statistics," *RG* 22 (June 13, 1890): 419–20, and untitled, *RG* 31(June 28, 1899): 541.

Appendix 2. Casualties and Accidents from Interstate Commerce Commission Statistics, 1888–1965

1. The commission never published a good history of its accident statistics. They are briefly discussed in "Memorandum Concerning the Railway Accident Statistics of the Interstate Commerce Commission," *Accident Bulletin 74, 1919* (Washington, 1920), 66–69.

2. The U.S. Railroad Retirement Board, *Reporting of Accidents & Casualties in the Railroad Industry* (Washington, 1959) notes that "nothing concrete could be found to explain the [history of] the '24-hour death' or the '3 day in 10' rule" (11). Because the ICC aggregated these and data presented earlier in *Statistics of Railways* I assume that this definition was employed from the outset but can find no direct evidence on the point.

3. For the impact of the new reporting rules see *Accident Bulletin 36, 1910* (Washington, 1911). The damage limit for train accidents obtained until 1948. It was then steadily raised to account for inflation. See the section on Train Accidents below.

4. The estimation of subsequent fatalities here and below employs the equation in the text and the "FORCST" routine in TSP.

5. Mark Aldrich, *Safety First: Technology, Labor, and Business in the Building of American Work Safety, 1870–1939* (Baltimore: Johns Hopkins University Press, 1997), Appendix 1.

6. There were 2,578 reported worker fatalities in 1920. Estimated subsequent fatalities that year were, therefore, $60.12 + 0.072(2,578)$, which is about 9 percent of the total.

7. Aldrich, *Safety First*, 287.

8. U.S. Railroad Retirement Board, *Reporting of Accidents,* Chart 4.6-A.

9. For a more detailed critique of railroad accident statistics, see Railroad Retirement Board, *Reporting of Accidents.*

10. For severity definitions see ICC *Accident Bulletin 103, 1934* (Washington, 1935), 8.

11. In the derailment equations I also experimented with a variable for train length but it is typically insignificant and often has the wrong sign, probably because, controlling for car miles, increases in train length implied longer trains, which may have increased the probability of derailment, but fewer trains, which reduced derailments.

12. See the quarterly accidents bulletins 71, 72, and 74. Railroad Retirement Board, *Reporting of Accidents.*

13. ICC, *Railroad Construction Indices 1914–1954* (Washington, 1955). John McCusker, *How Much is that in Real Money? A Historical Price Index for Use as a Deflator of Money Values in the Economy of the United States* (Worcester: American Antiquarian Society, 1992).

Essay on Sources

In writing this book I have relied mostly on primary printed and archival materials, especially the many railroad and engineering journals and proceedings. There is little secondary literature that focuses directly on railroad safety but much that discusses railroad technology, management, and operating practices that had an effect on safety. A germane modern literature also exists on the economics of safety, risk assessment, and regulation. This essay describes the primary sources I consulted as well as the most valuable secondary and contemporary materials.

Archives

The most valuable collection on railroad safety is undoubtedly the Pennsylvania Railroad Collection at the Hagley Museum and Library. Not only does it have the records of the company's safety department from its inception into the 1960s, but it is a rich source on other aspects of safety, including accident reports, rail and wheel technology, signaling and train control, and much else. A second set of Pennsylvania records at the Pennsylvania State Archives in Harrisburg is similarly valuable. The Philadelphia & Reading collection at Hagley contains information on accidents, operating practices, and medical care.

     Other collections were far less rich, but some contained important material on particular topics. The Delaware & Hudson materials at the New York State Archives contain small amounts of detail on that company's safety and medical work. Also at the state archives are the records of the New York railroad commission, but they contain little that was not published. The University of Connecticut at Storrs has early records of the Providence & Worcester that discuss construction and operating problems as well as the much larger New York, New Haven & Hartford collection, which yields small amounts of material on grade crossing problems, bridges, and rails. The records of the Union Pacific at the museum in Omaha reveal the concern of that company's top executives with safety problems around the turn of the twentieth century and are a rich source of photographs. The Railroad Museum of Pennsylvania is also a mine of photographs while the Lehigh & Hudson River records there document safety work on a small, impoverished carrier. At Cornell University the New York & Pennsylvania collection contains small amounts on early accidents and operating practices, but the Benjamin Johnson papers are a valuable window into the life of an engineman and air brake instructor on the Southern Pacific Railroad in the 1880s. Also at Cornell, the papers of Floyd Helmer contain diaries of W. C. Hartigan, who was a superintendent of motive power on the New York, Ontario & Western in the 1890s and reported details of many minor train accidents. I used the Chicago, Burlington & Quincy records at the Newberry Library in Chicago for insights into early decision making on air brake purchases while the Great Northern–Northern Pacific collection at the Minnesota Historical Society includes materials on early problems with steel rails.

The train robbery related records at the Bancroft library at Berkeley are a window into that topic while the Sierra Railroad collection contains only tidbits on accidents, operating practices, and engineering. The Virginia & Truckee records were of little value for my purposes. Stanford University's Letcher Railroad Collection is also a valuable trove of information on early conditions on the Southern Pacific. There are also Master Mechanics' files from that company at the California State Railroad Museum in Sacramento that I found useful for information on early operating conditions, while the museum also contains large numbers of photographs. I found little helpful material in the Lackawanna records at Syracuse University.

Railroad records at the National Archives are beneficially discussed in David Pfeiffer, *Records Relating to North American Railroads: Reference Information Paper 91* (Washington: National Archives, 2001). The records of the Interstate Commerce Commission at the National Archives (RG 134) are disappointing. The commissioners' minutes are largely a record of decisions while the files of the Bureau of Safety were mostly discarded, and the manuscript of the commission's hearing on automatic train control could not be found. However, Formal Docket 568 contains information on the carriers' efforts to postpone the original Safety Appliance Act. There are also useful inspection records of the Bureau of Locomotive Inspection. See my "Research Note," *Railroad History* 175 (Autumn 1996): 132–34. Locomotive inspection reports and accident photographs from that bureau are in the Museum of American History archives at the Smithsonian. Also at the National Archives are the records of the U.S. Railroad Administration (RG 14), which while thin help document its safety work. The Sperry Collection at the Smithsonian Museum of American History contains records relating to the legal wrangle between the railroads and Sperry over the development and use of his transverse fissure detector car.

The Vermont State Archives contain the records of the state railroad commission but they include little not published in the annual reports. Similarly, Connecticut's railroad commission records at the state library in Hartford largely duplicate published sources although they also contain a valuable field notebook of a state inspector of railroad bridges. Both the Vermont and Connecticut Historical Societies have a considerable collection of railroad photographs. Records of the Denver & Rio Grande at the Colorado Historical Society include a small amount of useful material on injury reports, engineering, and operating practices. Small collections there relating to the Colorado Midland and the Denver & Salt Lake contain little on safety.

The Kheel Center at Cornell University has the records of a number of the railroad brotherhoods. Those of the Brotherhood of Railroad Trainmen are voluminous but contain only glimpses of safety issues from the union viewpoint. I also found little on safety matters in the records of the Switchmen's Union of North America and those of the Railway Conductors and Brakemen. More helpful is the Brotherhood of Railroad Signalmen Collection containing the papers of William Doble, who was an employee of the New York Central and official in the Signalmen's Union from the 1930s through the 1960s; the papers reveal the many safety issues brought about by motorcars.

Primary Printed Materials

The published primary sources for this book fall into three broad categories. There are (1) state and federal reports, (2) trade journals and the publications of technical associations, and (3) books and annual reports. For the antebellum years no useful indices exist. Thomas Taber III, *Railroad Periodicals Index, 1831–1999* (Muncy, Pa.: privately printed, 2001), while it indexes much of this literature, is organized by railroad, which makes it problematic for subject research. Beginning in 1884, the *Engineering Index* covers the trade and technical literature as does the *Industrial Arts Index* from 1913 on. Government hearings are usefully indexed in Congressional Information Service (CIS) *Congressional Committee Hearing Index* (Washington, 1981). Lexis Nexis *Congressional Indices* online help navigate the maze of government documents. U.S. Commissioner of Labor, *Index of All Reports Issued by State Bureaus of Labor Statistics in the United States Prior to March 1, 1902* (Washington, 1902) is a valuable guide to that vast literature.

Government Reports

A number of state railroad commissions published valuable reports. I found most useful those from Connecticut, Massachusetts, New York, and Ohio while also consulting those from, Illinois, Iowa, Michigan, New Hampshire, and Vermont. For New York and Vermont both the corporate author and title of these reports vary over time; I have not chosen to follow all these twists and turns and instead have adopted a simplified citation. Together these reports contain various discussions germane to safety as well as statistics on bridges, fencing, locomotives, accidents, and much more. The data, however, need to be used with care, as they sometimes covered only the state in question and sometimes the whole railroad.

Occasionally states undertook special investigations of safety-related topics. Usually these are published with the annual reports but the following are important exceptions. Massachusetts State Senate, "Report of the Committee on 'Railroads and Canals' in their Examination in Relation to the Accidents Upon the Western Rail-road," *Document 55, 1842* (Boston, 1842). Ohio Legislature, *Report of the Joint Committee Concerning the Ashtabula Bridge Disaster* (Columbus, 1877). "Report of the Committee Appointed to Examine and Report the Causes of Railroad Accidents, the Means of Preventing their Recurrence, &c," New York State Senate *Document 13, 1853* (Albany, 1853). New York Board of Railroad Commissioners, *Report on Strains on Railroad Bridges of the State* (Albany, 1891).

A few state labor bureaus published reports on railway labor and these can be easily found in the index noted above. Two of the biggest and most helpful are Michigan Bureau of Labor and Industrial Statistics, *Eleventh Annual Report, 1894* (Lansing, 1895) and Kansas Bureau of Labor and Industry, *Sixteenth Annual Report, 1900* (Topeka, 1901). The Texas Bureau of Labor Statistics, *Second Annual Report, 1911–1912* (Austin, 1912) also contains a survey of wage earners. National Association of Railway Commissioners *Proceedings* (1889–) is an important source for following state efforts to deal with grade crossing accidents and other safety concerns. California also published several investigations of grade crossing problems, notably California Railroad Commission, *Bulletin of the Progress of the Commission's Investigation of Railroad Grade Crossings in the State* (Sacramento, 1917), and *Report on the Grade Crossing Situation of Public Streets, Roads and Highways with Steam and Electric Interurban Railroads in the State of California, December 1, 1932* (Sacramento, 1933).

The Interstate Commerce Commission supplanted the states for most railroad statistics after 1887. The commission's *Statistics on Railways of the United States* (Washington, 1888–) is the central source for most of these data while its *Annual Reports* reflect official thinking on events and policy and for a short time after 1900 have valuable reports from field inspectors. In the twentieth century the commission's Bureau of Safety published annual reports and reports of accidents; the former are mostly taken up with minutia while the latter provide important details on train accidents. These are now available online (minus most of the illustrations) at http://dotlibrary.specialcollection.net/. Its *Tabulation of Statistics Pertaining to Signals, Interlocking, Automatic Train Control . . . as Used on the Railroads of the United States* (Washington, 1908–) contains the basic data on these installations. The Bureau of Locomotive Inspection, *Annual Report of the Chief Inspector of Locomotive Boilers* (Washington, 1912–) includes data and analysis of locomotive safety as well as analysis of specific accidents. Beginning in 1902 the *Accident Bulletin* becomes the basic source of safety data.

The commission also published special reports from time to time, of which I found the following useful. Its "Railway Statistics Before 1890," *Statement No. 31251* (Washington, 1932) is a valuable summary taken largely from Poor's *Manual of Railroads*. Its *Railways in the United States, 1902* (Washington, 1903), part 4, "State Regulation of Railways," contains good summaries of the powers of state commissions. The *Report on Block Signal Systems* (Washington, 1907) is an important opening salvo in reformers' efforts to require the block system and later automatic train control. Its Block Signal and Train Control Board, *Reports*, 1908–12, follow up on the previous report and provide a wealth of information on both signaling practice and reformers' hopes. *Operation of Trains on Heavy Grades* (Washington, 1907) reveals the extent of noncompliance with safety appliance laws. The commission's *Report on the Formation of Transverse Fissures and Their Prevalence on Certain Railroads* (Washington, 1923) is a key survey of that safety problem. Train control may be followed in its various hearings. I found these valuable: Docket 13413, *69 ICC 258* (1922), *91 ICC 426* (1924), *148 ICC 188* (1928), and *186 ICC 131* (1932). The "Investigation of Power Brakes and Appliances for Operating Power-Brake Systems," Docket 13528, *91 ICC 481* (1924) summarizes that investigation. ICC, *Study of Railroad Motive Power* (Statement 5025, Washington, 1950) is useful on dieselization.

Congressional hearings on railroad safety contain much useful information. As noted, these may be found using the CIS index or Lexis Nexis online. A number of congressional special investigations as well as publications of executive agencies also provide useful glimpses into railroad safety.

There is a mine of information on railroad technology and safety in Tenth Census of the United States, 1880, 4, *Report on the Agencies of Transportation* (Washington, 1883). The U.S. Commissioner of Railroads, *Reports* occasionally have safety tidbits. U.S. Commissioner of Labor, *Fifth Annual Report, 1889, Railroad Labor* (Washington, 1890) has data on turnover. U.S. Industrial Commission, *Report on Transportation*, vol. 4 (Washington, 1901) contains much testimony on railroad accidents, employers' liability, and relief systems, while U.S. Commissioner of Labor, *Twenty-Third Annual Report, 1908 Workmen's Insurance and Benefit Funds in the United States* (Washington, 1909) is a crucial study of that topic. It should be supplemented by Emory Johnson, "Railway Relief Departments," U.S. Department of Labor *Bulletin* 8 (Washington, 1897): 39–57, Emory Johnson, "Brotherhood Relief and Insurance of Railway Employees," Bureau of Labor Statistics *Bulletin* 17 (Washington, 1898), Samuel Lindsay, "Report on Railway Labor in the United States," U.S. Industrial Commission, *Report on Labor Organizations, Labor Disputes, and Arbitration, and on Railway Labor* 17 (Washington, 1901), 709–1135, and Samuel Lindsay, "Railway Employees in the United States," U.S. Bureau of Labor Statistics *Bulletin* 37 (Washington, 1901). U.S. Railroad Administration, *Survey and Recommendations of the Committee on Health and Medical Relief* (Washington, 1920) reviews railway medical institutions as of World War I. U.S. Railroad Administration, *Annual Report of Walker D. Hines, Director General of Railroads, 1919, Automatic Train Control Committee* (Washington, 1920) is an important early statement. U.S. Federal Coordinator of Transportation, *Cost of Railroad Employee Accidents, 1932* (Washington, 1935) includes valuable data on the workings of the employers' liability system.

For British railroad safety the basic source for statistics and accident investigations is Great Britain Board of Trade, *General Report on Accidents* (London, various years), although the data are often presented in an extremely confusing fashion. British Parliamentary Papers, *Transport and Communication*, vols. 14–15 (Shannon: Irish University Press, 1968) contains many of the early investigations of railroad safety. I found the following especially useful: Captain Douglas Galton, *Report to the Lords of the Committee of Privy Council for Trade and Foreign Plantations on the Railways of the United States* (London, 1857); Royal Commission on Railway Accidents, *Report of the Commissioners* (London, 1877); Royal Commission on Accidents to Railway Servants, *Report into the Causes of the Accidents, Fatal and Non-fatal to Servants of Railway Companies and of Truck Owners* (London, 1900); Great Britain Board of Trade, *Report on a Visit to America September 19 to October 31, 1902 by Lieut.-Colonel H.A. Yorke, R.E.* (London, 1903); Great Britain Board of Trade, *Railway Employment, Safety Appliances: Third and Fourth Reports of the Committee Appointed by the Board of Trade* (London, 1908).

Trade Journals and Technical Associations

These are the most consistently valuable sources for railroad safety. As noted, they are quite thoroughly covered in the *Engineering Index* beginning in 1884 but because titles can be deceptive I found that there is no substitute for going through them page by page. Many of the journals have changed title, some of them several times. Thomas Taber's *Index* noted above is invaluable for keeping track of these twists and turns.

For the antebellum years the *American Railroad Journal* (1832–) and the *American Railway Times* (1849–) are virtually the only useful railroad journals, although the *Journal of the Franklin Institute* had a good deal of railroad material and is accessible through its cumulative index. Beginning in 1870, however, the literature becomes enormously rich. Far and away the most valuable single source is the *Railroad Gazette* (1870–), for it surveyed virtually all the technical and operating aspects of railroading and included some business and financial matters as well. Equally broad are the *Railway Age* (1870–1908) and *Railway Review* (1879–1926) and while there is considerable overlap among these three, all must be consulted. The *Engineering News* (1873–) was in fact a railroad journal from about 1884 to 1900. The British journal *Engineering* (1866–) also contained much useful information during the nineteenth century. *Iron Age* (1867–) includes details on rail manufacturing and technology and reflects the steel industry perspective on these matters. The *American Railroad Journal* also continued during these years but is usually less valuable until the twentieth century, when it also became more specialized. It changed names numerous times, becoming *Railway Mechanical Engineer* in 1916 and *Railway Locomotives and Cars* in 1953. *Modern Railroads* (1945–91) sometimes had feature stories on safety or new technologies.

There are a host of other specialty journals, some quite ephemeral and hard to find, but many of these provide detail not found in the broader works. The most valuable of these are as follows. *Locomotive Engineering,* which ran from 1892 to 1928, under various titles, *Railway Maintenance Engineer* (1916–), which also came out under various titles and after 1952 is *Railway Track and Structures. Railway Master Mechanic* lasted from 1889 to 1916, and *Railway Signaling* ran from 1908 to 1970 under a number of titles. *Railway Surgeon* (1894–) is hard to find but vital for railroad medical programs. The *Association of Engineering Societies Journal* (1881–) provided access to railroad information published by many of the regional engineering societies.

After World War I, as the railroads' financial position slowly crumbled and employment declined, many of these journals disappeared. In addition, especially after World War II, they become more celebratory and less candid sources of information on safety.

The proceedings or transactions of a number of professional societies are also important sources, although they are often excerpted in the journals noted above. None of the various antebellum railroad associations lasted long enough to publish much on railroad safety or technology. Thus the two earliest and most valuable are the *Proceedings* of the Master Car Builders and Master Mechanics, both of which began in 1867. In 1918 the organizations combined to become the Mechanical Section of the American Railway Association. The American Railway Engineering Association *Proceedings* (1900–) and its later *Bulletin* are required reading for anyone interested in road, track, signaling, and grade crossing technology. Its *Stresses in Railroad Track—the Talbot Reports* (Chicago, 1980) reprinted some of this important early work. For earlier years these developments may be followed in the American Society of Civil Engineers *Transactions* (1872–) for which the society published cumulative indices. For British developments and views on American railroads the Institution of Civil Engineers (London) *Proceedings* is equally valuable. Rail technology is often discussed in the American Institute of Mining Engineers *Transactions.* The American Society of Mechanical Engineers *Proceedings* (1880–) contains small amounts of material.

The publications of various railroad specialty organizations and regional railroad clubs are also vital. I consulted the Association of Railway Superintendents of Bridges and Buildings *Proceedings,* the Roadmasters Association *Proceedings,* and the Railway Signal Association *Journal.* I also found much useful information in the Western Railway Club *Proceedings,* and also those of the New York Railroad Club and the New England Railroad Club.

The American Railway Association (ARA) *Proceedings* (1886–) are especially valuable down to about World War I and contain information on freight car interchange, medical matters, and many other economic and technological topics. The association's Bureau of Explosives, *Annual Report of the Chief Inspector* (1908–) is hard to find but vital for that topic. Its rules and regulations are contained in *The American Railway Association Rules and the Interstate Commerce Commission Regulations for the Transportation of Explosives and the American Railway Association Regulations for the Transportation of Inflammable Articles and Acids* (1909 and subsequent years). In addition to the ARA Mechanical Section noted above, several other sections published important information in their proceedings. I consulted the Signal Section *Proceedings* (1918–). The ARA Safety Section *Proceedings* (1920–) is a key source for railroad safety activities although they become less valuable after World War II. Similarly, the National Safety Council Railroad Safety Section *Transactions* (1912–) are immensely valuable through the 1940s but thereafter become decreasingly useful.

Also important are ARA, *The Invention of the Track Circuit* (New York, 1922) and ARA Committee on Automatic Train Control, *Bulletin 1 Automatic Train Control, Development and Progress* (Nov. 1930). In 1934 the Association of American Railroads (AAR) replaced the ARA. Two of its recent studies of safety and accidents are A. E. Schulman and C. E. Taylor, *Analysis of Nine Years of Railroad Accident Data, 1966–1974* (Washington, 1976), and A. E. Schulman, *Analysis of Nine Years of Railroad Personal Casualty Data, 1966–1974* (Washington, 1976).

The Bureau of Railway Economics (BRE) publications contain a few items of value for this book. I consulted *The Arguments for and Against Train-Crew Legislation,* Miscellaneous Series Bulletin 18 (Washington, 1915) and *The Arguments for and Against Limitation of Length of Freight Trains,* Miscellaneous Series Bulletin 23 (Washington, 1916). BRE, *List of References on Grade Crossings* (Washington, 1915), *Grade Crossings: List of References to Material Published 1914–March 1927 Supplementing List of References of 1915* (Washington, 1927), and *Some References to Material in the Library on Separation of Railroad Grade Crossings, 1939–1954* (Washington, 1955) are invaluable bibliographies on that topic. Similarly, BRE, *Railway Hospital Service in North America: Memorandum Listing References to Material on the Subject in BRE Library, October 20, 1944* (Washington, 1945) is a useful guide for that topic.

All of the railroad brotherhoods published either journals or proceedings or both, but they are usually disappointing and extremely hard to use. An invaluable guide is Lloyd Reynolds and Charles Killingsworth, *Trade Union Publications: The Official Journals, Convention Proceedings, and Constitutions of International Unions and Federations, 1850–1941,* 3 vols. (Baltimore: Johns Hopkins Press, 1944–45). I searched the major journals through the 1930s and episodically thereafter. There is a small amount of information on Safety First work in the *Railroad Trainmen's Journal* from 1912 through the 1920s, and in both the *Railway Conductor* and the *Locomotive Fireman's and Engineman's Magazine* in the 1920s. The *Locomotive Engineer's Journal* comments occasionally on legislation and safety work up through the 1930s. The *Brotherhood of Maintenance of Way Employee's Journal, Railroad Telegrapher, Switchman's Journal,* and *Signalman's Journal* all had little information useful for this book.

Books and Articles

The most valuable early works are Franz Anton Ritter von Gerstner, *Early American Railroads (Die innern Communicationen [1842–1843])* ed. Frederick Gamst (Stanford, 1997), Dionysus Lardner, *Railway Economy* (New York: Harper, 1850), Zerah Colburn and Alexander Holley, *The Permanent Way and Coal Burning Locomotive Boilers of European Railways* (privately printed, 1858), and Alexander Holley, *American and European Railway Practice* (New York: D. Van Nostrand, 1860). All of them yield a comparative perspective on American practice as does William Acworth, *American and English Railways Compared* (Chicago: Railway News Bureau, 1898). See also Edward Bates Dorsey, "English and American Railroads Compared," *ASCE Transactions* 15 (January 1886): 1–70, and Robert Gordon, "On the Economical Construction and Operation of Railways in Countries Where Small Returns are Expected as Exemplified by American Practice," Institution of Civil Engineers *Proceedings* (London) 85 (1886): 54–85. The hard-to-find Wilhelm Hoff and Felix Schwabach, *North American Railroads* (New York, 1906) compares Prussian and American lines about 1900.

J. L. Ringwalt, *Development of Transportation Systems in the United States* (Philadelphia, 1888) contains much valuable detail on early practice. Charles Francis Adams Jr., *Notes on Railroad Accidents* (New York: Putnam's, 1879) is a sophisticated discussion of early safety problems. Francis Neison, "Analytical View of Railway Accidents," *Quarterly Journal of the Statistical Society* 16 (December 1853): 290–335 details accidents on early British lines.

Valuable discussions of technology, economics, operating practices, and much else are contained in Arthur Mellen Wellington, *Economic Theory of the Location of Railways,* rev. ed. (New York: John Wiley, 1887). Angus Sinclair, *Locomotive Engine Running and Management,* 23d ed. (New York: John Wiley, 1915) is helpful on that topic. Robert Thurston, *A Treatise on Iron and Steel* (New York: John Wiley, 1884) discusses metallurgy. A good discussion of switches is W. B. Parsons, *Track, A Complete Manual of Maintenance of Way* (New York: Engineering News, 1886). W. M. Camp, *Notes on Track* (Chicago: privately published, 1903) is definitive on that topic. Walter Berg, *Buildings and Structures of American Railroads* (New York: John Wiley, 1893) and his *American Railway Shop Systems* (New York: Railroad Gazette, 1904) contain enormous amounts of useful detail. John Droege, *Passenger Terminals and Trains* (New York: McGraw Hill, 1916) remains the most valuable discussion of the role of safety considerations in early station design. Harry Forman, *Rights of Trains,* rev. Peter Josserand (New York: Simmons-Boardman, 1945) is the bible on train rules.

On signal technology and practice I used Railroad Gazette, *American Practice in Block Signaling* (New York, 1891), Braman B. Adams, *The Block System* (New York, 1901), and Ralph Scott, *Automatic Block Signals and Signal Circuits* (New York: McGraw Hill, 1908). The development of the air brake is outlined in Westinghouse Air Brake Company, *Air Brake Tests* (Pittsburgh, 1905). See also Walter Turner and S. W. Dudley, *Development in Air Brakes for Railroads* (Westinghouse Air Brake Co., 1909). B. Joy Jeffries, *Color Blindness: Its Dangers and its Detection* (Boston, 1883) is an early call to regulate railroad workers' eyesight in the public interest. Railroad relief work is exhaustively treated in M. Riebenack, *Railway Provident Institutions in English Speaking Countries* (Philadelphia, 1905). For a later period I consulted Bryce Stewart and Murray Latimer, *Railroad Pensions in the United States* (Industrial Relations Counselors, 1929).

On bridges, Theodore Cooper, *General Specifications for Iron Railroad Bridges and Viaducts* (New York: Engineering News, 1884) is an important milestone in the development of specifications. Wolcott C. Foster, *A Treatise on Wooden Trestle Bridges According to the Present Practice on American Railroads* (New York: John Wiley, 1891) is a helpful guide. George Vose, *Bridge Disasters in America* (Boston: Lee & Shepherd, 1887) is hard to find and most of it appeared in the *Railroad Gazette.* Stephen Peet, *The Ashtabula Disaster* (Chicago, 1877) details that tragedy.

Several states and cities published reports on their efforts to remove grade crossings. I used Augustus Locke et al., *Report of an Investigation into the Subject of the Gradual Abolition of the Crossing of Highways by Railroads at Grade* (Boston, 1889) and Chicago Track Elevation Department, *Track Elevation Within the Corporate Limits of the City of Chicago to December 31st 1908* (Chicago, 1909). See also Fay,

Spofford and Thorndike Consulting Engineers, *Grade Crossing Elimination Syracuse, New York* (Boston, 1927).

A host of writings by railroad workers yield insight into job conditions, unionization, and technology. I consulted the following: Joseph Bromley, *Clear the Tracks* (New York: McGraw-Hill, 1943); Chauncey Del French, *Railroadman* (New York: Macmillan, 1938); J. Patrick Desmond, *Ace Carlson, Railroader, 1878–1960* (New York: Frederick Fell, 1964); Charles B. George, *Forty Years on the Rail* (Chicago: R. R. Donnelley & Sons, 1887); Roger Grant, ed., *Brownie the Boomer* (DeKalb: Northern Illinois University Press, 1991); Freeman Hubbard, *Railroad Avenue* (New York: McGraw Hill, 1945); Gilbert Lathrop, *Little Engines and Big Men* (Caldwell, Idaho, 1954); Stuart Leuthner, ed., *The Railroaders* (New York: Random House, 1983); Joseph Noble, *From Cab to Caboose: Fifty Years of Railroading* (Norman: University of Oklahoma Press, 1964); Herbert Pease, *Singing Rails* (New York: Thomas Y. Crowell, 1948); J. Harvey Reed, *Forty Years a Locomotive Engineer* (Prescott, Wash., 1913); James Ward, ed., *Southern Railroad Man: N.J. Bell's Recollections of the Civil War Era* (DeKalb: Southern Illinois University Press, 1994); Richard Reinhardt, ed., *Workin' on the Railroad* (Palo Alto: American West, 1970) reprints excerpts from many original sources. Benjamin Bodkin and Alvin Harlow, eds., *A Treasury of Railroad Folklore* (New York: Crown, 1953) and Robert Wayner, *Great Poems from Railroad Magazine* (New York: privately printed, 1968) contain illuminating and entertaining glimpses of railroad life and work. Youngstown State University Oral History Collection, available at http://www.maag.ysu.edu/oralhistory/oral_hist.html, contains valuable interviews.

Nineteenth- and early-twentieth-century railroads generated a rich ephemeral literature that included company annual reports, occasional reports of accident investigations, rulebooks, committee investigations, and company employee magazines. Although some rulebooks such as E. H. McHenry, *Engineering Rules and Instructions, Northern Pacific Railway* (New York, 1899) are insightful and a few of the early annual reports contain important comments on the adoption of steel rails or airbrakes, in general I found little of value in these materials. The B&O Employees Relief Association, *Annual Report* (Baltimore, 1882–) is invaluable for the origins of relief work. Origins of Safety First work are in Chicago & North Western Railway Co., *Report of the Central Safety Committee* (Chicago, 1913) and in the many talks given by Ralph Richards. See especially his "The Safety First Movement on American Railways," in *Proceedings of the Second Pan American Scientific Congress*, vol. 11 (Washington, 1917), 326–35, which has a good bibliography. The New York Central, *Safety* (1913–) and Pennsylvania Railroad System, *Information for Employees and the Public* (1913–) both contain information on company safety work.

Secondary Sources

The secondary writing that bears directly on railroad safety is limited, but the literature that contributed to an understanding of the subject is quite broad and spans a number of disciplines ranging from psychology to physics and includes economics as well as the many subdisciplines of history. In many of these fields the literature is immense and I have cited only what I have found to be the most relevant works. In the following I have organized these materials into general works (including those on technology, risk, railroads, and some other areas), and then works specifically on railroad safety, railroad technology, laws and regulations, and finally labor.

General Works

The evolution of railroad safety is part of the broader process of understanding economic and technological change through time and no one has written more perceptively on these subjects than Joseph Schumpeter in his *Theory of Economic Development* (Cambridge, Mass.: Harvard University Press, 1934) and later *Business Cycles,* 2 vols. (New York: McGraw Hill, 1939). Almost equally sweeping is the writing of Alfred Chandler. His many works, but especially *The Visible Hand: t he Managerial Revolution in American Business* (Cambridge, Mass.: Harvard University Press, 1977), and *Big Business and the Wealth of Nation* (New York: Cambridge University Press, 1997), place railroads in the broad course of organizational and technological history. Several scholars attempt to extend and modify Chandler: Naomi Lamoreaux, Daniel Raff, and Peter Temin, "Beyond Markets and Hierarchies: Toward a New Synthesis of American Business History," *American Historical Review* 108 (April 2003): 404–33; Richard Langlois, "The Vanishing Hand, the Changing Dynamics of Industrial Capitalism," *Industrial and Corporate Change* 12 (April 2003): 351–85.

I have found useful some of the writing on the ways institutions, organizations, information problems, and transactions costs shape economic behavior and outcomes. Key sources include Oliver Williamson, *The Economic Institutions of Capitalism* (New York: The Free Press, 1985), which is most valuable when read in tandem with Chandler's work. JoAnne Yates, *Control Through Communication* (Baltimore: Johns Hopkins University Press, 1989) helped me see common threads in much safety work. M. Norton Wise, *The Values of Precision* (Princeton: Princeton University Press, 1995), Samuel Krislov, *How Nations Choose Product Standards* (Pittsburgh: Pittsburgh University Press, 1997), and Trevor Pinch, 'Testing One, Two Three . . . Testing': Toward a Sociology of Testing," *Science, Technology, & Human Values* 18 (Winter 1993): 25–41 provide insights into the broader significance of standards, measurement, and testing. See also David Hemenway, *Industrywide Voluntary Product Standards* (Cambridge, Mass.: Ballinger, 1975) and Charles Kindleberger, "Standards as Public, Collective, and Private Goods," *Kyklos* 36 (June 1983): 377–96. The best introductions to economists' thinking about institutional change are Lance Davis and Douglass North, *Institutional Change and American Economic Growth* (Cambridge: Cambridge University Press, 1971) and especially Douglass North, *Institutions, Institutional Change, and Economic Performance* (Cambridge: Cambridge University Press, 1990).

The broader political developments shaping public policy toward railroads in the twentieth century are illuminated in a number of standard works. Samuel Hays, *Response to Industrialism* (Chicago: University of Chicago Press, 1957) remains a valuable guide to the forces shaping nineteenth-century political economy. Daniel Rogers, "In Search of Progressivism," *Reviews in American History* 10 (December 1982): 113–32 shaped my thinking about Progressivism. Robert Higgs, *Crisis and Leviathan* (New York: Oxford University Press, 1987) is insightful on when and why federal power over the economy expands. Ellis Hawley, "Herbert Hoover, the Commerce Secretariat and the Vision of an 'Associative State,' 1921–1928," *Journal of American History* 61 (June 1974): 116–40 is especially valuable.

Since railroad safety has reflected in part improvements in productivity and technology the general literature on this subject provides context. Two valuable overviews of technology and productivity are Frederic Scherer, *New Perspec-*

tives on Economic Growth and Technological Innovation (Washington: Brookings, 1999) and especially Vernon Ruttan, *Technology, Growth, and Development* (New York: Oxford University Press, 2001). Nathan Rosenberg's works are numerous and invariably insightful. A valuable sampling includes his *Technology and American Economic Growth* (New York: M. E. Sharpe, 1972) and *Inside the Black Box* (New York: Cambridge University Press, 1982). Similarly broad and perceptive are Joel Mokyr, *The Lever of Riches* (New York: Oxford University Press, 1990) and *The Gift of Athena* (Princeton: Princeton University Press, 2002). Recent work explaining technological change as path-dependent is in Paul David, "Cleo and the Economics of QWERTY," *American Economic Review* 75 (May 1985): 332–37, while Robert Allen, "Collective Invention," *Journal of Economic Behavior and Organization* 4 (1983): 1–24 is a useful explanation of the importance of learning networks, as is Gavin Wright, "Can a Nation Learn? American Technology as a Network Phenomenon," in *Learning By Doing in Markets, Firms and Countries,* ed. Naomi Lamoreux et al. (Chicago: University of Chicago Press, 1999), 295–326, and William Baumol, *The Free Market Innovation Machine* (Princeton: Princeton University Press, 2002). Henry Petroski, *Engineers of Dreams* (New York: Vintage, 1995) argues that engineers are by training disposed to improve technology.

In his *To Engineer Is Human: The Role of Failure in Successful Design* (New York: Vintage, 1985), Henry Petroski argues that technological improvements arise out of failure. Charles Perrow, *Normal Accidents* (New York: Basic, 1984) essentially reverses the claim, providing a sophisticated although unhistorical argument linking complex technologies to disasters. Jacob Feld and Kenneth Carper, *Construction Failure,* 2d ed. (New York: John Wiley, 1997) provides a technical assessment of why things fall down. Also valuable for those who need to know the technical details is Stephen Timoshenko, *History of the Strength of Materials* (New York: McGraw Hill, 1953) and H. F. Moore and J. B. Kommers, *The Fatigue of Metals* (New York: McGraw Hill, 1927), which has a good historical survey in chapter 2.

Grasping the history of safety requires understanding how people perceive risk and accidents, how and why these views evolve, and how firms and workers respond. I have found Thomas Haskell, "Capitalism and the Origins of Humanitarian Sensibility, Parts I and II," *American Historical Review* 90 (1985): 339–61, 547–66 especially insightful. Barbara Welke, *Recasting American Liberalism* (New York: Cambridge University Press, 2001) makes similar arguments. Haskell's *Emergence of Professional Social Science* (Champaign: University of Illinois Press, 1977) and Theodore Porter, *Trust in Numbers* (Princeton: Princeton University Press, 1995) also led me to think about how changing worldviews generated an emerging safety consciousness. Wolfgang Schivelbusch, *The Railway Journey* (New York: Berg, 1986) provides insights into the meaning of travel, risk, and much else. Vicki Goldberg, "Death Takes a Holiday, Sort of," in *Why We Watch: the Attractions of Violent Entertainment,* ed. Jeffrey Goldstein (New York: Oxford University Press, 1998), 27–51 also contains useful insights.

The work of Daniel Kahneman, Paul Slovic, and Amos Tversky discusses the ways people understand and respond to risk. Much of it is contained in their *Judgment Under Uncertainty: Heuristics and Biases* (Cambridge: Cambridge University Press, 1982). A number of works discuss the social and economic forces that shape risk and risk perception, including James Flynn et al., eds., *Risk, Media, and Stigma* (London: Earthscan, 2001), and Roger Kasperson and Jeanne Kasperson, "The Social Amplification and Attenuation of Risk," *Annals of the American Academy of*

Political and Social Science 545 (May 1996): 95–105. Aaron Wildavsky, *Searching for Safety* (New Brunswick: Transaction, 1988) also discusses the development of modern attitudes toward risk and the importance of wealth in improving safety.

The economic literature on safety and public policy is immense. Many of the issues involving risk trade-offs are raised in Walter Oi, "On the Economics of Industrial Safety," *Law and Contemporary Problems* 38 (4) (1974): 538–55. An excellent summary that extends Oi and reviews recent empirical findings is W. Kip Viscusi, *Fatal Tradeoffs* (New York: Oxford University Press, 1992). David Moss, *When All Else Fails: Government as the Ultimate Risk Manager* (Cambridge, Mass.: Harvard University Press, 2002) contains useful theoretical and historical insights.

Although there are probably thousands of books on railroads, most have little about safety, although many provide valuable details into operating conditions. Two early works are William Sipes, *The Pennsylvania Railroad* (Philadelphia, 1875) and James Dredge, *Pennsylvania Railroad* (London, 1879). Both document the operating practices and technology of that carrier. Seymour Dunbar, *A History of Travel in America* (Indianapolis: Bobbs-Merrill, 1915), vol. 3, provides a broad overview of travel and its dangers. Edward C. Kirkland, *Men, Cities, and Transportation* (Cambridge, Mass.: Harvard University Press, 1948) has an extended discussion of safety and much on the early commissions. Stephen Salsbury, *The State, the Investor, and the Railroad* (Cambridge, Mass.: Harvard University Press, 1967) discusses the political consequences of early accidents. Yehuda Grunfeld, "The Effect of the Per Diem Rate on the Efficiency and Size of the American Railroad Freight Car Fleet," *Journal of Business* 32 (Jan. 1959): 52–73 helped sharpen my understanding of how rates shaped safety incentives. Richard Saunders's several works, especially *Main Lines: Rebirth of the North American Railroads, 1970–2002* (DeKalb: Northern Illinois University Press, 2003), provide a valuable overview of the decline and rebirth of postwar railroading. A valuable economic analysis of British railroads containing much useful data is G. R. Hawke, *Railways and Economic Growth in England and Wales* (Oxford: Clarendon, 1970). There is little literature about the men who made the railroads safer as most have not merited a book-length biography. Two exceptions are Alfred Chandler, *Henry Varnum Poor, Business Editor Analyst, and Reformist* (Cambridge, Mass.: Harvard University Press, 1956) and Thomas P. Hughes, *Elmer Sperry, Inventor and Engineer* (Baltimore: Johns Hopkins University Press, 1971). There is no modern biography of Westinghouse and so I have relied on Henry G. Prout, *A Life of George Westinghouse* (New York: Arno, 1972) and Francis Leupp, *George Westinghouse: His Life and Achievements* (Boston: Little, Brown, 1918). Thomas McCraw, *Prophets of Regulation: Charles Francis Adams, Louis D. Brandeis, James M. Landis, Alfred E. Kahn* (Cambridge, Mass.: Harvard University Press, 1984) is the best source on Charles Francis Adams. Information on Charles B. Dudley is contained in ASTM, *Memorial Volume Commemorative of The Life and Life Work of Charles B. Dudley* (Philadelphia, 1911). Robert Paterson, *Robert Harvey Reed: A Sanitary Pioneer in Ohio* (Columbus: Ohio Public Health Association, 1935) is useful for this leader in railway medicine. For Arthur Mellen Wellington, Matthias Forney, and others I consulted Harold Lane, ed., *Biographical Dictionary of Railway Officials of America,* various editions (New York: McGraw Hill) and the *Biographical and Genealogical Master Index.*

Safety-Specific

There is a small secondary literature dealing specifically with railroad safety, most of it focusing on train accidents. For Great Britain I relied on Richard Blythe, *Danger Ahead* (London: Newman Neame, 1951) and L.T.C. Rolt, *Red For Danger*, 4th ed. (London: David & Charles, 1982). For the United States, Robert Shaw, *Down Brakes* (New York: P. R. Macmillan, 1961) remains the best survey but Robert Reed, *Train Wrecks* (New York: Bonanza, 1982) is also valuable. An early statistical analysis is C. Douglas Campbell, "The Business Cycle and Accidents to Railroad Employees in the United States," *Journal of the American Statistical Society* 26 (Sept. 1931): 295–302. Devra Golbe, "Product Safety in a Regulated Industry, Evidence from the Railroads," *Economic Inquiry* 21 (Jan. 1983): 39–52 relates several measures of safety to profitability.

More focused are the following. Warren Jacobs, "Early Rules and the Standard Code," Railway and Locomotive Historical Society *Bulletin* 50 (Oct. 1939): 29–55 is still valuable. Ian Bartky, "Running on Time," *Railroad History* 159 (Autumn 1988): 19–38 discusses early timetables and train rules. Charles Clark, "The Development of the Semi-Automatic Freight Car Coupler, 1863–1893," *Technology and Culture* 13 (Apr. 1972): 115–39 is definitive, while his "The Railroad Safety Movement, in the United States: Origins and Development, 1869–1913" (Ph.D. diss., University of Illinois, 1966) provides a broader discussion of early safety appliances. Steven Usselman, "Air Brakes for Freight Trains: Technological Innovation in the American Railroad Industry, 1869–1900," *Business History Review* 58 (Spring 1984): 30–50 provides a good discussion of this important development.

Mark Aldrich, "Safety First Comes to the Railroads, 1910–1939," *Railroad History* 162 (Spring 1992): 6–33 discusses the early Safety First movement. Work safety on the railroads and other industries is discussed in his *Safety First: Technology, Labor, and Business in the Building of American Work Safety, 1870–1939* (Baltimore: Johns Hopkins University Press, 1997). Various aspects of locomotive technology and safety are covered in more detail in Aldrich, "Safe and Suitable Boilers: The Railroads, the Interstate Commerce Commission, and Locomotive Safety, 1900–1945," *Railroad History* 171 (Autumn 1994): 23–44. Also see the following by the same author: "Combating the Collision Horror, The Interstate Commerce Commission and Automatic Train Control," *Technology and Culture* 34 (Jan. 1993): 49–77; "The Peril of the Broken Rail: The Carriers, the Steel Companies and Rail Technology, 1900–1945," *Technology and Culture* 40 (Apr. 1999): 263–91.

Individual disasters are discussed in a number of places. I used D. A. Gasparini and Melissa Fields, "Collapse of Ashtabula Bridge on December 29, 1876," *Journal of Performance of Constructed Facilities* 7 (May 1993): 109–25. Don Helmers, *Tragedy at Eden* (Pueblo, 1971) provides details of that tragedy. Marion Pinsdorf, "Engineering Dreams into Disaster: History of the Tay Bridge," *Railroad History* 179 (Aug. 1998): 89–116. I am indebted to Norman Turner for allowing me to use his *The Train Wreck Disaster at Meldrim, Georgia on June 28 1959* (privately printed, 2003).

The most valuable discussions of modern safety conditions during the ICC era are U.S. Railroad Retirement Board, *Work Injuries in the Railroad Industry, 1938–1940* (Chicago, 1947), *Reporting of Accidents & Casualties in the Railroad Industry* (Washington, 1959), and *Safety in the Railroad Industry* (Chicago, 1962). The postwar era is analyzed by Ian Savage, *The Economics of Railroad Safety* (Boston: Kluwer, 1998) and his "Trends in Transportation Employee Injuries since Economic

Deregulation," in *Research in Transportation Economics: Transportation Labor Issues and Regulatory Reform,* ed. James People and Wayne Talley (Amsterdam: Elsevier Science, 2004).

Railroad Technology

The writing on railroad technology is vast. John White's *American Locomotives: An Engineering History 1830–1880* (Baltimore: Johns Hopkins University Press, 1997), *The American Railroad Passenger Car* (Baltimore: Johns Hopkins University Press, 1978), and *The American Railroad Freight Car* (Baltimore: Johns Hopkins University Press, 1993) are the definitive works. ASME Rail Transportation Division, *Railway Mechanical Engineering, A Century of Progress: Car and Locomotive Design* (New York: ASME, 1979) is a good reference while August Mencken, *The Railroad Passenger Car* (Baltimore: Johns Hopkins University, 1957) is still useful. Volumes 1 and 2 of Anthony Bianculli, *Trains and Technology: The American Railroad in the Nineteenth Century* (Newark: University of Delaware Press, 2001–) are on locomotives and rolling stock, but add little to White. Lewis Sillcox, *Mastering Momentum* (New York: Simmons Boardman, 1936) is invaluable on air brakes. Al Krug, "North American Freight Train Brakes," http://www.railway-technical.com/brake2.html provides clear descriptions and diagrams. For British passenger cars I consulted Hamilton Ellis, *Nineteenth Century Railway Carriages* (London: Modern Transport, 1949). Frederick Westing, *Apex of the Atlantics* (New York: Bonanza, 1974) and *Pennsy Steam and Semaphores* (New York: Bonanza, 1974) provide excellent technical histories of these topics.

The technology of permanent way is less well covered. James Vance, *The North American Railroad: Its Origin, Evolution, and Geography* (Baltimore: Johns Hopkins University Press, 1995) is important for its emphasis on the shaping role of geography. William Middleton's *Landmarks on the Iron Road: Two Centuries of American Railroad Engineering* (Indianapolis: Indiana University Press, 1999) contains numerous case studies of construction projects. Volume 3 of Anthony Binaculli, *Trains and Technology: The American Railroad in the Nineteenth century* (Newark: University of Delaware Press, 2001–) is a helpful modern history of track while Earl Haytr, "The Fencing of Western Railroads," *Agricultural History* 19 (July 1945): 163–67 is still useful. A contemporary reference is William Hay, *Railroad Engineering,* 2d ed. (New York: Wiley, 1982). Robert Adam, "History of the Abolition of Railroad Grade Crossings in the City of Buffalo," Buffalo Historical Society *Proceedings* 8 (1905): 149–254 is a valuable case study. Railroad signaling is thoroughly discussed in Mary Brignano and Hax McCullough, *The Search for Safety: A History of Railroad Signals and the People Who Made them* (American Standards, 1981), which also has a valuable bibliography. An exhaustive treatment of signaling in Britain is Stanley Hall, *History and Development of Railway Signaling in the British Isles,* 2 vols. (York, England: Friends of the National Railway Museum, 2000). Carston Lunsden, "North American Signaling" (http://www.lundsten.dk/us_signaling/index.html), and "U.S. Railroad Signaling" (http://www.railway-technical.com/US-sig.html) provide good explanations of both modern and historical practice.

There is a considerable literature on railroad bridges. I found valuable material in the following. Carl Condit, *American Building Art, the Nineteenth Century* (New York: Oxford University Press, 1960). Llewellyn Edwards, *A Record of History and Evolution of Early American Bridges* (Orono: University of Maine Press, 1959). Henry Tyrrell, *History of Bridge Engineering,* 4th ed. (Chicago, 1911). J.A.L. Waddell,

Bridge Engineering (New York: John Wiley, 1916), vol. 1. Larry Kline and Anthony Thompson, "The Evolution of American Railroad Bridges, 1830–1994," in Anthony Thompson, ed., *Symposium on Railroad History* 3 (1994): 71–93. Donald Jackson, "Nineteenth Century American Bridge Failures: A Professional Perspective," *Proceedings of the Second Historic Bridges Conference, March 1988* (Columbus, 1988), 113–25.

The literature on other railroad structures (shops, stations, roundhouses) helps clarify a number of safety issues. An overview of architecture with many drawings and pictures is Carroll Meeks, *The Railroad Station: An Architectural History* (New Haven: Yale University Press, 1956). Edwin Alexander, *Down at the Depot* (New York: Clarkson Potter, 1970) and H. Roger Grant and Charles Bohi, *The Country Railroad Station in America* (Sioux Falls, S.D.: Center for Western Studies, 1988) are valuable for their pictures while the latter work also conveys a sense of life at the depot.

The best treatment of railroad technology in a broader context is Steven Usselman, *Regulating Railroad Innovation* (New York: Cambridge University Press, 2002), which has a superb discussion of the diffusion of the air brake. Brief but useful is David Jeremy and Darwin Stapleton, "Transfers Between Culturally Related Nations: The Movement of Textile and Railroad Technologies Between Britain and the United States, 1780–1840," in *International Technology Transfer, Europe, Japan and the USA, 1700–1914,* ed. David Jeremy (Brookfield, Vt.: Edward Elgar, 1991), 31–48. Nathan Rosenberg and Walter Vincenti, *The Britannia Bridge: The Generation and Diffusion of Technological Knowledge* (Cambridge, Mass.: MIT Press, 1978) provides a fascinating glimpse of interindustry technological diffusion. Of the works on organizations that helped develop and diffuse technology I found the following especially useful. Daniel H. Calhoun, *The American Civil Engineer: Origins and Conflict* (Cambridge, Mass.: MIT Press, 1960) remains the best study of that organization. Bruce Sinclair, *Early Research at the Franklin Institute: The Investigation into the Causes of Steam Boiler Explosions* (Philadelphia: Franklin Institute, 1966) is a valuable treatment of the Franklin Institute. Ira O. Baker and Everett King, *History of the College of Engineering of the University of Illinois, 1868–1945,* 2 vols. (Urbana: University of Illinois Press, 1947) discuss that school's engineering experiment station.

Economic aspects of railroad technology include Jeremy Atack and Jan Brueckner, "Steel Rails and American Railroads, 1867–1880," *Explorations in Economic History* 19 (1982): 339–59, which began a considerable exchange on the economic forces shaping diffusion of steel rails. Christopher Bell, "Charles Ellett Jr. and the Theory of Optimal Input Choice," *History of Political Economy* 18 (Fall 1986): 485–95 is one of the few writers to discuss the economic ideas of early railroad engineers. Reviews of modern developments that I found helpful are Kent Healy, "Regularization of Capital Investment in Railroads," in NBER, *Regularization of Business Investment* (Princeton, 1954), 137–212, and his *Performance of the U.S. Railroads since World War II* (New York: Vantage, 1985). An overview of postwar developments is National Research Council, *Science and Technology in the Railroad Industry* (Washington, 1963). Edwin Mansfield, "Innovation and Technical Change in the Railroad Industry," in NBER, *Transportation Economics* (New York: Columbia, 1965), 169–96 is an econometric assessment.

Early data on railroad productivity as well as an assessment of its sources are in Albert Fishlow, "Productivity and Technological Change in the Railroad Sector, 1840–1910," in *Output, Employment and Productivity in the United States After 1800*

(New York: NBER, 1966), 583–646. Productivity data for the railroads as well as the economy as a whole come from John W. Kendrick, *Productivity Trends in the United States* (Princeton: Princeton University Press, 1961) and, for the postwar years, his *Postwar Productivity Trends in the United States, 1948–1969* (New York: Columbia University Press, 1973). See also National Commission on Productivity, *Improving Railroad Productivity* (Washington, 1973).

Law and Regulation

Most work on railroad regulation has focused primarily on prices and quantities, but there are exceptions. Edward Glaeser and Andrei Schleifer, "The Rise of the Regulatory State," *Journal of Economic Literature* 41 (June 2003): 401–25 gives a broad explanation for the rise of regulation. Colleen Dunlavy, *Politics and Industrialization: Early Railroads in the United States and Prussia* (Princeton: Princeton University Press, 1994) provides both a historical analysis of how differing political regimes affected the development of railroad technology and regulation and a brief but excellent discussion of technology diffusion, but unfortunately her work stops about 1850. Henry Parris, *Government and the Railways in Nineteenth Century Britain* (London: Routledge and Kegan Paul, 1965) discusses British regulatory efforts. Thomas McCraw, "Regulation in America," *Business History Review* 49 (Summer 1975): 159–83 is still the best survey of that subject.

There are several valuable early summaries of railroad law. I consulted W. P. Gregg, *The Railroad Laws and Charters of the United States,* 2 vols. (Boston, 1851); Isaac Redfield, *Practical Treatise Upon the Law of Railroads,* 2d ed. (Boston: Little, Brown, 1858); and Christopher Patterson, *Railway Accident Law: The Liability of Railways for Injuries to the Person* (Philadelphia: T. & J. W. Johnson, 1886). An excellent modern treatment that emphasizes common law is James Ely, *Railroads and American Law* (Lawrence: University of Kansas Press, 2001). A number of recent legal histories have a major focus on railroads. I found the following valuable. Randolph Bergstrom, *Courting Danger: Injury and Law in New York City 1870–1910* (Ithaca: Cornell University Press, 1992). Lawrence Friedman, "Civil Wrongs: Personal Injury Law in the Late 19th Century," *American Bar Foundation Research Journal* 12 (1987): 351–76. Peter Karsten, *Heart vs. Head: Judge Made Law in Nineteenth Century America* (Chapel Hill: University of North Carolina Press, 1997). Frank Munger, "Social Change and Tort Legislation: Industrialization, Accidents and Trial Courts in Southern West Virginia," *Buffalo Law Review* 36 (1987): 75–118. Edward Purcell, *Litigation and Inequality: Federal Diversity Jurisdiction in Industrial America* (New York: Oxford University Press, 1992). William Thomas, *Lawyering for the Railroad: Business, Law, and Power in the New South* (Baton Rouge: Louisiana State University Press, 1999). John Witt, *The Accidental Republic* (Cambridge, Mass.: Harvard University Press, 2004).

Labor

The best overall study of railroad labor is Walter Licht, *Working for the Railroad* (Princeton: Princeton University Press, 1983). Excellent case studies are David Lightner, *Labor on the Illinois Central* (New York: Arno, 1977), Paul Black, "The Development of Management Personnel Policies on the Burlington Railroad, 1860–1900" (Ph.D. diss., University of Wisconsin, 1972). James Ducker, *Men of the Steel Rails* (Lincoln: University of Nebraska Press, 1983). Frederick Gamst, *The Hoghead: An Industrial Ethnology of the Locomotive Engineer* (New York: Holt, Rinehart and Winston, 1980) provides a fascinating glimpse into work life. The role of technological change in changing labor requirements is the topic of Robert Aldag, "Culture Clash: Diesel vs. Tradition," *Railroad History* (Millennium Special 2000): 89–99. An econometric analysis is Gilbert Yochum and Steven Rhiel, "Employment and Changing Technology in the Postwar Railroad Industry," *Industrial Relations* 29 (Fall 1990): 479–503.

Economists have looked for wage risk premiums among railroad workers, most notably Seung-Wook Kim and Price Fishback, "Institutional Change, Compensating Differentials, and Accident Risk in American Railroading, 1892–1945," *Journal of Economic History* 53 (Dec. 1993): 796–823. For historical evidence on wage premiums in general see Price Fishback and Shawn Kantor, *A Prelude to the Welfare State* (Chicago: University of Chicago Press, 2000). Skeptical modern assessments of this literature are William Dickens, "Assuming the Can Opener: Hedonic Wage Estimates and the Value of Life," NBER *Working Paper* 3446 (Cambridge, 1990) and Peter Dorman and Paul Hagstrom, "Wage Compensation for Dangerous Work Revisited," *Industrial and Labor Relations Review* 52 (Oct. 1998): 116–35.

There is a considerable literature on railroad beneficial organizations. I used the following. W. H. Baldwin, "Railroad Relief and Beneficiary Associations," American Economic Association *Papers and Proceedings of the Twelfth Annual Meeting* (1899): 213–30. See also W. H. Allport, "American Railway Relief Funds," *Journal of Political Economy* 20 (Jan. 1912): 49–78, and (Feb. 1912): 101–134. A survey of the extent of such programs as of 1930 is in Pierce Williams, *The Purchase of Medical Care Through Fixed Periodic Payment*, National Bureau of Economic Research 20 (New York, 1932).

J. B. Kennedy, "The Beneficiary Features of the Railway Unions," in *Studies in American Trade Unionism*, ed. Jacob Hollander and George Barnett (New York: Henry Holt, 1907), 321–50. Henry J. Short, *Railroad Doctors, Hospitals, and Associations: Pioneers in Comprehensive, Low Cost Medical Care* (Upper Lake, Calif., 1986). J. Roy Jones, *Memories, Men and Medicine* (Sacramento: Sacramento Society for Medical Improvement, 1950), and his *The Old Central Pacific Hospital* (San Francisco: Western Association of Railway Surgeons, 1960). Patricia Benoit, *Men of Steel, Women of Spirit: History of the Santa Fe Hospital 1891–1991* (Temple, Tex.: Scott & White Memorial Hospital, 1991). Paul Black, "Reluctant Paternalism: Employee Relief Activities of the Chicago, Burlington & Quincy Railroad in the 19th Century," *Business and Economic History* 6 (1977): 120–34.

Railway spine is a focus of Michael Trimble, *Post Traumatic Neurosis: From Railway Spine to Whiplash* (New York: Wiley, 1981). A broader discussion of its significance is in Eric Caplan, *Mind Games* (Berkeley: University of California Press, 1998). George McNeill, *A Study of Accidents and Accident Insurance* (Boston, 1900) contains a brief discussion of industrial life insurance, as does William F. Willoughby, *Workingmen's Insurance* (New York: Thomas Y. Crowell, 1898). A modern treatment of railroad medical care and benefit programs is Mark Aldrich, "Train Wrecks to Typhoid Fever: The Development of Railroad Medical Organizations, 1850–World War I," *Bulletin of History of Medicine* 75 (Summer 2001): 254–89. Don Hofsommer, "St. Louis Southwestern Railway's Campaign against Malaria in Arkansas and Texas," *Arkansas Historical Quarterly* 62 (Summer 2003): 182–93 is a good case study..

Index

accident liability: common law of,
24–26, 303; and company safety
work, 186, 189–93, 238–40, 276;
costs and effects of, 26, 53, 160,
168, 213, 218, 228, 234, 237–39,
280, 283–85, 400n3; Employers'
Liability Act of 1908, 186, 209,
237, 280; fraud, 170–72; for grade
crossing accidents, 25, 124, 213;
hazardous cargo, 216–18, 222–23,
228; new inventions, 25; for passengers, 24–26; for trespassers,
24–25, 121; for workers, 25, 108,
159–60, 190–91. *See also* medical
and benefit plans
Accident Report Act: of 1901, 182,
321; of 1910, 72, 186
accidents, 181; controllability of,
22; European and American,
10; as focusing devices, 7; from
hazardous substances, 217–19,
222–23, 227–28, 230–34, 301; and
induced innovation, 7, 10, 26; investigation of, 72, 191, 239; publicity and public concerns, 5, 26,
71–73, 96–97, 129, 153, 192, 196,
212, 222; social construction of,
97, 107, 117, 131, 140, 146–47, 152–
53, 266–67; views of causation,
24–25, 97, 107, 117–18, 265–67. *See
also* bridge and trestle accidents;
collisions; derailments; grade
crossings and accidents; little
accidents; locomotives and accidents; statistics; train accidents
Adams, Braman B., 187–88, 278
Adams, Charles Francis, Jr., 3, 71–
72, 116, 186, 189, 198, 277, 305–6;
and early regulation, 142; views
on personal responsibility, 117

agency problems, 16, 33–37, 59, 227;
as cause of accidents; 21, 89–90,
222–23; and labor markets, 8,
187–89. *See also* contracts and
contracting; management; personnel administration
air brakes, 64, 70–72, 82, 222, 225;
early development, 29–30, 40;
post–Civil War developments,
74–78, 83–85, 96, 110–11; in early
twentieth century, 195, 250–52;
after World War II, 293, 298–99.
See also brakes; locomotives and
accidents: brakes
Allen, Horatio, 12, 135
Allen, Robert, 48, 66
American Association of Railway
Surgeons, 165, 175
American Public Health Association, 174–75
American Railroad Journal, 12,
19–20, 25–31, 36, 38, 56, 79, 135
American Railway Association, 47,
89, 91, 105, 173, 181, 194, 204; explosives regulation, 216, 221–25;
mechanical division, 250–54;
research, 251–52, 255, 261–63, 298;
and safety work, 190–91, 210,
213–14, 237. *See also* safety section, ARA-AAR
American Railway Engineering Association, 47, 151; rail and track
research, 201, 203–8, 254, 256–58,
296; station design, 212; transverse fissures, 259–63
American Railway Times, 23, 25–26,
31, 38, 54, 56, 98, 136–37, 142
American Society of Civil Engineers, 26, 47, 57; and bridge
design, 132, 142–44, 149

American Society of Testing Materials, 203
American System of Railroading,
10, 14, 21; and bridge construction, 131–41
associationalism, 217, 226, 230, 234
Association of American Railways, 256, 267, 299; origins, 240;
research, 276–77, 293. *See also*
safety section, ARA-AAR
Association of Manufacturers of
Chilled Car Wheels, 199–201,
255–56
Atchison, Topeka and Santa Fe
Railroad, 52, 83, 114, 160, 165,
169, 172, 191, 209; accidents and
safety, 217, 279–80, 292
Atlantic Coast Line, 272, 280, 283,
289, 298
automatic block signals. *See* block
signals
automatic freight car couplers,
82–85, 109–12; incremental
improvement, 112–15. *See also*
couplers, nonautomatic
automatic train control, 186, 188;
in 1920s, 246–50, 404n32; after
WWII, 292–94
automobiles and trucks, 238,
270–71; and grade crossing risks,
8, 128, 213–14, 269, 290; and passenger safety, 263, 274, 292; and
trespassing, 212, 264, 274

Baltimore and Ohio Railroad, 10, 12,
21, 36–37, 48, 57, 59, 68, 100, 109,
114, 135–36, 191, 223, 226, 263; accidents and safety, 70–71, 210, 214,
279; medical and benefits plan,
91, 156, 161–65, 168–69, 172–73